CW01091009

Probability Theory and Stochastic Modelling

Volume 107

Probability Theory and Stochastic Modelling publishes cutting-edge research monographs in probability and its applications, as well as postgraduate-level textbooks that either introduce the reader to new developments in the field, or present a fresh perspective on fundamental topics.

Books in this series are expected to follow rigorous mathematical standards, and all titles will be thoroughly peer-reviewed before being considered for publication.

Probability Theory and Stochastic Modelling covers all aspects of modern probability theory including:

- Gaussian processes
- Markov processes
- Random fields, point processes, and random sets
- Random matrices
- Statistical mechanics, and random media
- Stochastic analysis
- High-dimensional probability

as well as applications that include (but are not restricted to) :

- Branching processes, and other models of population growth
- Communications, and processing networks
- Computational methods in probability theory and stochastic processes, including simulation
- Genetics and other stochastic models in biology and the life sciences
- Information theory, signal processing, and image synthesis
- Mathematical economics and finance
- Statistical methods (e.g. empirical processes, MCMC)
- Statistics for stochastic processes
- Stochastic control, and stochastic differential games
- Stochastic models in operations research and stochastic optimization
- Stochastic models in the physical sciences

Probability Theory and Stochastic Modelling is a merger and continuation of Springer's Stochastic Modelling and Applied Probability and Probability and Its Applications series.

Teemu Pennanen • Ari-Pekka Perkkiö

Convex Stochastic Optimization

Dynamic Programming and Duality in Discrete Time

Teemu Pennanen
Department of Mathematics
King's College London
London, UK

Ari-Pekka Perkkiö
Department of Mathematics
Ludwig-Maximilians-Universität München
Munich, Germany

ISSN 2199-3130 ISSN 2199-3149 (electronic)
Probability Theory and Stochastic Modelling
ISBN 978-3-031-76431-8 ISBN 978-3-031-76432-5 (eBook)
https://doi.org/10.1007/978-3-031-76432-5

Mathematics Subject Classification: 90C39, 46N10, 90C15, 90C46, 93E20, 91G80

This Springer imprint is published by the registered company Springer Nature Switzerland AG
The registered company address is: Gewerbestrasse 11, 6330 Cham, Switzerland

Preface

This book studies a general class of convex stochastic optimization (CSO) problems that unifies many common problem formulations from operations research, financial mathematics and stochastic optimal control. We extend the theory of dynamic programming and convex duality to allow for a unified and simplified treatment of various special problem classes found in the literature. The extensions allow also for significant generalizations to existing problem formulations. Both dynamic programming and duality have played crucial roles in the development of various optimality conditions and numerical techniques for the solution of convex stochastic optimization problems.

Several books on stochastic optimization (also known as "stochastic programming") have been published since the pioneering works of Beale [7] and Danzig [36]; see, e.g., [18, 25, 29, 43, 45, 71, 72, 76, 77, 81, 113, 117, 148]. The most influential to our work have been the articles [129–138] of Rockafellar and Wets who unified and extended most of the existing results in the field at the time. Their approach used the theory of measurable set-valued mappings, normal integrands and of convex duality to derive dual problems and optimality conditions in terms of Lagrange multipliers and other dual variables. The above references assume, however, that decisions are chosen from a locally convex function space of stochastic processes. While, in practice, such assumptions are harmless, they do exclude some popular models, e.g., in stochastic control and financial mathematics.

Much of the theory of stochastic optimization has been developed in the literature of optimal control where one separates the decision variables into a state and a control. Indeed, both the dynamic programming principle of Bellman [9] and the maximum principle of Pontryagin and his students [115] relied on such a format. On the other hand, various instances of stochastic optimization have been studied in financial mathematics ever since Markowitz [84] introduced his portfolio optimization model in the 1950s and Merton his optimal consumption model [85]. After the publication of the famous option pricing paper of Black and Scholes [24], mathematical finance focused largely on complete market models that can be analyzed with the techniques of stochastic analysis without relying much on optimization techniques. From the early 1990 (see [40] for more accurate historical account), however, the focus in financial mathematics started to move back toward incomplete markets and functional analytic techniques that are closely related to convex optimization. These works introduced some innovative measure theoretic techniques that were able to treat models beyond the reach of classical functional analysis in locally convex spaces.

Combining techniques from convex analysis and stochastics, the present book develops a unifying theoretical framework that covers and extends the classical models of Rockafellar and Wets, various models from financial mathematics as well as convex instances of stochastic optimal control in finite discrete time. The framework allows for dynamic programming, duality theory and optimality conditions that yield, for example, the optimality conditions derived by Rockafellar and Wets, the "fundamental theorem of asset pricing" of Dalang, Morton and Willinger [35] as well as the maximum principle in optimal stochastic control. Besides the classical models, the developed theory allows for many practically relevant extensions that have not been treated in the literature before.

This book builds on the theory of measurable set-valued mappings and normal integrands; see [139, Chapter 14] and its references. In particular, the first part of the book develops the dynamic programming principle in terms of conditional expectations of normal integrands introduced originally by Bismut [22] and, independently, by Rockafellar and Wets [131] and Evstigneev [53] in the context of stochastic optimization. Employing classical techniques of convex analysis, we relax the compactness assumptions made in [53, 131] in order to cover standard optimization problems in convex optimal control and financial mathematics. The theory of normal integrands allows for easy resolution of certain measurability problems that arise in some other formulations of stochastic optimal control; see, e.g., [14].

The second part of the book builds on the conjugate duality framework of Rockafellar [127] to develop a duality theory for CSO much like was done in [131, 136] in the case of stochastic programs with inequality constraints and in [20] for stochastic optimal control in continuous-time. However, as in the dynamic programming theory of the first part, we allow for unbounded objectives and decision strategies that take us beyond locally convex vector spaces. We also allow for general parameterizations that cover not only inequality constrained problems but also, e.g., optimal stochastic control where the parameter is the driving noise in the system equations. In financial problems, the parameter can be taken as the payout of, e.g., a contingent claim. Moreover, our approach is more direct than that of [131, 136] and it yields an explicit dual problem in the general formulation of CSO. Our analysis yields scenariowise optimality conditions where the decision strategies are sought from the space of general adapted stochastic processes without any boundedness or integrability conditions. Unless the probability space consists of finitely many atoms (scenarios), such a space is not locally convex so we are not strictly within the general duality framework of [127]. Allowing for general adapted decision strategies is important, e.g., in proving the celebrated "fundamental theorem of asset pricing" that provides the existence of martingale measures under the no-arbitrage condition; see, e.g., [35, 70, 144]. Our general results on CSO yields extensions of classical results of financial mathematics by allowing, e.g., for semi-static trading strategies, portfolio constraints and nonlinear illiquidity effects. In stochastic control, we obtain general formulations of the maximum principle with or without state constraints.

Each chapter starts by a section that reviews general concepts and techniques used in the analysis of stochastic optimization problems in that chapter. A reader interested only in stochastic optimization or its applications may wish to skip the first sections and only use them as reference when needed in the development of the theory. Chapters 1 and 2 start by presenting general results on measurable set-valued mappings and normal integrands which have numerous applications also beyond stochastic optimization. The preliminary sections of Chaprs. 3 and 5 review the theory of convex integral functionals on locally convex vector spaces of random variables. The beginning of Chap. 4 studies the L^0-space which fails to be locally convex but, nevertheless, allows convenient criteria for compactness-like properties for sequences. Appendices A and B review general convex analysis and probability theory, respectively. The sections on general theories constitute roughly a third of this book. The applications at the end of each chapter amount to roughly another third.

We are grateful to Jean-Philippe Chancelier, Michel De Lara, Patrick Louis Combettes and Igor Evstigneev for various corrections and suggestions on a preliminary version of this book.

London, UK Teemu Pennanen
Munich, Germany Ari-Pekka Perkkiö

Contents

Chapter 1
Convex Stochastic Optimization

Given a probability space $(\Omega, \mathcal{F}, P)$ with a filtration $(\mathcal{F}_t)_{t=0}^T$ (an increasing sequence of sub-σ-algebras of $\mathcal{F}$) and a sequence $n_0, \dots, n_T$ of integers, consider the problem

$$\text{minimize} \quad Eh(x) \quad \text{over } x \in \mathcal{N}, \qquad (P)$$

where $\mathcal{N}$ is the linear space of $(\mathcal{F}_t)_{t=0}^T$-adapted sequences $x = (x_t)_{t=0}^T$ of random vectors $x_t \in L^0(\Omega, P, \mathcal{F}_t; \mathbb{R}^{n_t})$ and Eh is an integral functional on the space $L^0(\Omega, \mathcal{F}, P; \mathbb{R}^n)$ of $\mathcal{F}$-measurable $\mathbb{R}^n$-valued functions given by

$$Eh(x) := \int_\Omega h(x(\omega), \omega) dP(\omega).$$

Here $n := n_0 + \ldots + n_T$ and h is a function from $\mathbb{R}^n \times \Omega$ to the extended reals $\overline{\mathbb{R}} := \mathbb{R} \cup \{\infty, -\infty\}$ such that $\omega \mapsto h(x(\omega), \omega)$ is measurable for all $x \in L^0(\Omega, \mathcal{F}, P; \mathbb{R}^n)$. Here and in what follows, we define the integral of an extended real-valued random variable as $+\infty$ unless its positive part is integrable. The integral of any extended real-valued measurable function is then a well-defined number in $\overline{\mathbb{R}}$. When the function h is convex in the first argument, the integral functional Eh will be a convex function on L^0. In this case, we say that (P) is a problem of *convex stochastic optimization.*

The measurable functions x_t may be thought of as decisions taken at time $t = 0, \dots, T$. The σ-algebra $\mathcal{F}_t$ represents the information available to the decision maker at time t. The requirement that x_t be $\mathcal{F}_t$-measurable means that the decision taken at time t can only depend on the information observed by time t; the decision maker does not know which scenario $\omega \in \Omega$ they live in but only to which elements of $\mathcal{F}_t$ the scenario belongs to. Since x_t takes values in $\mathbb{R}^{n_t}$, the paths of the process $x = (x_t)_{t=0}^T$ belong to $\mathbb{R}^n$ where $n = n_0 + \cdots + n_T$.

T. Pennanen, A.-P. Perkkiö, *Convex Stochastic Optimization*, Probability Theory and Stochastic Modelling 107, https://doi.org/10.1007/978-3-031-76432-5_1

Besides the measurability properties, we do not place other restrictions on the elements of $\mathcal{N}$. This is important e.g. when studying dynamic programming or existence of solutions where boundedness or even integrability requirements on the strategies often lead to complications. Allowing for general measurable strategies is essential, for example, in financial mathematics when proving various forms of the "fundamental theorem of asset pricing" that provide dual characterizations of the no-arbitrage conditions in terms of "martingale measures" or other kinds of "price systems".

Allowing for extended real-valued integrands h, we can incorporate various pointwise constraints into the objective as infinite penalties. This makes (P) a very general problem format that unifies and extends many classical optimization models from operations research, optimal stochastic control and financial mathematics. Examples will be given in Sect. 1.3. For the most part, we will study *convex* instances of (P) since that will give rise to the strongest theory in terms of existence results, duality and optimality conditions much of which is lost in nonconvex models involving e.g. probabilistic/chance constraints or integer variables.

The rest of this chapter is organized as follows. Section 1.1 reviews the basic theory of random closed sets and normal integrands (random lower semicontinuous functions). Section 1.2 then takes a closer look at problem (P) in the case where h is a convex normal integrand. Section 1.3 shows how some well-known problem classes and their extensions can be formulated as instances of (P) with an appropriate specification of h. Section 1.1 is somewhat technical and contains much more material than what is needed for the rest of this chapter. The material there, however, forms the basic building blocks for the theory of stochastic optimization needed for the later developments. Section 1.1 collects the employed theory that does not involve conditional expectations of random sets or normal integrands, a theory that will be covered in the first part of Chap. 2.

1.1 Normal Integrands and Integral Functionals

The objective of problem (P) is an integral functional associated with an extended real-valued function $h : \mathbb{R}^n \times \Omega \to \overline{\mathbb{R}}$. We will assume that h is a "normal integrand", which means that the epigraph of the function $h(\cdot, \omega)$ is closed and depends measurably on ω; see Sect. 1.1.2 below. Under quite general conditions, the measurability turns out to be equivalent to the measurability of h with respect to the product σ-algebra $\mathcal{F} \otimes \mathcal{B}(\mathbb{R}^n)$ where $\mathcal{B}(\mathbb{R}^n)$ is the Borel σ-algebra on $\mathbb{R}^n$. In general, however, the property of being a normal integrand is more convenient as it is often easy to verify in various applications and in the analysis of (P). This section reviews the basic theory of measurable set-valued mappings and normal integrands.

A *set-valued mapping* S from a space X to another U is a correspondence that assigns to each $x \in X$ a (possibly empty) set $S(x) \subseteq U$. We will write $S : X \rightrightarrows U$

to indicate that S is such a mapping. The *graph, domain* and the *range* of S are defined by

$$\begin{aligned}\operatorname{gph} S &:= \{(x, u) \in X \times U \mid u \in S(x)\},\\ \operatorname{dom} S &:= \{x \in X \mid S(x) \neq \emptyset\},\\ \operatorname{rge} S &:= \{u \in U \mid \exists x \in X : u \in S(x)\},\end{aligned}$$

respectively. The *inverse* $S^{-1} : U \rightrightarrows X$ of S is defined by

$$S^{-1}(u) := \{x \in X \mid u \in S(x)\}.$$

Clearly, $(S^{-1})^{-1} = S$, $\operatorname{dom} S^{-1} = \operatorname{rge} S$ and $\operatorname{rge} S^{-1} = \operatorname{dom} S$. The *image* of a set $C \subset X$ under S is the set

$$S(C) := \bigcup_{x \in C} S(x) = \{u \in U \mid S^{-1}(u) \cap C \neq \emptyset\}$$

while the *inverse image* of a $D \subset U$ under S is

$$S^{-1}(D) := \bigcup_{u \in D} S^{-1}(u) = \{x \in X \mid S(x) \cap D \neq \emptyset\}.$$

Sums and scalar multiples of sets in a vector space U are defined, as usual, by

$$D_1 + D_2 := \{u_1 + u_2 \in U \mid u_1 \in D_1,\ u_2 \in D_2\}$$

for $D_1, D_2 \subseteq U$ and

$$\lambda D := \{\lambda u \in U \mid u \in D\}$$

for $D \subseteq U$ and $\lambda \in \mathbb{R}$.

1.1.1 Random Sets

Let $(\Omega, \mathcal{F})$ be a measurable space and endow the n-dimensional Euclidean space $\mathbb{R}^n$ with the usual Euclidean topology. A set-valued mapping $S : \Omega \rightrightarrows \mathbb{R}^n$ is *measurable* if

$$S^{-1}(O) \in \mathcal{F}$$

for every open set $O \subset \mathbb{R}^n$. A *random set* in $\mathbb{R}^n$ is a measurable set-valued mapping from Ω to $\mathbb{R}^n$. Random variables can be identified with single-valued random sets.

The linear space of $\mathbb{R}^n$-valued random variables will be denoted by $\mathcal{L}^0(\mathbb{R}^n)$. If the dimension is clear from the context, we simply write $\mathcal{L}^0 := \mathcal{L}^0(\mathbb{R}^n)$.

Given a set $C \subset \mathbb{R}^n$, we denote its interior and closure by $\operatorname{int} C$ and $\operatorname{cl} C$, respectively. Given an open set $O \subset \mathbb{R}^n$, we have $O \cap C \neq \emptyset$ if and only if $O \cap \operatorname{cl} C \neq \emptyset$. This implies the following lemma where the *image-closure* of a mapping S is defined by

$$(\operatorname{cl} S)(\omega) := \operatorname{cl} S(\omega).$$

Lemma 1.1 *A set-valued mapping is measurable if any only if its image-closure is measurable.*

We will denote the *distance* of a point $x \in \mathbb{R}^n$ to a set $A \subset \mathbb{R}^n$ by

$$d(x, A) := \inf_{x' \in A} |x - x'|,$$

where $|\cdot|$ denotes the Euclidean norm on $\mathbb{R}^n$. We denote the closed Euclidean ball centered at x with radius r by $\mathbb{B}_r(x)$. When the ball is centered at the origin, we write it simply as $\mathbb{B}_r$. A set-valued mapping S is *closed-valued* if $S(\omega)$ is closed for each $\omega \in \Omega$. Similarly for convex-, cone-, linear- etc. valued mappings. The following theorem gives some useful criteria for measurability of closed-valued mappings.

Theorem 1.2 *Let $S : \Omega \rightrightarrows \mathbb{R}^n$ be closed-valued. The following are equivalent:*

1. *S is measurable;*
2. *$S^{-1}(C) \in \mathcal{F}$ for every compact set C;*
3. *$S^{-1}(C) \in \mathcal{F}$ for every closed set C;*
4. *$S^{-1}(B) \in \mathcal{F}$ for every closed ball B;*
5. *$S^{-1}(O) \in \mathcal{F}$ for every open ball O;*
6. *$\{\omega \in \Omega \mid S(\omega) \subset O\} \in \mathcal{F}$ for every open O;*
7. *$\{\omega \in \Omega \mid S(\omega) \subset C\} \in \mathcal{F}$ for every closed C;*
8. *$\omega \mapsto d(x, S(\omega))$ is measurable for every $x \in \mathbb{R}^n$.*

Proof Assume 1 and let $C \subset \mathbb{R}^n$ be compact. We have $S(\omega) \cap C \neq \emptyset$ if and only if $S(\omega) \cap (C + \operatorname{int} \mathbb{B}_{1/\nu}) \neq \emptyset$ for all $\nu \in \mathbb{N}$. Thus,

$$\begin{aligned} S^{-1}(C) &= \{\omega \in \Omega \mid S(\omega) \cap (C + \operatorname{int} \mathbb{B}_{1/\nu}) \neq \emptyset \ \forall \nu \in \mathbb{N}\} \\ &= \bigcap_{\nu \in \mathbb{N}} S^{-1}(C + \operatorname{int} \mathbb{B}_{1/\nu}) \\ &\in \mathcal{F}, \end{aligned}$$

so 2 holds. For a closed C, we have $C = \bigcup_{\nu \in \mathbb{N}} (C \cap \mathbb{B}_\nu)$ and thus,

$$S^{-1}(C) = \bigcup_{\nu \in \mathbb{N}} S^{-1}(C \cap \mathbb{B}_\nu) \in \mathcal{F}$$

so 2 implies 3. Clearly, 3 implies 4. Open balls in $\mathbb{R}^n$ are countable unions of closed balls so 4 implies 5. Open sets in $\mathbb{R}^n$ are countable unions of open balls, so 5 implies 1. For any $A \subseteq \mathbb{R}^n$, $\{\omega \in \Omega \mid S(\omega) \subseteq A\} = \Omega \setminus S^{-1}(A^C)$, so 3 and 6 are equivalent, and 1 and 7 are equivalent. For any $r \in \mathbb{R}_+$, we have $d(x, S(\omega)) < r$ if and only if $S(\omega) \cap \operatorname{int} \mathbb{B}_r(x) \neq \emptyset$, so 5 and 8 are equivalent. □

Recall that the *graph* of a set-valued mapping $S : \Omega \rightrightarrows \mathbb{R}^n$ is the set

$$\operatorname{gph} S := \{(\omega, x) \in \Omega \times \mathbb{R}^n \mid x \in S(\omega)\}.$$

We denote the Borel σ-algebra of $\mathbb{R}^n$ by $\mathcal{B}(\mathbb{R}^n)$.

Corollary 1.3 *The graph of a closed-valued measurable mapping is $\mathcal{F} \otimes \mathcal{B}(\mathbb{R}^n)$-measurable.*

Proof Given a closed-valued measurable mapping $S : \Omega \rightrightarrows \mathbb{R}^n$, we have $x \in S(\omega)$ if and only if, for every strictly positive $r \in \mathbb{Q}$, there exists a $q \in \mathbb{Q}^n$ such that $x \in \mathbb{B}_r(q)$ and $S(\omega) \cap \mathbb{B}_r(q) \neq \emptyset$. Thus

$$\operatorname{gph} S = \bigcap_{\substack{r \in \mathbb{Q} \\ r > 0}} \bigcup_{q \in \mathbb{Q}^n} [S^{-1}(\mathbb{B}_r(q)) \times \mathbb{B}_r(q)],$$

which is measurable by Theorem 1.2. □

In general, the measurability of the graph does not imply the measurability of a set-valued mapping. It does, however, if the underlying measure space is complete with respect to a probability measure; see Remark 1.4 below. It is natural to ask if the preimage of a general Borel-set is measurable. In general, the answer is no, unless the measure space is complete with respect to a probability measure.

Remark 1.4 Assume that the σ-algebra $\mathcal{F}$ is such that the projections of $\mathcal{F} \otimes \mathcal{B}(\mathbb{R}^n)$-measurable sets to Ω are $\mathcal{F}$-measurable, i.e.

$$\{\omega \in \Omega \mid \exists x \in \mathbb{R}^n : (\omega, x) \in A\} \in \mathcal{F} \quad \forall A \in \mathcal{F} \otimes \mathcal{B}(\mathbb{R}^n). \tag{1.1}$$

Given $S : \Omega \rightrightarrows \mathbb{R}^n$,

1. $\mathcal{F} \otimes \mathcal{B}(\mathbb{R}^n)$-measurability of $\operatorname{gph} S$ implies measurability of S,
2. measurability of S implies that $S^{-1}(B) \in \mathcal{F}$ for every $B \in \mathcal{B}(\mathbb{R}^n)$.

Indeed, the inverse image $S^{-1}(C)$ of any $C \subset \mathbb{R}^n$ is the projection of $\operatorname{gph} S \cap (\Omega \times C)$ to Ω so both claims hold under condition (1.1).

Condition (1.1) holds if there is a probability measure P such that $(\Omega, \mathcal{F}, P)$ is complete; see Theorem B.6 in the appendix. Without the extra assumption on $\mathcal{F}$, 1 and 2 above may fail. For a simple counter example to 2, let $(\Omega, \mathcal{F}) = (\mathbb{R}, \mathcal{B}(\mathbb{R}))$ and define $S : \Omega \rightrightarrows \mathbb{R}^2$ by $S(\omega) := \{\omega\} \times \mathbb{R}$. The inverse of S is the projection $(x_1, x_2) \mapsto x_1$ so $S^{-1}(O)$ is open for every open $O \subset \mathbb{R}^2$. Thus, S is closed-valued and measurable. However, a projection of a Borel set in $\mathbb{R}^2$ need not be a Borel set in $\mathbb{R}$. In other words, there exists an $A \in \mathcal{B}(\mathbb{R}^2)$ such that $S^{-1}(A) \notin \mathcal{F}$.

A *selection* of a set-valued mapping $S : \Omega \rightrightarrows \mathbb{R}^n$ is a function $x : \operatorname{dom} S \to \mathbb{R}^n$ such that $x(\omega) \in S(\omega)$ for all $\omega \in \operatorname{dom} S$. A sequence $(x^\nu)_{\nu\in\mathbb{N}}$ of measurable selections of S is said to be a *Castaing representation* of S if

$$S(\omega) = \operatorname{cl}\{x^\nu(\omega) \mid \nu \in \mathbb{N}\} \quad \forall \omega \in \operatorname{dom} S.$$

Here $\operatorname{dom} S$ is equipped with the σ-algebra $\{A \cap \operatorname{dom} S \mid A \in \mathcal{F}\}$. Note that $\operatorname{dom} S = S^{-1}(\mathbb{R}^n)$, so the domain of a measurable mapping is measurable.

Lemma 1.5 *A closed-valued mapping $S : \Omega \rightrightarrows \mathbb{R}^n$ is measurable if it has a Castaing representation and measurable domain.*

Proof For any Castaing representation $(x^\nu)_{\nu\in\mathbb{N}}$ and an open $O \subseteq \mathbb{R}^n$,

$$S^{-1}(O) = \operatorname{dom} S \cap \left(\bigcup_{\nu\in\mathbb{N}} (x^\nu)^{-1}(O) \right),$$

where the right side is measurable, by measurability of x^ν. □

Theorem 1.27 below shows that the converse to Lemma 1.5 holds as well, i.e. that any closed-valued measurable mapping admits a Castaing representation.

A set C is *solid* if $C = \operatorname{cl}\operatorname{int} C$. Solid-valued mappings have a simple characterization of measurability.

Theorem 1.6 *A solid-valued mapping $S : \Omega \rightrightarrows \mathbb{R}^n$ is measurable if and only if $S^{-1}(x)$ is measurable for every $x \in \mathbb{R}^n$.*

Proof Necessity follows from Theorem 1.2. To prove sufficiency, let $(x^\nu)_{\nu\in\mathbb{N}}$ be dense in $\mathbb{R}^n$. Since S is solid-valued, $(x^\nu)_{\nu\in\mathbb{N}} \cap S(\omega)$ is dense in $S(\omega)$ for every $\omega \in \operatorname{dom} S$, and $\operatorname{dom} S = \bigcup_\nu A^\nu$, where the sets $A^\nu := S^{-1}(x^\nu)$ are measurable, by assumption. In particular, $\operatorname{dom} S$ is measurable. The function

$$w(\omega) := \begin{cases} x^1 & \text{if } \omega \in A^1, \\ x^\nu & \text{if } \omega \in A^\nu \setminus \cup_{\mu \le \nu-1} A^\mu \end{cases}$$

is a measurable selection of S. The sequence $(\bar{x}^\nu)_{\nu\in\mathbb{N}}$ defined by

$$\bar{x}^\nu(\omega) := \begin{cases} x^\nu & \text{if } \omega \in A^\nu, \\ w(\omega) & \text{otherwise} \end{cases}$$

$$= \begin{cases} x^\nu & \text{if } x^\nu \in S(\omega), \\ w(\omega) & \text{otherwise} \end{cases}$$

is a Castaing representation of S. The claim thus follows from Lemma 1.5. □

Measurability is preserved under various algebraic operations.

Theorem 1.7 *Let J be a countable set and $S^j : \Omega \rightrightarrows \mathbb{R}^n$ a measurable mapping for each $j \in J$. Then*

1. $S(\omega) := \bigcap_{j\in J} S^j(\omega)$ *is measurable if each* S^j *is closed,*
2. $S(\omega) := \bigcup_{j\in J} S^j(\omega)$ *is measurable,*
3. $S(\omega) := \sum_{j\in J} \lambda^j S^j(\omega)$ *is measurable for finite* J *and* $\lambda^j \in \mathbb{R}$,
4. $S(\omega) := S^j(\omega)$ *for* $\omega \in A^j$ *is measurable for any measurable partition* $\{A^j\}_{j\in J}$ *of* Ω.

Given a finite collection $S^j : \Omega \rightrightarrows \mathbb{R}^{n_j}$, $j \in J$ *of measurable mappings,*

5. $S(\omega) := \prod_{j\in J} S^j(\omega)$ *is measurable.*

Proof Part 2 follows from the fact that $S^{-1}(O) = \bigcup_{j\in J}(S^j)^{-1}(O)$ for any open $O \subset \mathbb{R}^n$. Part 4 follows from the fact that $S^{-1}(O) = \bigcap_{j\in J}((S^j)^{-1}(O) \cap A^j)$ for any open $O \subset \mathbb{R}^n$. As to 5, every open $O \subset \prod_{j\in J} \mathbb{R}^{n_j}$ can be expressed as a countable union of sets of the form $\prod_{j\in J} O^j$ where $O^j \subset \mathbb{R}^{n_j}$ is open. Thus, $S^{-1}(O)$ is a countable union of sets of the form

$$S^{-1}(\prod_{j\in J} O^j) = \bigcap_{j\in J}(S^j)^{-1}(O^j),$$

where each $(S^j)^{-1}(O^j)$ is measurable by the assumption. Thus, S is measurable.

As to 3, let $O \subset \mathbb{R}^n$ be open. The set

$$O' = \{(x^j)_{j\in J} \in (\mathbb{R}^n)^J \mid \sum_{j\in J} \lambda^j x^j \in O\}$$

is open as a preimage of a continuous mapping. We get

$$\begin{aligned} S^{-1}(O) &= \{\omega \in \Omega \mid (\sum_{j\in J} \lambda^j S^j)(\omega) \cap O \neq \emptyset\} \\ &= \{\omega \in \Omega \mid (\prod_{j\in J} S^j(\omega)) \cap O' \neq \emptyset\}, \end{aligned}$$

which is measurable by part 5.

As to 1, assume first that $J = \{1, 2\}$, take a compact $C \subset \mathbb{R}^n$ and let $R^j(\omega) := S^j(\omega) \cap C$. We have

$$\begin{aligned} (S^1 \cap S^2)^{-1}(C) &= \{\omega \in \Omega \mid S^1(\omega) \cap S^2(\omega) \cap C \neq \emptyset\} \\ &= \{\omega \in \Omega \mid 0 \in R^1(\omega) - R^2(\omega)\} \\ &= (R^1 - R^2)^{-1}(\{0\}). \end{aligned}$$

Here $R^1 - R^2$ is measurable by part 3. Since R^j are compact-valued, $R^1 - R^2$ is compact-valued as well. Thus, $S^1 \cap S^2$ is measurable, by Theorem 1.2. The case of finite J follows by induction.

Suppose finally that $J = \{1, 2, 3, \dots\}$. The mappings $\tilde{S}^k := \bigcap_{j=1}^k S^j$ are measurable by the above. We have $\bigcap_{j\in\mathbb{N}} S^j(\omega) = \bigcap_{k\in\mathbb{N}} \tilde{S}^k(\omega)$. Since the sequence $\tilde{S}^k$ is nonincreasing, the finite intersection property implies that, for any compact $C \subset \mathbb{R}^n$,

$$\left(\bigcap_{\nu\in\mathbb{N}} S^j(\omega)\right) \cap C \neq \emptyset \quad \iff \tilde{S}^k(\omega) \cap C \neq \emptyset \quad \forall k$$

or, in other words,

$$\left(\bigcap_{\nu\in\mathbb{N}} S^j\right)^{-1}(C) = \bigcap_{k\in\mathbb{N}} (\tilde{S}^k)^{-1}(C).$$

The claim thus follows from Theorem 1.2. □

Given a set C, the *convex hull* $\operatorname{co} C$ of C is the smallest convex set containing C. Equivalently, $\operatorname{co} C$ is the set of all finite convex combinations of the points of C, i.e.

$$\operatorname{co} C = \left\{ \sum_{j\in J} \lambda^j C \;\middle|\; |J| < \infty,\ \sum_{j\in J} \lambda_j = 1,\ \lambda_j > 0 \right\}.$$

The *affine hull* $\operatorname{aff} C$ of C is the smallest affine set containing C. Equivalently, $\operatorname{aff} C$ is the set of all finite affine combinations of the points of C, i.e.

$$\operatorname{aff} C = \left\{ \sum_{j\in J} \lambda^j C \;\middle|\; |J| < \infty,\ \sum_{j\in J} \lambda_j = 1 \right\}.$$

Given a mapping $S : \Omega \rightrightarrows \mathbb{R}^n$, we define

$$(\operatorname{co} S)(\omega) := \operatorname{co} S(\omega),$$
$$(\operatorname{aff} S)(\omega) := \operatorname{aff} S(\omega).$$

Corollary 1.8 *Given a measurable mapping $S : \Omega \rightrightarrows \mathbb{R}^n$, the mappings* $\operatorname{co} S$ *and* $\operatorname{aff} S$ *are measurable.*

Proof The mapping $\operatorname{cl}\operatorname{co} S$ is the image-closure of a countable union of mappings of the form $\sum_{j\in J} \lambda^j S$, where J is finite and $\lambda^i > 0$ are rational with $\sum_{j\in J} \lambda^j = 1$. Thus, $\operatorname{cl}\operatorname{co} S$ is measurable by Theorem 1.7, and so is $\operatorname{co} S$, by Lemma 1.1. The proof of the second claim is analogous. □

A function $M : \mathbb{R}^n \times \Omega \to \mathbb{R}^m$ is a *Carathéodory mapping* if $M(\cdot, \omega)$ is continuous for all ω and $M(x, \cdot)$ is measurable for all $x \in \mathbb{R}^n$.

Theorem 1.9 *If $M : \mathbb{R}^n \times \Omega \rightrightarrows \mathbb{R}^m$ is such that*

$$(\operatorname{gph} M)(\omega) := \{(x, u) \mid u \in M(x, \omega)\}$$

is closed-valued and measurable, then the following mappings are measurable,

1. *$R(\omega) := M(S(\omega), \omega)$, where $S : \Omega \rightrightarrows \mathbb{R}^n$ is closed-valued and measurable,*
2. *$S(\omega) := M(\cdot, \omega)^{-1}(R(\omega))$, where $R : \Omega \rightrightarrows \mathbb{R}^m$ is closed-valued and measurable.*

If M is a Carathéodory mapping, then gph M *is closed-valued and measurable and $R(\omega) := M(S(\omega), \omega)$ is measurable for any measurable $S : \Omega \rightrightarrows \mathbb{R}^n$.*

Proof Let $\Pi : \mathbb{R}^n \times \mathbb{R}^m \to \mathbb{R}^m$ be the projection mapping. In 1, $R = \Pi \circ Q$, where $Q(\omega) := [S(\omega) \times \mathbb{R}^m] \cap (\operatorname{gph} M)(\omega)$ is measurable, by Theorem 1.2. Since $R^{-1}(O) = Q^{-1}(\Pi^{-1}(O))$, where $\Pi^{-1}(O)$ is open for any open O, R is measurable. The mapping $\Gamma(\cdot, \omega) := M^{-1}(\cdot, \omega)$ has the same graph as M, so 2 follows by applying 1 to Γ.

To prove the last claim, let $D \subset \mathbb{R}^n$ be countable and dense. When M is a Carathéodory mapping, the functions $\omega \mapsto (x, M(x, \omega))$ are measurable for each x and $\operatorname{gph} M(\cdot, \omega) = \operatorname{cl}\{(x, M(x, \omega)) \mid x \in D\}$. Thus, gph M is measurable, by Lemma 1.5. Continuity of $M(\cdot, \omega)$ implies that $\operatorname{cl} R(\omega) = \operatorname{cl} M(\operatorname{cl} S(\omega), \omega)$, so measurability of R follows from part 1 and Lemma 1.1. □

Example 1.10 Given a random matrix $A \in \mathcal{L}^0(\mathbb{R}^{m \times n})$, the mapping

$$M(x, \omega) := A(\omega)x$$

is Carathéodory. By Theorem 1.9, gph M is measurable. For any $u \in \mathbb{R}^m$, the mapping $\omega \mapsto \{x \in \mathbb{R}^m \mid A(\omega)x = u\}$ is measurable. This follows from Theorem 1.9.2 by choosing $R(\omega) \equiv u$. In particular, if $m = n$ and $A(\omega)$ is invertible for all $\omega \in \Omega$, then $A^{-1}(\omega) := A(\omega)^{-1}$ defines an element of $\mathcal{L}^0(\mathbb{R}^{n \times n})$ and $(u, \omega) \mapsto A^{-1}(\omega)u$ is a Carathéodory mapping.

1.1.2 Normal Integrands

Given extended real numbers $\alpha_1, \alpha_2 \in \overline{\mathbb{R}} := \mathbb{R} \cup \{+\infty, -\infty\}$, we define $\alpha_1 + \alpha_2 = +\infty$ if either $\alpha_1 = +\infty$ or $\alpha_2 = +\infty$, in accordance with the definition of the integral of an extended real-valued random variable at the beginning of this chapter. The product is defined as $\alpha_1 \alpha_2 = 0$ if either $\alpha_1 = 0$ or $\alpha_2 = 0$. The negative and positive parts of an $\alpha \in \overline{\mathbb{R}}$ will be denoted by α^- and α^+, respectively.

A function $h : \mathbb{R}^n \times \Omega \to \overline{\mathbb{R}}$ is a *normal integrand* if its *epigraphical mapping* $\operatorname{epi} h : \Omega \rightrightarrows \mathbb{R}^n \times \mathbb{R}$ defined by

$$(\operatorname{epi} h)(\omega) := \{(x, \alpha) \in \mathbb{R}^n \times \mathbb{R} \mid h(x, \omega) \le \alpha\}$$

is closed-valued and measurable. Note that the closed-valuedness means that the function $h(\cdot, \omega)$ is lower semicontinuous (lsc) for every $\omega \in \Omega$. A normal integrand is *convex/positively homogeneous/Lipschitz continuous* etc. if $h(\cdot, \omega)$ is convex/positively homogeneous/Lipschitz continuous etc. for all ω. Given a set-valued mapping $S : \Omega \rightrightarrows \mathbb{R}^n$, its indicator function

$$\delta_S(x, \omega) := \delta_{S(\omega)}(x) := \begin{cases} 0 & \text{if } x \in S(\omega), \\ +\infty & \text{otherwise} \end{cases}$$

defines a normal integrand if and only if S is closed-valued and measurable.

Recall that a set C is *solid* if $C = \operatorname{cl} \operatorname{int} C$.

Lemma 1.11 *If $h : \mathbb{R}^n \times \Omega \to \overline{\mathbb{R}}$ is such that* $\operatorname{epi} h$ *is solid-valued and $\omega \mapsto h(x, \omega)$ is measurable for every $x \in \mathbb{R}^n$, then h is a normal integrand.*

Proof For any $(x, \alpha) \in \mathbb{R}^n \times \mathbb{R}$,

$$(\operatorname{epi} h)^{-1}(x, \alpha) = \{\omega \mid (x, \alpha) \in (\operatorname{epi} h)(\omega)\} = \{\omega \mid h(x, \omega) \le \alpha\}$$

is measurable, by assumption. Since $\operatorname{epi} h$ is solid, the claim follows from Theorem 1.6. □

A real-valued function h on $\mathbb{R}^n \times \Omega$ is said to be a *Carathéodory integrand* if $h(\cdot, \omega)$ is continuous for each $\omega \in \Omega$ and $h(x, \cdot)$ is measurable for each $x \in \mathbb{R}^n$. In other words, a Carathéodory integrand is a real-valued Carathéodory mapping; see Theorem 1.9. The epigraph of a continuous function is solid, so Lemma 1.11 gives the following.

Example 1.12 A Carathéodory integrand is a normal integrand.

The *Euclidean inner product* of two vectors $x, v \in \mathbb{R}^n$ will be denoted by

$$x \cdot v := \sum_{i=1}^{n} x_i v_i.$$

Example 1.13 Given $Q \in \mathcal{L}^0(\mathbb{R}^{n \times n})$, $v \in \mathcal{L}^0(\mathbb{R}^n)$, $m \in \mathcal{L}^0(\mathbb{R})$, the function

$$h(x, \omega) := \frac{1}{2} x \cdot Q(\omega) x + x \cdot v(\omega) + m(\omega)$$

is a Carathéodory integrand and, in particular, a normal integrand.

Given a function $h : \mathbb{R}^n \times \Omega \to \overline{\mathbb{R}}$ and a random variable $\beta \in \mathcal{L}^0$, the associated *level-set mapping* is defined by

$$(\operatorname{lev}_\beta h)(\omega) := \{x \in \mathbb{R}^n \mid h(x,\omega) \le \beta(\omega)\}.$$

The *domain mapping* of h is defined by

$$(\operatorname{dom} h)(\omega) := \{x \in \mathbb{R}^n \mid h(x,\omega) < \infty\},$$

and its *strict epigraphical mapping* by

$$(\operatorname{epi}_s h)(\omega) := \{(x,\alpha) \in \mathbb{R}^n \times \mathbb{R} \mid h(x,\omega) < \alpha\}.$$

Since the preimages of open sets under $\operatorname{epi}_s h$ and $\operatorname{epi} h$ are the same, we have the following.

Lemma 1.14 *Given a function $h : \mathbb{R}^n \times \Omega \to \overline{\mathbb{R}}$, $\operatorname{epi} h$ is measurable if and only if $\operatorname{epi}_s h$ is measurable.*

Theorem 1.15 *Given a normal integrand h on $\mathbb{R}^n \times \Omega$, the domain mapping $\operatorname{dom} h$ is measurable and, for any random variable $\beta \in \mathcal{L}^0$, the level-set mapping $\operatorname{lev}_\beta h$ is closed-valued and measurable. A function $h : \mathbb{R}^n \times \Omega \to \overline{\mathbb{R}}$ is a normal integrand if and only if $\operatorname{lev}_{\le\beta} h$ is closed-valued and measurable for every $\beta \in \mathbb{R}$.*

Proof For a closed set $C \subset \mathbb{R}^n$, $R(\omega) := C \times \{\alpha \mid \alpha \le \beta(\omega)\}$ is closed-valued and measurable. We have

$$\begin{aligned}(\operatorname{lev}_\beta h)^{-1}(C) &= \{\omega \in \Omega \mid (\operatorname{epi} h)(\omega) \cap R(\omega) \ne \emptyset\}\\ &= \operatorname{dom}(\operatorname{epi} h \cap R).\end{aligned}$$

By Theorem 1.7, $\operatorname{epi} h \cap R$ is measurable which implies the measurability of $\operatorname{dom}(\operatorname{epi} h \cap R)$. Thus, $\operatorname{lev}_\beta h$ is measurable, by Theorem 1.2. Since

$$(\operatorname{dom} h)(\omega) = \bigcup_{\beta\in\mathbb{Q}} (\operatorname{lev}_\beta h)(\omega),$$

the domain mapping is measurable by Theorem 1.7.

To prove the third claim, recall first that the epigraph of a function is closed if and only if all its lower level-sets are closed. By Lemma 1.14, it thus suffices to show that $\operatorname{epi}_s h$ is measurable. It is easy to verify that

$$(\operatorname{epi}_s h)(\omega) = \bigcup_{\beta\in\mathbb{Q}} \left(\{x \in \mathbb{R}^n \mid h(x,\omega) < \beta\} \times [\beta,\infty)\right).$$

Thus, by Theorem 1.7, it suffices to show that the mappings

$$\omega \mapsto \{x \in \mathbb{R}^n \mid h(x,\omega) < \beta\}$$

are measurable. This follows by writing

$$\{x \in \mathbb{R}^n \mid h(x,\omega) < \beta\} = \bigcup_{\nu \in \mathbb{N}} (\operatorname{lev}_{\beta - 1/\nu} h)(\omega)$$

and using Theorem 1.7 again. □

Example 1.16 Given a Carathéodory integrand h, the set-valued mapping

$$\omega \mapsto \{x \in \mathbb{R}^n \mid h(x,\omega) = 0\}$$

is measurable.

Proof If h is a Carathéodory integrand, then both h and $-h$ are normal integrands, by Example 1.12. Since

$$\{x \in \mathbb{R}^n \mid h(x,\omega) = 0\} = \{x \in \mathbb{R}^n \mid h(x,\omega) \leq 0\} \cap \{x \in \mathbb{R}^n \mid -h(x,\omega) \leq 0\}$$

the claim follows from Theorems 1.15 and 1.7. □

The following simple fact will be crucial.

Theorem 1.17 *A normal integrand h on $\mathbb{R}^n \times \Omega$ is $\mathcal{B}(\mathbb{R}^n) \otimes \mathcal{F}$-measurable.*

Proof For any $\beta \in \mathbb{R}$, $\operatorname{lev}_{\leq \beta} h$ is closed-valued and measurable by Theorem 1.15, so $\{(x,\omega) \mid h(x,\omega) \leq \beta\} \in \mathcal{B}(\mathbb{R}^n) \otimes \mathcal{F}$, by Corollary 1.3. □

Given functions $h : \mathbb{R}^n \times \Omega \to \overline{\mathbb{R}}$ and $x : \Omega \to \mathbb{R}^n$, we will define $h(x) : \Omega \to \overline{\mathbb{R}}$ by

$$h(x)(\omega) := h(x(\omega), \omega).$$

Since the compositions of measurable functions are measurable, Theorem 1.17 gives the following.

Corollary 1.18 *Given a normal integrand h on $\mathbb{R}^n \times \Omega$, the function $h(x)$ is measurable for every $x \in \mathcal{L}^0(\mathbb{R}^n)$.*

Corollary 1.18 will be needed in much of what follows and it will be used without a mention. The following gives a converse of Theorem 1.17.

Remark 1.19 Assume, as in Remark 1.4, that

$$\{\omega \in \Omega \mid \exists x \in \mathbb{R}^n : (\omega, x) \in A\} \in \mathcal{F} \quad \forall A \in \mathcal{F} \otimes \mathcal{B}(\mathbb{R}^n).$$

If $h : \mathbb{R}^n \times \Omega \to \overline{\mathbb{R}}$ is $\mathcal{F} \otimes \mathcal{B}(\mathbb{R}^n)$-measurable and $h(\cdot, \omega)$ is lsc for every $\omega \in \Omega$, then h is a normal integrand.

Proof Given $\beta \in \mathbb{R}$, $\operatorname{gph}(\operatorname{lev}_{\leq\beta} h)$ is $\mathcal{F} \otimes \mathcal{B}(\mathbb{R}^n)$-measurable and $\operatorname{lev}_{\leq\beta} h$ is closed-valued. Thus, the claim follows from Remark 1.4 and Theorem 1.15. □

Given an extended real-valued function g and $\alpha \in \mathbb{R}_+$, we define

$$(\alpha g)(x) := \begin{cases} \alpha g(x) & \text{if } \alpha > 0, \\ \delta_{\operatorname{cl}\operatorname{dom} g}(x) & \text{if } \alpha = 0. \end{cases}$$

Note that, when $\alpha = 0$, $(\alpha g)(x) \neq \alpha g(x)$ when $x \notin \operatorname{cl}\operatorname{dom} g$ as the latter is defined as zero in the algebra of extended real numbers. Clearly, αg is lsc when g is so. The function $\delta_{\operatorname{cl}\operatorname{dom} g}$ is the lower semicontinuous hull of $\delta_{\operatorname{dom} g}$ which in turn is the pointwise limit of αg when $\alpha \searrow 0$. We will denote the *lower semicontinuous hull* of a function g by $\operatorname{lsc} g$.

Theorem 1.20 *The following are normal integrands:*

1. $h(x, \omega) := \sup_{j\in J} h^j(x, \omega)$, *where* $(h^j)_{j\in J}$ *is a countable collection of normal integrands;*
2. $h(x, \omega) := \inf_{j\in J} h^j(x, \omega)$, *where* $(h^j)_{j\in J}$ *is a finite collection of normal integrands;*
3. $h(x, \omega) := \sum_{i=1}^n h^i(x, \omega)$, *where* h^i *are normal integrands;*
4. $h(x, \omega) := (\alpha(\omega)h^0)(x, \omega)$, *where* h^0 *is a normal integrand and* $\alpha \in \mathcal{L}^0_+$*;*
5. $h(x, \omega) := f(x, u(\omega), \omega)$, *where* $f : \mathbb{R}^n \times \mathbb{R}^m \times \Omega \to \overline{\mathbb{R}}$ *is a normal integrand and* $u \in \mathcal{L}^0(\mathbb{R}^m)$ *is measurable;*
6. $h(x, \omega) := g(M(\omega, x), \omega)$, *where* g *is a normal integrand and* $M : \Omega \times \mathbb{R}^m \to \mathbb{R}^n$ *is a Carathéodory mapping.*

Proof In 1, $\operatorname{epi} h = \bigcap_{j\in J} \operatorname{epi} h^j$, so the claim follows directly from Theorem 1.7. In 2, $\operatorname{epi} h = \bigcup_{j\in J} \operatorname{epi} h^j$, so the claim follows from Theorem 1.7 and the fact that pointwise infimum of a finite collection of lsc functions is lsc.

To prove 3, let $S = \operatorname{epi} h^1 \times \cdots \times \operatorname{epi} h^n$ and

$$M(x_1, \alpha_1, \ldots, x_n, \alpha_n) = \begin{cases} (x, \alpha_1 + \cdots + \alpha_n) & \text{if } x_1 = \cdots = x_n = x, \\ \emptyset & \text{otherwise} \end{cases}$$

so that $\operatorname{epi} h = M \circ S$ and the claim follows from Theorem 1.9.1. Part 4, follows similarly since now $\operatorname{epi} h = M \circ S$ where

$$S(\omega) = \begin{cases} (\operatorname{epi} h^0)(\omega) & \text{if } \alpha(\omega) > 0, \\ \operatorname{epi} \delta_{\operatorname{cl}(\operatorname{dom} h^0)(\omega)} & \text{if } \alpha(\omega) = 0 \end{cases}$$

and $M(x,\alpha',\omega) = \{(x,\beta) \mid \alpha(\omega)\alpha' \le \beta\}$. Indeed, the measurability of S follows from Theorems 1.15 and 1.2 and the measurability of the set $\{\alpha > 0\}$ while the mapping M is closed-valued and has a measurable graph since α is measurable. In 5, we have

$$(\operatorname{epi} h)(\omega) = \{(x,\alpha) \mid M(x,\alpha,\omega) \in (\operatorname{epi} f)(\omega)\}$$

for the Carathéodory mapping $M(x,\alpha,\omega) := (x,u(\omega),\alpha)$. The claim thus follows from Theorem 1.9.2. In 6,

$$(\operatorname{epi} h)(\omega) = \{(x,\alpha) \mid (M(x,\omega),\alpha) \in (\operatorname{epi} g)(\omega)\},$$

so the claim follows from Theorem 1.9.2. □

The first part of the following result will be key in the study of the dynamic programming principle for stochastic optimization problems. We say that a mapping is *locally bounded* if every point in the domain space has a neighborhood whose image under the mapping is bounded.

Theorem 1.21 *Let $f : \mathbb{R}^n \times \mathbb{R}^m \times \Omega \to \overline{\mathbb{R}}$ be a normal integrand and define*

$$p(u,\omega) := \inf_{x\in\mathbb{R}^n} f(x,u,\omega).$$

The function $\operatorname{lsc} p$*, obtained by taking the lsc hull of $p(\cdot,\omega)$ for each ω, is a normal integrand on $\mathbb{R}^m$. In particular, if $p(\cdot,\omega)$ is lsc for every $\omega \in \Omega$, then p is a normal integrand.*

If $p(\cdot,\omega)$ is continuous for every $\omega \in \Omega$, then the solution mapping

$$S(u,\omega) := \operatorname*{argmin}_{x\in\mathbb{R}^n} f(x,u,\omega)$$

is such that $\omega \mapsto$ $\operatorname{gph} S(\cdot,\omega)$ is closed-valued and measurable. If, in addition, $S(\cdot,\omega)$ is locally bounded and single-valued, then it is continuous.

Proof Let $\Pi(x,u,\alpha) = (u,\alpha)$ be the projection from $\mathbb{R}^n \times \mathbb{R}^m \times \mathbb{R}$ to $\mathbb{R}^m \times \mathbb{R}$. It is easy to check that $\Pi(\operatorname{epi}_s f)(\omega) = (\operatorname{epi}_s p)(\omega)$ for all $\omega \in \Omega$, so

$$(\operatorname{epi}_s p)^{-1}(O) = (\operatorname{epi}_s f)^{-1}(\Pi^{-1}(O))$$

for every $O \subset \mathbb{R}^n \times \mathbb{R}$. Since Π is continuous, the right side is measurable, by Lemma 1.14 for every open O. Thus, by Lemma 1.14 again, $\operatorname{epi} p$ is measurable. By Lemma 1.1, this implies that $\omega \mapsto \operatorname{cl}\operatorname{epi} p(\cdot,\omega)$ is measurable so $(u,\omega) \mapsto (\operatorname{lsc} p)(u,\omega)$ is a normal integrand.

As to the last claim, we have

$$\operatorname{gph} S(\cdot,\omega) = \{(x,u) \in \mathbb{R}^n \times \mathbb{R}^m \mid f(x,u,\omega) \le p(u,\omega)\}.$$

The previous claim implies that p is a Carathéodory integrand so $-p$ is a normal integrand. The closedness and measurability of the graph now follow from Theorems 1.20 and 1.15. As to the last claim, it is easily shown that a locally bounded single-valued mapping with a closed graph in $\mathbb{R}^m \times \mathbb{R}^n$ is continuous. □

Remark 1.22 If, instead of continuity of $p(\cdot, \omega)$, we assume in Theorem 1.21 that $p(\cdot, \omega)$ is just lsc and $\mathcal{F}$ satisfies condition (1.1) in Remark 1.4, then the mapping $\omega \mapsto \operatorname{gph} S(\cdot, \omega)$ in Theorem 1.21 is still measurable but not necessarily closed-valued.

Proof By the first part of Theorem 1.21, p is a normal integrand so it is $\mathcal{B}(\mathbb{R}^n) \otimes \mathcal{F}$-measurable, by Theorem 1.17. This implies that the set

$$\{(x, u, \omega) \in \mathbb{R}^n \times \mathbb{R}^m \times \Omega \mid f(x, u, \omega) \le p(u, \omega)\}$$

is $\mathcal{B}(\mathbb{R}^n \times \mathbb{R}^m) \otimes \mathcal{F}$-measurable or, in other words, the graph of the mapping $\omega \mapsto \operatorname{gph} S(\cdot, \omega)$ is measurable. By Remark 1.4, this implies its measurability under condition (1.1). □

Corollary 1.23 *If h is a normal integrand, then the function*

$$p(\omega) := \inf_x h(x, \omega)$$

is measurable, and the mapping

$$S(\omega) := \operatorname*{argmin}_x h(x, \omega)$$

is closed-valued and measurable.

Proof Apply Theorem 1.21 to the normal integrand $f(x, u, \omega) := h(x, \omega)$. □

Corollary 1.24 *Given a closed-valued measurable $S : \Omega \rightrightarrows \mathbb{R}^n$, the projection mapping*

$$P_S(x, \omega) := \operatorname*{argmin}_{x' \in S(\omega)} |x' - x|$$

is such that the mapping $\omega \mapsto \operatorname{gph} P_S(\cdot, \omega)$ is closed-valued and $\mathcal{F}$-measurable. Moreover, for any $x \in \mathcal{L}^0(\mathbb{R}^n)$, the mapping $\omega \mapsto P_S(x(\omega), \omega)$ is closed-valued and measurable and its domain equals that of S. If S is convex-valued, then P_S is single-valued.

Proof The first claim follows by applying the last claim of Theorem 1.21 to the normal integrand $f(x, u, \omega) := \delta_{S(\omega)}(x) + |x - u|^2$. Indeed, by Theorem A.23, $p(\cdot, \omega)$ is continuous for every $\omega \in \Omega$. The second claim follows by combining the first claim with Theorem 1.9. When S is convex-valued, $f(\cdot, u, \omega)$ is strictly convex, which implies the last claim. □

Recall that the Moore-Penrose inverse of a matrix $A \in \mathbb{R}^{m\times n}$ is the matrix $A^\dagger \in \mathbb{R}^{n\times m}$ such that, for each $u \in \mathbb{R}^m$, the minimum-norm solution of the least squares problem of minimizing $|Ax - u|^2$ over $x \in \mathbb{R}^n$ is given by $A^\dagger u$; see Example A.84.

Example 1.25 Given a random matrix $A \in \mathcal{L}^0(\mathbb{R}^{m\times n})$, the scenariowise Moore-Penrose inverse $A^\dagger(\omega) := A(\omega)^\dagger$ is measurable.

Proof Fix a $u \in \mathbb{R}^m$. By Corollary 1.23, $\mathrm{argmin}_x |A(\omega)x - u|^2$ is measurable and then, by Corollary 1.24, $A^\dagger(\omega)u$ is measurable. Since $u \in \mathbb{R}^m$ was arbitrary, $A^\dagger$ is measurable. □

We say that a normal integrand h is *proper* if the function $h(\cdot, \omega)$ is proper for every $\omega \in \Omega$. Given $\rho \in \mathcal{L}^0$, we say that a normal integrand h is *ρ-Lipschitz continuous* if, for every $\omega \in \Omega$, the function $h(\cdot, \omega)$ is Lipschitz continuous with Lipschitz constant $\rho(\omega)$. Combining Theorem 1.21 and Lemma A.24 in the appendix gives the following.

Corollary 1.26 *Let $\rho, m \in \mathcal{L}^0_+$ with $\rho > 0$ and let h be a proper normal integrand such that $h(x, \omega) \ge -\rho(\omega)|x| - m$ for all $x \in \mathbb{R}^n$ and $\omega \in \Omega$. The functions*

$$h^\nu(x, \omega) := \inf_{x' \in \mathbb{R}^n} \{h(x', \omega) + \nu\rho(\omega)|x - x'|\} \quad \nu \in \mathbb{N}$$

are $(\nu\rho)$-Lipschitz continuous normal integrands with $h^\nu(x, \omega) \ge -\rho(\omega)|x| - m$ for all $x \in \mathbb{R}^n$ and $\omega \in \Omega$. As ν increases, the integrands h^ν increase pointwise to h.

1.1.3 Measurable Selections

Recall that a Castaing representation of a set-valued mapping $S : \Omega \rightrightarrows \mathbb{R}^n$ is a countable collection $\{x^\nu\}_{\nu\in\mathbb{N}}$ of measurable selections of S such that

$$S(\omega) = \mathrm{cl}\{x^\nu(\omega)\}_{\nu\in\mathbb{N}} \quad \forall \omega \in \mathrm{dom}\, S.$$

The following gives a converse of Lemma 1.5.

Theorem 1.27 (Castaing Representation) *A closed-valued measurable mapping admits a Castaing representation.*

Proof Let $S : \Omega \rightrightarrows \mathbb{R}^n$ be closed-valued and measurable. Let $(x^\nu)_{\nu\in\mathbb{N}}$ be dense in $\mathbb{R}^n$. Let $(v^i)_{i=1}^n \subset \mathbb{R}^n$ be orthogonal. For each ν, define recursively

$$S^{\nu,0}(\omega) := \mathrm{argmin}\{|x - x^\nu| \mid x \in S(\omega)\},$$

$$S^{\nu,i}(\omega) := \mathrm{argmin}\{x \cdot v_i \mid x \in S^{\nu,i-1}(\omega)\} \quad i = 1, \dots, n.$$

By Corollary 1.23, each $S^{\nu,i}$ is measurable. Moreover, $S^{\nu,n}(\omega)$ is the singleton $x^\nu(\omega) := \sum r_i(\omega)v_i$, where $r_i(\omega) := \inf\{x \cdot v_i \mid x \in S^{\nu,i-1}(\omega)\}$. Since $x^\nu \in S^{\nu,0}$, the point $x^\nu(\omega)$ is a vector in $S(\omega)$ nearest to x^ν. The family $\{\bar{x}^\nu\}_{\nu\in\mathbb{N}}$ is a Castaing representation of S. Indeed, if the family is not dense in $S(\omega)$, there exist $x \in S(\omega)$ and $\epsilon > 0$ such that $\bar{x}^\nu \notin \mathbb{B}_\epsilon(x)$ for any ν. But $x^\nu \in \mathbb{B}_x(\epsilon/3)$ for some ν, so $d(x, x^\nu) < d(x, \bar{x}^\nu(\omega))$, which is a contradiction. □

The following is an immediate consequence of Theorem 1.27.

Corollary 1.28 (Measurable Selection Theorem) *A closed-valued measurable mapping admits a measurable selection.*

Given set-valued mappings $S^1, S^2 : \Omega \rightrightarrows \mathbb{R}^n$, we write $S^1 \subseteq S^2$ if $S^1(\omega) \subseteq S^2(\omega)$ for every $\omega \in \Omega$.

Corollary 1.29 *Given closed-valued measurable mappings S^1 and S^2, the following are equivalent:*

1. $S^1 \subseteq S^2$;
2. $\{\omega \in \Omega \mid x(\omega) \in S^1(\omega)\} \subseteq \{\omega \in \Omega \mid x(\omega) \in S^2(\omega)\}$ *for every measurable function x;*
3. *every measurable selection of S^1 belongs to S^2 on* $\operatorname{dom} S^1$.

Proof The implication $1 \Rightarrow 2$ is clear. Let x be a measurable selection of S^1 and extend it to a measurable function on all of Ω by setting e.g. $x = 0$ outside of $\operatorname{dom} S^1$. Condition then 2 implies $x(\omega) \in S^2(\omega)$ for all $\omega \in \operatorname{dom} S^1$ so 3 holds. Let $\{x^\nu\}_{\nu\in\mathbb{N}}$ be a Castaing representation of S^1. Condition 3 implies that $\{x^\nu(\omega)\}_{\nu\in\mathbb{N}} \subseteq S^2(\omega)$ for every $\omega \in \operatorname{dom} S^1$. Taking closures on both sides gives $S^1 \subseteq S^2$ on $\operatorname{dom} S^1$. Thus, 1 holds. □

Combining Corollaries 1.23 and 1.28 gives the following.

Corollary 1.30 *Given a normal integrand h, the mapping $\omega \mapsto \operatorname{argmin} h(\cdot, \omega)$ admits a measurable selection.*

Given $C^1, C^2 \subseteq \mathbb{R}^n$, we use the notation $C^1 \setminus C^2 := \{x \in C^1 \mid x \notin C^2\}$.

Corollary 1.31 *Given closed-valued measurable mappings $S^j : \Omega \rightrightarrows \mathbb{R}^n$, the mapping*

$$(S^1 \setminus S^2)(\omega) := S^1(\omega) \setminus S^2(\omega),$$

is measurable.

Proof By Theorem 1.27, S^1 admits a Castaing representation $\{x^\nu\}_{\nu\in\mathbb{N}}$. For each x^ν, the closed-valued mapping

$$S^\nu(\omega) := \begin{cases} x^\nu(\omega) & \text{if } \omega \notin \operatorname{dom}(S^2 \cap \{x^\nu\}), \\ \emptyset & \text{otherwise} \end{cases}$$

is measurable by parts 2 and 4 of Theorem 1.7. Since $\{x^\nu(\omega)\}_{\nu\in\mathbb{N}}$ is dense in $S^1(\omega)$ and since $S^1(\omega)\setminus S^2(\omega)$ is open relative to $S^1(\omega)$, we have

$$\operatorname{cl}(S^1(\omega)\setminus S^2(\omega)) = \operatorname{cl}\bigcup_{\nu\in\mathbb{N}} S^\nu(\omega),$$

where the right side defines a measurable mapping by Theorem 1.7 and Lemma 1.1. By Lemma 1.1 again, $S^1\setminus S^2$ is measurable. □

The boundary of a set $C\subseteq\mathbb{R}^n$ will be denoted by

$$\operatorname{bdry} C := \operatorname{cl} C\setminus\operatorname{int} C.$$

Clearly, $\operatorname{bdry} C := \operatorname{cl} C\cap\operatorname{cl}(\mathbb{R}^n\setminus C)$.

Corollary 1.32 *Given a closed-valued measurable mapping $S:\Omega\rightrightarrows\mathbb{R}^n$, the complement mapping*

$$\omega\mapsto\mathbb{R}^n\setminus S(\omega),$$

the boundary mapping

$$(\operatorname{bdry} S)(\omega) := \operatorname{bdry} S(\omega)$$

and the interior mapping

$$(\operatorname{int} S)(\omega) := \operatorname{int} S(\omega)$$

are measurable.

Proof The complement mapping is measurable by Corollary 1.31. The boundary mapping is now measurable by Theorem 1.7. Since complement of the interior is the closure of the complement, the last claim follows from the first and Lemma 1.1. □

Given a random variable ξ with values in a measurable space $(\Xi,\mathcal{A})$, the smallest σ-algebra on Ω under which ξ is measurable is denoted by $\sigma(\xi)$.

Corollary 1.33 (Doob–Dynkin for Set-Valued Mappings) *Let ξ be a random variable with values in a measurable space $(\Xi,\mathcal{A})$. A closed-valued mapping S is $\sigma(\xi)$-measurable if and only if there exists closed-valued $\mathcal{A}$-measurable $\tilde{S}:\Xi\rightrightarrows\mathbb{R}^n$ such that $S(\omega)=\tilde{S}(\xi(\omega))$. If S is convex-valued, $\tilde{S}$ can be chosen convex-valued.*

Proof The sufficiency is clear. By Theorem 1.27, S has a $\sigma(\xi)$-measurable Castaing representation $(x^\nu)_{\nu\in\mathbb{N}}$. By Lemma B.7, there exists, for each ν, a Borel-measurable $g^\nu : \Xi \to \mathbb{R}^n$ such that $x^\nu(\omega) = g^\nu(\xi(\omega))$ for all $\omega \in \operatorname{dom} S$. Defining the mapping $\tilde{S}' : \Xi \rightrightarrows \mathbb{R}^n$ by

$$\tilde{S}'(\eta) := \operatorname{cl}\{g^\nu(\eta)\}_{\nu\in\mathbb{N}},$$

we have $S(\omega) = \operatorname{cl}\{g^\nu(\xi(\omega))\}_{\nu\in\mathbb{N}} = \tilde{S}'(\xi(\omega))$ on $\operatorname{dom} S$. Since $\operatorname{dom} S$ is $\sigma(\xi)$-measurable, there is a set $D \in \mathcal{A}$ such that $\operatorname{dom} S = \xi^{-1}(D)$. Defining

$$\tilde{S}(\eta) = \begin{cases} \tilde{S}'(\eta) & \eta \in D, \\ \emptyset & \eta \notin D \end{cases}$$

we have $S(\omega) = \tilde{S}(\xi(\omega))$ for all $\omega \in \Omega$. If S is convex-valued, $\tilde{S}(\eta)$ is convex for all $\eta \in \operatorname{rge} \xi := \{\xi(\omega) \mid \omega \in \Omega\}$. We can thus take the closed convex hull of $\tilde{S}(\eta)$ for all $\eta \in \Xi$ without interfering with the equality. By Corollary 1.8, the resulting mapping is still measurable. □

Corollary 1.34 (Doob–Dynkin for Normal Integrands) *Let ξ be a random variable with values in a measurable space $(\Xi, \mathcal{A})$. A function h is a $\sigma(\xi)$-normal integrand on $\mathbb{R}^n$ if and only if there exists a $\mathcal{A}$-normal integrand H on $\mathbb{R}^n$ such that*

$$h(x, \omega) = H(x, \xi(\omega)).$$

If h is convex, H can be chosen convex.

Proof Apply Corollary 1.33 to the epigraphical mapping of h. □

1.1.4 Convexity

The analysis of random sets and normal integrands simplifies in the presence of convexity. In particular, for a closed convex random set, the existence of measurable selections and a Castaing representation follow directly from Corollary 1.24.

Remark 1.35 (Castaing Representation, Convex Case) Assume that $S : \Omega \rightrightarrows \mathbb{R}^n$ is convex-valued and measurable and let $\{x^\nu\}_{\nu\in\mathbb{N}}$ be a dense set in $\mathbb{R}^n$. Let $\bar{x}^\nu(\omega)$ be the projection of x^ν to $S(\omega)$. By Corollary 1.24, $\bar{x}^\nu$ is well-defined and measurable. By Corollary A.82, convexity of $S(\omega)$ implies that every point in $S(\omega)$ is nearer to $\bar{x}^\nu(\omega)$ than x^ν. Since $S(\omega)$ is contained in the closure of $\{x^\nu\}_{\nu\in\mathbb{N}}$ it is contained in the closure of $\{\bar{x}(\omega)\}_{\nu\in\mathbb{N}}$ as well.

The following gives a version of Corollary 1.32 where the complement, boundary and interior are defined with respect to the affine hull of $S(\omega)$. Recall that the *relative interior* $\operatorname{rint} C$ of a set $C \subset \mathbb{R}^n$ is the interior of C with respect to the affine hull

$\operatorname{aff} C$ of C. The relative interior of a nonempty convex set in $\mathbb{R}^n$ is always nonempty. The *relative boundary* $\operatorname{rbdry} C$ of C is defined as the boundary of C with respect to $\operatorname{aff} C$, i.e.

$$\operatorname{rbdry} C := \operatorname{cl} C \setminus \operatorname{rint} C.$$

Note that $\operatorname{aff} C \setminus \operatorname{rint} C = \operatorname{cl}(\operatorname{aff} C \setminus C)$ so $\operatorname{rbdry} C = \operatorname{cl} C \cap \operatorname{cl}(\operatorname{aff} C \setminus C)$.

Corollary 1.36 *Given a measurable convex-valued $S : \Omega \rightrightarrows \mathbb{R}^n$, the relative complement mapping*

$$\omega \mapsto \operatorname{aff} S(\omega) \setminus S(\omega),$$

the relative boundary mapping

$$(\operatorname{rbdry} S)(\omega) := \operatorname{rbdry} S(\omega)$$

and the relative interior mapping

$$(\operatorname{rint} S)(\omega) := \operatorname{rint} S(\omega)$$

are measurable and they all admit measurable selections.

Proof By Theorem A.28, the sets $S(\omega)$ and $\operatorname{cl} S(\omega)$ have the same relative interiors, so the relative complements $\operatorname{aff} S(\omega) \setminus S(\omega)$ and $\operatorname{aff} S(\omega) \setminus \operatorname{cl} S(\omega)$ have the same closures. The last mapping is measurable, by Lemma 1.1 and Corollary 1.31, so the measurability of the relative complement follows from Lemma 1.1. The measurability of the relative boundary mapping now follows from Lemma 1.1 and Theorem 1.7. Since $\operatorname{cl} \operatorname{rint} S(\omega) = \operatorname{cl} S(\omega)$, by Theorem A.28, the measurability of the relative interior mapping follows from Lemma 1.1.

We show next that the relative interior mapping admits a measurable selection. By Lemma 1.1, the closed convex-valued mapping $\operatorname{cl} S$ is measurable. By Theorem 1.27, $\operatorname{cl} S$ admits a Castaing representation $\{x^\nu\}_{\nu \in \mathbb{N}}$. Let $\alpha^\nu = ce^{-|x^\nu|-\nu}$, where $c > 0$ is such that $\sum_{\nu=1}^\infty \alpha^\nu = 1$. It follows that $x := \sum_{\nu=1}^\infty \alpha^\nu x^\nu$ is well-defined and measurable. Since the Castaing representation is dense in $\operatorname{cl} S(\omega)$, the set $\{x^\nu(\omega)\}_{\nu \in \mathbb{N}}$ intersects $\operatorname{rint} S(\omega)$ for every $\omega \in \Omega$. Since $\alpha^\nu(\omega) > 0$ for all $\nu \in \mathbb{N}$ we have $x(\omega) \in \operatorname{rint} S(\omega)$, by Theorem A.28.

The relative boundary of S admits a measurable selection x^1, by Theorem 1.27. As observed above, the relative interior $\operatorname{rint} S$ admits a selection x^0. The domain of the relative complement mapping coincides with that of $\operatorname{bdry} S$. For any $\alpha > 1$, the measurable function $x^2 := \alpha x^1 + (1 - \alpha)x^0$ defined on $\operatorname{dom} \operatorname{bdry} S$ is a selection of the relative complement of S. Indeed, if we had $x^2(\omega) \in S(\omega)$, then, by Theorem A.28, $x^1(\omega) \in \operatorname{rint} S(\omega)$ which is impossible since $x^1(\omega) \in \operatorname{bdry} S(\omega)$. □

The following gives a simple sufficient condition, akin to the Carathéodory property, for measurability of convex normal integrands.

Theorem 1.37 *Let $h : \mathbb{R}^n \times \Omega \to \overline{\mathbb{R}}$ be such that, for every ω, the function $h(\cdot, \omega)$ is proper lsc convex and its domain has nonempty interior. Then h is a convex normal integrand if and only if $h(x, \cdot)$ is measurable for every $x \in \mathbb{R}^n$.*

Proof By Theorem A.33, $h(\cdot, \omega)$ is continuous on $\operatorname{int}\operatorname{dom} h(\cdot, \omega)$, which implies $\operatorname{int}(\operatorname{epi} h)(\omega) \neq \emptyset$. By Theorem A.28, a closed convex set with nonempty interior is solid, so the epigraphical mapping $\omega \mapsto (\operatorname{epi} h)(\omega)$ is solid-valued. Thus, by Theorem 1.6, $\omega \mapsto \operatorname{epi} h(\cdot, \omega)$ is a measurable mapping if and only if for each $(x, \alpha) \in \mathbb{R}^n \times \mathbb{R}$, the set

$$(\operatorname{epi} h)^{-1}(x, \alpha) = \{\omega \in \Omega \mid h(x, \omega) \leq \alpha\}$$

is measurable. This just means that $h(x, \cdot)$ is measurable for every $x \in \mathbb{R}^n$. □

The *recession cone* of a convex set $C \subseteq \mathbb{R}^n$ is the convex cone

$$C^\infty := \{x \in \mathbb{R}^n \mid \bar{x} + \lambda x \in C \ \forall \bar{x} \in C,\ \lambda > 0\};$$

see Sect. A.3. The recession cone describes the shape of a convex set infinitely far from the origin. For a convex-valued $S : \Omega \rightrightarrows \mathbb{R}^n$, we denote its scenariowise recession cone by

$$S^\infty(\omega) := S(\omega)^\infty \quad \forall \omega \in \Omega.$$

This defines a convex cone-valued mapping $S^\infty : \Omega \rightrightarrows \mathbb{R}^n$.

Theorem 1.38 *If $S : \Omega \rightrightarrows \mathbb{R}^n$ is closed convex-valued and measurable, then so too is S^∞.*

Proof By Theorem 1.27, there exists an $\bar{x} \in \mathcal{L}^0(\mathbb{R}^n)$ such that $\bar{x}(\omega) \in S(\omega)$ for all $\omega \in \operatorname{dom} S$. By Theorem A.15,

$$S^\infty(\omega) = \bigcap_{\nu \in \mathbb{N}} \frac{1}{\nu}(S(\omega) - \bar{x}(\omega))$$

for every $\omega \in \operatorname{dom} S$ so the measurability follows from Theorem 1.7. □

The *recession function* g^∞ of a convex function g is defined by

$$\operatorname{epi} g^\infty = (\operatorname{epi} g)^\infty;$$

see Sect. A.3. The recession function describes the growth of the function infinitely far from the origin. Given a convex normal integrand h, we denote its scenariowise recession function by

$$h^\infty(\cdot,\omega) := h(\cdot,\omega)^\infty \quad \forall\omega\in\Omega.$$

Theorem 1.39 *Given a convex normal integrand h, h^∞ is a sublinear normal integrand and, if h is proper, then*

$$h^\infty(x,\omega) = \sup_{\lambda>0} \frac{h(\bar{x}(\omega)+\lambda x,\omega) - h(\bar{x}(\omega),\omega)}{\lambda} \quad \forall(x,\omega)\in\mathbb{R}^n\times\Omega$$

for every $\bar{x}\in\mathcal{L}^0(\operatorname{dom} h)$.

Proof By definition, $(\operatorname{epi} h^\infty)(\omega) = (\operatorname{epi} h(\omega))^\infty$, so the first claim follows from Theorem 1.38. The formula follows from Theorem A.16 in the appendix. □

Convexity allows for convenient sufficient conditions for the lower semicontinuity condition in Theorem 1.21. Given a set $N\in\mathbb{R}^n$, we denote its *orthogonal complement* by

$$N^\perp := \{v\in\mathbb{R} \mid x\cdot v = 0 \quad \forall x\in N\}.$$

Theorem 1.40 *Let f be a proper convex normal integrand on $\mathbb{R}^n\times\mathbb{R}^m$ and assume that the set-valued mapping*

$$N(\omega) := \{x\in\mathbb{R}^n \mid f^\infty(x,0,\omega)\le 0\}$$

is linear-valued. Then

$$p(u,\omega) := \inf_{x\in\mathbb{R}^n} f(x,u,\omega)$$

is a convex normal integrand on $\mathbb{R}^m$ and

$$p^\infty(u,\omega) = \inf_{x\in\mathbb{R}^n} f^\infty(x,u,\omega).$$

Given $u\in\mathcal{L}^0(\mathbb{R}^m)$, there exists $x\in\mathcal{L}^0(\mathbb{R}^n)$ with $x(\omega)\in N(\omega)^\perp$ and

$$p(u(\omega),\omega) = f(x(\omega),u(\omega),\omega)$$

for all $\omega\in\Omega$.

Proof By Theorem A.77, the linearity condition implies for every $\omega \in \Omega$, that the set

$$S(u,\omega) := N(\omega)^{\perp} \cap \operatorname*{argmin}_{x\in\mathbb{R}^n} f(x,u,\omega)$$

is nonempty and that $p(\cdot,\omega)$ is a lower semicontinuous convex function with

$$p^{\infty}(u,\omega) = \inf_{x\in\mathbb{R}^n} f^{\infty}(x,u,\omega).$$

By Theorem 1.21, the lower semicontinuity implies that p is a normal integrand. Given a $u \in \mathcal{L}^0(\mathbb{R}^m)$, the function $h(x,\omega) := f(x,u(\omega),\omega)$ is a normal integrand, by Theorem 1.20.5. By Corollary 1.23, Theorems 1.39, 1.15 and 1.7, the mapping $\omega \mapsto S(u(\omega),\omega)$ is closed-valued and measurable so, by Theorem 1.27, it admits a measurable selection $x \in \mathcal{L}^0(\mathbb{R}^n)$. The same argument applied to f^{∞} gives the existence of $\tilde{x} \in \mathcal{L}^0(\mathbb{R}^n)$. □

Applying Theorem 1.40 to the indicator function of a measurable closed convex set gives the following.

Corollary 1.41 *Let $S : \Omega \rightrightarrows \mathbb{R}^n \times \mathbb{R}^m$ be closed convex-valued and measurable and such that*

$$N(\omega) = \{x \in \mathbb{R}^n \mid (x,0) \in S^{\infty}(\omega)\}$$

is linear-valued. Then the scenariowise projection

$$C(\omega) := \{u \in \mathbb{R}^m \mid \exists x \in \mathbb{R}^n : \ (x,u) \in S(\omega)\}$$

of S on $\mathbb{R}^m$ is closed convex-valued and measurable and

$$C^{\infty}(\omega) = \{u \in \mathbb{R}^m \mid \exists x \in \mathbb{R}^n : \ (x,u) \in S^{\infty}(\omega)\}.$$

Moreover, given a $u \in \mathcal{L}^0(C)$, there is an $x \in \mathcal{L}^0$ with $x(\omega) \perp N(\omega)$ and $(x,u) \in S$.

Combining Theorem 1.20 with the fact that the operations there preserve convexity gives the following.

Theorem 1.42 *The following are convex normal integrands:*

1. *$h(x,\omega) = \sup_{j\in J} h^j(x,\omega)$, where $(h^j)_{j\in J}$ is a countable collection of convex normal integrands;*
2. *$h(x,\omega) = \sum_{i=1}^n h^i(x,\omega)$, where h^i are convex normal integrands;*
3. *$h(\cdot,\omega) = (\alpha(\omega)h^0)(\cdot,\omega)$, where h^0 is a convex normal integrand and $\alpha \in \mathcal{L}^0_+$;*
4. *$h(x,\omega) = f(x,u(\omega),\omega)$, where $f : \mathbb{R}^n \times \mathbb{R}^m \times \Omega \to \overline{\mathbb{R}}$ is a convex normal integrand and $u \in \mathcal{L}^0(\mathbb{R}^m)$ is measurable.*

The notion of a vector-valued convex function from Sect. A.1 has a natural random counterpart. Given a convex cone-valued mapping $K : \Omega \rightrightarrows \mathbb{R}^m$, a *random K-convex function* from $\mathbb{R}^n$ to $\mathbb{R}^m$ is a family $H := \{H(\cdot,\omega)\}_{\omega\in\Omega}$ of $K(\omega)$-convex functions from $\mathbb{R}^n$ to $\mathbb{R}^m$ such that the set-valued mapping

$$(\operatorname{epi}_K H)(\omega) := \operatorname{epi}_{K(\omega)} H(\cdot,\omega)$$

is closed convex-valued and measurable. Here

$$\operatorname{epi}_{K(\omega)} H(\cdot,\omega) := \{(x,u) \in \mathbb{R}^n \times \mathbb{R}^m \mid x \in \operatorname{dom} H(\cdot,\omega),\ H(x,\omega) - u \in K(\omega)\},$$

the *$K(\omega)$-epigraph* of $H(\cdot,\omega)$; see Sect. A.1. Note that, for every $C \subseteq \mathbb{R}^n$,

$$(\operatorname{dom} H)^{-1}(C) = (\operatorname{epi}_K H)^{-1}(C \times \mathbb{R}^m)$$

so the domain of a random K-convex function is necessarily measurable.

Note that a convex normal integrand $h : \mathbb{R}^n \times \Omega \to \overline{\mathbb{R}}$ is a random $\mathbb{R}_-$-convex function. A normal integrand is affine if and only if it is a random $\{0\}$-convex function.

Example 1.43 Given a random matrix $A \in \mathcal{L}^0(\mathbb{R}^{m\times n})$, we obtain a random $\{0\}$-convex function H by defining $\operatorname{dom} H(\omega) := \mathbb{R}^n$ and $H(x,\omega) := A(\omega)x$. Indeed, we then have $(\operatorname{epi}_{\{0\}} H)(\omega) = \operatorname{gph} A(\omega)$ which is measurable by Example 1.10.

Example 1.44 If $f_1,\dots,f_m$ are convex normal integrands with f_j affine for j greater than some $l \le m$, then the function H defined by

$$\operatorname{dom} H(\cdot,\omega) := \bigcap_{j=1}^{m} \operatorname{dom} f_j(\cdot,\omega) \quad \text{and} \quad H(x,\omega) := (f_i(x,\omega))_{j=1}^{m}$$

is a random $\mathbb{R}^l_- \times \{0\}$-convex function from $\mathbb{R}^n$ to $\mathbb{R}^m$.

Proof We have

$$(\operatorname{epi}_{\mathbb{R}^l_-\times\{0\}} H)(\omega) = \{(x,u) \in \mathbb{R}^n \times \mathbb{R}^m \mid M(x,u) \in R(\omega)\},$$

where

$$R(\omega) := \prod_{j=1}^{l} (\operatorname{epi} f_j)(\omega) \times \prod_{j=l+1}^{m} (\operatorname{gph} f_j)(\omega)$$

and $M : \mathbb{R}^n \times \mathbb{R}^m \to \mathbb{R}^{(n+1)m}$ is the linear mapping given by

$$M(x,u) := (x,u^1,x,u^2,\dots,x,u^m).$$

By Theorem 1.9, the mappings $\omega \mapsto (\operatorname{gph} f_j)(\omega)$ are measurable for $j > l$, so by Theorem 1.7, R is measurable. The claim thus follows from Theorem 1.9. □

Under mild conditions, compositions of K-convex functions with convex normal integrands are again convex normal integrands.

Theorem 1.45 *Let g be a convex normal integrand on $\mathbb{R}^m$ and H a random K-convex function from $\mathbb{R}^n$ to $\mathbb{R}^M$ such that*

$$H(x,\omega) - u \in K(\omega) \implies g(H(x,\omega),\omega) \le g(u,\omega) \quad \forall x \in \operatorname{dom} H(\omega).$$

The function

$$h(x,\omega) := \begin{cases} g(H(x,\omega),\omega) & \text{if } x \in \operatorname{dom} H(\omega), \\ +\infty & \text{if } x \notin \operatorname{dom} H(\omega) \end{cases}$$

is a convex normal integrand as soon as $h(\cdot,\omega)$ is lsc for all $\omega \in \Omega$. A sufficient condition for the lower semicontinuity is that the random set

$$\omega \mapsto (-K(\omega)) \cap \{u \in \mathbb{R}^m \mid g^\infty(u,\omega) \le 0\}$$

be linear. In particular, given $A \in \mathcal{L}^0(\mathbb{R}^{m\times n})$, the function $h(x,\omega) = g(A(\omega)x,\omega)$ is a convex normal integrand.

Proof By Theorem 1.42, the function

$$f(x,u,\omega) := g(u,\omega) + \delta_{(\operatorname{epi}_K H)(\omega)}(x,u)$$

is a convex normal integrand on $\mathbb{R}^n \times \mathbb{R}^m \times \Omega$. The growth condition gives

$$h(x,\omega) = \inf_{u\in\mathbb{R}^m} f(x,u,\omega),$$

so the first two claims follow from Theorems 1.21 and 1.40. The last claim follows by choosing H as in Example 1.43. □

Given a normal integrand h, we denote its scenariowise *conjugate* by

$$h^*(v,\omega) := \sup_{x\in\mathbb{R}^n} \{x \cdot v - h(x,\omega)\}$$

and its scenariowise closed convex hull by $\operatorname{cl} \operatorname{co} h$; see Sect. A.8.

Theorem 1.46 *The conjugate and the closed convex hull of a normal integrand are convex normal integrands.*

Proof Given a normal integrand h,

$$
\begin{aligned}
h^*(v, \omega) &= \sup\{x \cdot v - \alpha \mid (x, \alpha) \in (\operatorname{epi} h)(\omega)\} \\
&= \sup\{x \cdot v - \alpha \mid (x, \alpha) \in \operatorname{cl} \operatorname{co}(\operatorname{epi} h)(\omega)\}.
\end{aligned}
$$

By Corollary 1.8 and Lemma 1.1, $\operatorname{cl} \operatorname{co} \operatorname{epi} h$ is measurable so, by Theorem 1.27, it has a Castaing representation $(x^\nu, \alpha^\nu)_{\nu \in \mathbb{N}}$. Since, for every $\omega \in \operatorname{dom} \operatorname{epi} h$, the Castaing representation is dense in $\operatorname{cl} \operatorname{co}(\operatorname{epi} h)(\omega)$, we get

$$
h^*(v, \omega) = \begin{cases} \sup_\nu \{x^\nu(\omega) \cdot v - \alpha^\nu(\omega)\} & \text{if } (\operatorname{epi} h)(\omega) \neq \emptyset, \\ -\infty & \text{otherwise.} \end{cases}
$$

The functions $(x, \omega) \mapsto x^\nu(\omega) \cdot v - \alpha^\nu(\omega)$ are Carathéodory integrands and thus, normal on $\mathbb{R}^n \times \operatorname{dom} \operatorname{epi} h$. Thus, by Theorem 1.20, h^* is a normal integrand on $\mathbb{R}^n \times \operatorname{dom} \operatorname{epi} h$. Since $\operatorname{dom} \operatorname{epi} h$ is measurable, h^* is a normal integrand on all of $\mathbb{R}^n \times \Omega$. The second claim follows from the first one and Theorem A.54. □

Given a measurable mapping $S : \Omega \rightrightarrows \mathbb{R}^n$, its *support function* is defined by

$$
\sigma_S(v, \omega) := \sigma_{S(\omega)}(v) := \sup_{x \in S(\omega)} x \cdot v.
$$

Corollary 1.47 *The support function of a measurable mapping is a normal integrand.*

Proof Given a measurable mapping $S : \Omega \rightrightarrows \mathbb{R}^n$, we have

$$
\sigma_S(v, \omega) = \delta^*_{\operatorname{cl} S(\omega)}(v),
$$

where $\omega \mapsto \operatorname{cl} S(\omega)$ is measurable, by Lemma 1.1. Thus, the claim follows from Theorem 1.46. □

Corollary 1.48 *Given a proper convex normal integrand h on $\mathbb{R}^n \times \Omega$,*

$$
H(x, \alpha, \omega) := \begin{cases} \alpha h(x/\alpha, \omega) & \text{if } \alpha > 0, \\ h^\infty(x, \omega) & \text{if } \alpha = 0, \\ +\infty & \text{otherwise} \end{cases}
$$

is a convex normal integrand on $\mathbb{R}^{n+1} \times \Omega$.

Proof By Corollary A.59,

$$
H(x, \alpha, \omega) = \sigma_{(\operatorname{epi} h^*)(\omega)}(x, -\alpha),
$$

so the claim follows from Corollary 1.47. □

Given a proper convex normal integrand h and $\alpha \in \mathcal{L}^0_+$, we define

$$(h\alpha)(x,\omega) := \begin{cases} \alpha(\omega)h(x/\alpha(\omega),\omega) & \text{if } \alpha(\omega) > 0, \\ h^\infty(x,\omega) & \text{if } \alpha(\omega) = 0. \end{cases}$$

By Corollary A.59, $(h\alpha)(\cdot,\omega)$ is conjugate to the function $(\alpha h^*)(\cdot,\omega)$. Recall that

$$(\alpha h^*)(x,\omega) := \begin{cases} \alpha(\omega)h^*(x,\omega) & \text{if } \alpha(\omega) > 0, \\ \delta_{\operatorname{cl}\operatorname{dom} h^*(\cdot,\omega)}(x) & \text{if } \alpha(\omega) = 0, \end{cases}$$

by definition. Combining Corollary 1.48 with Theorems 1.20.5 and 1.46 gives the following.

Corollary 1.49 *Given a proper convex normal integrand h and $\alpha \in \mathcal{L}^0_+$, the functions $h\alpha$ and αh^* are convex normal integrands conjugate to each other.*

If h is the indicator function of a closed random set C, then $h\alpha$ is the indicator function of the set

$$(\alpha_+ C)(\omega) := \begin{cases} \alpha(\omega)C(\omega) & \text{it } \alpha(\omega) > 0, \\ C^\infty(\omega) & \text{if } \alpha(\omega) = 0. \end{cases}$$

Corollary 1.49 thus gives the following.

Corollary 1.50 *Given a closed convex nonempty random set C and a nonnegative $\alpha \in \mathcal{L}^0$, the set $\alpha_+ C$ is closed convex and measurable.*

Given a convex function g on $\mathbb{R}^n$, a vector $v \in \mathbb{R}^n$ is a *subgradient* of g at $x \in \mathbb{R}^n$ if

$$g(x') \geq g(x) + (x' - x) \cdot v \quad \forall x' \in \mathbb{R}^n.$$

The set of such vectors v is known as the *subdifferential* of g at x and it is denoted by $\partial g(x)$. By the definition of the conjugate, $v \in \partial g(x)$ if and only if

$$g(x) + g^*(v) \leq x \cdot v.$$

The subdifferential defines a set-valued mapping $\partial g : \mathbb{R}^n \rightrightarrows \mathbb{R}^n$.

Theorem 1.51 *Given a convex normal integrand h, the set-valued mapping*

$$\omega \mapsto \operatorname{gph} \partial h(\cdot,\omega)$$

is closed-valued and measurable. In particular, given $x \in \mathcal{L}^0(\mathbb{R}^n)$, the mapping

$$\partial h(x)(\omega) := \partial h(x(\omega), \omega)$$

is closed convex-valued and measurable.

Proof We have

$$\operatorname{gph} \partial h(\cdot, \omega) = \{(x, v) \in \mathbb{R}^n \times \mathbb{R}^n \mid h(x, \omega) + h^*(v, \omega) - x \cdot v \leq 0\},$$

so the closedness follows from the lower semicontinuity of h and h^* while the measurability follows from Theorems 1.51 and 1.15. The second claim follows from Theorem 1.9 □

Applying Theorem 1.51 to the indicator function of a random closed convex set C shows that the *normal cone* (see Sect. A.6)

$$N_{C(\omega)}(x(\omega)) = \{v \in \mathbb{R}^n \mid (x' - x(\omega)) \cdot v \leq 0 \quad \forall x' \in C(\omega)\}$$

of C at an $x \in \mathcal{L}^0(\mathbb{R}^n)$ is measurable.

1.1.5 Integral Functionals

We now endow the measurable space $(\Omega, \mathcal{F})$ with a probability measure P. Given a normal integrand $h : \mathbb{R}^n \times \Omega \to \overline{\mathbb{R}}$, the associated *integral functional* $Eh : \mathcal{L}^0(\mathbb{R}^n) \to \overline{\mathbb{R}}$ is defined, for every $x \in \mathcal{L}^0(\mathbb{R}^n)$, by

$$Eh(x) := \int h(x(\omega), \omega) dP(\omega).$$

Here and in what follows, the expectation of an extended real-valued random variable is defined as $+\infty$ unless the positive part is integrable. Since, by Theorem 1.17, $\omega \mapsto h(x(\omega), \omega)$ is measurable for every $x \in \mathcal{L}^0$, the integral functional is a well-defined extended real-valued function on $\mathcal{L}^0$. Clearly, $Eh(x) = Eh(\tilde{x})$ if x and $\tilde{x}$ are *equivalent* in the sense that their values agree *almost surely* (a.s.), i.e. on a set $A \in \mathcal{F}$ with $P(A) = 1$. The integral functional Eh is thus a well-defined function on the space $L^0 := L^0(\Omega, \mathcal{F}, P; \mathbb{R}^n)$ of equivalence classes of $\mathbb{R}^n$-valued random variables.

We will say that a normal integrand h is *convex* if $h(\cdot, \omega)$ is so for almost every $\omega \in \Omega$. Similarly for sublinearity etc. Recall that a function g in a linear space is *sublinear* if

$$g(\alpha_1 x_1 + \alpha_2 x_2) \leq \alpha_1 g(x_1) + \alpha_2 g(x_2)$$

for all $x_i \in \mathbb{R}^n$ and $\alpha_i > 0$.

Theorem 1.52 *If h is a convex normal integrand, then Eh is convex on L^0. Similarly, if Eh is sublinear, then Eh is sublinear on L^0.*

Proof Given any $x_1, x_2 \in L^0$ and $\alpha_1, \alpha_2 > 0$ with $\alpha_1 + \alpha_2 = 1$, the convexity and Corollary B.4 give

$$\begin{aligned} Eh(\alpha_1 x_1 + \alpha_2 x_2) &\leq E[\alpha_1 h(x_1) + \alpha_2 h(x_2)] \\ &\leq E[\alpha_1 h(x_1)] + E[\alpha_2 h(x_2)] \\ &= \alpha_1 Eh(x_1) + \alpha_2 Eh(x_2), \end{aligned}$$

which proves the first claim. The proof of the second is analogous. □

The following is an immediate corollary of Lemma B.3.

Theorem 1.53 *Given normal integrands h_1 and h_2, we have*

$$E(h_1 + h_2) = Eh_1 + Eh_2$$

on a linear subspace $\mathcal{X}$ of L^0 under either of the following conditions:

1. *$h_1(x)_-, h_2(x)_- \in L^1$ for every $x \in \mathcal{X}$,*
2. *h_1 or h_2 is an indicator function of a closed-valued measurable mapping.*

We equip L^0 with the translation invariant metric

$$(x, x') \mapsto E\rho(|x' - x|),$$

where $\rho : \mathbb{R} \to [0, 1]$ is a nondecreasing continuous function vanishing only at the origin and such that $\rho(\alpha_1 + \alpha_2) \leq \rho(\alpha_1) + \rho(\alpha_2)$ for all $\alpha_1, \alpha_2 > 0$. Recall that a sequence $(x^\nu)_{\nu \in \mathbb{N}}$ in L^0 *convergences in probability* to an $x \in L^0$ if

$$\lim_{\nu \to \infty} P(\{|x^\nu - x| \geq \epsilon\}) = 0$$

for all $\epsilon > 0$. The following is Theorem B.1 from the appendix.

Theorem 1.54 *The space L^0 is a complete topological vector space where a sequence converges if and only if it converges in probability. A sequence converges in probability if and only if every subsequence has an almost surely convergent subsequence with a common limit.*

We say that a normal integrand h is *lower bounded* if there is an $m \in L^1$ such that

$$h(x, \omega) \geq m(\omega) \quad \forall x \in \mathbb{R}^n$$

for almost every $\omega \in \Omega$. The following is a simple consequence of Fatou's lemma.

Theorem 1.55 *The integral functional of a lower bounded normal integrand is L^0-lower semicontinuous.*

Proof Let h be a lower bounded normal integrand and $x^\nu \rightarrow x$ in L^0. By Theorem 1.54, we may assume, by passing to a subsequence if necessary, that $x^\nu \rightarrow x$ almost surely so that, by the lower semicontinuity of $h(\cdot, \omega)$,

$$\liminf h(x^\nu) \geq h(x)$$

almost surely. By Fatou's lemma,

$$\liminf Eh(x^\nu) \geq E[\liminf h(x^\nu)] \geq Eh(x),$$

which completes the proof. □

In the space L^0 of all random variables, the lower boundedness condition in Theorem 1.55 tends to be necessary for the integral functional Eh to be proper; see Remark 1.56 below. When restricted to appropriate locally convex subspaces of L^0, we can allow more general normal integrands; see Sect. 3.1. Recall that a measurable set A is an *atom* of the probability space $(\Omega, \mathcal{F}, P)$ if, for every measurable $B \subset A$, either $P(B) = 0$ or $P(B) = P(A)$. The usual measure-theoretic indicator function of a set $A \subseteq \Omega$ is denoted by

$$1_A(\omega) := \begin{cases} 1 & \text{if } \omega \in A, \\ 0 & \text{otherwise.} \end{cases}$$

Remark 1.56 If the probability space is atomless, then Eh is improper on L^0 unless h is lower bounded. If Ω is the union of finitely many atoms, then Eh is lsc if it is proper.

Proof The first claim holds trivially if $\operatorname{dom} Eh = \emptyset$. Otherwise, the positive part of $p(\omega) := \inf_x h(x, \omega)$ is integrable. If h fails to be lower bounded, we have $Ep = -\infty$. Assume that the probability space is atomless, let $\epsilon > 0$ and consider the set $A := \{\omega \mid p(\omega) = -\infty\}$. Define a random variable $\bar{p}$ as follows. If $P(A) = 0$, let $\bar{p} := p + \epsilon$, and if $P(a) > 0$, let

$$\bar{p} := 1_{A^C}(p + \epsilon) - \sum_{\nu \in \mathbb{N}} 1_{A^\nu} \frac{1}{P(A^\nu)},$$

where the sets $A^\nu \in \mathcal{F}$ are such that $A^\nu \subset A$ with $P(A^\nu) \searrow 0$. In both cases, we have $E\bar{p} = -\infty$ and $\bar{p} > p$ almost surely. By Corollary 1.28, there exists an $x \in L^0$ with $h(x) \leq \bar{p}$ almost surely. Clearly $Eh(x) = -\infty$, so Eh is improper. When the probability space is finite, Eh is a finite sum of lsc functions and thus, lsc. □

Recall that the indicator function $\delta_S(x,\omega) := \delta_{S(\omega)}(x)$ of a closed-valued measurable mapping $S : \Omega \rightrightarrows \mathbb{R}^n$ is a normal integrand. The associated integral functional can be expressed as $E\delta_S = \delta_{L^0(S)}$, where

$$L^0(S) := \{x \in L^0 \mid x \in S\ a.s.\}.$$

By Corollary 1.28, this set is nonempty if and only if $P(\operatorname{dom} S) = 1$ and then, $L^0(S)$ is the set of (equivalence classes of) measurable selections of S. Since the closedness of a set is equivalent to the lower semicontinuity of its indicator function, Theorem 1.55 yields the following.

Corollary 1.57 *Given a closed-valued measurable mapping $S : \Omega \rightrightarrows \mathbb{R}^n$, the set $L^0(S)$ is closed in L^0.*

The following expresses the recession function of an integral functional as the integral functional of the recession function of the normal integrand; see Sect. 1.1.4.

Theorem 1.58 *If h is a convex normal integrand such that Eh is lsc and proper on a topological vector space $\mathcal{X} \subseteq L^0$, then*

$$(Eh)^\infty(x) = Eh^\infty(x) \quad \forall x \in \mathcal{X}.$$

Proof Let $\bar{x} \in \operatorname{dom} Eh \cap \mathcal{X}$ and $x \in \mathcal{X}$. Since Eh is lsc on $\mathcal{X}$ and $h(\bar{x})$ is integrable, Theorem A.16 and Lemma B.3 give

$$(Eh)^\infty(x) = \sup_{\lambda>0} \frac{Eh(\lambda x + \bar{x}) - Eh(\bar{x})}{\lambda} = \sup_{\lambda>0} E\left[\frac{h(\lambda x + \bar{x}) - h(\bar{x})}{\lambda}\right].$$

By convexity, the difference quotients on the right side increase pointwise to $h^\infty(x(\omega),\omega)$ as $\lambda \nearrow \infty$. If $x + \bar{x} \notin \operatorname{dom} Eh$, then $(Eh)^\infty(x) = Eh^\infty(x) = +\infty$. If $x + \bar{x} \in \operatorname{dom} Eh$, we have $h(x+\bar{x}) \in L^1$ so the claim follows from the monotone convergence theorem. □

The following extension of the classical notion of homogeneity will be useful e.g. in the study of dynamic programming; see Sect. A.11 for further study of this notion. Given $p \in \mathbb{R}$, a function g on $\mathbb{R}^n$ is *p-homogeneous* if there is a constant C such that

$$g(\alpha x) = \begin{cases} \alpha^p g(x) - C\frac{\alpha^p - 1}{p} & \text{if } p \neq 0, \\ g(x) - C \ln \alpha & \text{if } p = 0 \end{cases}$$

for all $x \in \mathbb{R}^n$ and $\alpha > 0$; see Sect. A.11. Given $p \in \mathbb{R}$, a normal integrand h is *p-homogeneous* if $h(\cdot,\omega)$ is p-homogeneous for almost every $\omega \in \Omega$. This means that there exists a function $C : \Omega \to \mathbb{R}$ such that, for all ω outside a null set,

$$h(\alpha x,\omega) = \begin{cases} \alpha^p h(x,\omega) - C(\omega)\frac{\alpha^p - 1}{p} & \text{if } p \neq 0, \\ h(x,\omega) - C(\omega) \ln \alpha & \text{if } p = 0 \end{cases} \tag{1.2}$$

for all $x \in \mathbb{R}^n$ and $\alpha > 0$. If h is a p-homogeneous proper normal integrand, then C may be chosen measurable. Indeed, applying the measurable selection theorem to the epigraph of h gives a measurable selection $x \in L^0$ of $\operatorname{dom} h$ and then, the p-homogeneity implies that αx is a measurable selection of $\operatorname{dom} h$ as well. The measurability of C thus follows from Corollary 1.18.

The following shows that p-homogeneity of a normal integrand is inherited by the associated integral functional.

Lemma 1.59 *Let h be a p-homogeneous normal integrand. If the random variable C in* (1.2) *is integrable, then Eh is p-homogeneous with*

$$Eh(\alpha x) = \begin{cases} \alpha^p Eh(x) - E[C]\frac{\alpha^p - 1}{p} & \text{if } p \neq 0, \\ Eh(x) - E[C] \ln \alpha & \text{if } p = 0. \end{cases}$$

If there exists a linear subspace $\mathcal{X} \subseteq L^0$ such that $\operatorname{dom} Eh \cap \mathcal{X} \neq \emptyset$ *and $h(x)^- \in L^1$ for all $x \in \mathcal{X}$, then $C \in L^1$.*

Proof The first claim follows directly from Lemma B.3. Given $x \in \mathcal{X} \cap \operatorname{dom} Eh$ and $\alpha > 0$, we have $h(x) \in L^1$ and $h(\alpha x)^- \in L^1$, by assumption, so the last term in (1.2) has integrable negative part. Varying α, we get $C \in L^1$. □

The condition in Lemma 1.59 is a slightly stronger assumption than mere properness of Eh on $\mathcal{X}$. By Fenchel's inequality,

$$h(x, \omega) \geq x \cdot v - h^*(v, \omega)$$

so, if Eh is proper on $\mathcal{X}$, the condition holds if there exists a $v \in L^0$ such that $h^*(v) \in L^1$ and $x \cdot v \in L^1$ for all $x \in \mathcal{X}$. This is a standard assumption in the duality theory of integral functionals; see Sect. 3.1 below.

The theorem below gives a close connection between closed sets of random variables and measurable set-valued mappings. Given a collection $\mathcal{D}$ of random variables, its *essential infimum* $\operatorname{essinf} \mathcal{D}$ is a random variable such that $\operatorname{essinf} \mathcal{D} \leq \xi$ almost surely for every $\xi \in \mathcal{D}$ and if $\underline{\xi}$ is another random variable with this property, then $\underline{\xi} \leq \operatorname{essinf} \mathcal{D}$ almost surely. By Lemma B.21, every set of random variables has a unique essential infimum.

Theorem 1.60 *A set $\mathcal{D} \subseteq L^0$ can be expressed as $L^0(S)$ for a closed-valued measurable mapping $S : \Omega \rightrightarrows \mathbb{R}^n$ if and only if $\mathcal{D}$ is closed in L^0 and*

$$x 1_A + x' 1_{\Omega \setminus A} \in \mathcal{D} \quad \forall x, x' \in \mathcal{D},\ A \in \mathcal{F}. \tag{1.3}$$

Proof If $\mathcal{D} = L^0(S)$ for a closed-valued measurable mapping S, then it is clear that (1.3) holds while the closedness comes from Corollary 1.57. To prove the converse, let $\mathbb{Q}^n \subset \mathbb{R}^n$ be the set of rational vectors. For each $q \in \mathbb{Q}^n$, there exists, by Lemma B.21, a sequence $(x_q^j)_{j \in \mathbb{N}}$ in $\mathcal{D}$ such that $\inf_j |q - x_q^j| = \operatorname{essinf}_{x \in \mathcal{D}} |q - x|$

almost surely. Let $(x^\nu)_{\nu\in\mathbb{N}}$ be an enumeration of $(x_q^j)_{q\in\mathbb{Q}^n, j\in\mathbb{N}}$. The closed-valued mapping

$$S(\omega) := \operatorname{cl}\{x^\nu(\omega) \mid \nu \in \mathbb{N}\}.$$

is measurable, by Theorem 1.7. It suffices to show that $\mathcal{D} = L^0(S)$.

For any $x \in \mathcal{D}$ and $q \in \mathbb{Q}^n$,

$$|q - x| \geq \operatorname*{essinf}_{x'\in\mathcal{D}} |q - x'| = \inf_j |q - x_q^j| \geq \inf_\nu |q - x^\nu| = d(q, S) \quad a.s..$$

Since $S(\omega)$ is closed, $|q - x| \geq d(q, S(\omega))$ for all $q \in \mathbb{Q}^n$ implies that $x \in S(\omega)$. Thus, $x \in L^0(S)$ so $\mathcal{D} \subset L^0(S)$. On the other hand, let $x \in L^0(S)$ and $\epsilon > 0$. We construct a finite partition $(A^\nu)_{\nu=1}^\mu$ of Ω such that $E\rho(|x - \sum_{\nu=1}^\mu 1_{A^\nu} x^\nu|) < 2\epsilon$, where $\rho : \mathbb{R} \to [0, 1]$ is the function in the definition of the L^0-metric. We first define an infinite partition of Ω by setting $B_0 := \emptyset$ and

$$B^\nu := \{\rho(|x^\nu - x|) < \epsilon\} \setminus (\bigcup_{\nu'<\nu} B^{\nu'}).$$

For μ large enough, $P(\bigcup_{\nu\geq\mu} B^\nu) < \epsilon$, so we may choose $A^\nu := B^\nu$ for $\nu = 1, \ldots, \mu - 1$ and $A^\mu := \bigcup_{\nu\geq\mu} B^\nu$. Condition (1.3) implies $\sum_{\nu=1}^\mu 1_{A^\nu} x^\nu \in \mathcal{D}$. Since $\epsilon > 0$ was arbitrary and $\mathcal{D}$ is closed in L^0 we get $x \in \mathcal{D}$. Since $x \in L^0(S)$ was arbitrary, we have $L^0(S) \subseteq \mathcal{D}$. □

1.1.6 Indistinguishability

Given two set-valued mappings S and $\tilde{S}$, we say that $S = \tilde{S}$ *almost surely* if $S(\omega) = \tilde{S}(\omega)$ for almost every $\omega \in \Omega$. We say that two extended real-valued functions h and $\tilde{h}$ on $\mathbb{R}^n \times \Omega$ are *indistinguishable* if $\operatorname{epi} h = \operatorname{epi} \tilde{h}$ almost surely, i.e. if $h(\cdot, \omega) = \tilde{h}(\cdot, \omega)$ for almost every $\omega \in \Omega$. Clearly, the indistinguishability implies

$$Eh = E\tilde{h}.$$

We say that $h \leq \tilde{h}$ *almost surely everywhere* (a.s.e.) if $h(\cdot, \omega) \leq \tilde{h}(\cdot, \omega)$ for almost every $\omega \in \Omega$. In particular, h is lower bounded if there exists an $m \in L^1$ such that $h \geq m$ almost surely everywhere. Clearly, $h \leq \tilde{h}$ almost surely everywhere implies

$$Eh \leq E\tilde{h}.$$

Note that h and $\tilde{h}$ are indistinguishable if and only if $h = \tilde{h}$ almost surely everywhere.

Given a random variable x and a random set S, we will use the probabilistic abbreviation

$$\{x \in S\} := \{\omega \in \Omega \mid x(\omega) \in S(\omega)\}.$$

Similarly, $\{x^1 \leq x^2\} := \{\omega \in \Omega \mid x^1(\omega) \leq x^2(\omega)\}$ etc. The space of essentially bounded $\mathbb{R}^n$-valued random variables will be denoted by $L^\infty(\Omega, \mathcal{F}, P; \mathbb{R}^n)$ or simply L^∞ if the underlying spaces are clear from the context.

Theorem 1.61 *Given closed-valued measurable mappings $S, \tilde{S} : \Omega \rightrightarrows \mathbb{R}^n$, the following are equivalent:*

1. *$S \subseteq \tilde{S}$ almost surely;*
2. *$\{x \in S\} \subseteq \{x \in \tilde{S}\}$ almost surely for every $x \in L^0$;*
3. *$\{x \in S\} \subseteq \{x \in \tilde{S}\}$ almost surely for every $x \in L^\infty$.*

Proof The implications $1 \Rightarrow 2 \Rightarrow 3$ are clear. Let $x \in L^0$ and $x^\nu := x 1_{|x| \leq \nu}$. If 3 holds, we have

$$\{x \in S,\ |x| \leq \nu\} = \{x^\nu \in S,\ |x| \leq \nu\} \subseteq \{x^\nu \in \tilde{S},\ |x| \leq \nu\} = \{x \in \tilde{S},\ |x| \leq \nu\}$$

almost surely. Taking unions over $\nu \in \mathbb{N}$ gives 2. The proof that 2 implies 1 is analogous to the proof of Corollary 1.29. □

Applying Theorem 1.61 to epigraphical mappings of normal integrands gives the equivalences in the following.

Theorem 1.62 *Given normal integrands h and $\tilde{h}$, the following are equivalent:*

1. *$h \leq \tilde{h}$ almost surely everywhere;*
2. *$h(x) \leq \tilde{h}(x)$ almost surely for every $x \in L^0$;*
3. *$h(x) \leq \tilde{h}(x)$ almost surely for every $x \in L^\infty$*

and imply that $h(x, \cdot) \leq \tilde{h}(x, \cdot)$ almost surely for all $x \in \mathbb{R}^n$. The converse holds if h and $\tilde{h}$ are Carathéodory integrands.

Proof The equivalences follow by applying Theorem 1.61 to the mappings $S(\omega) = (\operatorname{epi} h)(\omega)$ and $\tilde{S}(\omega) = (\operatorname{epi} \tilde{h})(\omega)$. Assume now that the integrands are Carathéodory and let $D \subset \mathbb{R}^n$ be dense and countable. If $h(x, \cdot) \leq \tilde{h}(x, \cdot)$ almost surely for all $x \in \mathbb{R}^n$, then

$$h(x, \omega) \leq \tilde{h}(x, \omega) \quad \forall x \in D$$

for almost every $\omega \in \Omega$. By continuity, this extends from D to all of $\mathbb{R}^n$ so 1 holds. □

Theorem 1.62 yields the following criteria for scenariowise properties of normal integrand. The last part is concerned with p-homogeneity discussed in Lemma 1.59.

Corollary 1.63 *A normal integrand h is*

1. *convex if and only if*

$$h(\alpha_1 x_1 + \alpha_2 x_2) \leq \alpha_1 h(x_1) + \alpha_2 h(x_2) \quad a.s.$$

for every $x_1, x_2 \in L^\infty$ and $\alpha_1, \alpha_2 > 0$ with $\alpha_1 + \alpha_2 = 1$.

2. *sublinear if and only if*

$$h(\alpha_1 x_1 + \alpha_2 x_2) \leq \alpha_1 h(x_1) + \alpha_2 h(x_2) \quad a.s.$$

for every $x_1, x_2 \in L^\infty$ and $\alpha_1, \alpha_2 > 0$.

3. *the indicator of a closed-valued measurable mapping if and only if*

$$h(x) \in \{0, +\infty\} \quad a.s.$$

for every $x \in L^\infty$.

4. *p-homogeneous if and only if there exists $C \in L^0$ such that*

$$h(\alpha x) = \begin{cases} \alpha^p h(x) + C\frac{\alpha^p - 1}{p} & \text{if } p \neq 0, \\ h(x) - C \ln \alpha & \text{if } p = 0 \end{cases}$$

almost surely for every $x \in L^\infty$ and $\alpha > 0$.

Proof It is clear that convexity of h implies the condition in 1. By Theorem 1.62, the condition in 1 implies that, for any $\alpha^1, \alpha^2 > 0$ with $\alpha_1 + \alpha_2 = 1$,

$$h(\alpha_1 x_1 + \alpha_2 x_2) \leq \alpha_1 h(x_1) + \alpha_2 h(x_2) \quad \forall x^1, x^2 \in \mathbb{R}^n$$

almost surely. It follows that there exists a P-null set N such that, outside of N,

$$h(\alpha_1 x_1 + \alpha_2 x_2) \leq \alpha_1 h(x_1) + \alpha_2 h(x_2)$$

for all $x_1, x_2 \in \operatorname{dom} h$ and rational $\alpha_1, \alpha_2 > 0$ with $\alpha_1 + \alpha_2 = 1$. By lower semicontinuity, this extends to any $\alpha_1, \alpha_2 > 0$ with $\alpha_1 + \alpha_2 = 1$. Thus, h is a convex normal integrand. The proof of part 2 is analogous. It is clear that the condition in 3 holds if h is an indicator of a measurable set. Conversely, if the condition holds, we have

$$h(x) = \delta_{\operatorname{lev}_0 h}(x) \; a.s.$$

for all $x \in L^\infty$. Claim 3 now follows from Theorem 1.62. Necessity in 4 is clear. By Theorem 1.62, the condition in 4 gives, for any $\alpha > 0$,

$$h(\alpha x) = \begin{cases} \alpha^p h(x) - C\frac{\alpha^p - 1}{p} & \text{if } p \neq 0, \\ h(x) - C \ln \alpha & \text{if } p = 0 \end{cases}$$

for all $x \in \mathbb{R}^n$ almost surely. It follows that there exists a P-null set N such that, outside of N,

$$h(\alpha x) = \begin{cases} \alpha^p h(x) - C\frac{\alpha^p - 1}{p} & \text{if } p \neq 0, \\ h(x) - C \ln \alpha & \text{if } p = 0 \end{cases}$$

for all $x \in \mathbb{R}^n$ and all rational $\alpha > 0$. For any $x \in \mathbb{R}^n$, the functions involved are continuous on $(0, \infty)$, so the equality extends to all $\alpha > 0$. □

1.1.7 Decomposable Spaces

In many applications and e.g. in the duality theory of stochastic optimization (Chap. 3 below), one studies integral functionals on linear subspaces $\mathcal{X}$ of $L^0(\Omega, \mathcal{F}, P; \mathbb{R}^n)$. Throughout this section, we fix a linear space $\mathcal{X} \subseteq L^0$ which is *decomposable* in the sense that $L^\infty \subseteq \mathcal{X}$ and $1_A x \in \mathcal{X}$ whenever $A \in \mathcal{F}$ and $x \in \mathcal{X}$. Equivalently, $\mathcal{X}$ is decomposable if

$$1_A x + 1_{\Omega \setminus A} x' \in \mathcal{X}$$

whenever $A \in \mathcal{F}$, $x \in \mathcal{X}$ and $x' \in L^\infty$. In most applications, the relevant spaces of random variables are decomposable. This is the case e.g. with the familiar L^p, and Orlicz spaces; see Sect. 5.1.1. An example of a nondecomposable space is the space of continuous functions on a topological space. Another example is the space $\{x \in L^0 \mid \rho x \in L^\infty\}$, where $\rho \in L^0_+ \setminus L^\infty$. This one is not decomposable since it does not contain constant functions.

Decomposability is motivated by the following very useful result on integral functionals.

Theorem 1.64 (Interchange Rule) *Given a normal integrand $h : \mathbb{R}^n \times \Omega \to \overline{\mathbb{R}}$, we have*

$$\inf_{x \in \mathcal{X}} Eh(x) = E[\inf_{x \in \mathbb{R}^n} h(x)] \tag{1.4}$$

if the left side is less than $+\infty$ or if $\mathcal{X} = L^0$. If this common value is finite, then

$$\operatorname*{argmin}_{x \in \mathcal{X}} Eh = \{x \in \mathcal{X} \mid x \in \operatorname{argmin} h \ a.s.\}.$$

Proof The left side of (1.4) is clearly minorized by the right side. By Theorem 1.21, the function p defined by

$$p(\omega) := \inf_{x \in \mathbb{R}^m} h(x, \omega)$$

is measurable. When $Ep = +\infty$, the claim is trivial. Assume $Ep < \infty$ and let $\alpha > Ep$. By monotone convergence, there is an $\epsilon > 0$ such that $E\beta < \alpha$ where $\beta := \epsilon + \max\{p, -1/\epsilon\}$. The mapping

$$S(\omega) := \{x \in \mathbb{R}^n \mid h(x, \omega) \leq \beta(\omega)\}$$

is closed-valued and measurable and since $\beta > p$, we have $P(\operatorname{dom} S) = 1$. By Corollary 1.28, there exists $x \in L^0$ such that $h(x) \leq \beta$ almost surely and thus $Eh(x) \leq E\beta < \alpha$. Since $\alpha > Ep$ was arbitrary, this proves (1.4) in the case $\mathcal{X} = L^0$. Let $\bar{x} \in \mathcal{X}$ be such that $Eh(\bar{x}) < \infty$. Defining

$$x^\nu := 1_{|x| \leq \nu} x + 1_{|x| > \nu} \bar{x},$$

we have $x^\nu \in \mathcal{X}$, $h(x^\nu) \leq \max\{h(x), h(\bar{x})\}$ for all ν and $h(x^\nu) \to h(x)$ almost surely. By Fatou's lemma,

$$\limsup_\nu Eh(x^\nu) \leq Eh(x).$$

Since $\alpha > Ep$ was arbitrary, this proves (1.4).

It is clear that

$$\operatorname*{argmin}_{x \in \mathcal{X}} Eh \supseteq \{x \in \mathcal{X} \mid x \in \operatorname{argmin} h \ a.s.\}.$$

On the other hand, if $\bar{x} \in \mathcal{X}$ achieves the infimum on the left, the first claim implies $Eh(\bar{x}) = E[\inf_x h(x)]$. Since $h(\bar{x}) \geq \inf_x h(x)$, we must have $h(\bar{x}) = \inf_x h(x)$ almost surely when $E[\inf_x h(x)]$ is finite. □

Remark 1.65 Theorem 1.64 remains valid if the definition of decomposability is weakened by only requiring that, for every $x \in \mathcal{X}$, $1_A x + 1_{A^C} x' \in \mathcal{X}$ whenever $A \in \mathcal{F}$ and $x' \in \mathcal{X}$, and, for every $w \in L^0$, there exists an increasing covering $(A_\nu) \subset \mathcal{F}$ of Ω such that $1_{A_\nu} w + 1_{A_\nu^C} x \in \mathcal{X}$.

Remark 1.66 If an integral functional $Eh : \mathcal{X} \to \overline{\mathbb{R}}$ is proper, then $h(x)^- \in L^1$ for all $x \in \mathcal{X}$.

Proof Properness implies the existence of an $x^1 \in \mathcal{X}$ such that $h(x^1)^+ \in L^1$. If there existed an $x^2 \in \mathcal{X}$ with $h(x^2)^- \notin L^1$, then $Eh(\bar{x}) = -\infty$, where

$$\bar{x} := 1_{\Omega \setminus A} x^1 + 1_A x^2 \in \mathcal{X},$$

where $A := \{\omega \mid h(x^2) \leq 0\}$. □

We will denote by L^1 the linear space of random vectors with integrable components. For any $x \in L^1$ and any σ-algebra $\mathcal{G} \subset \mathcal{F}$, there exists a unique $\mathcal{G}$-measurable random vector $E^{\mathcal{G}}x \in L^1$ such that

$$E[x \cdot v] = E[(E^{\mathcal{G}}x) \cdot v]$$

for all $\mathcal{G}$-measurable $v \in L^1$; see Theorem B.11. The random vector $E^{\mathcal{G}}x$ is called the *$\mathcal{G}$-conditional expectation* of x. A normal integrand is said to be *$\mathcal{G}$-measurable* if its epigraphical mapping is $\mathcal{G}$-measurable.

We will say that a convex normal integrand h is *closed* if $h(\cdot, \omega)$ is closed in $\mathbb{R}^n$ for almost every $\omega \in \Omega$. Recall that closedness of h means that $h = h^{**}$; see Theorem A.54. Combining Theorem 1.64 with the biconjugate theorem (Theorem A.54) gives the following.

Theorem 1.67 (Jensen's Inequality) *Let h be a $\mathcal{G}$-measurable closed convex normal integrand such that* dom $Eh \cap L^1 \neq \emptyset$. *Then*

$$h(E^{\mathcal{G}}x) \leq E^{\mathcal{G}}[h(x)] \quad a.s. \tag{1.5}$$

for all $x \in L^1$ with $h(x)$ quasi-integrable. Moreover,

$$Eh(E^{\mathcal{G}}x) \leq Eh(x)$$

for all $x \in L^1$.

Proof Assume first that $h \geq 0$ so that $Eh^*(0) < \infty$. By Theorems A.54 and 1.64,

$$\begin{aligned} Eh(E^{\mathcal{G}}x) &= E \sup_{v \in \mathbb{R}^n} \{(E^{\mathcal{G}}x) \cdot v - h^*(v)\} \\ &= \sup_{v \in L^\infty(\mathcal{G})} E[(E^{\mathcal{G}}x) \cdot v - h^*(v)] \\ &= \sup_{v \in L^\infty(\mathcal{G})} E[x \cdot v - h^*(v)] \\ &\leq E \sup_{v \in \mathbb{R}^n} \{x \cdot v - h^*(v)\} \\ &= Eh(x). \end{aligned}$$

Consider now a general $\mathcal{G}$-measurable convex normal integrand h. Applying the above to the $\mathcal{G}$-measurable convex normal integrand $\delta_{\operatorname{epi} h}$ gives, $(E^{\mathcal{G}}x, E^{\mathcal{G}}\alpha) \in$ epi h or, in other words,

$$h(E^{\mathcal{G}}x) \leq E^{\mathcal{G}}\alpha \; a.s. \tag{1.6}$$

for any $(x, \alpha) \in L^1(\operatorname{epi} h)$. Assume now that $h(x)^+$ is integrable and, for $\nu \in \mathbb{N}$, let $\alpha^\nu := \max\{h(x), -\nu\}$. Since $(x, \alpha^\nu) \in L^1(\operatorname{epi} h)$, (1.6) gives $h(E^{\mathcal{G}} x) \leq E^{\mathcal{G}} \alpha^\nu$. Since $\alpha^\nu \searrow h(x)$ almost surely, Theorem B.12 gives (1.5).

Assume now that $h(x)^-$ is integrable and let $B \in \mathcal{G}$. To prove (1.5), it suffices, by Lemma B.5, to show that

$$E[h(E^{\mathcal{G}} x) 1_B] \leq E[E^{\mathcal{G}}[h(x)] 1_B]. \tag{1.7}$$

By Lemma B.14, $E[E^{\mathcal{G}}[h(x)]1_B] = E[h(x)1_B]$ so we may assume that $h(x)^+ 1_B$ is integrable since, otherwise, the right side of (1.7) equals $+\infty$. Since $L^1(\operatorname{epi} h) \neq \emptyset$, by assumption, there exists an $\bar{x} \in L^1$ such that $h(\bar{x})^+ \in L^1$. Let $\hat{x} := x 1_B + \bar{x} 1_{\Omega \setminus B}$. Since $h(\hat{x})^+ \in L^1$, the first part of the proof gives

$$h(E^{\mathcal{G}} \hat{x}) \leq E^{\mathcal{G}}[h(\hat{x})] \quad a.s.,$$

where, by Lemma B.13,

$$\begin{aligned} E^{\mathcal{G}}[\hat{x}] &= E^{\mathcal{G}}[x] 1_B + E^{\mathcal{G}}[\bar{x}] 1_{\Omega \setminus B}, \\ E^{\mathcal{G}}[h(\hat{x})] &= E^{\mathcal{G}}[h(x)] 1_B + E^{\mathcal{G}}[h(\bar{x})] 1_{\Omega \setminus B}. \end{aligned}$$

This implies (1.7). This proves the first claim. The second claim holds trivially unless $h(x)$ is quasi-integrable since then, $Eh(x) = +\infty$ by the definition of the extended real-valued integral. When $h(x)$ is quasi-integrable, it follows, by Lemma B.14, from the first claim by taking expectations on both sides of the inequality. □

Given a closed-valued measurable mapping $S : \Omega \rightrightarrows \mathbb{R}^n$, the integral functional associated with δ_S can be expressed on $\mathcal{X}$ as $E\delta_S = \delta_{\mathcal{X}(S)}$, where

$$\mathcal{X}(S) := \{x \in \mathcal{X} \mid x \in S \; a.s.\}$$

is the set of *measurable almost sure selections* of S in $\mathcal{X}$. Note that $\mathcal{X}(S) = \emptyset$ unless $P(\operatorname{dom} S) = 1$. The elements of $L^1(S)$ are called *integrable selections* of S.

Applying Theorem 1.67 to the indicator function of a random set gives the following.

Corollary 1.68 *If $S : \Omega \rightrightarrows \mathbb{R}^n$ is closed convex-valued and $\mathcal{G}$-measurable, then $E^{\mathcal{G}} x \in L^1(S)$ for every $x \in L^1(S)$.*

The following result is concerned with random vector-valued functions from Sect. 1.1.4. We say that a random K-convex function H from $\mathbb{R}^n$ to $\mathbb{R}^m$ is $\mathcal{G}$-*measurable* if its K-epigraph $\operatorname{epi}_K H$ is $\mathcal{G}$-measurable. Applying Corollary 1.68 $\operatorname{epi}_K H$ gives the following.

Corollary 1.69 *If H is a $\mathcal{G}$-measurable random K-convex function then*

$$H(E^{\mathcal{G}}x) - E^{\mathcal{G}}u \in K \ a.s.$$

for every $(x, u) \in L^1$ such that $H(x) - u \in K$ almost surely.

The remainder of this section gives further consequences of the measurable selection theorems from Sect. 1.1.3.

Lemma 1.70 *Let h and $\tilde{h}$ be normal integrands such that Eh and $E\tilde{h}$ are proper on $\mathcal{X}$. If $\bar{x} \in \operatorname{dom} Eh \cap \operatorname{dom} E\tilde{h} \cap \mathcal{X}$, then*

$$Eh(x) - Eh(\bar{x}) \geq E\tilde{h}(x) - E\tilde{h}(\bar{x})$$

for all $x \in \mathcal{X}$ if and only if

$$h - h(\bar{x}) \geq \tilde{h} - \tilde{h}(\bar{x}) \quad a.s.e.$$

Proof It is clear that the second condition implies the first one so assume the first one and denote $m := h(\bar{x}) - \tilde{h}(\bar{x})$. By Theorem 1.62, it suffices to show that $h(x) \geq \tilde{h}(x) + m$ almost surely for all $x \in L^\infty$. Assume, for contradiction, that there exists $x \in L^\infty$ such that the set $B := \{h(x) < \tilde{h}(x) + m\}$ has $P(B) > 0$. Since $h(x) < \infty$ on B, the set $A^\nu := B \cap \{h(x) \leq \nu\}$ has $P(A^\nu) > 0$ for ν large enough. Defining $x^\nu := x1_{A^\nu} + \bar{x}1_{\Omega\setminus A^\nu}$ we then have $x^\nu \in \operatorname{dom} Eh \cap \mathcal{X}$ and

$$\begin{aligned} Eh(x^\nu) &= E[1_{A^\nu}h(x)] + E[1_{\Omega\setminus A^\nu}h(\bar{x})] \\ &< E[1_{A^\nu}(\tilde{h}(x) + m)] + E[1_{\Omega\setminus A^\nu}(\tilde{h}(\bar{x}) + m)] \\ &= E\tilde{h}(x^\nu) + Eh(\bar{x}) - E\tilde{h}(\bar{x}), \end{aligned}$$

contradicting the first condition. □

The following is a simple consequence of Lemma 1.70.

Theorem 1.71 *Let h and $\tilde{h}$ be normal integrands such that Eh and $E\tilde{h}$ are proper on $\mathcal{X}$. Then $Eh = E\tilde{h}$ on $\mathcal{X}$ if and only if there exists $m \in L^1$ such that $Em = 0$ and*

$$h = \tilde{h} + m \quad a.s.e.$$

Proof Let $\bar{x} \in \operatorname{dom} Eh \cap \operatorname{dom} E\tilde{h} \cap \mathcal{X}$. If $Eh = E\tilde{h}$ on $\mathcal{X}$, the first inequality in Lemma 1.70 holds as an equality so the second one must hold as an equality too. In other words, $h = \tilde{h} + m$, where $m = h(\bar{x}) - h(\bar{x})$. The converse implication is immediate. □

Applying Lemma 1.70 to indicator functions of measurable mappings gives the following.

Theorem 1.72 *Let S and $\tilde{S}$ be closed-valued measurable mappings with $\mathcal{X}(S) \neq \emptyset$. Then $S \subseteq \tilde{S}$ almost surely if and only if $\mathcal{X}(S) \subseteq \mathcal{X}(\tilde{S})$.*

Proof It is clear that the first inclusion implies the second one so assume the second. Let $h = \delta_S$ and $\tilde{h} = \delta_{\tilde{S}}$. We have $Eh = \delta_{\mathcal{X}(S)}$ and $Eh = \delta_{\mathcal{X}(S)}$. Since $\mathcal{X}(S) \subseteq \mathcal{X}(\tilde{S})$, by assumption, the condition $\mathcal{X}(S) \neq \emptyset$ implies the existence of an $\bar{x} \in \operatorname{dom} Eh \cap \operatorname{dom} E\tilde{h} \cap \mathcal{X}$. The inclusion $\mathcal{X}(S) \subseteq \mathcal{X}(\tilde{S})$ means that h and $\tilde{h}$ satisfy the first inequality in Lemma 1.70 so they also satisfy the second one. Since the h and $\tilde{h}$ are indicators, the second inequality in Lemma 1.70 means that $S \subseteq \tilde{S}$ almost surely. □

The following gives a converse of Theorem 1.52.

Theorem 1.73 *If h is a normal integrand such that Eh is proper on $\mathcal{X}$, then*

1. *h is convex if and only if Eh is convex on $\mathcal{X}$,*
2. *h is sublinear if and only if Eh is sublinear on $\mathcal{X}$.*

Proof By Theorem 1.52, convexity of h implies that of Eh. If h is not convex, Corollary 1.63 gives the existence of $x_i \in L^\infty$, $\alpha_i > 0$ and $A \in \mathcal{F}$ such that $\alpha_1 + \alpha_2 = 1$, $P(A) > 0$ and

$$h(\alpha_1 x_1 + \alpha_2 x_2) > \alpha_1 h(x_1) + \alpha_2 h(x_2)$$

on A. By Remark 1.66, the negative part of $\alpha_1 h(x_1) + \alpha_2 h(x_2)$ is integrable. Choosing A small enough, we may assume that $1_A(\alpha_1 h(x_1) + \alpha_2 h(x_2)) \in L^1$. Let $x_0 \in \mathcal{X}$ be such that $h(x_0) \in L^1$ and define $\bar{x}_i := 1_A x_i + 1_{\Omega \setminus A} x_0$. By Lemma B.3,

$$\begin{aligned}
Eh(\alpha_1 \bar{x}_1 + \alpha_2 \bar{x}_2) &= E[1_A h(\alpha \bar{x}_1 + \alpha_2 \bar{x}_2)] + E[1_{\Omega \setminus A} h(x_0)] \\
&> E[1_A(\alpha_1 h(\bar{x}_1) + \alpha_2 h(\bar{x}_2))] \\
&\qquad + E[1_{\Omega \setminus A}(\alpha_1 h(x_0) + \alpha_1 h(x_0))] \\
&= \alpha_1 Eh(\bar{x}_1) + \alpha_2 h(\bar{x}_2)
\end{aligned}$$

so Eh is not convex. The proof of the second claim is analogous. □

Corollary 1.74 *A closed-valued measurable mapping S with $\mathcal{X}(S) \neq \emptyset$ is*

1. *convex-valued if and only if $\mathcal{X}(S)$ is convex,*
2. *cone-valued if and only if $\mathcal{X}(S)$ is a cone,*
3. *linear-valued if and only if $\mathcal{X}(S)$ is linear.*

Proof Applying Theorem 1.73 to δ_S gives parts 1 and 2. By 2, linearity of $\mathcal{X}(S)$ implies that S is a cone. It also implies $\mathcal{X}(S) = -\mathcal{X}(S) = \mathcal{X}(-S)$ so $S = -S$, by Theorem 1.72. Thus, linearity of $\mathcal{X}(S)$ implies that of S. The converse is clear. □

1.2 Convex Stochastic Optimization

Consider again the problem

$$\text{minimize} \quad Eh(x) \quad \text{over } x \in \mathcal{N} \tag{P}$$

from the introduction of this chapter. Recall that

$$\mathcal{N} := \{(x_t)_{t=0}^T \mid x_t \in L^0(\Omega, \mathcal{F}_t, P; \mathbb{R}^{n_t})\ t = 0, \ldots, T\},$$

the space of adapted decision strategies. We assume from now on that h is a normal integrand on $\mathbb{R}^n \times \Omega$, where $n = n_0 + \cdots + n_T$. As noted in Sect. 1.1.5, this implies, by Theorem 1.17, that the integral functional Eh is well-defined throughout L^0. We will also assume that h is convex. By Theorem 1.52, the integral functional $Eh : L^0 \to \overline{\mathbb{R}}$ is then convex and we say that (P) is a problem of *convex stochastic optimization*.

Many well-studied problem classes in operations research, optimal control and financial mathematics fit the above format. Problem (P) allows also for various generalizations to existing models. Some examples will be given in Sect. 1.3. More can be found in the references cited at the end of this chapter. In many stochastic programming models in the literature, the decision strategies x are restricted to certain subspaces of integrable or bounded functions in L^0.

Remark 1.75 (Integrable Strategies) Restricting the strategies $x \in \mathcal{N}$ to a decomposable subspace $\mathcal{X}$ of L^1 leads to the problem

$$\text{minimize} \quad Eh(x) \quad \text{over } x \in \mathcal{X}_a, \tag{$P_{\mathcal{X}}$}$$

where $\mathcal{X}_a := \mathcal{X} \cap \mathcal{N}$. Clearly, if $\operatorname{dom} Eh \subseteq \mathcal{X}$, the two problems coincide. This happens e.g. if $\operatorname{dom} h(\cdot, \omega)$ is contained in a bounded set of $\mathbb{R}^n$ for almost every $\omega \in \Omega$ since then, $\operatorname{dom} Eh \subseteq L^\infty$ (recall from Sect. 1.1.7 that decomposability implies $L^\infty \subseteq \mathcal{X}$). More generally, one could take $\mathcal{X}$ to be the Orlicz space associated with a Young function Φ (see Example 5.7) such that $\operatorname{dom} Eh$ is contained in the positive hull of $\operatorname{dom} E\Phi$ (see Sect. A.2). This holds, in particular, if there exist $\alpha > 0$ and $m \in L^1$ such that $\Phi(|x|/\alpha) \leq h(x) + m$.

In general, however, it may be difficult to specify an appropriate space $\mathcal{X}$ a priori or it may happen that optimal solutions can be found in L^0 but not in L^1. This is the case e.g. in many standard problems in financial mathematics; see Sect. 1.3.5 below. Integrability requirements on the strategies are cumbersome also in the context of dynamic programming where optimal strategies are obtained as scenariowise minimizers so that integrability properties may be difficult to guarantee; see Chap. 2. Chapters 2 and 4 will give sufficient conditions for the existence of solutions in L^0.

In most applications in practice, the uncertainties can be described by (finite-dimensional) random variables. The following gives a general description of such models.

Remark 1.76 (Canonical Representation) In most applications in practice, the uncertainty and information are represented by a sequence $\xi = (\xi_t)_{t=0}^{T+1}$ of random variables ξ_t each taking values in a measurable space $(\Xi_t, \mathcal{A}_t)$, typically a Euclidean space endowed with its Borel σ-algebra. Assume that

1. $\mathcal{F}_t = \sigma(\xi^t)$, where $\xi^t := (\xi_0, \ldots, \xi_t)$,
2. $h(x, \omega) = H(x, \xi(\omega))$ for an $\hat{\mathcal{F}}$-measurable convex normal integrand $H : \mathbb{R}^n \times \Xi \to \overline{\mathbb{R}}$,

where,

$$\Xi := \Xi_0 \times \cdots \times \Xi_{T+1} \quad \text{and} \quad \hat{\mathcal{F}} := \mathcal{A}_0 \otimes \cdots \otimes \mathcal{A}_{T+1}.$$

The first condition simply means that the new information the decision maker observes at time t is given by the realization of ξ_t. By Corollary 1.34, there exists a normal integrand H such that 2 holds if and only if h is $\sigma(\xi)$-measurable.

By Lemma B.7, we have $x \in \mathcal{N}$ if and only if, for each t, $x_t = \tilde{x}_t \circ \xi^t$ for some $\tilde{x}_t \in L^0(\Xi^t, \mathcal{A}^t, P_{\xi^t}; \mathbb{R}^{n_t})$, where

$$\Xi^t := \Xi_0 \times \cdots \times \Xi_t \quad \text{and} \quad \mathcal{A}^t := \mathcal{A}_0 \otimes \cdots \otimes \mathcal{A}_t$$

and P_{ξ^t} is the *distribution* of ξ^t, i.e. the probability measure on $(\Xi^t, \mathcal{A}^t)$ given by $P_{\xi^t}(A) = P(\{\omega \in \Omega \mid \xi^t(\omega) \in A\})$ for each $A \in \mathcal{A}^t$. Each such $\tilde{x}_t$ can be identified with a unique $\hat{x}_t \in L^0(\Xi, \hat{\mathcal{F}}_t, \hat{P}; \mathbb{R}^{n_t})$, where

$$\hat{\mathcal{F}}_t := \mathcal{A}^t \otimes \{\Xi_{t+1}, \emptyset\} \otimes \cdots \otimes \{\Xi_{T+1}, \emptyset\}$$

and $\hat{P}$ is the *distribution* of ξ. The identification is given by $\hat{x}_t = \tilde{x}_t \circ \pi^t$, where π^t denotes the projection $\xi \mapsto \xi^t$. The linear mapping $\tilde{x}_t \mapsto \tilde{x}_t \circ \pi^t$ from $L^0(\Xi^t, \mathcal{A}^t, P_{\xi^t}; \mathbb{R}^{n_t})$ to $L^0(\Xi, \hat{\mathcal{F}}_t, \hat{P}; \mathbb{R}^{n_t})$ is indeed a one-to-one mapping as its kernel is the origin.

We can then write (P) as

$$\text{minimize} \quad E^{\hat{P}} H(\hat{x}) \quad \text{over } \hat{x} \in \hat{\mathcal{N}}, \tag{$\hat{P}$}$$

where

$$\hat{\mathcal{N}} := \{(\hat{x}_t)_{t=0}^T \mid \hat{x}_t \in L^0(\Xi, \hat{\mathcal{F}}_t, \hat{P}; \mathbb{R}^{n_t}) \quad t = 0, \ldots, T\}$$

and $E^{\hat{P}}H : L^0(\Xi, \mathcal{A}, \hat{P}; \mathbb{R}^n) \to \overline{\mathbb{R}}$ is the convex integral functional given by

$$E^{\hat{P}}H(\hat{x}) := \int_{\Xi} H(\hat{x}(s), s)d\hat{P}(s).$$

We call $(\hat{P})$ the *canonical representation* of (P). The sequence of functions $s \mapsto s_t$ on the probability space $(\Xi, \hat{\mathcal{F}}, P_\xi)$ is known as the canonical process associated with ξ. In practice, one can of course choose $(\Omega, \mathcal{F}, P) := (\Xi, \hat{\mathcal{F}}, P_\xi)$ from the beginning.

As noted above, we can identify each component $\hat{x}_t$ of an $\hat{x} \in \hat{\mathcal{N}}$ with a function of ξ^t only. With a slight misuse of notation, we will denote the function by $\hat{x}_t$.

Remark 1.77 (Risk Aversion and Risk Measures) Even though problem (P) is about the minimization of an expectation, one should not think of it as a decision problem of a risk-neutral agent. For example, in problems of financial economics (see Sect. 1.3.5 below), risk preferences are often described by the expected utility of random wealth or cashflows. Such objectives fit directly into (P) while the utility function can be chosen to describe various levels of risk aversion.

One might also consider problems where the expectation in (P) is replaced by a more general convex function $\mathcal{V} : L^0 \to \overline{\mathbb{R}}$. Various "risk measures" can be expressed in the form

$$\mathcal{V}(u) = \inf_{\alpha \in \mathbb{R}} Ev(\alpha, u),$$

where v is a convex function on $\mathbb{R} \times \mathbb{R}$. For example, when

$$v(\alpha, u) = u + \lambda|u - \alpha|^2,$$

we obtain the classical mean-variance criterion. When

$$v(\alpha, u) = \alpha + V(u - \alpha)$$

for a nondecreasing convex function V on the real line, one obtains the *optimized certainty equivalent* that covers e.g. the Conditional Value at Risk and the entropic risk measure. Such problem formulations can be written in the format of (P) by incorporating the scalar variable α into the decision process x and requiring it to be measurable with respect to the trivial σ-algebra $\mathcal{F}_{-1} := \{\Omega, \emptyset\}$. This fits the format of (P) with time t running from -1 to T.

Remark 1.78 (Expectation Constraints) One could also consider problems of the form

$$\text{minimize} \quad Eh(x) + (g \circ EH)(x) \quad \text{over } x \in \mathcal{N},$$

where H is a random K-convex function (see Sect. 1.1.4) and EH is the K-convex function defined by

$$\operatorname{dom} EH := \{x \in L^0 \mid H(x) \in L^1(\Omega, \mathcal{F}, P; \mathbb{R}^m)\}$$

and $EH(x) := E[H(x)]$ for $x \in \operatorname{dom} EH$. If g is the indicator function of K, the problem can be written with expectation constraints as

$$\begin{aligned} &\text{minimize} \quad Eh(x) \quad \text{over } x \in \mathcal{N} \cap \operatorname{dom} EH, \\ &\text{subject to} \quad EH(x) \in K. \end{aligned}$$

Under appropriate "constraint qualifications", the objective can be expressed as the supremum over $y \in K^*$ of the function

$$L(x, y) := El(x, y),$$

where $l(x, y, \omega) := h(x, \omega) + y \cdot H(x, \omega) - g^*(y)$. The problem can then be treated with the theory of normal integrands and convex duality much like (P) in Chaps. 3–5 below.

The following simple remark illustrates some of the advantages of using normal integrands in the problem formulation. The features observed here will be taken full advantage of in Chap. 2 in the development of dynamic programming technique for the general format of (P).

Remark 1.79 (Two-Stage Problems) Assume that $T = 1$, h is $\mathcal{F}_T$-measurable and that $\mathcal{F}_0$ is the trivial σ-algebra $\{\emptyset, \Omega\}$. If the optimum value of (P) is finite, then an $x \in \mathcal{N}$ solves (P) if and only if

$$x_T(\omega) \in \operatorname*{argmin}_{x_T \in \mathbb{R}^{n_T}} h(x_0, x_T, \omega)$$

for almost every $\omega \in \Omega$ and x_0 solves the problem

$$\text{minimize} \quad E[Q(x_0)] \quad \text{over } x_0 \in \mathbb{R}^{n_0},$$

where

$$Q(x_0, \omega) := \inf_{x_T \in \mathbb{R}^{n_T}} h(x_0, x_T, \omega).$$

By Theorem 1.20.5, the function $(x_T, \omega) \mapsto h(x_0, x_T, \omega)$ is a normal integrand for every $x_0 \in \mathbb{R}^{n_0}$, so by Corollary 1.23, $Q(x_0, \cdot)$ is measurable for every $x_0 \in \mathbb{R}^{n_0}$.

Proof An x minimizes Eh over $\mathcal{N}$ if and only if x_0 minimizes the function

$$\mathcal{Q}(x_0) := \inf_{x_T \in L^0(\mathbb{R}^{n_T})} Eh(x_0, x_T)$$

over $\mathbb{R}^{n_0}$ and x_T minimizes the function $Eh(x_0, \cdot)$ over $L^0(\mathbb{R}^{n_T})$. By Theorem 1.64, $\mathcal{Q}(x_0) = E[Q(x_0)]$ and, since the optimum value is assumed finite, an x_T minimizes $Eh(x_0, \cdot)$ over $L^0(\mathbb{R}^{n_T})$ if and only if $x_T \in \operatorname{argmin} h(x_0, \cdot)$ almost surely. □

Another advantage of normal integrands is the L^0-lower semicontinuity of the associated integral functionals in Theorem 1.55. Combined with the lemma below, we then get that the essential objective $Eh + \delta_{\mathcal{N}}$ is lsc in the L^0-topology.

Lemma 1.80 *The set $\mathcal{N}$ is closed in L^0.*

Proof Let $(x^\nu)_{\nu \in \mathbb{N}}$ be a sequence in $\mathcal{N}$ converging to x in L^0. By Theorem 1.54, x is the almost sure limit of a subsequence of $(x^\nu)_{\nu \in \mathbb{N}}$ so x_t is $\mathcal{F}_t$-measurable for every $t = 0, \dots, T$. □

The L^0-lower semicontinuity implies lower semicontinuity on any space $\mathcal{X} \subset L^0$ whose topology is stronger than that of L^0. This will be crucial in development of the duality theory in Chap. 3. The lower semicontinuity yields also simple conditions for existence of solutions. In particular, if Eh is inf-compact in $\mathcal{X}$ with respect to a topology not weaker than that of L^0, then, by Lemma A.22, the problem $(P_{\mathcal{X}})$ in Remark 1.75 has a solution. Recall from Sect. A.4 that an extended real-valued function g in a topological space is *inf-compact* if the lower level sets $\operatorname{lev}_\alpha g$ are compact for every $\alpha \in \mathbb{R}$. In the case of (P), one can substitute topological compactness in L^0 for almost sure boundedness.

Theorem 1.81 *If h is a lower bounded convex normal integrand such that $h(\cdot, \omega)$ is inf-compact for almost every ω, then (P) has a solution.*

Proof Let $(x^\nu)_{\nu \in \mathbb{N}}$ be a minimizing sequence in $\mathcal{N}$ for Eh. Then $Eh(x^\nu) \le \gamma$ for some $\gamma \in \mathbb{R}$. Since h is lower bounded, the sequence $(h(x^\nu))_{\nu \in \mathbb{N}}$ is bounded in L^1. By Theorem B.25, there exists a sequence of convex combinations $\sum_{\mu \ge \nu} \alpha^\mu h(x^\mu)$ that converge almost surely. In particular, the sequence is almost surely bounded, so there exists $\beta \in L^0$ such that $\sum_{\mu \ge \nu} \alpha^\mu h(x^\mu) \le \beta$ almost surely. Defining, $\bar{x}^\nu = \sum_{\mu \ge \nu} \alpha^\mu x^\mu$, we have, by convexity, $h(\bar{x}^\nu) \le \beta$ almost surely for all ν. Since h is inf-compact almost surely, the sequence $(\bar{x}^\nu)$ is almost surely bounded. By Theorem B.25 again, there is a further sequence of convex combinations converging almost surely to an $x \in L^0$. By convexity and Theorem 1.55, $Eh(x) \le \liminf Eh(x^\nu) = \inf(P)$. □

A far reaching generalization of Theorem 1.81 will be given in Chap. 4.

1.3 Examples

This section presents five instances of the general stochastic optimization model (P). We will revisit the examples throughout the book as the theory develops. The five examples are somewhat an arbitrary choice as the list could be extended indefinitely. The examples considered below build on well-known problems from the literature so they allow for comparisons with existing results.

1.3.1 Mathematical Programming

Consider the problem

$$\begin{aligned} &\text{minimize} && Ef_0(x) \quad \text{over } x \in \mathcal{N}, \\ &\text{subject to} && f_j(x) \le 0 \quad j = 1, \dots, l \ a.s., \\ &&& f_j(x) = 0 \quad j = l+1, \dots, m \ a.s. \end{aligned} \tag{1.8}$$

where f_j are convex normal integrands with f_j affine for $j > l$. This fits the general format (P) with

$$h(x, \omega) = f_0(x, \omega) + \sum_{j=1}^{m} \delta_{C_j(\omega)}(x),$$

where

$$C_j(\omega) := \{x \in \mathbb{R}^n \mid f_j(x, \omega) \le 0\},$$

for $j = 1, \dots, l$ and

$$C_j(\omega) := \{x \in \mathbb{R}^n \mid f_j(x, \omega) = 0\}$$

for $j = l + 1, \dots, m$. Indeed, by Theorem 1.15 and Example 1.16, the set-valued mappings $\omega \mapsto C_j(\omega)$ are closed-valued and measurable so (since indicator functions of closed-valued measurable mappings are normal integrands) h is a normal integrand, by Theorem 1.20. By Theorem 1.53,

$$Eh(x) = \begin{cases} Ef_0(x) & \text{if } x \in C_j \ j = 1, \dots, m \text{ a.s.}, \\ +\infty & \text{otherwise.} \end{cases}$$

Thus, problem (1.8) is indeed an instance of (P).

Note that we can express h also as

$$h(x, \omega) = f_0(x, \omega) + \delta_{\text{epi}_K H}(x, 0, \omega),$$

where $K := \mathbb{R}^l_- \times \{0\}$ and H is the random K-convex function from $\mathbb{R}^n$ to $\mathbb{R}^m$ given by

$$\text{dom}\, H(\cdot, \omega) = \bigcap_{j=1}^{m} \text{dom}\, f_j(\cdot, \omega) \quad \text{and} \quad H(x, \omega) = (f_i(x, \omega))_{j=1}^{m};$$

see Example 1.44. Indeed, $(x, 0) \in (\text{epi}_K H)(\omega)$ means that $x \in \text{dom}\, H(\cdot, \omega)$ and $H(x, \omega) \in K$. That h is a normal integrand thus follows also from Example 1.44 and Theorem 1.42.

Example 1.82 (Composite Models) The above format can be extended to

$$h(x, \omega) = f_0(x, \omega) + g(H(x, \omega), \omega),$$

where H is a random K-convex function from $\mathbb{R}^n$ to $\mathbb{R}^m$ and g is a convex normal integrand on $\mathbb{R}^m$ satisfying the assumptions of Theorem 1.45. Problem (1.8) above is a special case of this with $K = \mathbb{R}^l_- \times \{0\}$ and $g = \delta_K$.

Example 1.83 (Linear Stochastic Programming) Consider (1.8) and assume that $f_0(x, \omega) = c(\omega) \cdot x$ and $f_j(x, \omega) = a_j(\omega) \cdot x - b_j(\omega)$ for random vectors c and a_j and random variables b_j. The problem can then be written in the matrix format as

$$\begin{aligned} &\text{minimize} \quad E[c \cdot x] \quad \text{over} x \in \mathcal{N} \\ &\text{subject to} \quad Ax + b \in K \quad a.s., \end{aligned}$$

where A is the random $m \times n$-matrix with the rows a_j, b is the random vector with components b_j and $K = \mathbb{R}^l_- \times \{0\}$.

1.3.2 *Optimal Stopping*

Let R be a real-valued adapted stochastic process and consider the problem

$$\text{maximize} \quad ER_\tau \quad \text{over } \tau \in \mathcal{T}, \tag{1.9}$$

where $\mathcal{T}$ is the set of *stopping times*, i.e. measurable functions $\tau : \Omega \to \{0, \dots, T+1\}$ such that $\{\omega \in \Omega \mid \tau(\omega) \leq t\} \in \mathcal{F}_t$ for each $t = 0, \dots, T$. This is the classical *optimal stopping problem* in finite discrete time. Choosing $\tau(\omega) = T + 1$ is interpreted as not stopping at all. Accordingly, we define $R_{T+1} := 0$.

Consider also the problem

$$\underset{x\in\mathcal{N}}{\text{maximize}} \quad E\left[\sum_{t=0}^{T} R_t x_t\right] \quad \text{subject to} \quad x \geq 0,\ \sum_{t=0}^{T} x_t \leq 1\ a.s., \tag{1.10}$$

where $\mathcal{N}$ is the linear space of adapted real-valued processes. This is a convex relaxation of (1.9) in the sense that feasible solutions x of (1.10) which take values in $\{0, 1\}$ are in one-to-one correspondence with stopping times via

$$x_t = \begin{cases} 1 & \text{if } t = \tau, \\ 0 & \text{if } t \neq \tau. \end{cases}$$

Under this correspondence, $\sum_{t=0}^{T} R_t x_t = R_\tau$. The feasible solutions of problem (1.10) are often referred to as *randomized stopping times*.

The relaxation is motivated by the following lemma. We leave its proof a little incomplete as it is not part of the theory developed in this book.

Lemma 1.84 *If* $R \in L^1$, *then* $\sup(1.9) = \sup(1.10)$ *and* argmax (1.10) *is the closed convex hull of strategies* x *that correspond to optimal solutions of* (1.9).

Proof It can be shown that the processes corresponding to stopping times are the extreme points of the feasible set of (1.10). By Banach-Alaoglu theorem, the feasible set of (1.10) is compact in the weak topology that L^∞ has as the dual of L^1. Thus, by Krein–Milman theorem, the feasible set of (1.10) is the closed convex hull of those x corresponding to stopping times.

If $R \in L^1$, the objective is weakly continuous so the relaxation does not affect the optimum value. It is easy to verify, by contradiction, that extreme points of the weakly compact argmax (1.10) are extreme points of the feasible set, which, by the Krein-Milman theorem again, proves the last claim. □

Problem (1.10) fits the general framework with $n_t = 1$ for all t and

$$h(x, \omega) = \begin{cases} -\sum_{t=0}^{T} R_t(\omega) x_t & \text{if } x \geq 0 \text{ and } \sum_{t=0}^{T} x_t \leq 1, \\ +\infty & \text{otherwise.} \end{cases}$$

Indeed, by Theorem 1.20, h is a normal integrand and, by Theorem 1.53,

$$Eh(x) = \begin{cases} -E\left[\sum_{t=0}^{T} R_t x_t\right] & \text{if } x \geq 0,\ \sum_{t=0}^{T} x_t \leq 1 \text{ a.s.}, \\ +\infty & \text{otherwise.} \end{cases}$$

1.3.3 Optimal Control

Consider the problem

$$\begin{aligned} &\text{minimize} \quad E\left[\sum_{t=0}^{T} L_t(X_t, U_t)\right] \quad \text{over } (X, U) \in \mathcal{N}, \\ &\text{subject to} \quad \Delta X_t = A_t X_{t-1} + B_t U_{t-1} + W_t \quad t = 1, \dots, T \text{ a.s.}, \end{aligned} \tag{1.11}$$

where the processes X and U take values in $\mathbb{R}^N$ and $\mathbb{R}^M$, respectively, A_t and B_t are $\mathcal{F}_t$-measurable random matrices of appropriate dimensions, W_t are $\mathcal{F}_t$-measurable random vectors and L_t are proper convex normal integrands on $\mathbb{R}^N \times \mathbb{R}^M \times \Omega$. Here and in what follows, $\Delta X_t := X_t - X_{t-1}$. The above is an *optimal control problem* with *state* X and *control* U. The linear constrains in (1.11) are called the *system equations*.

Problem (1.11) fits the format of (P) with $x = (X, U)$ and

$$h(x, \omega) = \sum_{t=0}^{T} L_t(X_t, U_t, \omega) + \sum_{t=1}^{T} \delta_{\{0\}}(\Delta X_t - A_t(\omega)X_{t-1} - B_t(\omega)U_{t-1} - W_t(\omega)).$$

Indeed, by Theorems 1.42 and 1.45, h is a normal integrand and, by Theorem 1.53,

$$Eh(x) = E\left[\sum_{t=0}^{T} L_t(X_t, U_t)\right]$$

if $\Delta X_t = A_t X_{t-1} + B_t U_{t-1} + W_t$ for all $t = 1, \dots, T$ almost surely while $Eh(x) = \infty$ otherwise. Thus, problem (1.11) is indeed the same as the problem of minimizing Eh over $\mathcal{N}$.

Remark 1.85 (Hidden State) Note that the system equations determine the state process $X = (X_t)_{t=0}^T$ scenariowise uniquely as a function of the initial state X_0, the control $U = (U_t)_{t=0}^T$ and the noise $W = (W_t)_{t=1}^T$. If X_0 is $\mathcal{F}_0$-measurable, the requirement that X_t be $\mathcal{F}_t$-measurable for $t \geq 1$ is thus redundant since it holds automatically when the sequences $(A_t)_{t=1}^T$, $(B_t)_{t=1}^T$, $(W_t)_{t=1}^T$ and $(U_t)_{t=0}^T$ are adapted. The problem is thus unaffected if we only require that the sequence $(X_t)_{t=1}^T$ be $\mathcal{F}$-measurable. This fits the general format of (P) with the same normal integrand h as above but with $x_0 = (X_0, U_0)$, $x_t = U_t$ for $t = 1, \dots, T$, $x_{T+1} = (X_t)_{t=1}^T$ and time t running from 0 to $T+1$ and $\mathcal{F}_{T+1} := \mathcal{F}$. This formulation makes good sense even for nonadapted A, B and W. Such generalizations are relevant in situations where the state X is not directly observable.

Remark 1.86 (Reduced Formulation) One can use the system equations to substitute out the state variables X_t for $t = 1, \dots, T$. Indeed, the state process

$X = (X_t)_{t=0}^T$ is a linear function of the initial state X_0, the control $U = (U_t)_{t=0}^T$ and the noise $W = (W_t)_{t=1}^T$. We call this scenariowise function the *solution mapping* and denote it by

$$R : \mathbb{R}^N \times \mathbb{R}^{(T+1)M} \times \mathbb{R}^{TN} \times \Omega \to \mathbb{R}^{(T+1)N},$$

so $X(\omega) = R(X_0(\omega), U(\omega), W(\omega), \omega)$ for every $\omega \in \Omega$. The function is given recursively by $R(X_0, U, W)_0 := X_0$ and

$$R(X_0, U, W)_t := (A_t + I)R(X_0, U, W)_{t-1} + B_t U_{t-1} + W_t$$

or directly in terms of (X_0, U, W) as

$$\begin{aligned} R(X_0, U, W)_t = &\left[\prod_{s=1}^{t}(A_s + I)\right] X_0 \\ &+ \sum_{s=1}^{t}\left[\prod_{r=s+1}^{t}(A_r + I)\right](B_s U_{s-1} + W_s), \end{aligned} \tag{1.12}$$

as is easily verified by substitution. Here, the products over empty sets are defined as identity mappings.

We can write the optimal control problem in the reduced form

$$\text{minimize } E\left[\sum_{t=0}^{T} L_t(R(X_0, U, W)_t, U_t)\right] \quad \text{over} \quad X_0 \in L^0(\mathcal{F}_0; \mathbb{R}^N),\ U \in \mathcal{N}',$$

where $\mathcal{N}'$ is the space of $\mathbb{R}^M$-valued adapted processes. An $(X, U) \in \mathcal{N}$ solves the optimal control problem if and only if (X_0, U) solves the reduced formulation and $X = R(X_0, U, W)$. Since R is an affine Carathéodory mapping, Theorem 1.45 implies that the function

$$(X_0, U) \mapsto \sum_{t=0}^{T} L_t(R(X_0, U, W)_t, U_t)$$

is a normal integrand so the reduced formulation is also an instance of (P).

Remark 1.87 (Controlled Volatility) Systems of the form

$$\Delta X_t = A_t X_{t-1} + B_t U_{t-1} + (C_t X_{t-1} + D_t U_{t-1})W_t \quad t = 1, \dots, T,$$

where C_t and D_t are random "tensors" of appropriate dimensions, can be written in the format of (1.11) as

$$\Delta X_t = (A_t + C_t^* W_t) X_{t-1} + (B_t + D_t^* W_t) U_{t-1} \quad t = 1, \dots, T,$$

where $C_t^*(\omega)$ and $D_t^*(\omega)$ are such that

$$(C_t(\omega)X)W = (C_t^*(\omega)W)X \quad \text{and} \quad (D_t(\omega)U)W = (D_t^*(\omega)W)U$$

for all $X \in \mathbb{R}^d$ and $W \in \mathbb{R}^r$.

1.3.4 Problems of Lagrange

Consider the problem

$$\text{minimize} \quad E\left[\sum_{t=0}^{T} K_t(x_t, \Delta x_t)\right] \quad \text{over } x \in \mathcal{N}, \tag{1.13}$$

where x is a process of fixed dimension d, K_t are proper convex normal integrands, $\Delta x_t := x_t - x_{t-1}$ and $x_{-1} := 0$. This is an instance of (P) with

$$h(x, \omega) = \sum_{t=0}^{T} K_t(x_t, \Delta x_t, \omega).$$

Indeed, by Theorems 1.42 and 1.45, h is a normal integrand. Problem (1.13) can be thought of as a discrete-time version of a problem studied in calculus of variations. Other problem formulations have $K_t(x_{t-1}, \Delta x_t)$ instead of $K_t(x_t, \Delta x_t)$ in the objective, or an additional term of the form $Ek(x_0, x_T)$, all of which fit the general format of (P).

If in the linear stochastic programming model Example 1.83, the constraints can be written as

$$T_t \Delta x_t + W_t x_t + b_t \in C_t \quad t = 0, \dots, T \tag{1.14}$$

for given random matrices T_t and W_t and cones C_t, then the problem in Example 1.83 is an instance of (1.13) with

$$K_t(\Delta x_t, x_t, \omega) = \begin{cases} c_t(\omega) \cdot x_t & \text{if } T_t(\omega)\Delta x_t + W_t(\omega)x_t + b_t(\omega) \in C_t(\omega), \\ +\infty & \text{otherwise.} \end{cases}$$

Note also that the problem of optimal control in Sect. 1.3.3 is a special case of problem (1.13) with $x_t = (X_t, U_t)$ and

$$K_t(x_t, \Delta x_t) = \begin{cases} L_t(X_t, U_t) & \text{if } \Delta X_t = A_t X_{t-1} + B_t U_{t-1} + W_t, \\ +\infty & \text{otherwise.} \end{cases}$$

The relaxed optimal stopping problem from Sect. 1.3.2 can be written as

$$\underset{x \in \mathcal{N}_+}{\text{maximize}} \quad E \sum_{t=0}^{T} R_t \Delta x_t \quad \text{subject to} \quad \Delta x \geq 0,\ x_T \leq 1\ a.s.$$

This is a problem of Lagrange with

$$K_t(x_t, u_t) = -R_t u_t + \delta_{\mathbb{R}_-}(x_t - 1) + \delta_{\mathbb{R}_+}(u_t).$$

Other examples can be found e.g. in financial mathematics; see Example 1.92 below.

1.3.5 *Financial Mathematics*

Let $s = (s_t)_{t=0}^T$ be an adapted $\mathbb{R}^J$-valued stochastic process describing the unit prices of a finite set J of traded assets in a perfectly liquid financial market. Consider the problem of finding a dynamic trading strategy $x = (x_t)_{t=0}^T$ that provides the "best hedge" against the financial liability of delivering a random amount $c \in L^0$ of cash at time T. The components of x_t are interpreted as the numbers of units of assets held over the time period $(t, t+1]$. We measure our risk preferences over random terminal cash positions by the "expected loss" associated with a nondecreasing convex "loss function" V. More precisely, we assume that $V : \mathbb{R} \times \Omega \to \overline{\mathbb{R}}$ is a convex normal integrand such that $V(\cdot, \omega)$ is nondecreasing and nonconstant for almost every ω. The optimal trading problem can be written as

$$\begin{aligned} &\text{minimize} \quad EV\left(c - \sum_{t=0}^{T-1} x_t \cdot \Delta s_{t+1}\right) \quad \text{over} \quad x \in \mathcal{N}, \\ &\text{subject to} \quad x_t \in D_t \quad t = 0, \ldots, T\ a.s., \end{aligned} \tag{1.15}$$

where $EV : L^0 \to \overline{\mathbb{R}}$ is the integral functional associated with V (see Sect. 1.1.5) and D_t is a random $\mathcal{F}_t$-measurable set describing possible portfolio constraints. We will assume that $0 \in D_t$ almost surely for all t and that $D_T = \{0\}$. The last assumption means that all positions have to be closed at the terminal date. Nondecreasing convex loss functions V are in one-to-one correspondence with nondecreasing concave utility functions U via $V(c) = -U(-c)$. It would be natural to assume non-random V, but most of what will be said about problem (1.15) apply to random loss functions as well.

Problem (1.15) is standard in financial mathematics although it is based on quite unrealistic assumptions about the financial market. In particular, it assumes that one can buy and sell arbitrary quantities of all assets at unit prices given by s. It also assumes that one can lend and borrow arbitrary amounts of cash at zero interest rate. Under these assumptions, the sum in the objective can be interpreted as the proceeds of the trading strategy x while the random variable c can be thought of as a payout of a financial product to be hedged. Implicit in (1.15) is the assumption that the purchase of portfolio x_t at time t can be financed by borrowing $x_t \cdot s_t$ units of cash at zero interest rate. The assumption of zero interest rates also allows us to describe nonzero initial wealth simply by subtracting it from the liability c.

Problem (1.15) is an instance of (P) with

$$h(x, \omega) = V\left(c(\omega) - \sum_{t=0}^{T-1} x_t \cdot \Delta s_{t+1}(\omega), \omega\right) + \sum_{t=0}^{T-1} \delta_{D_t(\omega)}(x_t, \omega).$$

Indeed, since indicator functions of measurable closed convex-valued mappings are convex normal integrands, h is a convex normal integrand by Theorems 1.42 and 1.45. By Theorem 1.53,

$$Eh(x) = \begin{cases} EV\left(c - \sum_{t=0}^{T-1} x_t \cdot \Delta s_{t+1}\right) & \text{if } x_t \in D_t \quad t = 0, \dots, T \text{ } a.s., \\ +\infty \quad \text{otherwise.} \end{cases}$$

Much of financial mathematics has revolved around the problem of assigning values to financial products that provide a random payout $c \in L^0$ at a future date T. A classical approach is to define such a value as the amount of money that would be required to finance a trading strategy whose liquidation value at time T equals c. In practice, however, there are few financial products whose payouts can be replicated by trading other assets. One possible extension of the approach is to relax exact replication to "superhedging".

Example 1.88 (Superhedging) Consider the problem

$$\begin{aligned} &\text{minimize} \quad \alpha \quad \text{over } \alpha \in \mathbb{R},\ x \in \mathcal{N} \\ &\text{subject to} \quad \alpha + \sum_{t=0}^{T-1} x_t \cdot \Delta s_{t+1} \geq c \quad a.s., \\ &\qquad\qquad x_t \in D_t \quad t = 0, \dots, T \text{ } a.s. \end{aligned} \tag{1.16}$$

This is the classical superhedging problem of finding the least amount of initial capital α that can finance a self-financing trading strategy x whose liquidation value at time T exceeds the liability c almost surely.

This is an instance of (P) but with time t running from -1 to T, $\mathcal{F}_{-1} = \{\Omega, \emptyset\}$, $x_{-1} = \alpha$ and

$$h(\alpha, x, \omega) = \alpha + \sum_{t=0}^{T} \delta_{D_t(\omega)}(x_t) + \delta_{\mathbb{R}_+}(\alpha + \sum_{t=0}^{T-1} x_t \cdot \Delta s_{t+1}(\omega) - c(\omega)).$$

Indeed, h is a normal integrand by Theorems 1.42 and 1.45.

Classical "risk-neutral" valuations in financial mathematics can be seen as a special case of superhedging. Indeed, we will see in Examples 3.74 and 4.32 that, when there are no portfolio constraints and the price process s satisfies the "no-arbitrage" condition, the superhedging cost can be expressed in terms of expectations of c under so called "equivalent martingale measures". In practice, however, the requirement of superhedging is often unreasonable and the associated cost is too high to be competitive. More practical approach is to use indifference pricing.

Remark 1.89 (Indifference Pricing) The *indifference selling price* of a claim $c \in L^0$ is defined by

$$\pi(\bar{c}; c) := \inf\{\alpha \in \mathbb{R} \mid \varphi(\bar{c} + c - \alpha) \leq \varphi(\bar{c})\},$$

where $\varphi(c)$ denotes the optimum value of (1.15). Here $\bar{c} \in L^0$ denotes the traders initial liability cashflows and α is the price she would receive in compensation of delivering an additional random cashflow $c \in L^0$ (a "contingent claim"). The indifference selling price $\pi(\bar{c}; c)$ is the least price at which it would make sense for the trader to sell the claim c for. The indifference buying price is defined analogously. If $\bar{c} = 0$, $V = \delta_{\mathbb{R}_-}$ and $\varphi(0) = 0$, the indifference price becomes the superhedging cost. When $V = \delta_{\mathbb{R}_-}$, the condition $\varphi(0) = 0$ means that one cannot turn a strictly negative initial wealth into a random terminal wealth that is nonnegative almost surely.

The indifference pricing principle makes good sense also in more general market models. The following extends problem (1.15) by allowing for investments in a finite set of contingent claims that can be traded at time $t = 0$ at a cost given by a convex function S_0.

Example 1.90 (Semi-Static Hedging) Consider the problem

$$\begin{aligned} &\text{minimize} \quad EV\left(c - \sum_{t=0}^{T-1} x_t \cdot \Delta s_{t+1} - \bar{c} \cdot \bar{x} + S_0(\bar{x})\right) \quad \text{over } x \in \mathcal{N},\ \bar{x} \in \mathbb{R}^{\bar{J}}, \\ &\text{subject to} \quad x_t \in D_t \quad t = 0, \ldots, T \text{ } a.s., \end{aligned} \tag{1.17}$$

where s, D and $\mathcal{N}$ are as in (1.15) and $\bar{J}$ is another finite set of assets that are traded only at time $t = 0$. The portfolio $\bar{x} \in \mathbb{R}^{\bar{J}}$ is bought before the dynamic trading of the assets J starts and it is held fixed (static) until time T. The random vector $\bar{c}$ gives the payouts of the statically held contingent claims and the function $S_0 : \mathbb{R}^{\bar{J}} \to \overline{\mathbb{R}}$ gives the cost of buying the portfolio $\bar{x}$ at the best available market prices. We assume that S_0 is a proper lsc convex function that vanishes at the origin.

Problem (1.17) fits the format of (P) with time t running from -1 to T, $\mathcal{F}_{-1} = \{\Omega, \emptyset\}$, $x_{-1} = \bar{x}$ and

$$h(\bar{x}, x, \omega) = V\left(c(\omega) - \sum_{t=0}^{T-1} x_t \cdot \Delta s_{t+1}(\omega) - \bar{c}(\omega) \cdot \bar{x} + S_0(\bar{x}), \omega\right) + \sum_{t=0}^{T-1} \delta_{D_t(\omega)}(x_t, \omega).$$

That h is a convex normal integrand follows from Theorems 1.42 and 1.45. Indeed, the conditions in Theorem 1.45 hold since S_0 is convex and V is nondecreasing, nonconstant and convex.

Convexity of S_0 arises naturally in practice. For example, if the buying and selling prices of the claims $\bar{c}$ are given by vectors $s^b \in \mathbb{R}^{\bar{J}}$ and $s^a \in \mathbb{R}^{\bar{J}}$, respectively, and if we assume that one can buy and sell infinite quantities at these prices, then

$$S_0(x) = \sup_{s \in [s^b, s^a]} x \cdot s.$$

If the bid and ask prices come with finite quantities given by vectors $q^b \in \mathbb{R}^{\bar{J}}$ and $q^a \in \mathbb{R}^{\bar{J}}$, respectively, then

$$S_0(x) = \sup_{s \in [s^b, s^a]} x \cdot s + \delta_{[-q^b, q^a]}(x).$$

More generally, the cost of buying a portfolio $\bar{x}$ in limit order markets always results in a proper lsc convex cost function S_0.

Example 1.91 (Optimal Control Formulation) Problem (1.15) can be written equivalently as

$$\begin{aligned} &\text{minimize} && EV(c - X_T) \quad \text{over} \quad (X, U) \in \mathcal{N}, \\ &\text{subject to} && \Delta X_t = U_{t-1} \cdot \Delta s_t \quad \forall t = 1, \ldots, T, \\ &&& U_t \in D_t \quad \forall t = 0, \ldots, T-1, \\ &&& X_0 = 0, \end{aligned}$$

where $(X_t)_{t=0}^T$ can be interpreted as the "gains process" associated with the trading strategy U. This is an instance of the optimal control problem (1.11) with $N = 1$, $M = |J|$, $A_t = 0$, $B_t = \Delta s_t$, $W_t = 0$ and

$$\begin{aligned} L_T(X_T, U_T) &= V(c - X_T), \\ L_t(X_t, U_t) &= \delta_{D_t}(U_t) \quad t = 1, \dots, T-1, \\ L_0(X_0, U_0) &= \delta_{\{0\} \times D_0}(X_0, U_0). \end{aligned}$$

If the price process s is componentwise almost surely nonzero, we can write (1.15) also as

$$\begin{aligned} &\text{minimize} && EV(c - X_T) \quad \text{over} \quad (X, U) \in \mathcal{N}, \\ &\text{subject to} && X_0 = 0, \\ &&& \Delta X_t = R_t \cdot U_{t-1} \quad \forall t = 1, \dots, T, \\ &&& U_t \in \tilde{D}_t \quad \forall t = 0, \dots, T, \end{aligned} \tag{1.18}$$

where $R_t^j := \Delta s_t^j / s_{t-1}^j$ is the *rate of return* on asset j, $U_t^j := s_t^j x_t^j$ is the amount of money invested in asset j over the period $(t, t+1]$ and

$$\tilde{D}_t(\omega) = \{U \in \mathbb{R}^J \mid (U^j / s_t^j(\omega))_{j \in J} \in D_t(\omega)\}.$$

This formulation will become convenient in the context of dynamic programming in Sect. 2.3.5 below. Formulation (1.18) also suggests extensions of form

$$\begin{aligned} &\text{minimize} && EV(c - X_T) \quad \text{over} \quad (X, U) \in \mathcal{N}, \\ &\text{subject to} && \Delta X_t = R_t \cdot U_{t-1} \quad \forall t = 1, \dots, T, \\ &&& (X_t, U_t) \in C_t \quad \forall t = 0, \dots, T-1, \end{aligned} \tag{1.19}$$

where C_t is an $\mathcal{F}_t$-measurable random set in $\mathbb{R} \times \mathbb{R}^J$. This formulation allows for pointwise constraints involving the wealth X_t. If $C_0 = \{0\} \times \tilde{D}_0$ and $C_t = \mathbb{R} \times \tilde{D}_t$, we recover (1.18). Constraints of the form

$$C_t = \{(X_t, U_t) \in \mathbb{R} \times \mathbb{R}^J \mid 1 \cdot U_t \leq \pi X_t\},$$

for example, describe the requirement that at most proportion π of the current wealth can be invested in the risky assets. The constraint

$$C_t = \{(X_t, U_t) \in \mathbb{R} \times \mathbb{R}^J \mid (U_t^j)^- \leq \pi X_t\}$$

would require that the short position in asset j is not allowed to be larger than proportion π of total wealth.

Example 1.92 (Currency Markets) Consider problem (1.13) with

$$K_t(x_{t-1}, \Delta x_t) = \delta_{D_{t-1}}(x_{t-1}, \omega) + \delta_{C_t}(\Delta x_t, \omega)$$

for adapted sequences $(D_t)_{t=0}^T$ and $(C_t)_{t=0}^T$ of closed convex random sets. This model can be used to describe trading in currency markets where $C_t(\omega)$ is the set of currency portfolios x_t that are freely available in the market at time t (in the sense that the negative components of x_t finance the positive components) while $D_t(\omega)$ describes portfolio constraints. In such a market model, none of the assets is taken as a numeraire unlike in traditional models like (1.15) where the costs and proceeds are measured in cash.

1.4 Bibliographical Notes

The study of multistage stochastic optimization problems in terms of normal integrands goes back to [131] and [53], where it was observed that many more traditional stochastic optimization models can be written in a unifying format. While [53] assumed merely that the integrand h be $\mathcal{B}(\mathbb{R}^n) \otimes \mathcal{F}$-measurable and lsc in the first argument, [131] assumed h to be a normal integrand. By Theorem 1.17, normality of h is a priori a stronger requirement but it is equivalent under the condition in Remark 1.4. In applications, normality of h usually comes for free and it provides us with all the machinery reviewed in Sect. 1.1. Early treatments of multistage convex stochastic optimization problems via the theory of normal integrands include [90–92, 129, 131, 135, 136, 138]. Continuous-time optimal control problems were studied with similar techniques already in [20, 21].

The concept of a convex normal integrand was introduced in [122] under a slightly different definition which turned out to be equivalent to the measurability of the epigraphical mapping. Section 1.1 covers only the part of the theory that is used in the analysis of stochastic optimization problems in this book. More general treatments and detailed commentaries on the historical development of the subject can be found e.g. in [139, Chapter 14], [30], [65] and [86]. Most of the material in Sect. 1.1 can be found in [139, Chapter 14]. When comparing the results, one should note, however, that our definition of a scalar multiple αg of a function g differs from that of [139] when $\alpha = 0$. While we define $0g$ as the indicator function of the set $\operatorname{cl}\operatorname{dom} g$, [139] defines $0g = 0$. Our convention is motivated mainly by the duality relation in Corollary 1.49, which was derived from Corollary 1.48. The operation in Corollary 1.48 arises in many situations in applied mathematics and statistics; see [33] for a detailed study and further references.

Results that are not found in [139] are: the measurability of nonlinear compositions in Theorem 1.45, the Doob–Dynkin-type representations in Corollary 1.33 and

Corollary 1.34 which seem to be new, the "section theorem" in Theorem 1.62 which can be found in e.g. in Truffert [153], the expression for the recession function of an integral functional in Theorem 1.58 which is [22, Proposition 1], the representation result in Theorem 1.60 which can be found for L^p-selections in [63, Theorem 3.1], the Jensen's inequality in Theorem 1.67 which can be found, under some additional conditions, e.g. in [153, page 139]. A different set of sufficient conditions (not requiring the lower semicontinuity in the first variable) for conditional Jensen's inequality can be found in [112] and [2, Appendix II.5]. The Jensen's inequality for random K-convex functions in Corollary 1.69 was given in [79, Section 5] in a Banach space setting. The necessity of lower boundedness in atomless probability spaces in Remark 1.56 is close to the necessity results for lower semicontinuity in [64].

Many formulations of stochastic optimization problems are given in terms of dynamic programming recursions similar to the formulations of the two-stage problem in Remark 1.79; see e.g. [7, 18, 36, 72, 91, 148, 155, 158] and their references. In this context, problem (P) is often referred to as the "deterministic equivalent". We have decided to follow [53, 131] by taking (P) as the basic model and then deriving the dynamic programming recursions afterwards in Chap. 2. This greatly simplifies the presentation as the rigorous treatment of dynamic programming requires more sophisticated tools than the formulation of (P).

In [131, 136] the decision strategies were restricted to be essentially bounded while in [49, 90] they were required to belong to a Lebesgue space L^p with $p \in (1, \infty)$. Such restrictions can often be justified via Remark 1.75. In computational applications in practice, the assumptions of Remark 1.75 are harmless but they become problematic when studying e.g. standard problems in financial mathematics where some important results only hold in models that allow for general adapted trading strategies. Representing uncertainties by a stochastic process $\xi = (\xi_t)_{t=0}^T$ as in Remark 1.76 is common in stochastic programming literature; see e.g. [29, 136, 148]. The optimized certainty equivalent in Remark 1.77 is from [10]; see also [11].

The mathematical programming models in Sect. 1.3.1 are essentially from [136] where they were analyzed through convex duality; see Chap. 3. We have extended the formulation by allowing for general adapted strategies and affine equality constraints. Linear stochastic programs in Example 1.83 seem to have first been studied in [36] and [7, Section 5], where results similar to Remark 1.79 were obtained; see also [76] and its references. Some instances of linear stochastic programs were considered also in [9]. The optimal stopping problem in Sect. 1.3.2 has been extensively analyzed in the literature of stochastic analysis. We refer to [111] for the history and applications of optimal stopping. Discrete-time stochastic optimal control has been studied through dynamic programming (see Chap. 2) e.g. in [12], [14] and [29]. Its reduced formulation in Remark 1.86 was studied in [136, Section 4] using convex duality. The problems of Lagrange in Sect. 1.3.4 are close to the stochastic problems of Bolza in [138]. The main difference is that [138] allows for a function of the form $l(x_0, x_T)$ in the objective. Our formulation already covers a variety of economic and financial applications. The problem of Lagrange can be seen as an extension of von Neumann–Gale model, whose stochastic extensions were studied in [44]. The special case in Example 1.92 extends the currency market model of [67] by adding portfolio constraints. Linear stochastic optimization models with constraints of the form (1.14) have been extensively studied in the stochastic programming literature; see e.g. [18, 148] and their references. The special form makes them convenient also in dynamic programming formulations; see Sect. 2.3.4 below. Sect. 1.3.5 gives only a small sample of stochastic optimization problems analyzed in the literature of mathematical finance. Problem (1.15) was studied in [120] and, implicitly, in [57, Section 8.2] neither of which, however, allowed for portfolio constraints. More general market models with nonlinear trading costs and constraints have been studied in [94, 97, 98, 102] using the techniques studied in this book.

Chapter 2
Dynamic Programming

This chapter studies the *dynamic programming* principle when applied to the general stochastic optimization problem (P) from Chap. 1. Our approach builds on the notion of *conditional expectation of a normal integrand*, which allows for significant extensions to many better known dynamic programming formulations while greatly simplifying the measurability questions that come up in dynamic programming recursions. More traditional forms of dynamic programming will be obtained as special cases in Sect. 2.3 below.

An extended real-valued random variable X is said to be *quasi-integrable* if either X^+ or X^- is integrable. Given a quasi-integrable X and a σ-algebra $\mathcal{G} \subseteq \mathcal{F}$, there exists an extended real-valued $\mathcal{G}$-measurable random variable $E^{\mathcal{G}}X$, almost surely unique, such that

$$E\left[\alpha(E^{\mathcal{G}}X)\right] = E\left[\alpha X\right] \quad \forall \alpha \in L^\infty_+(\Omega, \mathcal{G}, P);$$

see Theorem B.11. The random variable $E^{\mathcal{G}}X$ is known as the $\mathcal{G}$-conditional expectation of X.

Definition 2.1 below extends the notion of conditional expectation to normal integrands. Sufficient conditions for its existence will be given in Sect. 2.1 below. Recall that two normal integrands h_1 and h_2 are equal almost surely everywhere (a.s.e.) if $h_1(\cdot, \omega) = h_2(\cdot, \omega)$ for almost every $\omega \in \Omega$. A given condition defines a normal integrand almost surely everywhere uniquely if the normal integrands satisfying the condition coincide almost surely everywhere.

Definition 2.1 We say that a $\mathcal{G}$-measurable normal integrand $\bar{h}$ is a $\mathcal{G}$-*conditional expectation* of a normal integrand h if

$$\bar{h}(x) = E^{\mathcal{G}}[h(x)] \quad a.s.$$

T. Pennanen, A.-P. Perkkiö, *Convex Stochastic Optimization*, Probability Theory and Stochastic Modelling 107, https://doi.org/10.1007/978-3-031-76432-5_2

for all $x \in L^0(\mathcal{G})$ for which $h(x)$ is quasi-integrable. When the $\mathcal{G}$-conditional expectation of h exists and is almost surely everywhere unique, we denote it by $E^{\mathcal{G}}h$.

If the $\mathcal{G}$-conditional expectation of a normal integrand h exists, then by Lemma B.14,

$$Eh(x) = E(E^{\mathcal{G}}h)(x) \tag{2.1}$$

for every $x \in L^0(\mathcal{G})$ such that $h(x)$ is quasi-integrable. The random variable $h(x)$ is quasi-integrable, in particular, if h is lower-bounded or if x belongs to a decomposable space on which Eh is proper; see Remark 1.66.

Going back to problem (P), we will use the notations $x^t := (x_0, \dots, x_t)$, $n^t := n_0 + \cdots + n_t$ and $E_t := E^{\mathcal{F}_t}$. We say that an adapted sequence $(h_t)_{t=0}^T$ of normal integrands $h_t : \mathbb{R}^{n^t} \times \Omega \to \overline{\mathbb{R}}$ solves the generalized *Bellman equations for h* if

$$\begin{aligned} h_T &= E_T h, \\ h_t &= E_t \inf_{x_{t+1}} h_{t+1} \quad t = T-1, \dots, 0, \end{aligned} \tag{BE}$$

where the second equality means that

$$\tilde{h}_t(x^t, \omega) := \inf_{x_{t+1} \in \mathbb{R}^{n_{t+1}}} h_{t+1}(x^t, x_{t+1}, \omega)$$

is a normal integrand and $h_t = E_t \tilde{h}_t$. Here, and in what follows, all the equations are understood to hold almost surely everywhere. Equations (BE) are also known as *dynamic programming equations*. They extend and unify many more traditional formulations of Bellman equations, as we will see in the applications of Sect. 2.3. The formulation in terms of normal integrands simplifies, in particular, the proof of existence of solutions to Bellman equations. Indeed, the existence of solutions to (BE) boils down to preservation of normality under the scenariowise inf-projections and conditional expectations. The former holds under the general conditions given in Theorems 1.21 and 1.40 while the latter is the topic of Sect. 2.1 below. Section 2.2 will give convenient sufficient conditions for the existence of solutions to (BE) directly in terms of the problem data.

When they exist, solutions $(h_t)_{t=1}^T$ of the Bellman equations (BE) provide useful characterizations of optimal solutions and the optimum value of (P). We will denote the projection of the set $\mathcal{N}$ of adapted strategies to its first t components by

$$\mathcal{N}^t := \{x^t \mid x \in \mathcal{N}\} = \left\{ (x_{t'})_{t'=0}^t \,\middle|\, x_{t'} \in L^0(\Omega, \mathcal{F}_{t'}, P; \mathbb{R}^{n_{t'}}) \right\}.$$

The lower boundedness conditions in the following result will be relaxed in Sect. 2.2.2 below. Recall that a normal integrand is *lower bounded* if there exists an $m \in L^1$ such that $h \geq m$ almost surely everywhere; see Sect. 1.1.5.

Theorem 2.2 *Assume that h is lower bounded, (P) is feasible and that the Bellman equations* (BE) *admit a solution $(h_t)_{t=0}^T$ of lower bounded normal integrands. Then*

$$\inf(P) = \inf_{x^t \in \mathcal{N}^t} Eh_t(x^t)$$

for all $t = 0, \ldots, T$ and, moreover, an $\bar{x} \in \mathcal{N}$ solves (P) if and only if

$$\bar{x}_t(\omega) \in \operatorname*{argmin}_{x_t \in \mathbb{R}^{n_t}} h_t(\bar{x}^{t-1}(\omega), x_t, \omega) \tag{OP}$$

for almost every $\omega \in \Omega$ and all $t = 0, \ldots, T$. If, in addition, h_t are convex and

$$N_t(\omega) := \{x_t \in \mathbb{R}^{n_t} \mid h_t^\infty(0, x_t, \omega) \leq 0\}$$

is linear-valued for all $t = 0, \ldots, T$, then there exists an optimal $x \in \mathcal{N}$ with $x_t \perp N_t$ almost surely.

Proof Note first that

$$\inf_{x^t \in \mathcal{N}^t} Eh_t(x^t) = \inf_{x^{t-1} \in \mathcal{N}^{t-1}} \inf_{x_t \in L^0(\mathcal{F}_t)} Eh_t(x^{t-1}, x_t).$$

For any $x \in \mathcal{N}$, the function $(x_t, \omega) \mapsto h_t(x^{t-1}(\omega), x_t, \omega)$ is an $\mathcal{F}_t$-measurable normal integrand, by Theorem 1.20.5. Thus, Theorem 1.64 gives

$$\inf_{x_t \in L^0(\mathcal{F}_t)} Eh_t(x^{t-1}, x_t) = E \inf_{x_t \in \mathbb{R}^{n_t}} h_t(x^{t-1}, x_t) = E\tilde{h}_{t-1}(x^{t-1}).$$

Clearly, lower boundedness of h_t implies that of $\tilde{h}_{t-1}$, so (2.1) gives $E\tilde{h}_{t-1} = Eh_{t-1}$ on $\mathcal{N}^{t-1}$ and thus

$$\inf_{x^t \in \mathcal{N}^t} Eh_t(x^t) = \inf_{x^{t-1} \in \mathcal{N}^{t-1}} Eh_{t-1}(x^{t-1}).$$

The first claim now follows by induction on t.

To prove the second claim, note first that, by (2.1), an $\bar{x} \in \mathcal{N}$ solves (P) if and only if $\bar{x}$ minimizes Eh_T. We have

$$\bar{x}^t \in \operatorname*{argmin}_{x^t \in \mathcal{N}^t} Eh_t(x^t)$$

if and only if

$$\bar{x}^{t-1} \in \operatorname*{argmin}_{x^{t-1} \in \mathcal{N}^{t-1}} Eh_{t-1}(x^{t-1}) \quad \text{and} \quad \bar{x}_t \in \operatorname*{argmin}_{x_t \in L^0(\mathcal{F}_t)} Eh_t(\bar{x}^{t-1}, x_t).$$

Since (P) is feasible and h is lower bounded, the first claim implies that $\inf_{x^t \in L^0(\mathcal{F}_t)} Eh_t(\bar{x}^{t-1}, x_t)$ is finite. Thus, by the second part of Theorem 1.64, the second inclusion means that

$$\bar{x}_t \in \operatorname*{argmin}_{x_t \in \mathbb{R}^{n_t}} h_t(\bar{x}^{t-1}, x_t) \quad a.s.$$

A backward recursion shows that optimal solutions satisfy (OP). The converse follows from a forward recursion.

Assume now that h_t are convex. Applying Theorem 1.40 recursively forward in time shows that, when N_t is linear-valued, (OP) has an $\mathcal{F}_t$-measurable solution $\bar{x}_t \perp N_t$ almost surely for all $t = 0, \ldots, T$. The last claim thus follows from the second one. □

Condition (OP) in Theorem 2.2 characterizes optimal solutions $\bar{x} \in \mathcal{N}$ of (P) in terms of optimal solutions of optimization problems over $\mathbb{R}^{n_t}$. In the optimal control format, condition (OP) reduces to the famous *Bellman's optimality principle* which states that the optimal control depends on the previous decisions only through the current state; see Theorem 2.91 below. In the literature of optimal control, results like Theorem 2.2 are often called "verification theorems". The existence of solutions to the Bellman equations (BE) will be studied in Sect. 2.2 below.

Instead of normal integrands, the dynamic programming equations are sometimes formulated in terms of "essential infimums" of random variables; see Sect. B.5. The following provides a connection between the two notions.

Remark 2.3 In the setting of Theorem 2.2,

$$h_t(x^t) = \operatorname*{essinf}_{\tilde{x} \in \mathcal{N}} \{E_t h(\tilde{x}) \mid \tilde{x}^t = x^t\} \quad \forall x^t \in L^0(\mathcal{F}_t).$$

Proof Given $x^t \in L^0(\mathcal{F}_t)$, the definition of $(h_t)_{t=0}^T$ and Lemma 2.4 below give

$$\begin{aligned} h_t(x^t) &= E_t[\inf_{x_{t+1} \in \mathbb{R}^{n_{t+1}}} h_{t+1}(x^t, x_{t+1})] \\ &= \operatorname*{essinf}_{x_{t+1} \in L^0(\mathcal{F}_{t+1})} E_t[h_{t+1}(x^t, x_{t+1})] \\ &= \operatorname*{essinf}_{x_{t+1} \in L^0(\mathcal{F}_{t+1})} \operatorname*{essinf}_{x_{t+2} \in L^0(\mathcal{F}_{t+2})} E_t h_{t+2}(x^t, x_{t+1}, x_{t+2}), \end{aligned}$$

where the last equality follows by repeating the argument that gives the second one and by using Lemma B.14. The claim now follows by induction from this and the fact that if we have a doubly indexed collection $C = \{\xi_{\alpha,\alpha'} \mid \alpha \in \mathcal{J}, \alpha' \in \mathcal{J}'\}$ of random variables, then

$$\operatorname{essinf} C = \operatorname*{essinf}_{\alpha \in \mathcal{J}} \{\operatorname*{essinf}_{\alpha' \in \mathcal{J}'} \xi_{\alpha,\alpha'}\}$$

as is easily verified. □

The following can be viewed as a conditional version of the interchange rule in Theorem 1.64. It was used in the proof of Remark 2.3 but will not be needed later.

Remark 2.4 If h is a lower bounded normal integrand, then

$$E^{\mathcal{G}}[\inf_{x\in\mathbb{R}^n} h(x)] = \operatorname*{essinf}_{x\in L^0(\mathcal{F})} E^{\mathcal{G}}[h(x)].$$

Proof Clearly,

$$E^{\mathcal{G}}[\inf_{x\in\mathbb{R}^n} h(x)] \leq \operatorname*{essinf}_{x\in L^0(\mathcal{F})} E^{\mathcal{G}}[h(x)].$$

Let $\epsilon > 0$ and

$$S(\omega) := \{x' \in \mathbb{R}^n \mid h(x', \omega) \leq \inf_{x\in\mathbb{R}^n} h(x, \omega) + \epsilon\}.$$

Since h is a normal integrand, S is closed-valued and measurable, so, by the measurable selection theorem (Corollary 1.28), there exists $\bar{x} \in L^0(\mathcal{F})$ such that $h(\bar{x}) \leq \inf_{x\in\mathbb{R}^n} h(x, \omega) + \epsilon$ almost surely. Thus,

$$E^{\mathcal{G}} h(\bar{x}) \leq E^{\mathcal{G}}[\inf_{x\in\mathbb{R}^n} h(x)] + \epsilon$$

and

$$\operatorname*{essinf}_{x\in L^0(\mathcal{F})} E^{\mathcal{G}}[h(x)] \leq E^{\mathcal{G}}[\inf_{x\in\mathbb{R}^n} h(x)] + \epsilon$$

almost surely. Since $\epsilon > 0$ was arbitrary, this completes the proof. □

The rest of this chapter is organized as follows. Section 2.2 will give sufficient conditions for the existence of solutions to (BE) and (P) and it extends Theorem 2.2 by relaxing the lower boundedness assumption on h. The arguments will build on the general properties of conditional expectations of normal integrands that are developed in Sect. 2.1 below. Section 2.3 illustrates the main results of this chapter by applying them to the examples given in Sect. 1.3.

2.1 Conditional Expectation of a Normal Integrand

Section 2.1.1 below gives sufficient conditions for the existence of the conditional expectation of a normal integrand. Section 2.1.2 gives some of its basic properties and Sect. 2.1.3 studies conditional expectations in convexity preserving operations. Section 2.1.4 is devoted to the notions of conditional essential infimum and essential supremum of normal integrands. They will be used in Sect. 2.2.3 to study "induced constraints" of problem (P). Section 2.1.5 studies representations of conditional expectations of normal integrands in terms of regular conditional distributions.

2.1.1 Existence

We say that a function $h : \mathbb{R}^n \times \Omega \to \overline{\mathbb{R}}$ is *L-bounded* if there exist $\rho, m \in L^1$ such that

$$h(x, \omega) \geq -\rho(\omega)|x| - m(\omega) \quad \forall x \in \mathbb{R}^n$$

for almost every $\omega \in \Omega$.

Lemma 2.5 *Given a convex normal integrand h, the following are equivalent:*

1. *h is L-bounded;*
2. *there exist $v \in L^1(\mathbb{R}^n)$ and $m \in L^1$ such that*

$$h(x, \omega) \geq x \cdot v(\omega) - m(\omega) \quad \forall x \in \mathbb{R}^n$$

for almost every $\omega \in \Omega$;

3. $\operatorname{dom} Eh^* \cap L^1 \neq \emptyset$.

Proof The inequality in 2 means that $h^*(v) \leq m$ almost surely so 2 is equivalent to 3. Assume 3 and let $v \in L^1$ such that $Eh^*(v) < \infty$. By Fenchel's inequality,

$$h(x, \omega) \geq x \cdot v(\omega) - h^*(v, \omega) \geq -|v(\omega)||x| - h^*(v, \omega),$$

so h 1 holds with $\rho = |v|$ and $m = h^*(v)^+$. On the other hand, the inequality $h \geq -\rho|\cdot| - m$ can be written as $(h + \rho|\cdot|)^*(0) \leq m$. This means that

$$\inf_{v \in \mathbb{R}^n} \{h^*(v, \omega) + \delta_{\mathbb{B}}(v/\rho(\omega))\} \leq m(\omega),$$

where the infimum is attained. Indeed, if $h(\cdot, \omega)$ is proper, this follows from Corollary A.80 in the appendix while otherwise, this is trivial. By Corollary 1.28, there exists a $v \in L^0$ with $|v| \leq \rho$ and $h^*(v) \leq m$ and thus, 3 holds. □

If a normal integrand h is L-bounded, then $h(x)$ is quasi-integrable and, by Theorem B.11, $E^{\mathcal{G}}[h(x)]$ is well-defined for every $x \in L^\infty(\mathcal{G})$. The following lemma shows that, for L-bounded normal integrands, it suffices to test with elements of $L^\infty(\mathcal{G})$ in Definition 2.1.

Lemma 2.6 *Given an L-bounded normal integrand h, a $\mathcal{G}$-measurable normal integrand $\bar{h}$ is a $\mathcal{G}$-conditional expectation of h if and only if*

$$\bar{h}(x) = E^{\mathcal{G}}[h(x)] \quad a.s.$$

for all $x \in L^\infty(\mathcal{G})$. When it exists, the conditional expectation of an L-bounded normal integrand is almost surely everywhere unique. In particular, if h is $\mathcal{G}$-measurable and L-bounded, then $E^{\mathcal{G}}h = h$.

Proof It is clear that the condition holds if $\bar{h}$ is a $\mathcal{G}$-conditional expectation of h. On the other hand, if $\bar{h}$ is not a conditional $\mathcal{G}$-expectation of h, then there exists an $x \in L^0(\mathcal{G})$ such that $h(x)$ is quasi-integrable and $P(\{\bar{h}(x) \neq E^{\mathcal{G}}[h(x)]\}) > 0$. For ν large enough,

$$P(\{1_{\{|x|\le\nu\}}\bar{h}(x) \neq 1_{\{|x|\le\nu\}}E^{\mathcal{G}}[h(x)]\}) > 0.$$

We have $1_{\{|x|\le\nu\}}\bar{h}(x) = 1_{\{|x|\le\nu\}}\bar{h}(1_{\{|x|\le\nu\}}x)$ while, by Lemma B.13,

$$\begin{aligned}1_{\{|x|\le\nu\}}E^{\mathcal{G}}[h(x)] &= 1_{\{|x|\le\nu\}}E^{\mathcal{G}}[1_{\{|x|\le\nu\}}h(x)]\\ &= 1_{\{|x|\le\nu\}}E^{\mathcal{G}}[1_{\{|x|\le\nu\}}h(1_{\{|x|\le\nu\}}x)]\\ &= 1_{\{|x|\le\nu\}}E^{\mathcal{G}}[h(1_{\{|x|\le\nu\}}x)].\end{aligned}$$

Thus, $\bar{h}(1_{\{|x|\le\nu\}}x) \neq E^{\mathcal{G}}[h(1_{\{|x|\le\nu\}}x)]$ for ν large enough. Since $1_{\{|x|\le\nu\}}x \in L^\infty(\mathcal{G})$, the condition fails. By Theorem 1.62, the condition defines $\bar{h}$ uniquely almost surely everywhere. □

Example 2.7 Given $Q \in L^1(\mathbb{R}^{n\times n})$, $v \in L^1(\mathbb{R}^n)$, $m \in L^1$, the function

$$h(x,\omega) := \frac{1}{2}x \cdot Q(\omega)x + x \cdot v(\omega) + m(\omega)$$

is a normal integrand, by Example 1.12. If Q is almost surely positive semidefinite, then h is convex, L-bounded and

$$(E^{\mathcal{G}}h)(x,\omega) = \frac{1}{2}x \cdot E^{\mathcal{G}}[Q](\omega)x + x \cdot (E^{\mathcal{G}}v)(\omega) + (E^{\mathcal{G}}m)(\omega).$$

Proof Given $x \in L^\infty$, Lemma B.13 gives

$$E^{G}[h(x)] = \frac{1}{2}x \cdot E^{\mathcal{G}}[Q](\omega)x + x \cdot (E^{\mathcal{G}}v)(\omega) + (E^{\mathcal{G}}m)(\omega),$$

so the claim follows from Lemma 2.6. □

Argument similar to that in Example 2.7 gives the following.

Example 2.8 Given $\rho, m \in L^1_+$ and an lsc function g on $\mathbb{R}^n$ such that $g(x) \geq -\gamma|x| - l$ for some $\gamma, l \in \mathbb{R}$, the function

$$h(x,\omega) = \rho(\omega)g(x) - m(\omega)$$

is an L-bounded normal integrand with the $\mathcal{G}$-conditional expectation

$$(E^{\mathcal{G}}h)(x,\omega) = (E^{\mathcal{G}}\rho)(\omega)g(x) - (E^{\mathcal{G}}m)(\omega).$$

The following gives sufficient conditions for the existence of the conditional expectation for scenariowise Lipschitz continuous normal integrands.

Theorem 2.9 *Let h be a real-valued normal integrand such that there exist $\bar{x} \in \mathbb{R}^n$ and $\rho \in L^1$ with $h(\bar{x}) \in L^1$ and*

$$|h(x) - h(x')| \leq \rho|x - x'| \quad \forall x, x' \in \mathbb{R}^n$$

almost surely. Then $E^{\mathcal{G}}h$ exists and it is characterized by

$$(E^{\mathcal{G}}h)(x) = E^{\mathcal{G}}[h(x)] \quad \forall x \in \mathbb{R}^n.$$

Moreover,

$$|(E^{\mathcal{G}}h)(x - x')| \leq (E^{\mathcal{G}}\rho)|x - x'| \quad \forall x, x' \in \mathbb{R}^n \tag{2.2}$$

almost surely.

Proof The assumption on h implies that $h(x) \in L^1$ for all $x \in \mathbb{R}^n$ and that, by Jensen's inequality Theorem 1.67,

$$|E^{\mathcal{G}}[h(x)] - E^{\mathcal{G}}[h(x')]| \leq (E^{\mathcal{G}}\rho)|x - x'| \quad a.s.$$

for all $x, x' \in \mathbb{R}^n$. Let D be a countable dense set in $\mathbb{R}^n$ and define $\tilde{h}(x,\omega) := E^{\mathcal{G}}[h(x)](\omega)$ for each $x \in D$. By countability of D, there is a P-null set $N \in \mathcal{G}$ such that, for all $\omega \in \Omega \setminus N$,

$$|\tilde{h}(x,\omega) - \tilde{h}(x',\omega)| \leq (E^{\mathcal{G}}\rho)(\omega)|x - x'| \quad \forall x, x' \in D.$$

The function $\tilde{h}$ has a unique extension to $\mathbb{R}^n \times (\Omega \setminus N)$ such that $\tilde{h}(\cdot,\omega)$ is continuous for all $\omega \in \Omega \setminus N$. Finally, we extend the definition of $\tilde{h}$ to all of $\mathbb{R}^n \times \Omega$ by setting $\tilde{h}(\cdot,\omega) = 0$ for $\omega \in N$. The function thus constructed is a $\mathcal{G}$-measurable Carathéodory integrand and thus, normal. It is clear that it satisfies (2.2) as well.

If $x = \sum_{i=1}^{\nu} x^i 1_{A^i}$, where $x^i \in D$ and $A^i \in \mathcal{G}$ form a disjoint partition of Ω, then

$$E^{\mathcal{G}}[h(x)] = E^{\mathcal{G}}[h(\sum_{i=1}^{\nu} x^i 1_{A^i})] = E^{\mathcal{G}}[\sum_{i=1}^{\nu} 1_{A^i} h(x^i)] = \sum_{i=1}^{\nu} 1_{A^i}\tilde{h}(x^i) = \tilde{h}(x).$$

Any $x \in L^\infty(\mathcal{G})$ is the almost sure limit of such simple random variables x^ν bounded by $\|x\|_{L^\infty}$. By dominated convergence for conditional expectations (see

Theorem B.12), the continuity of $h(\cdot,\omega)$ and $\tilde{h}(\cdot,\omega)$ imply

$$E^{\mathcal{G}}[h(x)] = \lim_{\nu} E^{\mathcal{G}}[h(x^\nu)] = \lim_{\nu} \tilde{h}(x^\nu) = \tilde{h}(x) \quad a.s.$$

so the claim follows from Lemma 2.6. □

Counterexample 2.10 The claim of Theorem 2.9 fails if ρ is not integrable. Indeed, let ξ be a random variable uniformly distributed on (0, 1), let $\mathcal{G} = \sigma(\xi)$ and let $\eta > 1$ be a nonintegrable random variable independent of ξ. Let $h(x,\omega) = \eta(\omega)|x - \xi(\omega)|$. Then $E^{\mathcal{G}}[h(x)] = +\infty$ for every constant $x \in \mathbb{R}$. On the other hand, choosing $x = \xi$, we have $E^{\mathcal{G}}[h(x)] = 0$, so $E^{\mathcal{G}}h$ is not characterized by the condition in Theorem 2.9 even though h is scenariowise Lipschitz and L-bounded.

Lemma 2.11 *Let h^1 and h^2 be L-bounded normal integrands with $h^1 \le h^2$ almost surely everywhere. Then $E^{\mathcal{G}}h^1 \le E^{\mathcal{G}}h^2$ almost surely everywhere whenever the conditional expectations exist.*

Proof For any $x \in L^\infty(\mathcal{G})$, $h^1(x)$ and $h^2(x)$ are quasi-integrable, so the monotonicity of conditional expectation on random variables implies

$$(E^{\mathcal{G}}h^1)(x) = E^{\mathcal{G}}[h^1(x)] \le E^{\mathcal{G}}[h^2(x)] = (E^{\mathcal{G}}h^2)(x)$$

and the result follows from Theorem 1.62. □

Recall that countable pointwise supremums of normal integrands are normal integrands; see Theorem 1.20. The following result is a monotone convergence theorem for conditional expectations of integrands.

Theorem 2.12 *Let $(h^\nu)_{\nu\in\mathbb{N}}$ be an almost surely everywhere nondecreasing sequence of L-bounded normal integrands and let*

$$h := \sup_{\nu} h^\nu.$$

If each $E^{\mathcal{G}}h^\nu$ exists, then $E^{\mathcal{G}}h$ exists and

$$E^{\mathcal{G}}h = \sup_{\nu} E^{\mathcal{G}}h^\nu \quad a.s.e.$$

Proof For any $x \in L^\infty(\mathcal{G})$ and $\alpha \in L^\infty(\mathcal{G};\mathbb{R}_+)$, monotone convergence and Lemmas B.14 and 2.11 imply that

$$\begin{aligned} E[\alpha h(x)] &= E[\alpha \sup_{\nu} h^\nu(x)] = \sup_{\nu} E[\alpha h^\nu(x)] = \sup_{\nu} E[\alpha E^{\mathcal{G}}[h^\nu(x)]] \\ &= E[\alpha \sup_{\nu} E^{\mathcal{G}}[h^\nu(x)]] = E[\alpha \sup_{\nu}(E^{\mathcal{G}}h^\nu)(x)]. \end{aligned}$$

Since $\alpha \in L^\infty(\mathcal{G};\mathbb{R}_+)$ was arbitrary, $E^{\mathcal{G}}[h(x)] = \sup_\nu(E^{\mathcal{G}}h^\nu)(x)$. Since h is L-bounded and $x \in L^\infty(\mathcal{G})$ was arbitrary, the claim follows from Lemma 2.6. □

The following is our main result on the existence of conditional normal integrands.

Theorem 2.13 *An L-bounded normal integrand h with*

$$h(x) \geq -\rho|x| - m \quad \forall x \in \mathbb{R}^n \ a.s.$$

for $\rho, m \in L^1$ admits an almost surely everywhere unique conditional expectation with

$$(E^{\mathcal{G}}h)(x) \geq -(E^{\mathcal{G}}\rho)|x| - E^{\mathcal{G}}m \quad \forall x \in \mathbb{R}^n \ a.s.$$

Proof Assume first that $h(\bar{x}) \in L^1$ for some $\bar{x} \in \mathbb{R}^n$. Clearly,

$$h(x) \geq -\bar{\rho}|x| - m \quad \forall x \in \mathbb{R}^n \ a.s.,$$

where $\bar{\rho} := \rho^+ + 1 > 0$ almost surely. By Corollary 1.26, the functions

$$h^\nu(x,\omega) := \inf_{x' \in \mathbb{R}^n}\{h(x',\omega) + \nu\bar{\rho}(\omega)|x - x'|\}$$

satisfy the assumptions of Theorems 2.9 and they increase pointwise to h. Thus, by Theorems 2.9 and 2.12, $E^{\mathcal{G}}h$ exists. To remove the assumption $h(\bar{x}) \in L^1$, consider the nondecreasing sequence of functions $h^\mu(x) := \min\{h(x), \mu\}$ (which are normal integrands, by Theorem 1.20) and apply Theorem 2.12 again. Uniqueness follows from Lemma 2.6. The lower bound follows from Lemma 2.11 and Example 2.8. □

Given a set-valued mapping $S : \Omega \rightrightarrows \mathbb{R}^n$, the largest $\mathcal{G}$-measurable mapping contained in S almost surely will be denoted by $\operatorname{essinf}^{\mathcal{G}} S$ and called the $\mathcal{G}$-*conditional essential infimum of* S. We will use the notation

$$L^0(\mathcal{G}; S) := \{x \in L^0(\Omega, \mathcal{G}, P; \mathbb{R}^n) \mid x \in S \ a.s.\}$$

for the set of $\mathcal{G}$-measurable selections of S. Applying Theorem 2.13 to the indicator function of a closed-valued measurable mapping gives the following.

Theorem 2.14 *A closed-valued measurable $S : \Omega \rightrightarrows \mathbb{R}^n$ admits a $\mathcal{G}$-conditional essential infimum* $\operatorname{essinf}^{\mathcal{G}} S$ *and it satisfies*

$$L^0(\mathcal{G}; S) = L^0(\mathcal{G}; \operatorname{essinf}^{\mathcal{G}} S).$$

It is the almost surely unique, closed-valued $\mathcal{G}$-measurable mapping S' satisfying the following equivalent conditions:

1. $E^{\mathcal{G}}\delta_S = \delta_{S'}$;
2. *For any $x \in L^0(\mathcal{G})$ and $A \in \mathcal{G}$, $x \in S$ almost surely on A if and only if $x \in S'$ almost surely on A.*

If $L^0(\mathcal{G}; S) \neq \emptyset$, the above conditions are equivalent to

3. $L^0(\mathcal{G}; S) = L^0(\mathcal{G}; S')$.

Proof By Theorem 2.13, the normal integrand δ_S has an almost surely everywhere unique conditional expectation $E^{\mathcal{G}}\delta_S$. Since the conditional expectation of a $\{0, +\infty\}$-valued random variable is almost surely $\{0, +\infty\}$-valued as well, Corollary 1.63 gives the existence of a closed-valued $\mathcal{G}$-measurable mapping S' satisfying condition 1. We claim that $S' = \operatorname{essinf}^{\mathcal{G}} S$.

By Theorem 1.27, S' has a Castaing representation $\{x^\nu\}_{\nu\in\mathbb{N}}$. Extending the definition of each $x^\nu : \operatorname{dom} S' \to \mathbb{R}^n$ by setting e.g. $x^\nu(\omega) := 0$ for $\omega \notin \operatorname{dom} S'$, the $\mathcal{G}$-measurability of $\operatorname{dom} S'$ and condition 1 give

$$E^{\mathcal{G}}[1_{\operatorname{dom} S'}\delta_S(x^\nu)] = 1_{\operatorname{dom} S'}\delta_{S'}(x^\nu) = 0 \quad a.s.$$

so $x^\nu(\omega) \in S(\omega)$ for $\omega \in \operatorname{dom} S'$. Since $S'(\omega) = \operatorname{cl}\{x^\nu(\omega)\}_{\nu\in\mathbb{N}}$, this implies $S' \subseteq S$ almost surely.

Let $\tilde{S}$ be a closed-valued $\mathcal{G}$-measurable mapping contained in S and let $x \in L^0(\mathcal{G})$. Condition 1 gives

$$1_{\{x\in\tilde{S}\}}\delta_{S'}(x) = E^{\mathcal{G}}[1_{\{x\in\tilde{S}\}}\delta_S(x)] = 0 \quad a.s.$$

so $x \in S'$ almost surely on $\{x \in \tilde{S}\}$. Since $x \in L^0(\mathcal{G})$ was arbitrary, Theorem 1.61 gives $\tilde{S} \subseteq S'$ almost surely so $S' = \operatorname{essinf}^{\mathcal{G}} S$. The uniqueness of $\operatorname{essinf}^{\mathcal{G}}$ follows from that of S' which, in turn, follows from Theorem 2.13.

As to the equivalence of 1 and 2, recall that 1 means that

$$E[\delta_S(x)1_A] = E[\delta_{S'}(x)1_A]$$

for every $x \in L^0(\mathcal{G})$ and $A \in \mathcal{G}$. This just means that condition 2 holds.

Since $E^{\mathcal{G}}\delta_S = \delta_{\operatorname{essinf}^{\mathcal{G}} S}$, we have $E\delta_S = E\delta_{\operatorname{essinf}^{\mathcal{G}} S}$, by (2.1). This just means that

$$L^0(\mathcal{G}; \operatorname{essinf}^{\mathcal{G}} S) = L^0(\mathcal{G}; S).$$

This proves the second claim. That this fact characterizes $\operatorname{essinf}^{\mathcal{G}} S$ when $L^0(\mathcal{G}; S) \neq \emptyset$ follows from Theorem 1.72. Indeed, if $\tilde{S}$ is any $\mathcal{G}$-measurable mapping with $L^0(\mathcal{G}; \tilde{S}) = L^0(\mathcal{G}; S)$, then $\tilde{S} = \operatorname{essinf}^{\mathcal{G}} S$, by Theorem 1.72. $\square$

2.1.2 General Properties

Many properties of conditional expectations on random variables (see Sect. B.4) pass on to conditional expectations of normal integrands. Some operations on extended real-valued normal integrands, however, require special attention. Recall that the product of two extended real numbers is defined as zero if one of them is zero while the nonnegative scalar multiple of a function h is defined by

$$(\alpha h)(x) := \begin{cases} \alpha h(x) & \text{if } \alpha > 0, \\ \delta_{\mathrm{cl\,dom}\,h}(x) & \text{if } \alpha = 0, \end{cases}$$

or equivalently,

$$(\alpha h)(x) := \alpha h(x) + \delta_{\mathrm{cl\,dom}\,h}(x) \quad \alpha \geq 0.$$

The third part of the following statement involves the notion of $\mathcal{G}$-conditional essential infimum characterized by Theorem 2.14.

Theorem 2.15 *Let h, h^1 and h^2 be L-bounded normal integrands.*

1. *$h^1 + h^2$ is L-bounded and*

$$E^{\mathcal{G}}(h^1 + h^2) = E^{\mathcal{G}}h^1 + E^{\mathcal{G}}h^2.$$

2. *If $\alpha \in L^1_+$, h is $\mathcal{G}$-measurable and αh is L-bounded, then*

$$E^{\mathcal{G}}(\alpha h) = E^{\mathcal{G}}[\alpha]h.$$

3. *If $\alpha \in L^0_+(\mathcal{G})$ and αh is L-bounded, then*

$$E^{\mathcal{G}}(\alpha h) = \alpha E^{\mathcal{G}}h$$

   if either α is strictly positive or $\operatorname{essinf}^{\mathcal{G}}[\mathrm{cl\,dom}\,h] = \mathrm{cl\,dom}\,E^{\mathcal{G}}h$.
4. *If f is an L-bounded normal integrand on $\mathbb{R}^n \times \mathbb{R}^m \times \Omega$ and $u \in L^0(\mathcal{G})$ is such that $h(x, \omega) := f(x, u(\omega), \omega)$ is L-bounded, then*

$$(E^{\mathcal{G}}h)(x, \omega) = (E^{\mathcal{G}}f)(x, u(\omega), \omega).$$

5. *If $M : \mathbb{R}^m \times \Omega \to \mathbb{R}^n$ is a $\mathcal{G}$-measurable Carathéodory mapping and $h \circ M$ is L-bounded, then*

$$E^{\mathcal{G}}(h \circ M) = (E^{\mathcal{G}}h) \circ M.$$

Proof Let $x \in L^\infty(\mathcal{G})$. Since $h^1(x)^-$ and $h^2(x)^-$ are integrable, claim 1 follows from Lemma 2.6 and the first part of Lemma B.13. In claim 2,

$$\begin{aligned} E^{\mathcal{G}}[(\alpha h)(x)] &= E^{\mathcal{G}}[\alpha h(x) + \delta_{\mathrm{cl\,dom}\, h}(x)] \\ &= E^{\mathcal{G}}[\alpha]h(x) + \delta_{\mathrm{cl\,dom}\, h}(x) \\ &= (E^{\mathcal{G}}[\alpha]h)(x), \end{aligned}$$

by Lemma B.13 so the claim follows from Lemma 2.6 again. In claim 3,

$$\begin{aligned} E^{\mathcal{G}}[(\alpha h)(x)] &= E^{\mathcal{G}}[\alpha h(x) + \delta_{\mathrm{cl\,dom}\, h}(x)] \\ &= \alpha E^{\mathcal{G}}[h(x)] + E^{\mathcal{G}}[\delta_{\mathrm{cl\,dom}\, h}(x)], \end{aligned}$$

by Lemma B.13. If α is strictly positive or if $\operatorname{essinf}^{\mathcal{G}}[\mathrm{cl\,dom}\, h] = \mathrm{cl\,dom}\, E^{\mathcal{G}}h$, we thus get

$$\begin{aligned} E^{\mathcal{G}}[(\alpha h)(x)] &= \alpha(E^{\mathcal{G}}h)(x) + \delta_{\mathrm{cl\,dom}\, E^{\mathcal{G}}h}(x) \\ &= (\alpha E^{\mathcal{G}}h)(x). \end{aligned}$$

Part 4 follows from part 5 with $M(x,\omega) = (x, u(\omega))$. As to 5, by Theorem 1.20, $h \circ M$ is a normal integrand. By Theorem 2.13, both h and $h \circ M$ have almost surely everywhere unique conditional expectations. The integrand $h \circ M$ is quasi-integrable at $x \in L^0(\mathcal{G})$ if and only if h is quasi-integrable at $M(x)$. Thus, 5 follows from the definition of a conditional expectation of a normal integrand. □

The following shows that part 3 of Theorem 2.15 may fail without the extra assumptions.

Counterexample 2.16 Let $\mathcal{G}$ be trivial, $\alpha = 0$ almost surely and $h(x,\omega) := \eta|x|$, where $\eta \notin L^1$ is nonnegative. Then $\alpha h = 0$ while $E^{\mathcal{G}}h = \delta_{\{0\}}$, so

$$E^{\mathcal{G}}(\alpha h) \neq \alpha E^{\mathcal{G}}h.$$

Lemma 2.17 *Let h be an L-bounded normal integrand, $A \in \mathcal{G}$ and $\tilde{h}(x,\omega) := h(x,\omega)1_A(\omega)$. Then $(E^{\mathcal{G}}\tilde{h})(x,\omega) = (E^{\mathcal{G}}h)(x,\omega)1_A(\omega)$.*

Proof Given $x \in L^\infty(\mathcal{G})$, the L-boundedness implies that $h(x)$ is quasi-integrable, so Lemma B.13 gives

$$E^{\mathcal{G}}[\tilde{h}(x)] = E^{\mathcal{G}}[h(x)1_A] = (E^{\mathcal{G}}h)(x)1_A.$$

Thus, the claim follows from Lemma 2.6. □

The class of normal integrands is not a linear space, so one cannot hope for linearity of the conditional expectation, in general.

Remark 2.18 The conditional expectation is a linear operator on the linear space of normal integrands that satisfy the assumptions of Theorem 2.9.

Proof By Theorem 2.15, addition and nonnegative scalar multiplication preserve normality. If h satisfies the assumptions of Theorem 2.9, then so does $-h$, and $h(x) \in L^1$ for every $x \in \mathbb{R}^n$. Thus,

$$E^{\mathcal{G}}(-h)(x) = E^{\mathcal{G}}[-h(x)] = -E^{\mathcal{G}}[h(x)] = -(E^{\mathcal{G}}h)(x) \quad \forall x \in \mathbb{R}^n,$$

so $E^{\mathcal{G}}(-h) = -E^{\mathcal{G}}h$, by Theorem 2.9. □

Recall that if ξ is a quasi-integrable random variable and $\mathcal{G}'$ is a sub-σ-algebra of $\mathcal{G}$, then

$$E^{\mathcal{G}'}[E^{\mathcal{G}}\xi] = E^{\mathcal{G}'}\xi \quad \text{and} \quad E[E^{\mathcal{G}}\xi] = E\xi;$$

see Lemma B.14. This extends to normal integrands as follows.

Theorem 2.19 (Tower Property) *Assume that h is an L-bounded normal integrand, and $\mathcal{G}'$ is a sub-σ-algebra of $\mathcal{G}$. Then*

$$E^{\mathcal{G}'}(E^{\mathcal{G}}h) = E^{\mathcal{G}'}h$$

for all $x \in L^0(\mathcal{G})$ such that $h(x)$ is quasi-integrable.

Proof By Theorem 2.13, all the conditional expectations exist and are L-bounded. For any $x \in L^\infty(\mathcal{G}')$, the tower property of conditional expectation (see Lemma B.14) gives

$$(E^{\mathcal{G}'}(E^{\mathcal{G}}h))(x) = E^{\mathcal{G}'}[(E^{\mathcal{G}}h)(x)] = E^{\mathcal{G}'}[E^{\mathcal{G}}[h(x)]] = E^{\mathcal{G}'}[h(x)] = (E^{\mathcal{G}'}h)(x).$$

The claim thus follows from Lemma 2.6. □

Given $\mathcal{H} \subset \mathcal{F}$, σ-algebras $\mathcal{G}$ and $\mathcal{G}'$ are *$\mathcal{H}$-conditionally independent* if

$$E^{\mathcal{H}}[1_{A'}1_A] = E^{\mathcal{H}}[1_{A'}]E^{\mathcal{H}}[1_A]$$

for every $A \in \mathcal{G}$ and $A' \in \mathcal{G}'$. A random variable w is *$\mathcal{H}$-conditionally independent* of $\mathcal{G}$ if $\sigma(w)$ and $\mathcal{G}$ are $\mathcal{H}$-conditionally independent. Likewise, we say that a normal integrand h is *$\mathcal{H}$-conditionally independent* of $\mathcal{G}$ if $\sigma(h)$ and $\mathcal{G}$ are $\mathcal{H}$-conditionally independent. Here $\sigma(h)$ is the smallest σ-algebra under which epi h is measurable. In other words, $\sigma(h)$ is generated by the family

$$\{(\operatorname{epi} h)^{-1}(O) \mid O \subset \mathbb{R}^{n+1} \text{ open}\}.$$

Example 2.20 Let ξ be a random variable with values in a measurable space $(\Xi, \mathcal{A})$, H a $\mathcal{A}$-measurable normal integrand on $\mathbb{R}^n$ and

$$h(x, \omega) = H(x, \xi(\omega)).$$

If ξ is $\mathcal{H}$-conditionally independent of $\mathcal{G}$, then h is so too. Indeed, given an open $O \subset \mathbb{R}^{n+1}$, we have $(\operatorname{epi} h)^{-1}(O) = \xi^{-1}((\operatorname{epi} H)^{-1}(O))$, so $\sigma(h) \subseteq \sigma(\xi)$. To conclude, it suffices to note that sub-σ-algebras inherit conditional independence.

If an integrable random variable w is $\mathcal{H}$-conditionally independent of $\mathcal{G}$, then, by Lemma B.16,

$$E^{\mathcal{G} \vee \mathcal{H}}[w] = E^{\mathcal{H}}[w],$$

where $\mathcal{G} \vee \mathcal{H}$ denotes the smallest σ-algebra containing both $\mathcal{G}$ and $\mathcal{H}$. This extends to normal integrands as follows.

Theorem 2.21 *Let h be an L-bounded normal integrand $\mathcal{H}$-conditionally independent of $\mathcal{G}$. Then*

$$E^{\mathcal{G} \vee \mathcal{H}} h = E^{\mathcal{H}} h.$$

In particular, if h is independent of $\mathcal{G}$, then $E^{\mathcal{G}} h$ is deterministic.

Proof Assume first that h satisfies the assumptions of Theorem 2.9. Then $E^{\mathcal{H}} h$ is characterized by

$$E^{\mathcal{H}}(h(x)) = (E^{\mathcal{H}} h)(x) \quad \forall x \in \mathbb{R}^n$$

and likewise for $E^{\mathcal{G} \vee \mathcal{H}} h$. Thus, $E^{\mathcal{G} \vee \mathcal{H}} h = E^{\mathcal{H}} h$, by Lemma B.16. As to the general case, we use Lipschitz regularizations and truncations as in the proof of Theorem 2.13. By Lemma 2.11, the L-boundedness implies

$$h(x) \geq -E^{\sigma(h)}[\rho]|x| - E^{\sigma(h)}[m]$$

so we may assume, without loss of generality, that ρ and m are $\sigma(h)$-measurable. Theorem 1.21 then implies that the Lipschitz regularizations h^ν are $\sigma(h)$-measurable as well so $E^{\mathcal{G} \vee \mathcal{H}} h^\nu = E^{\mathcal{H}} h^\nu$. The first claim now follows from Theorem 2.12. The second claim follows by taking $\mathcal{H}$ the trivial σ-algebra. □

Recall that a normal integrand h is p-homogeneous if there exists $C \in L^0$ such that

$$h(\alpha x, \omega) = \begin{cases} \alpha^p h(x, \omega) - C(\omega) \frac{\alpha^p - 1}{p} & \text{if } p \neq 0, \\ h(x, \omega) - C(\omega) \ln \alpha & \text{if } p = 0. \end{cases} \tag{2.3}$$

Lemma 2.22 *Let h be an L-bounded p-homogeneous normal integrand. If the random variable C in (2.3) is integrable, then the conditional expectation of h is p-homogeneous with*

$$(E^{\mathcal{G}}h)(\alpha x) = \begin{cases} \alpha^p (E^{\mathcal{G}}h)(x) - E^{\mathcal{G}}[C]\frac{\alpha^p - 1}{p} & \text{if } p \neq 0, \\ (E^{\mathcal{G}}h)(x) - E^{\mathcal{G}}[C]\ln \alpha & \text{if } p = 0. \end{cases}$$

Proof Let $\alpha > 0$ and $x \in L^\infty$. By Lemmas 2.6 and B.13,

$$\begin{aligned} (E^{\mathcal{G}}h)(\alpha x) &= E^{\mathcal{G}}[h(\alpha x)] \\ &= \begin{cases} E^{\mathcal{G}}[\alpha^p h(x) - C(\alpha^p - 1)] & \text{if } p \neq 0, \\ E^{\mathcal{G}}[h(x) - C \ln \alpha] & \text{if } p = 0 \end{cases} \\ &= \begin{cases} \alpha^p E^{\mathcal{G}}[h(x)] - E^{\mathcal{G}}[C](\alpha^p - 1)] & \text{if } p \neq 0, \\ E^{\mathcal{G}}[h(x)] - E^{\mathcal{G}}[C]\ln \alpha & \text{if } p = 0 \end{cases} \\ &= \begin{cases} \alpha^p (E^{\mathcal{G}}h)(x) - E^{\mathcal{G}}[C](\alpha^p - 1)] & \text{if } p \neq 0, \\ (E^{\mathcal{G}}h)(x) - E^{\mathcal{G}}[C]\ln \alpha & \text{if } p = 0. \end{cases} \end{aligned}$$

Thus, $E^{\mathcal{G}}h$ is p-homogeneous, by Corollary 1.63.4. □

2.1.3 Convexity

Conditional expectations behave particularly well in the presence of convexity.

Theorem 2.23 *The conditional expectation of an L-bounded convex normal integrand is convex.*

Proof Assume that h is an L-bounded convex normal integrand and let $\alpha^1, \alpha^2 > 0$ with $\alpha^1 + \alpha^2 = 1$ and $x^1, x^2 \in L^\infty(\mathcal{G})$. By Lemmas 2.6 and B.13,

$$\begin{aligned} (E^{\mathcal{G}}h)(\alpha^1 x^1 + \alpha^2 x^2) &= E^{\mathcal{G}}[h(\alpha^1 x^1 + \alpha^2 x^2)] \\ &\leq E^{\mathcal{G}}[\alpha^1 h(x^1) + \alpha^2 h(x^2)] \\ &= \alpha^1 E^{\mathcal{G}}[h(x^1)] + \alpha^2 E^{\mathcal{G}}[h(x^2)] \\ &= \alpha^1 (E^{\mathcal{G}}h)(x^1) + \alpha^2 (E^{\mathcal{G}}h)(x^2) \end{aligned}$$

so $E^{\mathcal{G}}h$ is convex by Corollary 1.63. □

Under convexity, we can add one more algebraic operation to the list in Theorem 2.15. The following result gives a simple expression for the conditional expectation of composite integrands studied in Sect. 1.1.4. It assumes that the inner function in the composition is $\mathcal{G}$-measurable.

Theorem 2.24 *Let g be an L-bounded convex normal integrand and H a $\mathcal{G}$-measurable random K-convex function satisfying the assumptions of Theorem 1.45. If $g \circ H$ is L-bounded, then*

$$E^{\mathcal{G}}(g \circ H) = (E^{\mathcal{G}} g) \circ H.$$

The integrand $g \circ H$ is L-bounded, in particular, if there exists $y \in \operatorname{dom} Eg^* \cap L^1$ such that $y \cdot H$ is L-bounded.

Proof By Theorem 1.45, $g \circ H$ is a normal integrand. By Theorem 2.13, both g and $g \circ H$ have almost surely everywhere unique conditional expectations. The integrand $g \circ H$ is quasi-integrable at $x \in L^0(\mathcal{G})$ if and only if g is quasi-integrable at $H(x)$. Thus, the first claim follows from the definition of a conditional expectation of a normal integrand. As to the second claim, by Fenchel's inequality, the extra condition implies

$$g(H(x, \omega), \omega) \geq y(\omega) \cdot H(x, \omega) - g^*(y),$$

so h is L-bounded, by Lemma 2.5. □

Corollary 2.25 *Let g be an L-bounded convex normal integrand on $\mathbb{R}^m$ and $A \in L^0(\mathbb{R}^{m \times n})$ a $\mathcal{G}$-measurable random matrix. If $g \circ A$ is L-bounded, then*

$$E^{\mathcal{G}}(g \circ A) = (E^{\mathcal{G}} g) \circ A.$$

The integrand $g \circ A$ is L-bounded, in particular, if there exists a $y \in \operatorname{dom} Eg^* \cap L^1$ such that $A^* y \in L^1$.

Proof This follows from Theorem 2.24 with $\operatorname{dom} H = \mathbb{R}^n$, $H(x) = Ax$ and $K = \{0\}$. □

The following result is concerned with measurable selections of the scenariowise subdifferential $\partial h(x)$ of a convex normal integrand; see Theorem 1.51. It says that conditional expectations of integrable selections of $\partial h(x)$ are selections of $\partial(E^{\mathcal{G}} h)(x)$. Sufficient conditions for the converse will be given in Theorems 3.79 and 5.62. Given a set-valued mapping $S : \Omega \rightrightarrows \mathbb{R}^n$, the set of integrable selections of S will be denoted by

$$L^1(S) := \{x \in L^1 \mid x \in S \ a.s.\}.$$

Since L^1 convergence implies convergence in probability, Corollary 1.57 implies that, if S is closed-valued, then $L^1(S)$ is closed in the L^1-topology.

Theorem 2.26 *Let h be an L-bounded convex normal integrand and let $x \in L^0(\mathcal{G})$ be such that $h(x)$ is quasi-integrable. Then $E^{\mathcal{G}} v \in \partial(E^{\mathcal{G}} h)(x)$ almost surely for all $v \in L^1(\partial h(x))$.*

Proof Let $v \in L^1(\partial h(x))$ so that

$$h(x', \omega) \geq h(x(\omega), \omega) + (x' - x(\omega)) \cdot v(\omega) \quad \forall x' \in \mathbb{R}^n.$$

Let $\alpha := e^{-|x|}$ so that $\alpha x \in L^\infty(\mathcal{G})$ and

$$\alpha(\omega) h(x', \omega) \geq \alpha(\omega) h(x(\omega), \omega) + \alpha(\omega)(x' - x(\omega)) \cdot v(\omega) \quad \forall x' \in \mathbb{R}^n.$$

Since h is L-bounded, the definition of the conditional expectation and Theorem 2.15 give,

$$\alpha(E^{\mathcal{G}} h)(x') \geq \alpha(E^{\mathcal{G}} h)(x) + \alpha(x' - x) \cdot E^{\mathcal{G}} v \quad a.s.$$

for all $x' \in L^\infty(\mathcal{G})$. Thus, by Theorem 1.62,

$$\alpha(\omega)(E^{\mathcal{G}} h)(x', \omega) \geq \alpha(\omega)(E^{\mathcal{G}} h)(x(\omega), \omega) + \alpha(\omega)(x' - x(\omega)) \cdot (E^{\mathcal{G}} v)(\omega).$$

Dividing by α gives $E^{\mathcal{G}} v \in \partial(E^{\mathcal{G}} h)(x)$ almost surely. □

Given a set-valued mapping $S : \Omega \rightrightarrows \mathbb{R}^n$, the set of $\mathcal{G}$-measurable integrable selections of S will be denoted by

$$L^1(\mathcal{G}; S) := \{x \in L^1(\mathcal{G}) \mid x \in S \; a.s.\}.$$

A *$\mathcal{G}$-conditional expectation* of an $\mathcal{F}$-measurable set-valued mapping $S : \Omega \rightrightarrows \mathbb{R}^n$ is a closed-valued $\mathcal{G}$-measurable mapping $E^{\mathcal{G}} S$ such that

$$L^1(\mathcal{G}; E^{\mathcal{G}} S) = \mathrm{cl}\{E^{\mathcal{G}} x \mid x \in L^1(S)\},$$

where the closure is taken in the L^1-topology.

The following lemma establishes the existence and uniqueness in the convex case.

Theorem 2.27 *If S is a closed convex-valued mapping with $L^1(S) \neq \emptyset$, then $E^{\mathcal{G}} S$ exists and it is the almost surely unique closed convex-valued mapping with*

$$\sigma_{E^{\mathcal{G}} S} = E^{\mathcal{G}} \sigma_S.$$

Proof If $x \in L^1(S)$, then $\sigma_S(v) \geq x \cdot v$ so σ_S is L-bounded. By Theorem 2.23 and Lemma 2.22, $E^{\mathcal{G}} \sigma_S$ is a positively homogeneous convex normal integrand. By Theorem 1.46 and Corollary A.56, there is a convex-valued $\mathcal{G}$-measurable mapping Γ such that $E^{\mathcal{G}} \sigma_S = \sigma_\Gamma$.

Let $C := \{E^{\mathcal{G}}x \mid x \in L^1(S)\}$. For every $v \in L^\infty(\mathcal{G})$,

$$\begin{aligned}\sigma_C(v) &= \sup\{E[v \cdot E^{\mathcal{G}}x] \mid x \in L^1(S)\}\\ &= \sup\{E[v \cdot x] \mid x \in L^1(S)\}\\ &= E\sigma_S(v)\\ &= E[E^{\mathcal{G}}\sigma_S(v)]\\ &= E[\sigma_\Gamma(v)]\\ &= \sigma_{L^1(\Gamma)}(v),\end{aligned}$$

where the third and the last equality follow from Theorem 1.64 and the fourth from Theorem B.11 since $\sigma_S(v)$ is quasi-integrable. Applying Theorem A.54 to δ_C gives $\operatorname{cl} C = L^1(\Gamma)$ so $\Gamma = E^{\mathcal{G}}S$. □

Conditional expectation of a closed convex cone-valued mapping is again convex cone-valued while the $\mathcal{G}$-conditional expectation of the indicator function of a closed measurable set S is the indicator function of the $\mathcal{G}$-conditional essential infimum $\operatorname{essinf}^{\mathcal{G}} S$ of S; see Theorem 2.14. Thus, when S is a cone, Theorem 2.27 gives following relation between polar cones; see Sect. A.6.

Corollary 2.28 *If S is a closed convex cone-valued mapping such that $L^1(S) \neq \emptyset$, then $E^{\mathcal{G}}S$ exists. It is the almost surely unique closed convex cone-valued mapping with*

$$(E^{\mathcal{G}}S)^\circ = \operatorname{essinf}^{\mathcal{G}}(S^\circ).$$

Recall from Corollary 1.50 that if C is a closed convex random set and $\alpha \in L^0$ is nonnegative, then the set

$$(\alpha_+ C)(\omega) := \begin{cases}\alpha(\omega)C(\omega) & \text{it } \alpha(\omega) > 0,\\ C^\infty(\omega) & \text{if } \alpha(\omega) = 0\end{cases}$$

is measurable. Note that the support function of $\alpha_+ C$ is given by

$$\sigma_{\alpha_+ C} = \begin{cases}\alpha\sigma_C & \text{if } \alpha > 0,\\ \operatorname{cl}\operatorname{dom}\sigma_C & \text{if } \alpha = 0,\end{cases}$$

in accordance with the definition of positive scalar multiple of a function; see Corollary A.59. The sets $\operatorname{cl}\operatorname{dom}\sigma_C$ and C^∞ are polar to each other; see Example A.43.

Corollary 2.29 *Let $\alpha \in L^0_+$ and let S, S_1 and S_2 be closed convex-valued mappings each admitting an integrable selection.*

1. *If $S_1 + S_2$ and $E^{\mathcal{G}} S_1 + E^{\mathcal{G}} S_2$ are closed-valued, then*

$$E^{\mathcal{G}}[S_1 + S_2] = E^{\mathcal{G}} S_1 + E^{\mathcal{G}} S_2.$$

2. *If $\alpha \in L^1_+$, S is $\mathcal{G}$-measurable and $\alpha_+ S$ admits an integrable selection, then*

$$E^{\mathcal{G}}[\alpha_+ S] = E^{\mathcal{G}}[\alpha]_+ S.$$

3. *If $\alpha \in L^0_+(\mathcal{G})$ and $\alpha_+ S$ admits an integrable selection, then*

$$E^{\mathcal{G}}[\alpha_+ S] = \alpha_+ E^{\mathcal{G}} S$$

if either α is strictly positive or $E^{\mathcal{G}}(S^\infty) = (E^{\mathcal{G}} S)^\infty$.

Proof Note first that if a set-valued mapping has an integrable selection, then its support function is L-bounded. Recall that the support function of the sum of two sets is the sum of the support functions of the sets. Thus, by Theorems 2.27 and 2.15.1,

$$\begin{aligned}
\sigma_{E^{\mathcal{G}}(S_1+S_2)} &= E^{\mathcal{G}}[\sigma_{(S_1+S_2)}] \\
&= E^{\mathcal{G}}[\sigma_{S_1} + \sigma_{S_2}] \\
&= E^{\mathcal{G}}\sigma_{S_1} + E^{\mathcal{G}}\sigma_{S_2} \\
&= \sigma_{E^{\mathcal{G}} S_1} + \sigma_{E^{\mathcal{G}} S_2} \\
&= \sigma_{(E^{\mathcal{G}} S_1 + E^{\mathcal{G}} S_2)}.
\end{aligned}$$

Taking conjugates, gives 1. In 2,

$$\sigma_{E^{\mathcal{G}}(\alpha_+ S)} = E^{\mathcal{G}}[\sigma_{\alpha_+ S}] = E^{\mathcal{G}}[\alpha\sigma_S] = E^{\mathcal{G}}[\alpha]\sigma_S = \sigma_{E^{\mathcal{G}}[\alpha]_+ S},$$

by Theorems 2.15 and 2.27. Taking conjugates, gives 2. By Corollary 2.28, the additional condition in 3 means that either α is strictly positive or $\operatorname{essinf}^{\mathcal{G}}[(S^\infty)^*] = ((E^{\mathcal{G}} S)^\infty)^*$. The latter can be written as $\operatorname{essinf}^{\mathcal{G}}[\operatorname{cl}\operatorname{dom}\sigma_S] = \operatorname{cl}\operatorname{dom}\sigma_{E^{\mathcal{G}} S}$, so

$$\sigma_{E^{\mathcal{G}}(\alpha_+ S)} = E^{\mathcal{G}}[\sigma_{\alpha_+ S}] = E^{\mathcal{G}}[\alpha\sigma_S] = \alpha E^{\mathcal{G}}[\sigma_S] = \alpha\sigma_{E^{\mathcal{G}} S} = \sigma_{\alpha_+ E^{\mathcal{G}} S},$$

by Theorems 2.15 and 2.27. Taking conjugates, gives 3. □

Applying Lemma 2.17 to $\tilde{h}(x, \omega) = \sigma_S(x, \omega)1_A(\omega)$ and using Theorem 2.27 gives the following.

Lemma 2.30 *Let S be a closed convex-valued mapping, $A \in \mathcal{G}$ and*

$$\tilde{S}(\omega) := \begin{cases} S(\omega) & \text{if } \omega \in A, \\ \{0\} & \text{otherwise.} \end{cases}$$

If S admits an integrable selection, then

$$(E^{\mathcal{G}}\tilde{S})(\omega) = \begin{cases} (E^{\mathcal{G}}S)(\omega) & \text{if } \omega \in A, \\ \{0\} & \text{otherwise.} \end{cases}$$

It is natural to ask how the conjugate of the conditional expectation of a normal integrand h is related to the conjugate of h. Given a normal integrand h, the mapping $E^{\mathcal{G}} \operatorname{epi} h$, whenever it exists, is the epigraph of a $\mathcal{G}$-measurable normal integrand. We denote this $\mathcal{G}$-measurable normal integrand by ${}^{\mathcal{G}}h$ and call it the *$\mathcal{G}$-conditional epi-expectation* of h.

The conditional epi-expectation can be seen as the dual operation of the conditional expectation.

Theorem 2.31 *Assume that h is a convex normal integrand such that there exists $x \in \operatorname{dom} Eh \cap L^0(\mathcal{G})$ and $v \in \operatorname{dom} Eh^* \cap L^1$. Then*

$$(E^{\mathcal{G}}h)^* = {}^{\mathcal{G}}(h^*).$$

Proof By assumption, $(v, h^*(v)^+)$ is an integrable selection of $\operatorname{epi} h^*$. Applying Theorem 2.27 to $\operatorname{epi} h^*$, we get that $E^{\mathcal{G}} \operatorname{epi} h^*$ exists and is characterized by $\sigma_{E^{\mathcal{G}} \operatorname{epi} h^*} = E^{\mathcal{G}}\sigma_{\operatorname{epi} h^*}$. By Lemma 2.5 and Theorem 2.13, the assumptions imply that $E^{\mathcal{G}}h$ is well-defined and proper. By Corollary A.59, a closed proper convex function g can be expressed as $g = \sigma_{\operatorname{epi} g^*} \circ H$, where $H(x) = (x, -1)$. Thus, by Theorem 2.24,

$$\begin{aligned} E^{\mathcal{G}}h = E^{\mathcal{G}}[\sigma_{\operatorname{epi} h^*} \circ H] = (E^{\mathcal{G}}\sigma_{\operatorname{epi} h^*}) \circ H \\ = \sigma_{E^{\mathcal{G}} \operatorname{epi} h^*} \circ H = \sigma_{\operatorname{epi}({}^{\mathcal{G}}(h^*))} \circ H = ({}^{\mathcal{G}}(h^*))^*. \end{aligned}$$

Since $E^{\mathcal{G}}h$ is proper, the claim thus follows from the biconjugate theorem. □

The following gives an expression for the conditional expectation of the sublinear normal integrand from Corollary 1.48.

Corollary 2.32 *Let h be as in Theorem 2.31. Then*

$$H(x, \alpha, \omega) := \begin{cases} \alpha h(x/\alpha, \omega) & \text{if } \alpha > 0, \\ h^\infty(x, \omega) & \text{if } \alpha = 0, \\ +\infty & \text{otherwise} \end{cases}$$

is an L-bounded convex normal integrand on $\mathbb{R}^{n+1} \times \Omega$ *with*

$$(E^{\mathcal{G}} H)(x, \alpha, \omega) = \begin{cases} \alpha (E^{\mathcal{G}} h)(x/\alpha, \omega) & \text{if } \alpha > 0, \\ (E^{\mathcal{G}} h)^\infty (x, \omega) & \text{if } \alpha = 0, \\ +\infty & \text{otherwise.} \end{cases}$$

Proof By Corollary A.59,

$$H(x, \alpha, \omega) = \sigma_{\operatorname{epi} h^*(\omega)}(x, -\alpha).$$

By assumption, $(v, h^*(v)^+)$ is an integrable selection of $\operatorname{epi} h^*$, so H is L-bounded with $H(x, \alpha, \omega) \ge x \cdot v(\omega) + \alpha h^*(v)(\omega)^+$. By Theorem 2.27, $E^{\mathcal{G}} H(x, \alpha, \omega) = \sigma_{E^{\mathcal{G}} \operatorname{epi} h^*(\omega)}(x, -\alpha)$, where $E^{\mathcal{G}} \operatorname{epi} h^* = \operatorname{epi}(E^{\mathcal{G}} h)^*$, by Theorem 2.31. The claim thus follows from Corollary A.59. □

Recall the normal integrand $h\alpha$ from Corollary 1.49. Combining Corollary 2.32 with Theorem 2.15.4 gives the following.

Corollary 2.33 *Let h be as in Theorem 2.31 and let $\alpha \in L^0(\mathcal{G})$ such that $h\alpha$ is L-bounded. Then*

$$E^{\mathcal{G}}(h\alpha) = (E^{\mathcal{G}} h)\alpha.$$

Taking $\alpha = 0$ in Corollary 2.33 gives the following.

Corollary 2.34 *Let h be as in Theorem 2.31. Then*

$$(E^{\mathcal{G}} h)^\infty = E^{\mathcal{G}} h^\infty.$$

Proof By Lemma 2.5, the assumptions of Theorem 2.31 imply that h is L-bounded. This implies that $h0 = h^\infty$ is L-bounded as well, so the claim follows from Corollary 2.33. □

The conditional epi-expectation of a normal integrand has the following property reminiscent of Jensen's inequality. Besides its statement, the proof of the following is remarkably similar to that of Theorem 1.67.

Theorem 2.35 (Jensen's Inequality) *Given a convex normal integrand h with* $\operatorname{dom} Eh \cap L^1 \neq \emptyset$, *we have*

$$^{\mathcal{G}}h(E^{\mathcal{G}} x) \le E^{\mathcal{G}}[h(x)] \quad a.s. \tag{2.4}$$

for all $x \in L^1$ such that $h(x)$ is quasi-integrable. Moreover,

$$^{\mathcal{G}}h \le E^{\mathcal{G}} h \quad a.s.e.$$

as soon as h is L-bounded.

Proof Given $(x, \alpha) \in L^1(\operatorname{epi} h)$, we have

$$(E^{\mathcal{G}}x, E^{\mathcal{G}}\alpha) \in E^{\mathcal{G}}[\operatorname{epi} h] = \operatorname{epi}({}^{\mathcal{G}}h),$$

by the definitions of conditional expectation of a measurable mapping and of conditional epi-expectation, so

$${}^{\mathcal{G}}h(E^{\mathcal{G}}x) \le E^{\mathcal{G}}\alpha \quad a.s. \tag{2.5}$$

Assume now that $h(x)^+$ is integrable and, for $\nu \in \mathbb{N}$, let $\alpha^\nu := \max\{h(x), -\nu\}$. Since $(x, \alpha^\nu) \in L^1(\operatorname{epi} h)$, (2.5) gives ${}^{\mathcal{G}}h(E^{\mathcal{G}}x) \le E^{\mathcal{G}}\alpha^\nu$. Since $\alpha^\nu \searrow h(x)$ almost surely, Theorem B.12 gives (2.4).

Assume now that $h(x)^-$ is integrable and let $B \in \mathcal{G}$. To prove (2.4), it suffices, by Lemma B.5, to show that

$$E[{}^{\mathcal{G}}h(E^{\mathcal{G}}x)1_B] \le E[E^{\mathcal{G}}[h(x)]1_B]. \tag{2.6}$$

By Lemma B.14, $E[E^{\mathcal{G}}[h(x)]1_B] = E[h(x)1_B]$ so we may assume that $h(x)^+1_B$ is integrable since, otherwise, the right side of (2.6) equals $+\infty$. Since $L^1(\operatorname{epi} h) \neq \emptyset$, by assumption, there exists an $\bar{x} \in L^1$ such that $h(\bar{x})^+ \in L^1$. Let $\hat{x} := x1_B + \bar{x}1_{\Omega\setminus B}$. Since $h(\hat{x})^+ \in L^1$, the first part of the proof gives

$${}^{\mathcal{G}}h(E^{\mathcal{G}}\hat{x}) \le E^{\mathcal{G}}[h(\hat{x})] \quad a.s.,$$

where, by Lemma B.13,

$$E^{\mathcal{G}}[\hat{x}] = E^{\mathcal{G}}[x]1_B + E^{\mathcal{G}}[\bar{x}]1_{\Omega\setminus B},$$

$$E^{\mathcal{G}}[h(\hat{x})] = E^{\mathcal{G}}[h(x)]1_B + E^{\mathcal{G}}[h(\bar{x})]1_{\Omega\setminus B}.$$

This implies (2.6) thus finishing the proof of the first claim. Since $E^{\mathcal{G}}[h(x)] = (E^{\mathcal{G}}h)(x)$ for every $x \in L^\infty(\mathcal{G})$, the second claim follows from the first and Theorem 1.62. □

The following shows, in particular, that $\mathcal{G}$-measurability of a set-valued mapping S is, in fact, equivalent to the property given in Corollary 1.68.

Theorem 2.36 *Given a closed convex-valued random set S with $L^1(S) \neq \emptyset$, the following are equivalent:*

1. *S is $\mathcal{G}$-measurable;*
2. *$E^{\mathcal{G}}x \in L^1(S)$ for every $x \in L^1(S)$;*
3. *$E^{\mathcal{G}}S \subseteq S$ almost surely.*

Proof By Corollary 1.68, 1 implies 2. Since $L^1(S)$ is closed, 2 implies 3. To prove that 3 implies 1, it suffices, by Theorem 1.2, to show that $\omega \mapsto d(x, S(\omega))$ is $\mathcal{G}$-measurable for every $x \in \mathbb{R}^n$. By Corollary 1.24, there exists $y \in L^0(S)$ such that

$$d(x, S(\omega)) = |x - y(\omega)|.$$

Since $L^1(S) \neq \emptyset$, $d(x, S)$ is integrable, so $y \in L^1(S)$. By Jensen's inequality,

$$E[d(x, S)] = E|x - y| \geq E|x - E^{\mathcal{G}} y|.$$

When $E^{\mathcal{G}} S \subseteq S$, we have $E^{\mathcal{G}} y \in S$, so $|x - E^{\mathcal{G}} y| \geq d(x, S)$ and thus, $d(x, S) = |x - E^{\mathcal{G}} y|$, which is $\mathcal{G}$-measurable. □

The following gives a converse to Theorem 1.67 by showing that, if a convex normal integrand h satisfies Jensen's inequality with respect to σ-algebra $\mathcal{G} \subset \mathcal{F}$, then it is necessarily $\mathcal{G}$-measurable.

Corollary 2.37 *A convex normal integrand h with* $\operatorname{dom} Eh \cap L^1 \neq \emptyset$ *is $\mathcal{G}$-measurable if and only if*

$$h(E^{\mathcal{G}} x) \leq E^{\mathcal{G}}[h(x)] \quad a.s.$$

for all $x \in L^1$ with $h(x)$ quasi-integrable.

Proof The $\mathcal{G}$-measurability of h means that its epigraph is $\mathcal{G}$-measurable. By Theorem 2.36, this holds if and only if $h(E^{\mathcal{G}} x) \leq E^{\mathcal{G}} \alpha$ almost surely for every $(x, \alpha) \in L^1(\operatorname{epi} h)$ or, in other words,

$$h(E^{\mathcal{G}} x) \leq E^{\mathcal{G}} \alpha \quad a.s.$$

for every $(x, \alpha) \in L^1$ with $h(x) \leq \alpha$. This clearly holds if the condition in the statement holds. The first claim of Theorem 1.67 gives the converse implication.

□

2.1.4 Conditional essinf and esssup

Recall from Theorem 2.14 that any closed-valued measurable mapping $S : \Omega \rightrightarrows \mathbb{R}^n$ admits a *$\mathcal{G}$-conditional essential infimum* $\operatorname{essinf}^{\mathcal{G}} S$ which is the largest closed-valued $\mathcal{G}$-measurable mapping contained in S almost surely. The scenariowise interior of $\operatorname{essinf}^{\mathcal{G}} S$ is the largest open-valued $\mathcal{G}$-measurable mapping contained in S.

Lemma 2.38 *Given a closed-valued measurable mapping $S : \Omega \rightrightarrows \mathbb{R}^n$,* $\operatorname{int} \operatorname{essinf}^{\mathcal{G}} S$ *is the largest open-valued $\mathcal{G}$-measurable mapping contained in S.*

Proof By Corollary 1.32, $\operatorname{int}\operatorname{essinf}^{\mathcal{G}} S$ is $\mathcal{G}$-measurable. Given an arbitrary open-valued $\mathcal{G}$-measurable mapping contained in S, its image closure is $\mathcal{G}$-measurable by Lemma 1.1 and contained in $\operatorname{essinf}^{\mathcal{G}} S$, by definition of the essential infimum. □

Theorem 2.39 below says that, for any measurable mapping $S : \Omega \rightrightarrows \mathbb{R}^n$, there also exists a smallest closed-valued $\mathcal{G}$-measurable set-valued mapping containing S almost surely. This mapping will be denoted by $\operatorname{esssup}^{\mathcal{G}} S$ and called the $\mathcal{G}$-*conditional essential supremum* of S. Note that

$$\operatorname{esssup}^{\mathcal{G}} S = \operatorname{esssup}^{\mathcal{G}} \operatorname{cl} S,$$

where $(\operatorname{cl} S)(\omega) := \operatorname{cl} S(\omega)$ is measurable, by Lemma 1.1. The $\mathcal{G}$-conditional supremum of a random set is a natural generalization of the $\mathcal{G}$-conditional support of a random variable; see Example 2.40 below.

Theorem 2.39 *A measurable set-valued mapping admits a $\mathcal{G}$-conditional essential supremum.*

Proof It suffices to prove the existence for a closed-valued measurable mapping S. Assume first that S is solid-valued. By Corollary 1.32 and Lemma 1.1, the mapping $\operatorname{cl}(S^C)$, where $S^C(\omega) := \mathbb{R}^n \setminus S(\omega)$, is measurable. By Lemma 2.38, the mapping $\operatorname{int}\operatorname{essinf}^{\mathcal{G}} \operatorname{cl}(S^C)$ is the largest $\mathcal{G}$-measurable open-valued mapping contained in $\operatorname{cl} S^C$. Since S is solid-valued, we have $S^C = \operatorname{int}(\operatorname{cl} S^C)$, so $\operatorname{int}\operatorname{essinf}^{\mathcal{G}}(\operatorname{cl} S^C)$ is the largest $\mathcal{G}$-measurable open valued mapping contained in S^C. In other words, $(\operatorname{int}\operatorname{essinf}^{\mathcal{G}}(\operatorname{cl} S^C))^C = \operatorname{esssup}^{\mathcal{G}} S$.

As to the general case, the mappings $S^\nu := S + \mathbb{B}_{1/\nu}$ are solid-valued and measurable, by Theorem 1.7, so they admit $\mathcal{G}$-conditional essential supremums $\bar{S}^\nu$. We claim that $\bar{S} := \bigcap_{\nu=1}^{\infty} \bar{S}^\nu$ is the $\mathcal{G}$-conditional essential supremum of S. Let $\tilde{S}$ be a closed-valued $\mathcal{G}$-measurable mapping containing S. Then $\tilde{S}^\nu := \tilde{S} + \mathbb{B}_{1/\nu}$ is a closed-valued $\mathcal{G}$-measurable mapping containing S^ν. Thus, $\bar{S}^\nu \subseteq \tilde{S}^\nu$ so

$$\bar{S} = \bigcap_{\nu=1}^{\infty} \bar{S}^\nu \subseteq \bigcap_{\nu=1}^{\infty} \tilde{S}^\nu = \tilde{S},$$

which completes the proof. □

Example 2.40 Let ξ be an $\mathbb{R}^n$-valued random variable with regular $\mathcal{G}$-conditional distribution μ (see Theorem B.18) and let $\operatorname{supp}\mu(\omega, \cdot)$ be the support of the measure $\mu(\omega, \cdot)$. The mapping

$$\omega \mapsto \operatorname{supp}\mu(\omega, \cdot)$$

is $\mathcal{G}$-measurable and almost surely equal to the $\mathcal{G}$-conditional essential supremum $\operatorname{esssup}^{\mathcal{G}}\{\xi\}$ of the single-valued mapping $\omega \mapsto \{\xi(\omega)\}$.

Proof The mapping $S(\omega) := \operatorname{supp} \mu(\omega, \cdot)$ is $\mathcal{G}$-measurable by Theorem 1.2. Indeed, given a closed $C \subseteq \mathbb{R}^n$, the inclusion $S(\omega) \subseteq C$ means that $\mu(\omega, C) = 1$, so $\{\omega \in \Omega \mid S(\omega) \subseteq C\} = \mu(\cdot, C)^{-1}(\{1\})$, where the right side is $\mathcal{G}$-measurable.

Given a $\mathcal{G} \otimes \mathcal{B}(\mathbb{R}^n)$-measurable set $A \subseteq \Omega \times \mathbb{R}^n$, Theorem B.19 implies that

$$E^{\mathcal{G}}[1_A(\cdot, \xi(\cdot))](\omega) = \int_{\mathbb{R}^n} 1_A(\omega, s)\mu(\omega, ds)$$

for almost every ω. By Corollary 1.3, graphs of measurable maps are measurable. Choosing $A = \operatorname{gph} \operatorname{esssup}^{\mathcal{G}}\{\xi\}$, the left side equals 1 almost surely, so $S \subseteq \operatorname{esssup}^{\mathcal{G}}\{\xi\}$. Choosing $A = \operatorname{gph} S$, the right side equals 1 almost surely, which gives the opposite inclusion. □

Remark 2.41 By Theorem 2.23, the first characterization in Theorem 2.14 implies that the $\mathcal{G}$-conditional essential infimum of a closed convex-valued measurable mapping is convex-valued. However, the $\mathcal{G}$-conditional essential supremum of a closed convex-valued mapping need not be convex; see Example 2.40.

Applying Theorem 2.15 to indicator functions of measurable mappings yields the following calculus rules for essential infimums.

Theorem 2.42 (Essential Infimum in Algebraic Operations)

1. *If $\{S^\nu\}_{\nu \in \mathbb{N}}$ are closed-valued measurable mappings from Ω to $\mathbb{R}^n$, then*

$$\operatorname{essinf}^{\mathcal{G}} \bigcap_{\nu \in \mathbb{N}} S^\nu = \bigcap_{\nu \in \mathbb{N}} \operatorname{essinf}^{\mathcal{G}} S^\nu.$$

2. *If $C : \Omega \rightrightarrows \mathbb{R}^n$ and $D : \Omega \rightrightarrows \mathbb{R}^m$ are closed-valued and measurable, then*

$$\operatorname{essinf}^{\mathcal{G}}(C \times D) = (\operatorname{essinf}^{\mathcal{G}} C) \times (\operatorname{essinf}^{\mathcal{G}} D).$$

3. *If $\tilde{S}(\omega) := \{x \in \mathbb{R}^n \mid (x, u(\omega)) \in S(\omega)\}$, where $S : \Omega \rightrightarrows \mathbb{R}^n \times \mathbb{R}^m$ is closed-valued and measurable and $u \in L^0(\mathcal{G}; \mathbb{R}^m)$, then*

$$(\operatorname{essinf}^{\mathcal{G}} \tilde{S})(\omega) = \{x \in \mathbb{R}^n \mid (x, u(\omega)) \in (\operatorname{essinf}^{\mathcal{G}} S)(\omega)\}.$$

4. *If $D : \Omega \rightrightarrows \mathbb{R}^m$ is closed-valued and measurable and $M : \mathbb{R}^n \times \Omega \to \mathbb{R}^m$ is a $\mathcal{G}$-measurable Carathéodory mapping, then*

$$\operatorname{essinf}^{\mathcal{G}}(M^{-1}(D)) = M^{-1}(\operatorname{essinf}^{\mathcal{G}} D).$$

5. *If $S : \Omega \rightrightarrows \mathbb{R}^n$ is closed convex-valued and measurable with $L^0(\mathcal{G}; S) \neq \emptyset$, then*

$$(\operatorname{essinf}^{\mathcal{G}} S)^\infty = \operatorname{essinf}^{\mathcal{G}}(S^\infty).$$

Proof By Theorem 2.14, $E^{\mathcal{G}}\delta_C = \delta_{\text{essinf}^{\mathcal{G}} C}$, so parts 3 and 4 follow by applying Theorem 2.15 to indicator functions of random sets. As to 1, Theorem 2.15.1 implies

$$\text{essinf}^{\mathcal{G}} \bigcap_{\nu'=0,\dots,\nu} S^{\nu'} = \bigcap_{\nu'=0,\dots,\nu} \text{essinf}^{\mathcal{G}} S^{\nu'}$$

for every $\nu \in \mathbb{N}$. Part 1 follows by applying Theorem 2.12 to the indicators of sets $\tilde{S}^\nu := \bigcap_{\nu'=0,\dots\nu} S^{\nu'}$. Part 2 is clear if C or D is the whole space almost surely. In general,

$$C \times D = (C \times \mathbb{R}^m) \cap (\mathbb{R}^n \times D),$$

so part 2 follows from part 1. To prove 5, let $x \in L^0(\mathcal{G}; S)$ and $\tilde{S}(\omega) := S(\omega) - x(\omega)$. By 1,

$$\text{essinf}^{\mathcal{G}}(\tilde{S}^\infty) = \text{essinf}^{\mathcal{G}}(\bigcap_{\nu\in\mathbb{N}} \frac{1}{\nu}\tilde{S}) = \bigcap_{\nu\in\mathbb{N}} \frac{1}{\nu}\text{essinf}^{\mathcal{G}} \tilde{S} = (\text{essinf}^{\mathcal{G}} \tilde{S})^\infty.$$

Defining $M(x', \omega) := x' + x(\omega)$, part 4 gives $\text{essinf}^{\mathcal{G}} \tilde{S} = \text{essinf}^{\mathcal{G}} S - x$. Thus, by Theorem A.15, $\tilde{S}^\infty = S^\infty$ and $(\text{essinf}^{\mathcal{G}} \tilde{S})^\infty = (\text{essinf}^{\mathcal{G}} S)^\infty$. □

Given a measurable $S : \Omega \rightrightarrows \mathbb{R}^n$ and an $A \in \mathcal{F}$, the mapping

$$(S_A)(\omega) := \begin{cases} S(\omega) & \text{if } \omega \in A, \\ \emptyset & \text{otherwise} \end{cases}$$

is measurable.

Theorem 2.43 (Essential Supremum in Algebraic Operations)

1. *If $S : \Omega \rightrightarrows \mathbb{R}^n$ is measurable and A is $\mathcal{G}$-measurable, then*

$$(\text{esssup}^{\mathcal{G}} S)_A = \text{esssup}^{\mathcal{G}} S_A.$$

2. *If $\{S^\nu\}_{\nu\in\mathbb{N}}$ are set-valued measurable mappings from Ω to $\mathbb{R}^n$, then*

$$\text{esssup}^{\mathcal{G}} \bigcup_{\nu\in\mathbb{N}} S^\nu = \text{cl} \bigcup_{\nu\in\mathbb{N}} \text{esssup}^{\mathcal{G}} S^\nu.$$

3. *If $C : \Omega \rightrightarrows \mathbb{R}^n$ is closed-valued and measurable and $D : \Omega \rightrightarrows \mathbb{R}^m$ is closed-valued and $\mathcal{G}$-measurable, then*

$$\text{esssup}^{\mathcal{G}}(C \times D) = (\text{esssup}^{\mathcal{G}} C) \times D.$$

4. *If* $M : \mathbb{R}^n \times \Omega \to \mathbb{R}^m$ *is a* $\mathcal{G}$*-measurable Carathéodory mapping and* $S : \Omega \rightrightarrows \mathbb{R}^n$ *is closed-valued and measurable, then*

$$\operatorname{esssup}^{\mathcal{G}} M(S) = \operatorname{cl} M(\operatorname{esssup}^{\mathcal{G}} S).$$

5. *If* $C : \Omega \rightrightarrows \mathbb{R}^n$ *is measurable and* $C' : \Omega \rightrightarrows \mathbb{R}^n$ *is* $\mathcal{G}$*-measurable, then*

$$\operatorname{esssup}^{\mathcal{G}}(C + C') = \operatorname{cl}(\operatorname{esssup}^{\mathcal{G}} C + C').$$

Proof In part 2, it is clear that $\operatorname{esssup}^{\mathcal{G}} \bigcup_\nu S^\nu \subseteq \bigcup_\nu \operatorname{esssup}^{\mathcal{G}} S^\nu$. If $\tilde{S}$ is a closed-valued $\mathcal{G}$-measurable mapping containing $\bigcup_\nu S^\nu$, then it clearly contains $\bigcup_\nu \operatorname{esssup}^{\mathcal{G}} S^\nu$. This proves part 2. In part 1, it is clear that $\operatorname{esssup}^{\mathcal{G}} S_{\Omega\setminus A} \subseteq (\operatorname{esssup}^{\mathcal{G}} S)_{\Omega\setminus A}$, so part 2 implies

$$\begin{aligned}
(\operatorname{esssup}^{\mathcal{G}} S)_A &= (\operatorname{esssup}^{\mathcal{G}}(S_A \cup C_{\Omega\setminus A}))_A \\
&= (\operatorname{esssup}^{\mathcal{G}}(S_A) \cup \operatorname{esssup}^{\mathcal{G}}(S_{\Omega\setminus A}))_A \\
&= \operatorname{esssup}^{\mathcal{G}} S_A,
\end{aligned}$$

which proves part 1.

In part 4 it is clear that

$$\begin{aligned}
M(S) &\subseteq M(\operatorname{esssup}^{\mathcal{G}} S), \\
S &\subseteq M^{-1}(\operatorname{esssup}^{\mathcal{G}} M(S)).
\end{aligned}$$

By Theorem 1.9, the $\mathcal{G}$-measurability of M implies that of $M(\operatorname{esssup}^{\mathcal{G}} S)$ and $M^{-1}(\operatorname{esssup}^{\mathcal{G}} M(S))$ so

$$\begin{aligned}
\operatorname{esssup}^{\mathcal{G}} M(S) &\subseteq \operatorname{cl} M(\operatorname{esssup}^{\mathcal{G}} S), \\
\operatorname{esssup}^{\mathcal{G}} S &\subseteq \operatorname{cl} M^{-1}(\operatorname{esssup}^{\mathcal{G}} M(S)).
\end{aligned}$$

If M is continuous, then the second closure is superfluous and we get

$$\begin{aligned}
\operatorname{esssup}^{\mathcal{G}} M(S) &\subseteq \operatorname{cl} M(\operatorname{esssup}^{\mathcal{G}} S), \\
&\subseteq \operatorname{cl} M(M^{-1}(\operatorname{esssup}^{\mathcal{G}} M(S))) \subseteq \operatorname{esssup}^{\mathcal{G}} M(S),
\end{aligned}$$

which proves the claim.

In part 3, Theorem 1.27 gives the existence of a collection $\{u^\nu\}_{\nu\in\mathbb{N}}$ of $\mathcal{G}$-measurable functions $u^\nu : \operatorname{dom} D \to \mathbb{R}^m$ such that

$$D(\omega) = \operatorname{cl}\{u^\nu(\omega) \mid \nu \in \mathbb{N}\}$$

for every $\omega \in \operatorname{dom} D$. Let

$$\bar{u}^\nu(\omega) := \begin{cases} u^\nu(\omega) & \text{if } \omega \in \operatorname{dom} D, \\ 0 & \text{otherwise.} \end{cases}$$

Applying 3 with $M(x, \omega) = (x, \bar{u}^\nu(\omega))$ and $S = C$, we get

$$\operatorname{esssup}^{\mathcal{G}}(C \times \{\bar{u}^\nu\}) = (\operatorname{esssup}^{\mathcal{G}} C) \times \{\bar{u}^\nu\}.$$

By part 1,

$$\begin{aligned} \operatorname{esssup}^{\mathcal{G}}((C \times \{u^\nu\})) &= \operatorname{esssup}^{\mathcal{G}}((C \times \{\bar{u}^\nu\})_{\operatorname{dom} D}) \\ &= (\operatorname{esssup}^{\mathcal{G}}(C \times \{\bar{u}^\nu\}))_{\operatorname{dom} D} \\ &= ((\operatorname{esssup}^{\mathcal{G}} C) \times \{\bar{u}^\nu\})_{\operatorname{dom} D} \\ &= (\operatorname{esssup}^{\mathcal{G}} C) \times \{u^\nu\}. \end{aligned}$$

Thus, part 2 gives

$$\begin{aligned} \operatorname{esssup}^{\mathcal{G}}(C \times D) &= \operatorname{esssup}^{\mathcal{G}} \operatorname{cl} \bigcup_{\nu \in \mathbb{N}} (C \times \{u^\nu\}) \\ &= \operatorname{cl} \bigcup_{\nu \in \mathbb{N}} \operatorname{esssup}^{\mathcal{G}}(C \times \{u^\nu\}) \\ &= \operatorname{cl} \bigcup_{\nu \in \mathbb{N}} (\operatorname{esssup}^{\mathcal{G}} C \times \{u^\nu\}) \\ &= \operatorname{esssup}^{\mathcal{G}} C \times D, \end{aligned}$$

which proves part 3. Part 5 follows from parts 3 and 4 with $D = C'$ and $M(x, x') = x + x'$. □

Given an extended real-valued random variable ξ, Corollary B.23 gives the existence of a $\mathcal{G}$-measurable random variable $\operatorname{esssup}^{\mathcal{G}} \xi$ that dominates ξ and is minimal in the sense that if $\bar{\xi}$ is another $\mathcal{G}$-measurable random variable with $\bar{\xi} \geq \xi$ almost surely, then $\bar{\xi} = \operatorname{esssup}^{\mathcal{G}} \xi$. The $\mathcal{G}$-conditional essential infimum is defined analogously. When ξ is a real-valued random variable, the random set $\operatorname{co} \operatorname{esssup}^{\mathcal{G}}\{\xi\}$ coincides with the random interval $[\operatorname{essinf}^{\mathcal{G}} \xi, \operatorname{esssup}^{\mathcal{G}} \xi]$.

Given a normal integrand h, the lowest $\mathcal{G}$-measurable normal integrand that majorizes h will be called the *$\mathcal{G}$-conditional essential supremum* of h. Similarly, the greatest $\mathcal{G}$-measurable normal integrand that minorizes h will be called the *$\mathcal{G}$-conditional essential infimum* of h. If h does not depend on x, it can be identified with a random variable and then its essential supremum and infimum can be identified with those of the random variable. The last part of the following gives a more general correspondence between the two notions.

Theorem 2.44 *The $\mathcal{G}$-conditional essential supremum and infimum of a normal integrand h exist and they satisfy*

$$\operatorname{epi} \operatorname{esssup}^{\mathcal{G}} h = \operatorname{essinf}^{\mathcal{G}} \operatorname{epi} h$$

and

$$\operatorname{epi} \operatorname{essinf}^{\mathcal{G}} h = \operatorname{esssup}^{\mathcal{G}} \operatorname{epi} h.$$

The $\mathcal{G}$-conditional essential supremum of h is the unique normal integrand h' satisfying

$$h'(x) = \operatorname{esssup}^{\mathcal{G}}[h(x)] \quad \forall x \in L^0(\mathcal{G}).$$

If h is convex, then $\operatorname{esssup}^{\mathcal{G}} h$ *is convex.*

Proof For existence, it suffices to observe that $\operatorname{essinf}^{\mathcal{G}} \operatorname{epi} h$ and $\operatorname{esssup}^{\mathcal{G}} \operatorname{epi} h$ are epigraphical mappings of normal integrands. That $\operatorname{essinf}^{\mathcal{G}} \operatorname{epi} h$ is an epigraphical mapping follows directly from its definition. That $\operatorname{esssup}^{\mathcal{G}} \operatorname{epi} h$ is the epigraphical mapping follows by observing that the construction of the essential supremum in the proof of Theorem 2.39 preserves the property of being an epigraph.

As to the third claim, let $x \in L^0(\mathcal{G})$. By the definition of $\operatorname{esssup}^{\mathcal{G}} h$, $(\operatorname{esssup}^{\mathcal{G}} h)(x) \geq h(x)$ so $(\operatorname{esssup}^{\mathcal{G}} h)(x) \geq \operatorname{esssup}^{\mathcal{G}}[h(x)]$ almost surely. On the other hand, if this inequality were strict with positive probability, then the closed-valued $\mathcal{G}$-measurable mapping

$$\omega \mapsto (\operatorname{essinf}^{\mathcal{G}} \operatorname{epi} h)(\omega) \cup \{(x(\omega), \operatorname{esssup}^{\mathcal{G}}[h(x)](\omega))\}$$

would be strictly larger than $\operatorname{essinf}^{\mathcal{G}} \operatorname{epi} h$ but still contained in $\operatorname{epi} h$ almost surely, contradicting the definition of $\operatorname{esssup}^{\mathcal{G}} \operatorname{epi} h$. The last claim follows from the first one and Remark 2.41. □

Unlike the conditional essential supremum in the last claim of Theorem 2.44, the conditional essential infimum of a normal integrand does not allow similar characterization in terms of conditional essential infimums of random variables; see Counterexample 2.54.

Remark 2.45 Given a measurable mapping S, we have $\operatorname{essinf}^{\mathcal{G}} S \subseteq S \subseteq \operatorname{esssup}^{\mathcal{G}} S$, so

$$\operatorname{essinf}^{\mathcal{G}} S \subseteq E^{\mathcal{G}} S \subseteq \operatorname{esssup}^{\mathcal{G}} S$$

whenever the conditional expectation $E^{\mathcal{G}} S$ exists; see Theorem 2.27. Applying this to the epigraph of a normal integrand h gives

$$\operatorname{esssup}^{\mathcal{G}} h \geq {}^{\mathcal{G}}h \geq \operatorname{essinf}^{\mathcal{G}} h,$$

whenever the conditional epi-expectation ${}^{\mathcal{G}}h$ of h exists; see Theorem 2.31. By definition, $\operatorname{essinf}^{\mathcal{G}} h \leq h \leq \operatorname{esssup}^{\mathcal{G}} h$. Thus, as soon as $\operatorname{essinf}^{\mathcal{G}} h$ is L-bounded, Lemma 2.11 gives $\operatorname{essinf}^{\mathcal{G}} h \leq E^{\mathcal{G}} h$. If, in addition, h is convex, Theorem 2.35 says that ${}^{\mathcal{G}}h \leq E^{\mathcal{G}} h$ so

$$\operatorname{essinf}^{\mathcal{G}} h \leq E^{\mathcal{G}} h \leq {}^{\mathcal{G}}h \leq \operatorname{esssup}^{\mathcal{G}} h.$$

The $\mathcal{G}$-conditional essential supremum and infimum of an indicator function are given in terms of the $\mathcal{G}$-conditional essential infimum and supremum, respectively, of the underlying mapping.

Example 2.46 Given a closed-valued measurable mapping $C : \Omega \rightrightarrows \mathbb{R}^n$,

1. $\operatorname{esssup}^{\mathcal{G}} \delta_C = \delta_{\operatorname{essinf}^{\mathcal{G}} C}$,
2. $\operatorname{essinf}^{\mathcal{G}} \delta_C = \delta_{\operatorname{esssup}^{\mathcal{G}} C}$.

Proof Clearly $\operatorname{essinf}^{\mathcal{G}} \operatorname{epi} \delta_C = \operatorname{essinf}^{\mathcal{G}}(C \times \mathbb{R}_+) = \operatorname{essinf}^{\mathcal{G}} C \times \mathbb{R}_+ = \operatorname{epi} \delta_{\operatorname{essinf}^{\mathcal{G}} C}$, so the first claim follows from Theorem 2.44. To prove the second claim, let $\hat{C} = \operatorname{lev}_0(\operatorname{essinf}^{\mathcal{G}} \delta_C)$. Since $\operatorname{essinf}^{\mathcal{G}} \delta_C \leq \delta_C$ almost surely, we must have $C \subseteq \hat{C}$ almost surely and thus, $\delta_{\hat{C}} \leq \delta_C$. Since $\hat{C}$ is $\mathcal{G}$-measurable and $\operatorname{essinf}^{\mathcal{G}} \delta_C \leq \delta_{\hat{C}}$, we must have $\operatorname{essinf}^{\mathcal{G}} \delta_C = \delta_{\hat{C}}$. Since $\operatorname{epi} \operatorname{essinf}^{\mathcal{G}} \delta_C = \hat{C} \times \mathbb{R}_+$ is the smallest closed $\mathcal{G}$-measurable set containing $\operatorname{epi} \delta_C = C \times \mathbb{R}_+$, we must have $\hat{C} = \operatorname{esssup}^{\mathcal{G}} C$. □

Theorems 2.42 and 2.44 yield the following.

Theorem 2.47 *Given normal integrands h and h' on $\mathbb{R}^n \times \Omega$, we have*

1. $\operatorname{esssup}^{\mathcal{G}}(h \vee h') = (\operatorname{esssup}^{\mathcal{G}} h) \vee (\operatorname{esssup}^{\mathcal{G}} h')$,
2. $\operatorname{esssup}^{\mathcal{G}}(h \circ A) = (\operatorname{esssup}^{\mathcal{G}} h) \circ A$ *for any $\mathcal{G}$-measurable $A \in L^0(\mathbb{R}^{n \times m})$,*
3. $\operatorname{esssup}^{\mathcal{G}}(h + h') \leq \operatorname{esssup}^{\mathcal{G}} h + \operatorname{esssup}^{\mathcal{G}} h'$, *where equality holds if h' is $\mathcal{G}$-measurable,*
4. $\operatorname{essinf}^{\mathcal{G}}(\operatorname{lev}_\alpha h) = \operatorname{lev}_\alpha(\operatorname{esssup}^{\mathcal{G}} h)$ *for any $\alpha \in L^0(\mathcal{G})$.*

Proof Part 1 follows by applying Theorem 2.42.1 to the epigraphs of h and h' and using Theorem 2.44. Given $x \in L^0(\mathcal{G})$, we have

$$(\operatorname{esssup}^{\mathcal{G}} h \circ A)(x) = (\operatorname{esssup}^{\mathcal{G}} h)(Ax) = \operatorname{esssup}^{\mathcal{G}}[h(Ax)] = \operatorname{esssup}^{\mathcal{G}}[(h \circ A)(x)],$$

where the second equality holds by the second last claim of Theorem 2.44. The same claim now implies part 2.

The inequality in part 3 obvious. Assume now that h' is $\mathcal{G}$-measurable and let $x \in L^0(\mathcal{G})$. We have

$$\begin{aligned}(\operatorname{esssup}^{\mathcal{G}} h + h')(x) &= (\operatorname{esssup}^{\mathcal{G}} h)(x) + h'(x) \\ &= \operatorname{esssup}^{\mathcal{G}}[h(x)] + h'(x) \\ &= \operatorname{esssup}^{\mathcal{G}}[h(x) + h'(x)] \\ &= \operatorname{esssup}^{\mathcal{G}}[(h + h')(x)],\end{aligned}$$

where the second equality holds by the second last claim of Theorem 2.44 and the third equality by Corollary B.23 in the appendix. Part 3 thus follows from the second last claim of Theorem 2.44.

To prove 4, let $x \in L^0(\mathcal{G})$ and $A \in \mathcal{G}$. We have, almost surely on A that

$$\begin{aligned}& x \in \operatorname{lev}_\alpha \operatorname{esssup}^{\mathcal{G}} h \\ \iff & (x, \alpha) \in \operatorname{epi} \operatorname{esssup}^{\mathcal{G}} h \\ \iff & (x, \alpha) \in \operatorname{essinf}^{\mathcal{G}} \operatorname{epi} h \\ \iff & (x, \alpha) \in \operatorname{epi} h \\ \iff & x \in \operatorname{lev}_\alpha h,\end{aligned}$$

where the second equivalence holds by Theorem 2.44 and the third by Theorem 2.14. By Theorem 2.14 again, $\operatorname{essinf}^{\mathcal{G}}(\operatorname{lev}_\alpha h) = \operatorname{lev}_\alpha(\operatorname{esssup}^{\mathcal{G}} h)$. □

Much like Theorem 2.47 gives calculus rules for essential supremums of normal integrands, one could derive calculus rules for essential infimums by applying Theorem 2.43 to epigraphs of normal integrands.

The $\mathcal{G}$-conditional essential supremum and infimum of a normal integrand can be seen as dual operations when taking conjugates.

Theorem 2.48 *Given a normal integrand h,*

$$\begin{aligned}(\operatorname{essinf}^{\mathcal{G}} h)^* &= \operatorname{esssup}^{\mathcal{G}} h^*, \\ (\operatorname{esssup}^{\mathcal{G}} \operatorname{cl} \operatorname{co} h)^* &= \operatorname{cl} \operatorname{co}(\operatorname{essinf}^{\mathcal{G}} h^*).\end{aligned}$$

Proof Since $\operatorname{esssup}^{\mathcal{G}} h^* \geq h^*$, we have $(\operatorname{esssup}^{\mathcal{G}} h^*)^* \leq \operatorname{cl} \operatorname{co} h$, by Theorem A.54. Thus, by the definition of the essential infimum,

$$(\operatorname{esssup}^{\mathcal{G}} h^*)^* \leq \operatorname{essinf}^{\mathcal{G}} \operatorname{cl} \operatorname{co} h \leq \operatorname{essinf}^{\mathcal{G}} h \leq h.$$

Taking conjugates and applying Theorem A.54 to $\text{esssup}^{\mathcal{G}} h^*$, gives

$$\text{cl co esssup}^{\mathcal{G}} h^* \geq (\text{essinf}^{\mathcal{G}} \text{cl co}\, h)^* \geq (\text{essinf}^{\mathcal{G}} h)^* \geq h^*.$$

Since $\text{esssup}^{\mathcal{G}} h^*$ is the smallest $\mathcal{G}$-measurable normal integrand majorizing h^* and since $\text{esssup}^{\mathcal{G}} h^* \geq \text{cl co esssup}^{\mathcal{G}} h^*$, we must have

$$\text{esssup}^{\mathcal{G}} h^* = \text{cl co esssup}^{\mathcal{G}} h^* = (\text{essinf}^{\mathcal{G}} \text{cl co}\, h)^* = (\text{essinf}^{\mathcal{G}} h)^*.$$

This proves the first claim. The second one follows by applying the first one to h^* and using Theorem A.54 again. □

Remark 2.49 In the second formula in Theorem 2.48, one cannot omit the closed convex hulls in general. Indeed, let $\mathcal{G} = \{\Omega, \emptyset\}$ and $h := \delta_S$ where $S := \{\xi\} \bigcup \{-\xi\}$ and ξ is normally distributed with nonzero variance. Then $\text{esssup}^{\mathcal{G}} h = +\infty$ while $\text{esssup}^{\mathcal{G}} \text{cl co}\, h = \delta_{\{0\}}$, so they have different conjugates.

When applied to support functions of measurable mappings, Theorem 2.48 yields the following.

Corollary 2.50 *Given a measurable mapping C,*

$$\text{esssup}^{\mathcal{G}} \sigma_C = \sigma_{\text{esssup}^{\mathcal{G}} C},$$
$$\text{cl co essinf}^{\mathcal{G}} \sigma_C = \sigma_{\text{essinf}^{\mathcal{G}} \text{cl co}\, C}.$$

Proof Since $\sigma_C = \sigma_{\text{cl}\, C}$ and $\text{esssup}^{\mathcal{G}} C = \text{esssup}^{\mathcal{G}} \text{cl}\, C$, we may assume that C is closed-valued. Applying Theorem 2.48 to the indicator function of C gives

$$(\text{essinf}^{\mathcal{G}} \delta_C)^* = \text{esssup}^{\mathcal{G}} \sigma_C,$$
$$(\text{esssup}^{\mathcal{G}} \text{cl co}\, \delta_C)^* = \text{cl co essinf}^{\mathcal{G}} \sigma_C,$$

where $\text{essinf}^{\mathcal{G}} \delta_C = \delta_{\text{esssup}^{\mathcal{G}} C}$, by Example 2.46. Since $\text{cl co}\, \delta_C = \delta_{\text{cl co}\, C}$, Example 2.46 also gives $\text{esssup}^{\mathcal{G}} \text{cl co}\, \delta_C = \delta_{\text{essinf}^{\mathcal{G}} \text{cl co}\, C}$. □

We will denote the scenariowise polar of a measurable mapping C by

$$C^\circ(\omega) := \{v \in \mathbb{R}^n \mid x \cdot v \leq 1 \quad \forall x \in C(\omega)\}.$$

Note that $C^\circ = \text{lev}_1\, \sigma_C$, so C° is measurable by Corollary 1.47 and Theorem 1.15. Recall that if C is cone-valued, then

$$C^\circ(\omega) = \{v \in \mathbb{R}^n \mid x \cdot v \leq 0 \quad \forall x \in C(\omega)\}.$$

Corollary 2.51 *Given a measurable mapping C containing the origin,*

$$\operatorname{essinf}^{\mathcal{G}}(C^\circ) = (\operatorname{esssup}^{\mathcal{G}} C)^\circ,$$
$$\operatorname{cl}\operatorname{co}\operatorname{esssup}^{\mathcal{G}}(C^\circ) = (\operatorname{essinf}^{\mathcal{G}} \operatorname{cl}\operatorname{co} C)^\circ.$$

If C is cone-valued, then $\operatorname{esssup}^{\mathcal{G}} C$ *and* $\operatorname{essinf}^{\mathcal{G}} \operatorname{cl} C$ *are cone-valued as well.*

Proof Theorem 2.47.4 and Corollary 2.50 imply

$$\begin{aligned}\operatorname{essinf}^{\mathcal{G}}(C^\circ) = \operatorname{essinf}^{\mathcal{G}} \operatorname{lev}_1 \sigma_C &= \operatorname{lev}_1 \operatorname{esssup}^{\mathcal{G}} \sigma_C \\ &= \operatorname{lev}_1 \sigma_{\operatorname{esssup}^{\mathcal{G}} C} = (\operatorname{esssup}^{\mathcal{G}} C)^\circ.\end{aligned}$$

By Corollary A.58, $C^{\circ\circ} = \operatorname{cl}\operatorname{co} C$, so the second claim follows from the first when applied to C°. Assume now that C is cone-valued. Since $\delta_{\operatorname{essinf}^{\mathcal{G}} \operatorname{cl} C} = E^{\mathcal{G}} \delta_{\operatorname{cl} C}$, by definition, Lemma 2.22 implies that $\operatorname{essinf}^{\mathcal{G}} \operatorname{cl} C$ is cone-valued. Given any $\alpha > 0$, we have $C = \alpha C$, so by Theorem 2.43.4, $\operatorname{esssup}^{\mathcal{G}} C = \alpha(\operatorname{esssup}^{\mathcal{G}} C)$. Thus, $\operatorname{esssup}^{\mathcal{G}} C$ is cone-valued as well. □

Corollary 2.52 *Given a measurable mapping C,*

$$\operatorname{cl}\operatorname{pos}\operatorname{esssup}^{\mathcal{G}} C = \operatorname{esssup}^{\mathcal{G}} \operatorname{pos} C.$$

Proof Since $C \subseteq \operatorname{esssup}^{\mathcal{G}} \operatorname{pos} C$ where, by Corollary 2.51, the right side is $\mathcal{G}$-measurable and closed cone-valued, we get

$$\operatorname{cl}\operatorname{pos}\operatorname{esssup}^{\mathcal{G}} C \subseteq \operatorname{esssup}^{\mathcal{G}} \operatorname{pos} C.$$

On the other hand, $C \subseteq \operatorname{esssup}^{\mathcal{G}} C$, so $\operatorname{esssup}^{\mathcal{G}} \operatorname{pos} C \subseteq \operatorname{cl}\operatorname{pos}\operatorname{esssup}^{\mathcal{G}} C$. □

Combining Theorem 2.47.4 with Corollary 2.50 gives the following.

Example 2.53 Given a measurable mapping C and a measurable α,

$$(\operatorname{essinf}^{\mathcal{G}} \operatorname{lev}_\alpha \sigma_C)(\omega) = \{x \in \mathbb{R}^n \mid \sigma_{\operatorname{esssup}^{\mathcal{G}}(C \times \{-\alpha\})}(x, 1, \omega) \le 0\}.$$

If α is $\mathcal{G}$-measurable, then

$$(\operatorname{essinf}^{\mathcal{G}} \operatorname{lev}_\alpha \sigma_C)(\omega) = \{x \in \mathbb{R}^n \mid \sigma_{\operatorname{esssup}^{\mathcal{G}} C}(x, \omega) \le \alpha(\omega)\}.$$

If, in addition, $C(\omega) := \{\xi^1 + \xi^2\}$, where $\xi^1 \in L^0(\mathcal{G}; \mathbb{R}^n)$ and $\xi^2 \in L^0(\mathbb{R}^n)$, then

$$(\operatorname{essinf}^{\mathcal{G}} \operatorname{lev}_\alpha \sigma_C)(\omega) = \{x \in \mathbb{R}^n \mid x \cdot \xi^1(\omega) + \sigma_{\operatorname{esssup}^{\mathcal{G}} \{\xi^2\}}(x, \omega) \le \alpha(\omega)\}.$$

Proof Defining

$$S(\omega) := \{(x, u) \in \mathbb{R}^n \times \mathbb{R} \mid \sigma_{C\times\{-\alpha\}}(x, u) \leq 0\},$$

we have $(\operatorname{lev}_\alpha \sigma_C)(\omega) = \{x \in \mathbb{R}^n \mid (x, 1) \in S(\omega)\}$. By Theorem 2.42,

$$(\operatorname{essinf}^{\mathcal{G}} \operatorname{lev}_\alpha \sigma)(\omega) = \{x \in \mathbb{R}^n \mid (x, 1) \in (\operatorname{essinf}^{\mathcal{G}} S)(\omega)\},$$

where

$$(\operatorname{essinf}^{\mathcal{G}} S)(\omega) = \{(x, u) \in \mathbb{R}^n \mid \sigma_{\operatorname{esssup}^{\mathcal{G}}(C\times\{-\alpha\})}(x, u, \omega) \leq 0\},$$

by Theorem 2.47.4 and Corollary 2.50. The second claim follows from Theorem 2.43.3 and the third from Theorem 2.43.5. □

Counterexample 2.54 In contrast to the formula for essential supremum in Theorem 2.44, essential infimum does not satisfy

$$(\operatorname{essinf}^{\mathcal{G}} h)(x) = \operatorname{essinf}^{\mathcal{G}}[h(x)] \quad \forall x \in L^0(\mathcal{G})$$

in general. Indeed, given a random variable ξ, Example 2.46 gives

$$\operatorname{essinf}^{\mathcal{G}} \delta_{\{\xi\}} = \delta_{\operatorname{esssup}^{\mathcal{G}}\{\xi\}}.$$

When $\mathcal{G}$ is the trivial σ-algebra and ξ is, e.g., normally distributed on the real line, we have $\operatorname{esssup}^{\mathcal{G}}\{\xi\} = \mathbb{R}$, so $\delta_{\operatorname{esssup}^{\mathcal{G}}\{\xi\}} = 0$, while $\operatorname{essinf}^{\mathcal{G}}[\delta_{\{\xi\}}(x)] = \infty$ for any $x \in \mathbb{R}$.

2.1.5 Regular Conditional Distributions

Section 2.2.4 below studies dynamic programming in the canonical representation of problem (P) given in Remark 1.76. The random variables ξ_t in the problem formulation give rise to regular conditional distributions which are "random measures" that allow for scenariowise representations of conditional expectations of random variables. This section extends such representations to normal integrands. When available, such scenariowise representations can be very useful computationally. This is most notable in applications of (P) where the underlying risk factors are modelled by a Markov process and the problem has an appropriate time separable structure; see Sect. 2.3. The current section can be safely skipped if the reader is not interested in regular conditional expectations and related computational aspects. The material presented here will not be needed in the later chapters.

Given a measurable space $(\Xi, \mathcal{A})$, a function $\mu : \Omega \times \mathcal{A} \to \overline{\mathbb{R}}$ is a *probability kernel* from $(\Omega, \mathcal{F})$ to $(\Xi, \mathcal{A})$ if for each $\omega \in \Omega$, $\mu(\omega, \cdot)$ is a probability measure on $(\Xi, \mathcal{A})$ and, for each $A \in \mathcal{A}$, $\mu(\cdot, A)$ is a $\mathcal{F}$-measurable function on Ω. Given a random variable $\xi : (\Omega, \mathcal{F}) \to (\Xi, \mathcal{A})$ and a σ-algebra $\mathcal{G} \subseteq \mathcal{F}$, a probability kernel μ from $(\Omega, \mathcal{G})$ to $(\Xi, \mathcal{A})$ is a *regular $\mathcal{G}$-conditional distribution* of ξ if, for every $\mathcal{A}$-measurable X such that $X(\xi)$ is quasi-integrable, we have

$$E^{\mathcal{G}}[X(\xi)](\omega) = \int_{\Xi} X(s)\mu(\omega, ds);$$

see Theorem B.18. Given a regular $\mathcal{G}$-conditional distribution μ of ξ and a normal integrand of the form $h(x, \omega) = H(x, \xi(\omega))$ it is natural to expect that

$$E^{\mathcal{G}} h = E^{\mu} H,$$

where

$$(E^{\mu} H)(x, \omega) := \int_{\Xi} H(x, s)\mu(\omega, ds) \quad \forall x \in \mathbb{R}^n.$$

This would characterize the conditional expectation $E^{\mathcal{G}} h$ pointwise in $\mathbb{R}^n \times \Omega$ as opposed to Definition 2.1 where x varies in $L^0(\mathcal{G})$.

Lemma 2.55 *Let $H : \mathbb{R}^n \times \Xi \to \overline{\mathbb{R}}$ be an $\mathcal{A}$-measurable normal integrand and μ be a probability kernel from $(\Omega, \mathcal{G})$ to $(\Xi, \mathcal{A})$. Assume that there exist $R, M : \Xi \to \mathbb{R}_+$ with $E^{\mu} M$ and $E^{\mu} R$ real-valued and*

$$H(x, s) \ge -R(s)|x| - M(s) \quad \forall x \in \mathbb{R}^n \ \mu(\omega, \cdot)\text{-a.e.}$$

almost surely. Then $E^{\mu} H$ is a $\mathcal{G}$-measurable normal integrand and

$$E^{\mu} H(x, \omega) \ge -(E^{\mu} R)(\omega)|x| - (E^{\mu} M)(\omega) \quad \forall x \in \mathbb{R}^n$$

almost surely.

If μ is the regular conditional distribution of a random variable $\xi : (\Omega, \mathcal{F}) \to (\Xi, \mathcal{A})$, then the normal integrand

$$h(x, \omega) := H(x, \xi(\omega))$$

is L-bounded and its $\mathcal{G}$-conditional expectation can be expressed as

$$E^{\mathcal{G}} h = E^{\mu} H.$$

Proof Without loss of generality, we may assume that

$$H(x, s) \ge -R(s)|x| - M(s) \quad \forall x \in \mathbb{R}^n \ \mu(\omega, \cdot)\text{-a.e.}$$

for all ω. Indeed, if the lower bound holds on a set A of full measure, then one can extend the normal integrand to the whole Ω by defining it e.g. as zero on $\mathbb{R}^n \times (\Omega \setminus A)$. Adding a positive constant to R, if necessary, we can also assume that $R(s) > 0$ for all $s \in \Xi$.

By Lemma B.10, $E^\mu H$ is $\mathcal{B}(\mathbb{R}^n) \otimes \mathcal{G}$-measurable. For $j \in \mathbb{N}$, let

$$H^j(x, s) := \inf_{x' \in \mathbb{R}^n} \{H(x', s) + jR(s)|x - x'|\},$$

$$(E^\mu H^j)(x, \omega) := \int_S H^j(x, s)\mu(\omega, ds).$$

By Lemma A.24 in the appendix, we have for $\mu(\omega, \cdot)$-almost every s that each $H^j(\cdot, s)$ is $(jR(s))$-Lipschitz, $H^j(x, s)$ increase pointwise to $H(x, s)$ and

$$H^j(x, s) \geq -R(s)|x| - M(s).$$

By Jensen's inequality (Theorem 1.67) and the Lipschitz property, we have for every $x, x' \in \mathbb{R}^n$ that

$$\begin{aligned} |(E^\mu H^j)(x, \omega) - (E^\mu H^j)(x', \omega)| &\leq \int_S |H^j(x, s) - H^j(x', s)|\mu(\omega, ds) \\ &\leq \int_S jR(s)|x - x'|\mu(\omega, ds) \\ &\leq j(E^\mu R)(\omega)|x - x'|. \end{aligned}$$

Thus, since $E^\mu H^j$ is measurable, it is a Carathéodory integrand and hence normal, by Example 1.12. Since H^j have a common lower bound, monotone convergence gives

$$(E^\mu H)(x, \omega) = \sup_j (E^\mu H^j)(x, \omega),$$

so $E^\mu H$ is a normal integrand by Theorem 1.20. Integrating the lower bound for H with respect to $\mu(\omega, \cdot)$, gives the lower bound for $E^\mu H$.

Assume now that μ is the regular conditional distribution of ξ and let $h(x, \omega) = H(x, \xi(\omega))$. The lower bound on H implies that h is L-bounded. Let $x \in L^\infty(\mathcal{G})$. Applying Theorem B.19 to the function $X(s, \omega) := H(x(\omega), s)$ gives $E^\mathcal{G}[h(x)] = E^\mu H(x)$. Lemma 2.6 now gives $E^\mathcal{G} h = E^\mu H$. □

The following shows that, the lower bound on the Doob–Dynkin representation in Lemma 2.55 is equivalent to L-boundedness of a normal integrand.

Lemma 2.56 *A $\sigma(\xi)$-measurable normal integrand h is L-bounded if and only if*

$$h(x, \omega) = H(x, \xi(\omega)),$$

where $H : \mathbb{R}^n \times \Xi \to \overline{\mathbb{R}}$ *is an* $\mathcal{A}$*-measurable normal integrand such that there exist* $\mathcal{A}$*-measurable* $M, R : \Xi \to \mathbb{R}_+$ *such that* $R(\xi), M(\xi) \in L^1$ *and*

$$H(x, \xi(\omega)) \geq -R(\xi(\omega))|x| - M(\xi(\omega)) \quad \forall x \in \mathbb{R}^n$$

almost surely. If μ *is a regular* $\mathcal{G}$*-conditional distribution of* ξ*, then the L-boundedness implies*

$$H(x, s) \geq -R(s)|x| - M(s) \quad \forall x \in \mathbb{R}^n \ \mu(\omega, \cdot)\text{-a.e.}$$

for almost every ω*.*

Proof It is clear that the stated conditions imply L-boundedness of h. On the other hand, $\sigma(\xi)$-measurability of h implies, by Corollary 1.34, that there exists an $\mathcal{A}$-measurable normal integrand $H : \mathbb{R}^n \times \Xi \to \overline{\mathbb{R}}$ such that $h(x, \omega) = H(x, \xi(\omega))$. L-boundedness of h means that there exist $\rho, m \in L^1_+$ such that

$$h(x, \omega) \geq -\rho(\omega)|x| - m(\omega) \quad \forall x \in \mathbb{R}^n.$$

Since h is $\sigma(\xi)$-measurable, $E^{\sigma(\xi)}h = h$ and

$$h(x, \omega) \geq -(E^{\sigma(\xi)}\rho)(\omega)|x| - (E^{\sigma(\xi)}m)(\omega) \quad \forall x \in \mathbb{R}^n$$

almost surely. By Lemma B.7, there exist measurable $R, M : \Xi \to \mathbb{R}$ such that $(E^{\sigma(\xi)}\rho)(\omega) = R(\xi(\omega))$ and $(E^{\sigma(\xi)}m)(\omega) = M(\xi(\omega))$.

Assume now that h is L-bounded. Then $X(\xi) \geq 0$ almost surely, where

$$X(s) := \inf_{x \in \mathbb{R}^n} \{H(x, s) + R(s)|x| + M(s)\}$$

is $\mathcal{A}$-measurable, by Corollary 1.23. Thus, $1_{\{X<0\}}(\xi) = 0$ almost surely, so

$$E^{\mu} 1_{\{X<0\}} = E^{\mathcal{G}}[1_{\{X<0\}}(\xi)] = 0$$

almost surely. In other words,

$$X(s) \geq 0 \quad \mu(\omega, \cdot)\text{-a.e.}$$

for almost every ω, which proves the last claim. □

Combining Lemmas 2.55 and 2.56 gives the following.

Theorem 2.57 *Let*

$$h(x, \omega) := H(x, \xi(\omega)),$$

where $H : \mathbb{R}^n \times \Xi \to \overline{\mathbb{R}}$ *is an* $\mathcal{A}$*-measurable normal integrand and* $\xi : (\Omega, \mathcal{F}) \to (\Xi, \mathcal{A})$ *is a random variable with a regular* $\mathcal{G}$*-conditional distribution* μ*. If* h *is L-bounded, then*

$$E^{\mathcal{G}} h = E^{\mu} H.$$

The following corollary is a special of case Theorem 2.57 with $(\Xi, \mathcal{A}) = (\Omega, \mathcal{F})$ and $\xi(\omega) = \omega$. It implies that, in most applications, the conditional expectation of a normal integrand can be expressed scenariowise.

Corollary 2.58 *Let* h *be an L-bounded normal integrand. If the identity mapping on* Ω *has a regular* $\mathcal{G}$*-conditional distribution* μ*, then*

$$E^{\mathcal{G}} h = E^{\mu} h.$$

Given another random variable $\hat{\xi}$ with values in a measurable space $(\hat{\Xi}, \hat{\mathcal{A}})$, a probability kernel ν from $(\hat{\Xi}, \hat{\mathcal{A}})$ to $(\Xi, \mathcal{A})$ is called a *regular* $\hat{\xi}$*-conditional distribution of* ξ if $\mu(\omega, A) := \nu(\hat{\xi}(\omega), A)$ defines a $\sigma(\hat{\xi})$-conditional regular distribution of ξ. This means that, for every $\mathcal{A}$-measurable function X on Ξ such that $X(\xi)$ is quasi-integrable, we have

$$E^{\sigma(\hat{\xi})}[X(\xi)](\omega) = (E^{\nu} X)(\hat{\xi}(\omega)), \tag{2.7}$$

for P-almost every $\omega \in \Omega$. Here

$$(E^{\nu} X)(\hat{s}) := \int_{\Xi} X(s)\nu(\hat{s}, ds).$$

Corollary 2.59 below extends this to normal integrands. Given a random variable ξ, we denote its distribution by P_{ξ}. That is, P_{ξ} is a probability measure on $(\Xi, \mathcal{A})$ given, for each $A \in \mathcal{A}$, by

$$P_{\xi}(A) = P(\{\omega \mid \xi(\omega) \in A\}).$$

Corollary 2.59 *Let*

$$h(x, \omega) := H(x, \xi(\omega))$$

where $H : \mathbb{R}^n \times \Xi \to \overline{\mathbb{R}}$ *is a* $\mathcal{A}$*-measurable normal integrand. If* h *is L-bounded and* ν *is a regular* $\hat{\xi}$*-conditional distribution of* ξ*, then*

$$(E^{\sigma(\hat{\xi})} h)(x, \omega) = (E^{\nu} H)(x, \hat{\xi}(\omega)),$$

where

$$(E^{\nu}H)(x,\hat{s}) := \int_{\Xi} H(x,s)\nu(\hat{s},ds).$$

Moreover, $E^{\nu}H$ *is* $P_{\hat{\xi}}$*-indistinguishable from a normal integrand on* $\mathbb{R}^n \times \hat{\Xi}$*.*

Proof Applying Theorem 2.57 with $\mu(\omega, A) := \nu(\hat{\xi}(\omega), A)$, gives

$$\begin{aligned}(E^{\sigma(\hat{\xi})}h)(x,\omega) &= (E^{\mu}H)(x,\omega)\\ &= \int H(x,s)d\mu(\omega,ds)\\ &= \int H(x,s)d\nu(\hat{\xi}(\omega),ds),\end{aligned}$$

which proves the first claim.

By definition, μ is a regular $\sigma(\hat{\xi})$-conditional distribution of ξ so, by Lemma 2.56, there exist $\mathcal{A}$-measurable functions $M, R : \Xi \to \mathbb{R}_+$ such that $R(\xi), M(\xi) \in L^1$ and

$$H(x,s) \geq -R(s)|x| - M(s) \quad \forall x \in \mathbb{R}^n \ \nu(\hat{\xi}(\omega),\cdot)\text{-a.s.}$$

for P-almost every ω. This means that $\hat{\xi}$ belongs P-almost surely to the set $B = \{\hat{s} \in \hat{\Xi} \mid \nu(\hat{s}, A) = 1\}$, where

$$A := \{s \in \Xi \mid \inf_x\{H(x,s) + R(s)|x| + M(s)\} \geq 0\} \in \mathcal{A}.$$

Since ν is a probability kernel, B is $\hat{\mathcal{A}}$-measurable with $P(\hat{\xi} \in B) = 1$. In summary,

$$H(x,s) \geq -R(s)|s| - M(s) \quad \forall x \in \mathbb{R}^n \ \nu(\hat{s},\cdot)\text{-a.e.}$$

for all $\hat{s} \in B$, where $B \in \hat{\mathcal{A}}$ has $P_{\hat{\xi}}(B) = 1$. Thus, the claim follows from Lemma 2.55 with $(\Omega, \mathcal{G}, P) := (\hat{\Xi}, \hat{\mathcal{A}}, P_{\hat{\xi}})$. □

Lemma B.20 yields the following reformulation of Corollary 2.59 which will be convenient when studying dynamic programming in the canonical representation of (P); see Sect. 2.2.4 below.

Corollary 2.60 *Let* ξ_1 *and* ξ_2 *be random variables with values in measurable spaces* $(\Xi_1, \mathcal{A}_1)$ *and* $(\Xi_2, \mathcal{A}_2)$*, respectively, and let*

$$h(x,\omega) := H(x,\xi_1(\omega),\xi_2(\omega))$$

where $H : \mathbb{R}^n \times \Xi_1 \times \Xi_2 \to \overline{\mathbb{R}}$ is a $\mathcal{A}_1 \otimes \mathcal{A}_2$-measurable normal integrand. If h is L-bounded and ν is a regular ξ_1-conditional distribution of ξ_2, then

$$(E^{\sigma(\xi_1)}h)(x,\omega) = (E^{\nu}H)(x,\xi_1(\omega)),$$

where

$$(E^{\nu}H)(x,s_1) := \int_{\Xi_2} H(x,s_1,s_2)\nu(s_1,ds_2).$$

Moreover, $E^{\nu}H$ is P_{ξ_1}-indistinguishable from a normal integrand on $\mathbb{R}^n \times \Xi_1$.

If, in addition, $\sigma(\xi_1) \subseteq \mathcal{G}$ and $\sigma(\xi_2)$ is $\sigma(\xi_1)$-conditionally independent of $\mathcal{G}$, then

$$(E^{\mathcal{G}}h)(x,\omega) = (E^{\nu}H)(x,\xi_1(\omega)) \quad a.s.e.$$

Proof Applying Corollary 2.59 with $\xi := (\xi_1, \xi_2)$ and $\hat{\xi} := \xi_1$ gives

$$(E^{\sigma(\xi_1)}h)(x,\omega) = (E^{\hat{\nu}}H)(x,\hat{\xi}(\omega)) = \int_{\Xi_2} H(x,\xi_1(\omega),s_2)\nu(\xi_1(\omega),ds_2),$$

where the second inequality comes from Lemma B.20. Corollary 2.59 also says that $E^{\nu}H$ is P_{ξ_1}-indistinguishable from a normal integrand.

The last claim follows from the first and Theorem 2.21 with $\mathcal{H} := \sigma(\xi_1)$. Indeed, by Lemma B.17, $\sigma(\xi_1, \xi_2)$ is $\sigma(\xi_1)$-conditionally independent of $\mathcal{G}$, so by Example 2.20, the normal integrand h is $\sigma(\xi_1)$-conditionally independent of $\mathcal{G}$. □

2.2 Existence of Solutions

Consider again problem (P) and recall from the introduction of this chapter that an adapted sequence $(h_t)_{t=0}^T$ of normal integrands $h_t : \mathbb{R}^{n^t} \times \Omega \to \overline{\mathbb{R}}$ solves the generalized Bellman equations for a normal integrand h if

$$\begin{aligned} h_T &= E_T h, \\ h_t &= E_t \inf_{x_{t+1}} h_{t+1} \quad t = T-1, \ldots, 0, \end{aligned} \tag{BE}$$

where the second equality means that

$$\tilde{h}_t(x^t,\omega) := \inf_{x_{t+1}\in\mathbb{R}^{n_{t+1}}} h_{t+1}(x^t, x_{t+1}, \omega)$$

is a normal integrand and $h_t = E_t\tilde{h}_t$. As observed in Theorem 2.2, solutions of the Bellman equation (BE) can be used to characterize the optimal solutions and optimum value of (P).

Section 2.2.1 below gives sufficient conditions for the existence of solutions to (BE) when h is lower bounded in the sense that there exists an $m \in L^1$ such that $h \geq m$ almost surely everywhere; see Sect. 1.1.5. For most applications, the lower boundedness assumption is harmless but it does fail e.g. in some utility maximization problems where the utility function is unbounded from above. Section 2.2.2 gives more general conditions without lower boundedness but under which problem (P) is equivalent to another problem that satisfies the assumptions of Sect. 2.2.1. Besides the existence of solutions to the Bellman equations, Sect. 2.2.2 also extends Theorem 2.2 beyond lower bounded normal integrands. In the context of utility maximization, the generalized lower bounds of Sect. 2.2.1 hold for utility functions that satisfy well-known asymptotic elasticity conditions; see Remark 2.121.

2.2.1 Lower Bounded Objectives

This section establishes the existence of a unique solution $(h_t)_{t=0}^T$ to (BE) when h is lower bounded and convex, (P) is feasible and the set

$$\mathcal{L} := \{x \in \mathcal{N} \mid h^\infty(x) \leq 0 \; a.s.\}$$

is linear. The definition of $\mathcal{L}$ involves the scenariowise recession function h^∞ of h; see Theorem 1.39. Note also that lower boundedness of h implies that h^∞ is nonnegative; see Remark A.20. The sublinearity of h^∞ implies that the set $\mathcal{L}$ is a convex cone. Thus, the linearity of $\mathcal{L}$ simply means that $\mathcal{L} = -\mathcal{L}$. This holds trivially if h is inf-compact since then, h^∞ is strictly positive except at the origin; see Theorems A.27 and A.17. The inf-compactness condition was used in the preliminary existence result in Theorem 1.81 but it precludes many important applications satisfying the above conditions.

Whenever h_t is well-defined and convex, h_t^∞ is a sublinear normal integrand by Theorem 1.39 so, by Theorem 1.15,

$$N_t(\omega) := \{x_t \in \mathbb{R}^{n_t} \mid h_t^\infty(0, x_t, \omega) \leq 0\}$$

is a closed cone-valued measurable mapping and

$$\mathcal{L}^t := \{x^t \in \mathcal{N}^t \mid h_t^\infty(x^t) \leq 0 \; a.s.\}$$

is a convex cone.

Lemma 2.61 *Consider* (BE) *and assume that* $\tilde{h}_t$ *is a well-defined lower bounded convex normal integrand and that* dom $E\tilde{h}_t \cap \mathcal{N}^t \neq \emptyset$. *Then* h_t *is well-defined lower* *bounded convex normal integrand with*

$$h_t^\infty = E_t \tilde{h}_t^\infty \tag{2.8}$$

and

$$\mathcal{L}^t = \{x^t \in \mathcal{N}^t \mid \tilde{h}_t^\infty(x^t) \leq 0 \, a.s.\}.$$

If N_t is linear-valued almost surely, then $\tilde{h}_{t-1}$ and h_{t-1} are well-defined lower bounded convex normal integrands,

$$\tilde{h}_{t-1}^\infty(x^{t-1}, \omega) = \inf_{x_t \in \mathbb{R}^{n_t}} h_t^\infty(x^{t-1}, x_t, \omega) \tag{2.9}$$

and

$$\mathcal{L}^{t-1} = \{x^{t-1} \in \mathcal{N}^{t-1} \mid \exists \hat{x}^t \in \mathcal{L}^t, \ \hat{x}^{t-1} = x^{t-1}\}. \tag{2.10}$$

Proof By Theorems 2.13 and 2.23, h_t is well-defined lower bounded convex normal integrand. By Corollary 2.34, $h_t^\infty = (E_t \tilde{h}_t)^\infty = E_t \tilde{h}_t^\infty$, so

$$\begin{aligned}
\mathcal{L}^t &= \{x^t \in \mathcal{N}^t \mid (E_t \tilde{h}_t^\infty)(x^t) \leq 0 \, a.s.\} \\
&= \{x^t \in \mathcal{N}^t \mid E_t[\tilde{h}_t^\infty(x^t)] \leq 0 \, a.s.\} \\
&= \{x^t \in \mathcal{N}^t \mid \tilde{h}_t^\infty(x^t) \leq 0 \, a.s.\},
\end{aligned}$$

where the second equality follows from the definition of the conditional expectation of a normal integrand and the third one from the nonnegativity of $\tilde{h}_t^\infty$.

Assume now that N_t is linear-valued. By Theorem 1.40, $\tilde{h}_{t-1}$ is a well-defined convex normal integrand with

$$\tilde{h}_{t-1}^\infty(x^{t-1}, \omega) = \inf_{x_t \in \mathbb{R}^{n_t}} h_t^\infty(x^{t-1}, x_t, \omega),$$

and, for every $x^{t-1} \in \mathcal{N}^{t-1}$, there is an $\hat{x}_t \in L^0(\mathcal{F}_t; \mathbb{R}^{n_t})$ with

$$\tilde{h}_{t-1}^\infty(x^{t-1}) = h_t^\infty(x^{t-1}, \hat{x}_t) \quad a.s. \tag{2.11}$$

Lower boundedness of h_t implies that of $\tilde{h}_{t-1}$. By Theorem 2.13, h_{t-1} is well-defined and lower bounded. It is clear that $\operatorname{dom} E\tilde{h}_t \cap \mathcal{N}^t \neq \emptyset$ implies $\operatorname{dom} E\tilde{h}_{t-1} \cap \mathcal{N}^{t-1} \neq \emptyset$. Thus,

$$\begin{aligned}
\mathcal{L}^{t-1} &= \{x^{t-1} \in \mathcal{N}^{t-1} \mid \tilde{h}_{t-1}^\infty(x^{t-1}) \leq 0 \, a.s.\} \\
&= \{x^{t-1} \in \mathcal{N}^{t-1} \mid \exists \hat{x}_t \in L^0(\mathcal{F}_t) \, h_t^\infty(x^{t-1}, \hat{x}_t) \leq 0 \, a.s.\}
\end{aligned}$$

$$= \{x^{t-1} \in \mathcal{N}^{t-1} \mid \exists \hat{x}^t \in \mathcal{N}^t : h_t^\infty(\hat{x}^t) \le 0,\ \hat{x}^{t-1} = x^{t-1}\ a.s.\}$$
$$= \{x^{t-1} \in \mathcal{N}^{t-1} \mid \exists \hat{x}^t \in \mathcal{L}^t : \hat{x}^{t-1} = x^{t-1}\ a.s.\},$$

where the first equality follows from the third claim while the second equality follows from (2.11). □

Theorem 2.62 below is our first existence result for the existence of solutions to the Bellman equations (BE). The observation about the linearity of the recession cones gives the sufficient condition for the existence of solutions of (P) used in Theorem 2.2. The last claim will be used in the proof of the main result of Chap. 4.

Theorem 2.62 *Assume that h is lower bounded and convex and that (P) is feasible. Then $\mathcal{L}$ is linear if and only if* (BE) *has an almost surely everywhere unique solution $(h_t)_{t=0}^T$ of lower bounded convex normal integrands and N_t are linear-valued almost surely for all t. In this case, if $x \in \mathcal{L}$ is such that $x^{t-1} = 0$, then $x_t \in N_t$ almost surely.*

Proof The feasibility means that there is an $\tilde{x} \in \operatorname{dom} Eh \cap \mathcal{N}$. By Theorem 2.13, h_T is well-defined and lower bounded and $\tilde{x}^T \in \operatorname{dom} Eh_T$. Denoting $\tilde{h}_T := h$, Lemma 2.61 gives $\mathcal{L}^T = \mathcal{L}$.

Assume first that $\mathcal{L}$ is linear. We make the induction hypotheses that h_t is well-defined and lower bounded with $\tilde{x}^t \in \operatorname{dom} Eh_t$ and that $\mathcal{L}^t$ is linear. By definition

$$L^0(\mathcal{F}_t; N_t) = \{x_t \in L^0(\mathcal{F}_t; \mathbb{R}^{n_t}) \mid (0, x_t) \in \mathcal{L}^t\}, \tag{2.12}$$

so the linearity of $\mathcal{L}^t$ implies that of $L^0(\mathcal{F}_t; N_t)$ which in turn, by Corollary 1.74, implies that N_t is linear-valued. Lemma 2.61 now implies that $\tilde{h}_{t-1}$ and h_{t-1} are well-defined and lower bounded and that (2.10) holds, so the linearity of $\mathcal{L}^t$ implies that of $\mathcal{L}^{t-1}$. Since $\mathcal{L}$ is linear, the induction hypotheses holds for $t = T$ so, by backward induction on t, (BE) has a solution and N_t are linear-valued.

To prove the converse, assume that (BE) has a solution and that N_t are linear-valued. The linear-valuedness of N_0 implies that $\mathcal{L}^0$ is linear. Assume now that $\mathcal{L}^{t-1}$ is linear and let $x^t \in \mathcal{L}^t$. By (2.10), $x^{t-1} \in \mathcal{L}^{t-1}$, so $-x^{t-1} \in \mathcal{L}^{t-1}$. By (2.10) again, there exists an $\hat{x}^t \in \mathcal{L}^t$ with $\hat{x}^{t-1} = -x^{t-1}$. Since $\mathcal{L}^t$ is a cone, $x^t + \hat{x}^t \in \mathcal{L}^t$. Since $x^{t-1} + \hat{x}^{t-1} = 0$, the linear-valuedness of N_t gives $-x^t - \hat{x}^t \in \mathcal{L}^t$. Since $\mathcal{L}^t$ is a cone, we get $-x^t = \hat{x}^t + (-x^t - \hat{x}^t) \in \mathcal{L}^t$ so $\mathcal{L}^t$ is linear. By forward induction on t, $\mathcal{L}^T$ is linear. As noted at the beginning of the proof, $\mathcal{L} = \mathcal{L}^T$.

To prove the last claim, let $x \in \mathcal{L}$ with $x^{t-1} = 0$. Since $\mathcal{L} = \mathcal{L}^T$, recursive application of (2.10) gives $x^t \in \mathcal{L}^t$, so (2.12) implies $x_t \in N_t$ almost surely. □

The following corollary says that, even if h fails to be lower bounded, the linearity of $\mathcal{L}$ in Theorem 2.62 can be characterized also in terms of the solution of (BE) associated with h^∞. When h is lower bounded, the last claim says that the recession functions of the solutions of (BE) solve the Bellman equations associated with the recession function h^∞ of h.

Corollary 2.63 *Assume that h is a proper and convex and that h^∞ is lower bounded. Then $\mathcal{L}$ is linear if and only if* (BE) *associated with h^∞ has an almost surely everywhere unique solution $((h^\infty)_t)_{t=0}^T$ of lower bounded sublinear normal integrands and the set*

$$\{x_t \in \mathbb{R}^{n_t} \mid (h^\infty)_t(0, x_t, \omega) \le 0\}$$

is linear for almost every ω for all t. If the above hold, h is lower bounded and (P) is feasible, then $h_t^\infty = (h^\infty)_t$ for all t, where $(h_t)_{t=0}^T$ is the solution of (BE) *associated with h.*

Proof By Theorem 1.39, the properness and convexity of h imply that h^∞ is a proper sublinear normal integrand. The equivalence follows by applying Theorem 2.62 to h^∞. When h is lower bounded and (P) is feasible, Corollary 2.34 gives $(h^\infty)_T = h_T^\infty$, so the last claim follows from (2.8) and (2.9) in Lemma 2.61. □

Theorem 2.65 below is an immediate corollary of Theorems 2.2 and 2.62.

Assumption 2.64 *The normal integrand h is convex and lower bounded, the associated problem (P) is feasible, and the set*

$$\mathcal{L} := \{x \in \mathcal{N} \mid h^\infty(x) \le 0 \; a.s.\}$$

is linear.

Theorem 2.65 *Under Assumption 2.64,* (BE) *has an almost surely everywhere unique solution $(h_t)_{t=0}^T$ of lower bounded convex normal integrands,*

$$\inf(P) = \inf_{x^t \in \mathcal{N}^t} Eh_t(x^t)$$

for all $t = 0, \dots, T$ and (P) has a solution $x \in \mathcal{N}$ with $x_t \perp N_t$ almost surely. The solutions $\bar{x} \in \mathcal{N}$ of (P) are characterized by

$$\bar{x}_t(\omega) \in \operatorname*{argmin}_{x_t \in \mathbb{R}^{n_t}} h_t(\bar{x}^{t-1}(\omega), x_t, \omega) \tag{OP}$$

for almost every $\omega \in \Omega$ and all $t = 0, \dots, T$.

2.2.2 L-Bounded Objectives

This section extends the existence result in Theorem 2.65 and the optimality conditions in Theorem 2.2 by relaxing the lower boundedness assumption on h.

The extension is based on the interplay of the space $\mathcal{N}$ of adapted strategies with the set

$$\mathcal{N}^\perp := \{v \in L^1 \mid E[x \cdot v] = 0 \; \forall x \in \mathcal{N}^\infty\},$$

where $\mathcal{N}^\infty := \mathcal{N} \cap L^\infty$. The following gives an alternative expression for $\mathcal{N}^\perp$.

Lemma 2.66 $\mathcal{N}^\perp = \{v \in L^1 \mid E_t[v_t] = 0 \quad t = 0, \dots, T\}$.

Proof We have

$$E[x \cdot v] = \sum_{t=0}^{T} E[x_t \cdot v_t]$$

so $v \in \mathcal{N}^\perp$ if and only if $E[x_t \cdot v_t] = 0$ for all $x_t \in L^\infty(\mathcal{F}_t)$. The claim now follows from Lemma B.13. □

Since the elements of $\mathcal{N}$ are not integrable in general, we do not necessarily have $E[x \cdot v] = 0$ for all $x \in \mathcal{N}$ and $v \in \mathcal{N}^\perp$ as the scenariowise inner product $x \cdot v$ does not have to be integrable. However, we have the following very useful result.

Lemma 2.67 *Let $x \in \mathcal{N}$ and $v \in \mathcal{N}^\perp$. If $[x \cdot v]^+ \in L^1$, then $E[x \cdot v] = 0$.*

Proof Assume first that $T = 0$. Defining $x^\nu := 1_{\{|x| \le \nu\}} x$, we have $x^\nu \in \mathcal{N}^\infty$, so $E[x^\nu \cdot v] = 0$ and thus, $E[x^\nu \cdot v]^- = E[x^\nu \cdot v]^+$. Since $[x^\nu \cdot v]^+ \le [x \cdot v]^+$, Fatou's lemma gives

$$E[x \cdot v]^- \le \liminf_{\nu \to \infty} E[x^\nu \cdot v]^- = \liminf_{\nu \to \infty} E[x^\nu \cdot v]^+ \le E[x \cdot v]^+$$

so if $[x \cdot v]^+ \in L^1$ we have $x \cdot v \in L^1$. Since $|x^\nu \cdot v| \le |x \cdot v|$, dominated convergence theorem gives $E[x \cdot v] = \lim E[x^\nu \cdot v] = 0$.

Assume now that the claim holds for every $(T-1)$-period model. Defining $x^\nu := 1_{\{|x_0| \le \nu\}} x$, we have

$$[\sum_{t=1}^{T} x_t^\nu \cdot v_t]^+ = [x^\nu \cdot v - x_0^\nu \cdot v_0]^+ \le [x^\nu \cdot v]^+ + [x_0^\nu \cdot v_0]^- \le [x \cdot v]^+ + [x_0^\nu \cdot v_0]^-,$$

where the right side is integrable. Thus, $E[\sum_{t=1}^{T} x_t^\nu \cdot v_t] = 0$, by the induction hypothesis. Since $x_0^\nu \in L^\infty$ and $v \in \mathcal{N}^\perp$, we also have $E[x_0^\nu \cdot v_0] = 0$ so $E[x^\nu \cdot v] = 0$. Since $[x^\nu \cdot v]^+ \le [x \cdot v]^+$, we conclude $E[x \cdot v] = 0$ just like in the case $T = 0$ above. □

The following illustrates Lemma 2.67 with the stochastic integral of an adapted process with respect to a martingale.

Example 2.68 Assume that $n_t = d$ for all t and let s be a d-dimensional martingale, i.e. an adapted integrable stochastic process such that $E_t[\Delta s_{t+1}] = 0$ for all t. If $x \in \mathcal{N}$ is such that

$$E[\sum_{t=0}^{T-1} x_t \cdot \Delta s_{t+1}]^+ < \infty,$$

then $E[\sum_{t=0}^{T-1} x_t \cdot \Delta s_{t+1}] = 0$. This follows from Lemma 2.67 with $v \in \mathcal{N}^\perp$ defined by $v_t = \Delta s_{t+1}$.

The extensions of Theorems 2.62 and 2.2 below are based on the following lemma that allows us to reduce a more general problem to one with a lower bounded integrand.

Lemma 2.69 *Assume that there exists $p \in \mathcal{N}^\perp$ such that the normal integrand $k(x,\omega) := h(x,\omega) - x \cdot p(\omega)$ is lower bounded. Then* (BE) *has a solution $(h_t)_{t=0}^T$ for h if and only if* (BE) *has a solution $(k_t)_{t=0}^T$ for k. In this case, the solutions are almost surely everywhere unique and related by*

$$k_t(x^t,\omega) = h_t(x^t,\omega) - x^t \cdot (E_t p_t)(\omega). \tag{2.13}$$

Proof Assume that $(h_t)_{t=0}^T$ solves (BE) for h. Since k is lower bounded, there exists $m \in L^1$ such that

$$h(x,\omega) \geq x \cdot p(\omega) - m(\omega).$$

We have

$$h_t(x^t,\omega) \geq x^t \cdot (E_t p^t)(\omega) - (E_t m)(\omega) \tag{2.14}$$

for $t = T, \dots, 0$ and

$$\tilde{h}_t(x^t,\omega) \geq x^t \cdot (E_{t+1} p^t)(\omega) - (E_{t+1} m)(\omega) \tag{2.15}$$

for $t = T-1, \dots, 0$. Indeed, since $h_T = E_T h$, (2.14) holds for $t = T$. If (2.14) holds for t, then, since $E_t p_t = 0$, we get (2.15) for $t-1$. If (2.15) holds a given t, then by Lemma 2.11 and Example 2.7, (2.14) holds for t as well.

Define $(k_t)_{t=0}^T$ by (2.13) and $(\tilde{k}_t)_{t=0}^{T-1}$ by

$$\tilde{k}_t(x^t,\omega) := \tilde{h}_t(x^t,\omega) - x^t \cdot (E_{t+1} p^t)(\omega).$$

By Theorem 2.15 and Example 2.7, the lower bound (2.15) implies $E_t\tilde{k}_t = k_t$. Since $E_t p_t = 0$,

$$\begin{aligned}\tilde{k}_{t-1}(x^{t-1},\omega) &= \tilde{h}_{t-1}(x^{t-1},\omega) - x^{t-1}\cdot(E_t p^{t-1})(\omega),\\ &= \inf_{x_t\in\mathbb{R}^{n_t}}\{h_t(x^t,\omega) - x^t\cdot(E_t p^t)(\omega)\}\\ &= \inf_{x_t\in\mathbb{R}^{n_t}} k_t(x^{t-1},x_t,\omega).\end{aligned}$$

Thus, $(k_t)_{t=1}^T$ solves (BE) for k. Conversely, assume that $(k_t)_{t=0}^T$ and $(\tilde{k}_t)_{t=0}^{T-1}$ satisfy (BE) for k. The lower boundedness of k implies that of k_t and $\tilde{k}_t$. Similar argument as above, then shows that the functions

$$\begin{aligned}\tilde{h}_t(x^t,\omega) &:= \tilde{k}_t(x^t,\omega) + x^t\cdot(E_{t+1}p^t)(\omega),\\ h_t(x^t,\omega) &:= k_t(x^t,\omega) + x^t\cdot(E_t p^t)(\omega)\end{aligned}$$

satisfy (BE) for h. This completes the proof. □

The following two theorems extend Theorems 2.2 and 2.62 on the solutions to (P) and to the Bellman equations, respectively. The theorems below relax the lower boundedness assumptions made earlier. The sufficient conditions may seem somewhat technical but they are implied by the more concrete conditions given in Lemmas 2.74 and 2.75 below. The conditions hold trivially in the case of lower bounded integrands.

The following is a generalization of Theorem 2.2.

Theorem 2.70 *Assume that there exists $p \in \mathcal{N}^\perp$ such that the normal integrand $k(x,\omega) := h(x,\omega) - x\cdot p(\omega)$ is lower bounded and $Ek(x) = Eh(x)$ for all $x \in \mathcal{N}$. If (P) is feasible and the Bellman equations* (BE) *admit a solution $(h_t)_{t=0}^T$ of normal integrands, then*

$$\inf(P) = \inf_{x^t\in\mathcal{N}^t} Eh_t(x^t)$$

for all $t = 0,\dots,T$ and, moreover, an $\bar{x}\in\mathcal{N}$ solves (P) if and only if

$$\bar{x}_t(\omega) \in \operatorname*{argmin}_{x_t\in\mathbb{R}^{n_t}} h_t(\bar{x}^{t-1}(\omega), x_t, \omega) \tag{OP}$$

for almost every $\omega\in\Omega$ and all $t = 0,\dots,T$. If, in addition, h_t are convex and

$$N_t(\omega) := \{x_t \in \mathbb{R}^{n_t} \mid h_t^\infty(0,x_t,\omega) \le 0\}$$

is linear-valued for all $t = 0,\dots,T$, then there exists an optimal $x\in\mathcal{N}$ with $x_t \perp N_t$ almost surely.

Proof Since $Eh = Ek$ on $\mathcal{N}$, the optimum value and solutions of (P) coincide with those of the problem

$$\text{minimize} \quad Ek(x) \quad \text{over } x \in \mathcal{N}.$$

By Lemma 2.69, the sequence $(k_t)_{t=0}^T$ given by

$$k_t(x^t, \omega) = h_t(x^t, \omega) - x^t \cdot (E_t p_t)(\omega)$$

solves (BE) for k. Since k is lower bounded, Theorem 2.2 implies that

$$\inf(P) = \inf_{x^t \in \mathcal{N}^t} Ek_t(x^t) \tag{2.16}$$

for all $t = 0, \dots, T$ and that an $\bar{x} \in \mathcal{N}$ solves (P) if and only if

$$\bar{x}_t \in \operatorname*{argmin}_{x_t \in \mathbb{R}^{n_t}} k_t(\bar{x}^{t-1}, x_t) \quad a.s.$$

for all $t = 0, \dots, T$.

By definition of $(h_t)_{t=0}^T$, we always have $\inf(P) \geq \inf_{x^t \in \mathcal{N}^t} Eh_t(x^t)$. The lower boundedness of k implies that of k_t, so there exist $m_t \in L^1$ such that

$$h_t(x^t) = k_t(x^t) + x^t \cdot E_t p^t \geq x^t \cdot E_t p^t - m_t.$$

If $Eh_t(x^t) < \infty$, then $E[x^t \cdot E_t p^t] < \infty$, so Lemma 2.67 gives $Ek_t(x^t) = Eh_t(x^t)$. Thus,

$$\inf(P) = \inf_{x^t \in \mathcal{N}^t} Eh_t(x^t),$$

which proves the first claim. Since $E_t p_t = 0$, we have

$$\operatorname*{argmin}_{x_t} k_t(x^{t-1}(\omega), x_t, \omega) = \operatorname*{argmin}_{x_t} h_t(x^{t-1}(\omega), x_t, \omega),$$

which proves the second claim. When h_t are convex, the normal integrands k_t are convex as well so the last claim of Theorem 2.2 says that, if the sets

$$\hat{N}_t(\omega) := \{x_t \in \mathbb{R}^{n_t} \mid k_t^\infty(0, x_t, \omega) \leq 0\}$$

are linear valued, then there exists a solution $x \in \mathcal{N}$ with $x_t \perp \hat{N}_t$ almost surely. It now suffices to note that, since $E_t p_t = 0$, we have $\hat{N}_t = N_t$ almost surely. □

The following is a generalization of Theorem 2.62.

Theorem 2.71 *Assume that there exists $p \in \mathcal{N}^\perp$ such that the normal integrand $k(x, \omega) := h(x, \omega) - x \cdot p(\omega)$ satisfies Assumption 2.64. Then* (BE) *has an almost* *surely everywhere unique solution* $(h_t)_{t=0}^T$ *of L-bounded convex normal integrands and*

$$N_t(\omega) := \{x_t \in \mathbb{R}^{n_t} \mid h_t^\infty(0, x_t, \omega) \le 0\}$$

are linear-valued for all t.

Proof By Theorem 2.62, the Bellman equations associated with k have an a.s.e. unique solution $(k_t)_{t=0}^T$ of lower bounded normal integrands, and the measurable mappings

$$\hat{N}_t(\omega) := \{x_t \in \mathbb{R}^{n_t} \mid k_t^\infty(0, x_t, \omega) \le 0\}$$

are linear-valued for all t. By Lemma 2.69,

$$h_t(x^t, \omega) = k_t(x^t, \omega) + x^t \cdot (E_t p_t)(\omega)$$

defines a unique solution of (BE) for h. The lower boundedness of k_t implies the L-boundedness of h_t. Since $E_t p_t = 0$, we have $N_t = \hat{N}_t$ almost surely. □

We combine the assumptions of Theorems 2.70 and 2.71 into the following.

Assumption 2.72 *Problem* (P) *is feasible and there exists* $p \in \mathcal{N}^\perp$ *such that*

1. $k(x, \omega) := h(x, \omega) - x \cdot p(\omega)$ *is convex and lower bounded,*
2. $\{x \in \mathcal{N} \mid k^\infty(x) \le 0 \text{ a.s.}\}$ *is linear,*
3. $Ek(x) = Eh(x)$ *for all* $x \in \mathcal{N}$.

Note that, if h is lower bounded, one can take $p = 0$ so Assumption 2.72 reduces to Assumption 2.64. More general sufficient conditions are given in Lemmas 2.74 and 2.75 below. Applications in Sect. 2.3 illustrate the conditions in specific examples. In particular, in the financial model of Sect. 2.3.5, the assumption is related to the existence of a martingale measure for the price process.

The following combines Theorem 2.70 and 2.71.

Theorem 2.73 *Under Assumption 2.72,* (BE) *has an almost surely everywhere unique solution* $(h_t)_{t=0}^T$ *of L-bounded normal integrands,*

$$\inf(P) = \inf_{x^t \in \mathcal{N}^t} Eh_t(x^t)$$

for all $t = 0, \dots, T$, (P) *has solutions* $\bar{x} \in \mathcal{N}$ *and they are characterized by*

$$\bar{x}_t(\omega) \in \operatorname*{argmin}_{x_t \in \mathbb{R}^{n_t}} h_t(\bar{x}^{t-1}(\omega), x_t, \omega) \tag{OP}$$

for almost every $\omega \in \Omega$ and all $t = 0, \dots, T$. Moreover, the mapping

$$N_t(\omega) := \{x_t \in \mathbb{R}^{n_t} \mid h_t^\infty(0, x_t, \omega) \le 0\}$$

is linear-valued for all $t = 0, \dots, T$ and there exists an optimal $x \in \mathcal{N}$ with $x_t \perp N_t$ almost surely.

The following gives sufficient conditions for Assumption 2.72. Recall that the scenariowise conjugate

$$h^*(v, \omega) := \sup_{x \in \mathbb{R}^n} \{x \cdot v - h(x, \omega)\}$$

of h is a normal integrand; see Theorem 1.46.

Lemma 2.74 *Assumption 2.72 holds if (P) is feasible, h is convex and*

1. $\{x \in \mathcal{N} \mid h^\infty(x) \le 0\ a.s.\}$ *is linear,*
2. *there exist $p \in \mathcal{N}^\perp$, $m \in L^1$ and $\epsilon > 0$ such that, for every $\lambda \in [1 - \epsilon, 1 + \epsilon]$,*

$$h(x, \omega) \ge \lambda x \cdot p(\omega) - m(\omega) \quad \forall x \in \mathbb{R}^n$$

and almost every $\omega \in \Omega$.

Condition 2 can be written equivalently as

2'. *there exist $p \in \mathcal{N}^\perp$ and $\epsilon > 0$ such that*

$$\lambda p \in \operatorname{dom} Eh^* \quad \forall \lambda \in [1 - \epsilon, 1 + \epsilon].$$

Proof The lower bound in condition 2 gives

$$\begin{aligned} h(x, \omega) &\ge x \cdot p(\omega) - m(\omega), \\ h(x, \omega) - x \cdot p(\omega) &\ge \epsilon x \cdot p(\omega) - m(\omega). \end{aligned}$$

Since h is convex, the first inequality means that Assumption 2.72.1 holds. Let $x \in \mathcal{N}$. If either $Eh(x) < \infty$ or $E[h(x) - x \cdot p] < \infty$, the above inequalities and Lemma 2.67 give $E[x \cdot p] = 0$, so

$$Eh(x) = E[h(x) - x \cdot p],$$

which is Assumption 2.72.3. By Remark A.20, the above inequalities also give

$$\begin{aligned} h^\infty(x, \omega) &\ge x \cdot p(\omega), \\ h^\infty(x, \omega) - x \cdot p(\omega) &\ge \epsilon x \cdot p(\omega). \end{aligned}$$

If either $h^\infty(x,0) \le 0$ or $h^\infty(x,0) - x \cdot p \le 0$ almost surely, then $x \cdot p \le 0$ almost surely. Lemma 2.67 then implies $x \cdot p = 0$ almost surely, so

$$\{x \in \mathcal{N} \mid h^\infty(x) \le 0 \text{ a.s.}\} = \{x \in \mathcal{N} \mid h^\infty(x) - x \cdot p \le 0 \text{ a.s.}\}.$$

Thus, under condition 2, condition 1 is equivalent to Assumption 2.72.2.

The lower bound in 2 can be written as

$$h^*(\lambda p) \le m \; a.s.$$

Thus, 2 implies 2'. Assuming 2', there exist, for every $\lambda \in [1-\epsilon, 1+\epsilon]$, an $m_\lambda \in L^1$ such that $h^*(\lambda p) \le m_\lambda$. By convexity,

$$h^*(\lambda p) \le \max\{m_{1-\epsilon}, m_{1+\epsilon}\},$$

so 2 holds with $m := \max\{m_{1-\epsilon}, m_{1+\epsilon}\}$. □

The above proof shows that, under condition 2 of Lemma 2.74, the linearity conditions in Assumptions 2.64 and 2.72 coincide. Note that condition 2 of Lemma 2.74 can also be written as

$$h(x,\omega) \ge x \cdot p(\omega) + \epsilon |x \cdot p(\omega)| - m(\omega) \quad \forall x \in \mathbb{R}^n$$

for almost every $\omega \in \Omega$.

The following gives an alternative set of sufficient conditions for Assumption 2.72.

Lemma 2.75 *Assumption 2.72 holds if (P) is feasible and there exists $p \in \mathcal{N}^\perp$ such that*

1. *$k(x,\omega) := h(x,\omega) - x \cdot p(\omega)$ is convex and lower bounded,*
2. *$\{x \in \mathcal{N} \mid k^\infty(x) \le 0 \; a.s.\}$ is linear,*
3. *$\mathcal{N}(\operatorname{dom} h) \subseteq \operatorname{dom} Eh$.*

Proof Let $x \in \mathcal{N}$ be such that either $Ek(x)$ or $Eh(x)$ is finite. Then $x \in \operatorname{dom} h$ almost surely, so $x \in \operatorname{dom} Eh$ by 2. By 1, there exists $m \in L^1$ such that $h(x) = k(x) + x \cdot p \ge x \cdot p - m$, so $E[x \cdot p] = 0$ by Lemma 2.67. Thus, $Ek(x) = Eh(x)$. □

The following simple observation will be useful in certain applications of dynamic programming. It is inspired by the solution of the classical Merton problem in dynamic portfolio optimization; see Example 2.127 below.

Remark 2.76 Let $(h_t)_{t=0}^T$ be an L-bounded solution of (BE). If h is p-homogeneous such that the random variable C in (2.3) is integrable (see Lemma 2.22), then h_t are p-homogeneous with

$$h_t(\alpha x) = \begin{cases} \alpha^p h_t(x) - E_t[C] \frac{\alpha^p - 1}{p} & \text{if } p \ne 0, \\ h_t(x) - E_t[C] \ln \alpha & \text{if } p = 0. \end{cases}$$

Proof This follows from Lemma 2.22 and Theorem A.86 and by noting that the constant C is preserved in infimal projections. □

Counterexample 2.77 Without Assumption 2.72, condition (OP) is not sufficient for optimality, in general. Consider, for example, the normal integrand $h(x, \omega) = x \cdot p$ where $p \in \mathcal{N}^\perp$. Since $E_t p_t = 0$ for all t, Example 2.7 implies

$$h_t(x, \omega) = x^t \cdot E_t p^t$$

so every $x \in \mathcal{N}$ satisfies (OP) but, by Lemma 2.67, the optimal solutions $x \in \mathcal{N}$ are characterized by the condition $E[x \cdot p] < \infty$. Here, h satisfies conditions 1 and 2 of Assumption 2.72 but fails 3, in general.

2.2.3 *Induced Constraints and Relatively Complete Recourse*

When making a decision x_t at time t, the requirement that one has to be able to extend $x^t := (x_{t'})_{t'=0}^t$ to an adapted feasible solution of (P) may induce constraints on x^t that cannot be specified by inspecting the domain of the normal integrand h (or that of the associated integral functional Eh) alone.

Any feasible solution x of (P) belongs to the set

$$\mathcal{N}(\operatorname{cl}\operatorname{dom} h) := \{x \in \mathcal{N} \mid x \in \operatorname{cl}\operatorname{dom} h \quad a.s.\}.$$

In particular, if x is feasible, then x^t belongs to the set $\mathcal{D}_t$ defined recursively by

$$\begin{aligned}
\mathcal{D}_T &:= L^0(\mathcal{F}_T; \operatorname{cl}\operatorname{dom} h),\\
\tilde{\mathcal{D}}_t &:= \{x^t \in L^0(\mathcal{F}_{t+1}; \mathbb{R}^{n^t}) \mid x^{t+1} \in \mathcal{D}_{t+1}\},\\
\mathcal{D}_t &:= \{x^t \in L^0(\mathcal{F}_t; \mathbb{R}^{n^t}) \mid x^t \in \tilde{\mathcal{D}}_t\}.
\end{aligned}$$

It turns out that the constraint $x^t \in \mathcal{D}_t$ admits a scenariowise representation as soon as the mappings $\tilde{\mathcal{D}}_t$ are closed in L^0 for all $t = 0, \ldots, T$.

Given a closed-valued measurable mapping $S : \Omega \rightrightarrows \mathbb{R}^n$, we say that a sequence $(S_t)_{t=0}^T$ of closed-valued measurable mappings $S_t : \Omega \rightrightarrows \mathbb{R}^{n^t}$ solves the *Bellman equations* for S if the mappings

$$\tilde{S}_t(\omega) := \{x^t \in \mathbb{R}^{n^t} \mid x^{t+1} \in S_{t+1}(\omega)\}$$

are closed-valued and measurable and

$$\begin{aligned}
S_T &= \operatorname{essinf}_T S,\\
S_t &= \operatorname{essinf}_t \tilde{S}_t \quad \forall t = 0, \ldots, T-1.
\end{aligned} \tag{2.17}$$

Here, and in what follows, we use the shorthand notation $\text{essinf}_t := \text{essinf}^{\mathcal{F}_t}$ for the $\mathcal{F}_t$-conditional essential infimum; see Sect. 2.1.4. Note that (2.17) is an instance of the Bellman equations (BE) applied to the indicator function $h = \delta_S$. Indeed, by Theorem 2.14, sequence $(S_t)_{t=0}^T$ satisfies Bellman equations (2.17) if and only if $h_t = \delta_{S_t}$ satisfies Bellman equation (BE) with $h = \delta_S$.

Theorem 2.78 *Assume that* (P) *is feasible. The following are equivalent:*

1. *the sets* $\tilde{\mathcal{D}}_t$ *are closed in* L^0 *for all* t*;*
2. *there exist closed-valued* $\mathcal{F}_t$*- and* $\mathcal{F}_{t+1}$*-measurable mappings* D_t *and* $\tilde{D}_t$*, respectively, such that*

$$\mathcal{D}_t = L^0(\mathcal{F}_t; D_t) \quad \text{and} \quad \tilde{\mathcal{D}}_t = L^0(\mathcal{F}_{t+1}; \tilde{D}_t)$$

   for all $t = 0, \ldots, T$*;*
3. *the Bellman equations admit a solution* $(S_t)_{t=0}^T$ *for* $S = \text{cl dom}\, h$*,*

and in this case, $D_t = S_t$ *almost surely for all* $t = 0, \ldots, T$*. The above conditions hold, in particular, if* cl dom h *is convex-valued and the set*

$$\{x \in \mathcal{N} \mid x \in (\text{cl dom}\, h)^\infty \ a.s.\}$$

is linear.

Proof By Corollary 1.57, 2 implies 1. To prove the converse, we apply Theorem 1.60. If $\tilde{\mathcal{D}}_t$ is closed in L^0, then $\mathcal{D}_t$ is closed as well. If

$$x^{t+1}1_A + \bar{x}^{t+1}1_{\Omega\setminus A} \in \mathcal{D}_{t+1} \quad \forall x^{t+1}, \bar{x}^{t+1} \in \mathcal{D}_{t+1}, A \in \mathcal{F}_{t+1},$$

then, by definition of $\tilde{\mathcal{D}}_t$ and $\mathcal{D}_t$,

$$\begin{aligned} x^t 1_A + \bar{x}^t 1_{\Omega\setminus A} \in \tilde{\mathcal{D}}_t \quad \forall x^t, \bar{x}^t \in \tilde{\mathcal{D}}_t,\ A \in \mathcal{F}_{t+1}, \\ x^t 1_A + \bar{x}^t 1_{\Omega\setminus A} \in \mathcal{D}_t \quad \forall x^t, \bar{x}^t \in \mathcal{D}_t,\ A \in \mathcal{F}_t. \end{aligned} \tag{2.18}$$

It is clear that $\mathcal{D}_T$ satisfies (2.18), so by induction, (2.18) holds for every t. Thus, by Theorem 1.60, 1 implies 2.

Condition 2 means that

$$L^0(\mathcal{F}_{t+1}, \tilde{D}_t) = \{x^t \in L^0(\mathcal{F}_{t+1}; \mathbb{R}^{n^t}) \mid x^{t+1} \in L^0(\mathcal{F}_{t+1}; D_{t+1})\}, \tag{2.19}$$

$$L^0(\mathcal{F}_t; D_t) = \{x^t \in L^0(\mathcal{F}_t; \mathbb{R}^{n^t}) \mid x^t \in L^0(\mathcal{F}_{t+1}; \tilde{D}_t)\}. \tag{2.20}$$

It is clear that

$$\{x^t \in L^0(\mathcal{F}_{t+1}; \mathbb{R}^{n^t}) \mid x^{t+1} \in L^0(\mathcal{F}_{t+1}; D_{t+1})\} \subseteq L^0(\mathcal{F}_{t+1}; D_{t+1}^t), \tag{2.21}$$

where $D^t_{t+1}(\omega) := \{x^t \in \mathbb{R}^{n^t} \mid x^{t+1} \in D_{t+1}(\omega)\}$. On the other hand, let $x^t \in L^0(\mathcal{F}_{t+1}; D^t_{t+1})$. Applying Corollary 1.28, to the mapping $S(\omega) := \{x_{t+1} \in \mathbb{R}^{n_{t+1}} \mid (x^t(\omega), x_{t+1}) \in D_{t+1}\}$ shows that (2.21) holds as an equality. Thus, (2.19) means that

$$L^0(\mathcal{F}_{t+1}, \tilde{D}_t) = L^0(\mathcal{F}_{t+1}; D^t_{t+1})$$

while (2.20) means that

$$L^0(\mathcal{F}_t; D_t) = L^0(\mathcal{F}_t; \tilde{D}_t).$$

By Theorems 1.72 and 2.14, condition 2 thus means that $\tilde{D}_t = D^t_{t+1}$ and $D_t = \operatorname{essinf}_t \tilde{D}_t$. This proves the equivalence of 2 and 3.

As noted earlier, the Bellman equations (2.17) for $S = \operatorname{cl}\operatorname{dom} h$ is an instance of (BE) applied to the normal integrand $\delta_{\operatorname{cl}\operatorname{dom} h}$. Thus, the last claim follows from Theorem 2.62. □

The random sets D_t in Theorem 2.78 give a pointwise description of the requirement that one has to be able to extend $x^t \in \mathcal{N}^t$ to an adapted selection of $\operatorname{cl}\operatorname{dom} h$. In other words,

$$\mathcal{N}^t(D_t) = \{x^t \mid x \in \mathcal{N}(\operatorname{cl}\operatorname{dom} h)\}.$$

The random sets D_t will be called *induced constraints* of problem (P). Note that D_t may be strictly smaller than the scenariowise projection

$$(\operatorname{cl}\operatorname{dom} h)^t(\omega) := \{x^t \in \mathbb{R}^{n^t} \mid x \in \operatorname{cl}\operatorname{dom} h(\cdot, \omega)\}$$

of $\operatorname{cl}\operatorname{dom} h(\cdot, \omega)$; see Examples 2.97, 2.111 and 2.119 below. If $D_t = (\operatorname{cl}\operatorname{dom} h)^t$ for all t, problem (P) is said to have *relatively complete recourse*. Relatively complete recourse plays a fundamental role in Chap. 5 on the existence of dual solutions.

We say that a set-valued mapping $S : \Omega \rightrightarrows \mathbb{R}^{n_0} \times \cdots \times \mathbb{R}^{n_T}$ is $(\mathcal{F}_t)_{t=0}^T$*-adapted* if, for each $t = 0, \ldots, T$, the projection

$$S^t(\omega) := \{x^t \in \mathbb{R}^{n^t} \mid x \in S(\omega)\}$$

is $\mathcal{F}_t$-measurable. Note that a stochastic process $s = (s_t)_{t=0}^T$ is adapted if and only if the set-valued mapping $S(\omega) = \{s(\omega)\}$ is adapted in the above sense.

Corollary 2.79 *Problem* (P) *has relatively complete recourse if and only if* $\operatorname{cl}\operatorname{dom} h$ *is adapted and each* $(\operatorname{cl}\operatorname{dom} h)^t$ *is closed-valued.*

Proof The sets $S_t(\omega) := (\mathrm{cl}\,\mathrm{dom}\,h)^t(\omega)$ can be expressed recursively as

$$S_T(\omega) = (\mathrm{cl}\,\mathrm{dom}\,h)(\omega),$$
$$S_t(\omega) = \tilde{S}_t(\omega) := \{x^t \in \mathbb{R}^{n^t} \mid x^{t+1} \in S_{t+1}(\omega)\}.$$

Thus, $(S_t)_{t=0}^T$ solves (2.17) if and only if $S_t = \mathrm{essinf}_t \tilde{S}_t$, which means that S_t is $\mathcal{F}_t$-measurable. □

When it exists, the solution to the Bellman equations (2.17) for S allows us to construct the largest closed-valued adapted mapping contained in S.

Theorem 2.80 *If $(S_t)_{t=0}^T$ solves the Bellman equations for a closed-valued measurable mapping $S : \Omega \rightrightarrows \mathbb{R}^n$, then*

$$(\mathrm{essinf}_a S)(\omega) := \{x \in \mathbb{R}^n \mid x^t \in S_t(\omega)\ t = 0, \dots, T\}$$

is the largest closed-valued adapted mapping contained in S and

$$\mathcal{N}(\mathrm{essinf}_a S) = \mathcal{N}(S).$$

Proof Let $\hat{S}$ be a closed-valued adapted mapping contained in S. Since $\hat{S}$ is $\mathcal{F}_T$-measurable, $\hat{S}^T \subseteq S_T$ almost surely. Assume now that $\hat{S}^{t+1} \subseteq S_{t+1}$. It follows that $\hat{S}^t \subseteq \tilde{S}_t$ so $\hat{S}^t \subseteq S_t$ since $\hat{S}^t$ is $\mathcal{F}_t$-measurable. By induction, $\hat{S}^t \subseteq S_t$ almost surely for all $t = 0, \dots, T$. Thus, if $x \in \hat{S}(\omega)$, we have $x^t \in S_t(\omega)$ for every $t = 0, \dots, T$ so $x \in (\mathrm{essinf}_a S)(\omega)$.

Since $\mathrm{essinf}_a S \subseteq S$ almost surely, we have $\mathcal{N}(\mathrm{essinf}_a S) \subseteq \mathcal{N}(S)$. On the other hand, if $x \in \mathcal{N}(S)$, then the mapping $\tilde{S}(\omega) := (\mathrm{essinf}_a S)(\omega) \cup \{x(\omega)\}$ is closed-valued and adapted. Indeed, the projection of the union of two sets is the union of the projections, so $\tilde{S}^t$ is $\mathcal{F}_t$-measurable for all $t = 0, \dots, T$. This contradicts the first claim unless $x \in \mathrm{essinf}_a S$ almost surely. □

Consider a convex stochastic optimization problem (P) associated with a convex normal integrand h. When it exists, the solution to the Bellman equations (2.17) for $S = \mathrm{cl}\,\mathrm{dom}\,h$ allows us to construct an equivalent convex stochastic optimization problem that has relatively complete recourse.

Corollary 2.81 *Assume that $S := \mathrm{dom}\,h$ is closed-valued and that the Bellman equations* (2.17) *admit a solution $(S_t)_{t=0}^T$ for S, and let*

$$\bar{h}(x, \omega) := h(x, \omega) + \delta_{\mathrm{essinf}_a S}(x, \omega),$$

where $\mathrm{essinf}_a S$ is as in Theorem 2.80. The stochastic optimization problem associated with the normal integrand $\bar{h}$ has relatively complete recourse and

$$E\bar{h}(x) = Eh(x) \quad \forall x \in \mathcal{N}.$$

Proof We have $\text{essinf}_a S \subseteq \text{dom}\, h$, so $\text{dom}\, \bar{h} = \text{dom}\, h \cap \text{essinf}_a S = \text{essinf}_a S$, which is adapted by Theorem 2.80. Thus, the first claim follows from Corollary 2.79. The second claim holds by the last claim of Theorem 2.80. □

2.2.4 Canonical Representation

Consider again the canonical representation

$$\text{minimize} \quad \int_{\Xi} H(\hat{x}(s), s) d\hat{P} \quad \text{over } \hat{x} \in \hat{\mathcal{N}} \tag{$\hat{P}$}$$

of (P) from Remark 1.76. Recall that $\hat{P}$ is the distribution of a stochastic process $\xi = (\xi_t)_{t=0}^T$ where ξ_t takes values in a measurable space $(\Xi_t, \mathcal{A}_t)$. We assume throughout that ξ_{t+1} has a regular ξ^t-conditional distribution ν_t; see Sect. 2.1.5. As pointed out in Remark 1.76, one may identify ξ with its associated canonical process so that $(\Omega, \mathcal{F}, P) = (\Xi, \hat{\mathcal{F}}, \hat{P})$. Note that the conditional distributions of a process coincide with the conditional distributions of the associated canonical process.

Example 2.82 In many familiar stochastic process models, the conditional distributions ν_t have simple expressions. For example, if $\Xi_t = \mathbb{R}^d$ for all t and ξ follows the autoregressive model

$$\xi_t = A\xi_{t-1} + a + \epsilon_t,$$

where $A \in \mathbb{R}^{d\times d}$ and $a \in \mathbb{R}^d$ are fixed and ϵ_t are independent and identically distributed Gaussian random variables with zero mean and positive definite variance Σ, then the regular ξ^t-conditional distributions $\nu_t(\xi^t, \cdot)$ are Gaussian with mean $A\xi_t + a$ and variance Σ. Indeed, denoting the probability density of ϵ_{t+1} by ϕ, we have for every $\alpha \in L^\infty_+(\Omega, \sigma(\xi^t), P)$ and every $X \in L^0(\Xi_{t+1}, \mathcal{A}_{t+1})$ such that $X(\xi_{t+1})$ is quasi-integrable,

$$\begin{aligned} E[\alpha X(\xi_{t+1})] &= E[\alpha X(A\xi_t + a + \epsilon_{t+1})] \\ &= E[\int_{\mathbb{R}^n} \alpha X(A\xi_t + a + s)\phi(s) ds] \\ &= E[\alpha \int_{\mathbb{R}^n} X(s)\phi(s - A\xi_t - a) ds], \end{aligned}$$

where the second equality follows, by Fubini, from the independence of ξ^t and ϵ_{t+1}. Thus,

$$E_t[X] = \int_{\mathbb{R}^n} X(s)\phi(s - A\xi_t - a) ds,$$

so $\nu_t(\xi^t, \cdot)$ is indeed Gaussian.

We say that a sequence $(H_t)_{t=0}^T$ of normal integrands $H_t : \mathbb{R}^{n^t} \times \Xi^t \to \overline{\mathbb{R}}$ solves the *Bellman equations in canonical form* for H if the functions

$$\tilde{H}_t(x^t, s^{t+1}) := \inf_{x_{t+1} \in \mathbb{R}^{n_{t+1}}} H_{t+1}(x^t, x_{t+1}, s^{t+1}),$$

are normal integrands on $\mathbb{R}^{n^t} \times \Xi^{t+1}$ and

$$\begin{aligned} H_T(x^T, s^T) &= (E^{\nu_T} H)(x^T, s^T), \\ H_t(x^t, s^t) &= (E^{\nu_t} \tilde{H}_t)(x^t, s^t) \end{aligned} \tag{2.22}$$

for $t = T-1, \dots, 0$ and P_{ξ^t}-almost every s^t. Here P_{ξ^t} is the distribution of ξ^t and

$$(E^{\nu_t} \tilde{H}_t)(x^t, s^t) := \int_{\Xi_{t+1}} \tilde{H}_t(x^t, s^t, s_{t+1}) \nu_t(s^t, ds_{t+1});$$

see Corollary 2.60. In the context of Example 2.82, it would be straightforward to construct quadrature approximations of $E^{\nu_t} \tilde{H}_t$ e.g. with Monte Carlo averages.

Let

$$\hat{\mathcal{N}}^{\perp} := \{\hat{p} \in L^1(\Xi, \mathcal{A}, P_\xi) \mid E_{P_\xi}[\hat{x} \cdot \hat{p}] = 0 \quad \forall \hat{x} \in \hat{\mathcal{N}}^\infty\},$$

where

$$\hat{\mathcal{N}}^\infty := \hat{\mathcal{N}} \cap L^\infty(\Xi, \mathcal{A}, P_\xi).$$

The following is the canonical representation of Assumption 2.72.

Assumption 2.83 *Problem $(\hat{P})$ is feasible and*

1. *there exists $V \in \hat{\mathcal{N}}^\perp$ such that*

$$K(x, s) := H(x, s) - x \cdot V(s) \quad \forall x \in \mathbb{R}^n,$$

is convex and lower bounded,

2. *$\{\hat{x} \in \hat{\mathcal{N}} \mid K^\infty(\hat{x}) \leq 0 P_\xi\text{-a.e.}\}$ is linear,*
3. *$E_{P_\xi} K(x) = E_{P_\xi} H(x)$ for all $x \in \hat{\mathcal{N}}$.*

As pointed out at the end of Remark 1.76, we can view the components $\hat{x}_t$ of $\hat{x} \in \hat{\mathcal{N}}$ as functions of ξ^t instead of the whole path ξ. This interpretation is used in the following statement.

Theorem 2.84 *Under Assumption 2.83, the Bellman equations* (2.22) *have a solution* $(H_t)_{t=0}^T$ *where* H_t *is* P_{ξ^t}*-almost surely everywhere unique,*

$$\inf(\hat{P}) = \inf_{\hat{x}^t \in \hat{\mathcal{N}}^t} \int_{\Xi^t} H_t(\hat{x}^t(s^t), s^t) dP_{\xi^t}(s^t)$$

for all $t = 0, \ldots, T$, $(\hat{P})$ *has solutions* $\bar{x} \in \hat{\mathcal{N}}$ *and they are characterized by*

$$\bar{x}_t(s^t) \in \operatorname*{argmin}_{x_t \in \mathbb{R}^{n_t}} H_t(\bar{x}^{t-1}(s^{t-1}), x_t, s^t) \qquad (\hat{OP})$$

for P_{ξ^t}*-almost every* $s^t \in \Xi^t$ *and all* $t = 0, \ldots, T$.

Proof We apply Theorem 2.73 with $(\Omega, \mathcal{F}, P) = (\Xi, \hat{\mathcal{F}}, \hat{P})$ to $(\hat{P})$. This gives the existence of almost surely everywhere unique sequences of L-bounded convex $\hat{\mathcal{F}}_t$-measurable normal integrands $h_t : \mathbb{R}^{n^t} \times \Xi \to \overline{\mathbb{R}}$ and $\tilde{h}_{t-1} : \mathbb{R}^{n^{t-1}} \times \Xi$ satisfying (BE). Applying the Doob–Dynkin representation in Corollary 1.34 with the random variable $\xi \mapsto \xi^t$ gives the existence of $\mathcal{A}^t$-measurable convex normal integrands $H_t : \mathbb{R}^{n^t} \times \Xi^t \to \overline{\mathbb{R}}$ and $\tilde{H}_{t-1} : \mathbb{R}^{n^{t-1}} \times \Xi^t \to \overline{\mathbb{R}}$ such that

$$h_t(x^t, s) = H_t(x^t, s^t)$$

and

$$\tilde{h}_{t-1}(x^{t-1}, s) = \tilde{H}_{t-1}(x^{t-1}, s^t)$$

for all t. The P_{ξ^t}-almost surely everywhere uniqueness of H_t follows from the almost surely everywhere uniqueness of h_t.

Applying Corollary 2.60 with $(\Omega, \mathcal{F}, P) = (\Xi, \hat{\mathcal{F}}, \hat{P})$, $\xi_1 = s^t$ and $\xi_2 = s_{t+1}$ gives

$$(E_t \tilde{h}_t)(x^t, s) = (E^{\nu_t} \tilde{H}_t)(x^t, s^t).$$

Thus, the Bellman equations (BE) can be written as (2.22) so the claims follow from Theorem 2.73. □

2.3 Applications

This section applies some of the main results of this chapter to the examples presented in Sect. 1.3. We will see that the time-separable structure in some of the examples allows for writing the dynamic programming recursions in more explicit form.

2.3.1 Mathematical Programming

Consider again problem (1.8)

$$\begin{aligned} &\text{minimize} && Ef_0(x) \quad \text{over} \quad x \in \mathcal{N}, \\ &\text{subject to} && f_j(x) \le 0 \quad j = 1, \dots, l \; a.s., \\ &&& f_j(x) = 0 \quad j = l+1, \dots, m \; a.s. \end{aligned}$$

from Sect. 1.3.1. As observed there, this is an instance of (P) with

$$h(x,\omega) := \begin{cases} f_0(x,\omega) & \text{if } f_j(x,\omega) \le 0 \; j = 1 \dots, l, \\ & \quad\;\; f_j(x,\omega) = 0 \; j = l+1, \dots, m, \\ +\infty & \text{otherwise.} \end{cases}$$

This section applies the main results of Sect. 2.2.2 to give sufficient conditions for the existence of solutions of (1.8). In the general format above, the mathematical programming model does not allow for significant simplifications to the dynamic programming recursion.

Assumption 2.85 *Problem* (1.8) *is feasible and*

1. $\{x \in \mathcal{N} \mid f_j^\infty(x) \le 0 \; j = 0, \dots l, \; f_j^\infty(x) = 0 \; j = l+1, \dots, m \; a.s.\}$ *is linear,*
2. *there exist* $p \in \mathcal{N}^\perp$, $m \in L^1$ *and* $\epsilon > 0$ *such that, for every* $\lambda \in [1-\epsilon, 1+\epsilon]$,

$$f_0(x,\omega) \ge \lambda x \cdot p(\omega) - m(\omega)$$

for almost every $\omega \in \Omega$ *and every* $x \in \mathbb{R}^n$ *with* $f_j(x,\omega) \le 0$ *for all* $j = 1, \dots, l$ *and* $f_j(x,\omega) = 0$ *for all* $j = l+1, \dots, m$.

Assumption 2.85 is a reformulation of the conditions in Lemma 2.74 which imply Assumption 2.72. The last condition in Assumption 2.85 holds, in particular, if f_0 is lower bounded as one can then take $p = 0$.

Theorem 2.86 *Under Assumption 2.85,* (1.8) *has a solution.*

Proof By Theorem 2.73, it suffices to show that Assumption 2.85 implies the conditions of Lemma 2.74. We have

$$h(x,\omega) = f_0(x,\omega) + \sum_{j=1}^m \delta_{C_j(\omega)}(x),$$

where

$$C_j(\omega) := \{x \in \mathbb{R}^n \mid f_j(x,\omega) \le 0\},$$

for $j = 1, \dots, l$ and

$$C_j(\omega) := \{x \in \mathbb{R}^n \mid f_j(x, \omega) = 0\}$$

for $j = l + 1, \dots, m$. Applying Theorem A.17, we get

$$h^\infty(x,\omega) := \begin{cases} f_0^\infty(x,\omega) & \text{if } f_j^\infty(x,\omega) \le 0 \; j = 1 \dots, l, \\ & \quad f_j^\infty(x,\omega) = 0 \; j = l+1, \dots, m, \\ +\infty & \text{otherwise.} \end{cases}$$

Thus, the linearity condition in Assumption 2.85 implies that of Lemma 2.74 while the lower bound in Assumption 2.85 gives that in Lemma 2.74. □

Example 2.87 (Linear Programming) In Example 1.83, Assumption 2.85 means that problem (1.8) is feasible and

1. $\{x \in \mathcal{N} \mid c \cdot x \le 0, \; Ax \in K \; a.s.\}$ is linear,
2. there exist $p \in \mathcal{N}^\perp$, $m \in L^1$ and an $\epsilon > 0$ such that, almost surely,

$$c \cdot x \ge x \cdot p + \epsilon |x \cdot p| - m \quad a.s.$$

for all $x \in \mathbb{R}^n$ with $Ax - b \in K$.

The above are merely examples how the results of Sect. 2.2.2 can be used. In some applications, the conditions of Lemma 2.75 could be more convenient than Lemma 2.74 which was used above.

2.3.2 *Optimal Stopping*

Recall the relaxed optimal stopping problem (1.10)

$$\begin{aligned} &\text{maximize} && E\sum_{t=0}^T R_t x_t \quad \text{over } x \in \mathcal{N}, \\ &\text{subject to} && x \ge 0, \; \sum_{t=0}^T x_t \le 1 \quad a.s. \end{aligned}$$

from Sect. 1.3.2. This fits the general framework with

$$h(x,\omega) = \begin{cases} -\sum_{t=0}^T R_t(\omega) x_t & \text{if } x \ge 0 \text{ and } \sum_{t=0}^T x_t \le 1, \\ +\infty & \text{otherwise.} \end{cases}$$

Assume that $\max_t R_t \in L^1$ and let S be the *Snell envelope* of the reward process R, i.e. the adapted stochastic process given by

$$S_{T+1} := 0$$
$$S_t := \max\{R_t, E_t S_{t+1}\}.$$

The Snell envelope is a *supermartingale* in the sense that

$$E_t[\Delta S_{t+1}] \leq 0 \quad \forall t.$$

It is the smallest supermartingale that dominates the positive part R^+ of the reward process R. Indeed, let $\tilde{S}$ be another supermartingale that dominates R^+. Then $\tilde{S}_T \geq S_T$ and

$$\tilde{S}_t \geq \max\{R_t, E_t \tilde{S}_{t+1}\}$$

so $\tilde{S}_t \geq S_t$ for all t, by induction.

Theorem 2.88 *Assume that* $\max_t R_t \in L^1$. *The optimum value of* (1.10) *coincides for all* $t = 0, \dots, T$ *with that of*

$$\begin{aligned} &\underset{x^t \in \mathcal{N}^t}{\text{maximize}} && E\left[\sum_{s=0}^{t} R_s x_s + E_t[S_{t+1}](1 - \sum_{s=0}^{t} x_s)\right] \\ &\text{subject to} && x^t \geq 0,\ \sum_{s=0}^{t} x_s \leq 1 \quad a.s. \end{aligned}$$

In particular, the optimum value of (1.10) *is* ES_0. *An* $x \in \mathcal{N}$ *is optimal if and only if*

$$x_t \in \underset{x_t \in \mathbb{R}}{\operatorname{argmax}} \left\{ (R_t - E_t[S_{t+1}])x_t \;\middle|\; x_t \in [0, 1 - \sum_{s=0}^{t-1} x_s] \right\} \quad a.s. \quad t = 0, \dots, T.$$

Optimum values of (1.9) *and* (1.10) *coincide and* (1.9) *admits optimal solutions. A stopping time* $\tau \in \mathcal{T}$ *is optimal if and only if* $R_\tau = S_\tau$ *and the stopped process* $S^{(\tau)}$ *defined by*

$$S_t^{(\tau)} = \begin{cases} S_t & \text{if } t \leq \tau, \\ S_\tau & \text{if } t > \tau \end{cases}$$

is a martingale.

Proof The domain of h is contained in the unit simplex, so $h^\infty = \delta_{\{0\}}$ and, since $\max_{t=0,\dots,T} R_t \in L^1$, h is lower bounded. Thus, by Theorem 2.71, (BE) has a solution $(h_t)_{t=0}^T$. To prove the claims concerning (1.10), it suffices, by Theorem 2.2, to show that

$$\begin{aligned} h_t(x^t, \omega) = &\sum_{s=0}^{t} \left(-R_s(\omega)x_s + \delta_{\mathbb{R}_+}(x_s)\right) \\ &- E_t[S_{t+1}](\omega)(1 - \sum_{s=0}^{t} x_s) + \delta_{\mathbb{R}_+}(1 - \sum_{s=0}^{t} x_s). \end{aligned} \tag{2.23}$$

Since R is adapted, h is $\mathcal{F}_T$-measurable so $h_T = h$ and (2.23) holds for $t = T$ since $S_{T+1} = 0$. Assume that (2.23) holds for t. Since

$$\begin{aligned} &\sup_{x_t \in \mathbb{R}} \left\{ E_t[S_{t+1}](\omega)(1 - \sum_{s=0}^{t} x_s) \;\middle|\; 0 \le x_t \le 1 - \sum_{s=0}^{t-1} x_s \right\} \\ &= \sup_{x_t \in \mathbb{R}} \left\{ R_t(\omega)x_t + E_t[S_{t+1}](\omega)(1 - \sum_{s=0}^{t} x_s) \;\middle|\; 0 \le x_t \le 1 - \sum_{s=0}^{t-1} x_s \right\} \\ &= \max\{R_t(\omega), E_t[S_{t+1}](\omega)\}(1 - \sum_{s=0}^{t-1} x_s) + \delta_{\mathbb{R}_+}(1 - \sum_{s=0}^{t-1} x_s) \\ &= S_t(\omega)(1 - \sum_{s=0}^{t-1} x_s) + \delta_{\mathbb{R}_+}(1 - \sum_{s=0}^{t-1} x_s), \end{aligned}$$

we get

$$\begin{aligned} \tilde{h}_{t-1}(x_{t-1}, \omega) := &\inf_{x_t \in \mathbb{R}} h_t(x^t, \omega) \\ = &\sum_{s=0}^{t-1} (-R_s(\omega)x_s + \delta_{\mathbb{R}_+}(x_s)) \\ &- S_t(\omega)(1 - \sum_{s=0}^{t-1} x_s) + \delta_{\mathbb{R}_+}(1 - \sum_{s=0}^{t-1} x_s). \end{aligned}$$

By Example 2.7, the $\mathcal{F}_{t-1}$-conditional expectation of the second last term is $E_{t-1}[S_t](\omega)(\sum_{s=0}^{t-1} x_s - 1)$, so (2.23) holds for $t-1$. It is clear that the argmax over x_t always contains either 0 or $1 - \sum_{s=0}^{t-1} x_s$. Thus, if $\sum_{s=0}^{t-1} x_s$ takes values in $\{0, 1\}$, we can choose an optimal x_t such that $\sum_{s=0}^{t} x_s$ takes values in $\{0, 1\}$. We can thus construct recursively an optimal strategy taking values in $\{0, 1\}$. As observed in

Sect. 1.3.2, such strategies are in one-to-one correspondence with optimal stopping times.

An integer-valued feasible process $x \in \mathcal{N}$ satisfies the optimality condition if and only if $R_t = S_t$ when $x_t = 1$ and $R_t \leq E_t[S_{t+1}]$ for $t \leq \tau$. Here, τ is the stopping time corresponding to x, i.e. $\tau = t \iff x_t = 1$. By the definition of S, we have $R_t \leq E_t[S_{t+1}]$ for $t \leq \tau$ if and only if $S_t = E_t[S_{t+1}]$ for $t \leq \tau$. This means that $S^{(\tau)}$ is a martingale. □

A process $(R_t)_{t=0}^T$ is said to be a *Markov process* if, for all t, $(R_{t+1}, \dots, R_T)$ is R_t-conditionally independent of $\mathcal{F}_t$.

Remark 2.89 If $R \in L^1$ is a Markov process, then its Snell envelope S can be expressed as $S_t = \psi_t(R_t)$ for some measurable functions $\psi_t : \mathbb{R} \to \mathbb{R}$. In particular, the optimal stopping times $\tau \in \mathcal{T}$ are characterized by the condition $R_\tau = \psi_\tau(R_\tau)$.

Proof Clearly, $S_{T+1} := 0$ is of the required form. Assume that $S_{t+1} = \psi_{t+1}(R_{t+1})$ so that $S_t = \max\{R_t, E_t[\psi_{t+1}(R_{t+1})]\}$. By the conditional independence and Lemma B.7, $E_t[\psi_{t+1}(R_{t+1})] = E^{\sigma(R_t)}[\psi_{t+1}(R_{t+1})] = \hat{\psi}_t(R_t)$ for some measurable function $\hat{\psi}_t$. Defining $\psi_t(x) := \max\{x, \hat{\psi}_t(x)\}$, we get $S_t = \psi_t(R_t)$, so the claim follows by induction. □

2.3.3 Optimal Control

Consider the optimal control problem (1.11)

$$\begin{aligned} &\text{minimize} \quad && E\left[\sum_{t=0}^{T} L_t(X_t, U_t)\right] \quad \text{over } (X, U) \in \mathcal{N}, \\ &\text{subject to} \quad && \Delta X_t = A_t X_{t-1} + B_t U_{t-1} + W_t \quad t = 1, \dots, T \ a.s. \end{aligned}$$

from Sect. 1.3.3. The special structure in (1.11) allows us to express the solution $(h_t)_{t=0}^T$ of the general Bellman equations (BE) in terms of normal integrands $J_t : \mathbb{R}^N \times \Omega \to \overline{\mathbb{R}}$ and $I_t : \mathbb{R}^N \times \mathbb{R}^M \times \Omega \to \overline{\mathbb{R}}$ that solve the following dynamic programming equations

$$\begin{aligned} I_T &= 0, \\ J_t(X_t, \omega) &= \inf_{U_t \in \mathbb{R}^M} E_t(L_t + I_t)(X_t, U_t, \omega), \\ I_{t-1}(X_{t-1}, U_{t-1}, \omega) &= J_t(X_{t-1} + A_t(\omega) X_{t-1} + B_t(\omega) U_{t-1} + W_t(\omega), \omega). \end{aligned} \tag{2.24}$$

Note that J_t is a function only of X_t and ω while the functions h_t in the general Bellman equations (BE) may depend on the whole path x^t of x up to time t. The equations (2.31) are also known as *Bellman equations* and the functions J_t are often called the 'value functions', 'cost-to-go functions' or 'Bellman functions'. Theorem 2.91 below gives optimality conditions for (1.11) in terms of the functions J_t. Theorem 2.94 below will give sufficient conditions for the existence of solutions to (2.24).

Informally, (2.24) can be written in the more familiar form

$$\begin{aligned} J_T(X_T) &= \inf_{U_T \in \mathbb{R}^M} E_T[L_T(X_T, U_T)], \\ J_t(X_t) &= \inf_{U_t \in \mathbb{R}^M} E_t[L_t(X_t, U_t) + J_{t+1}(X_t + A_{t+1}X_t + B_{t+1}U_t + W_{t+1})], \end{aligned}$$

but this should only be taken as shorthand for (2.24) which is interpreted in terms of conditional expectations of normal integrands. The formulation in (2.24) allows us to use the results of the previous sections to establish the existence of the cost-to-go functions J_t. The shorter formulation without conditional expectations of normal integrands, leads to complicated measurability problems in general.

Compared to the general Bellman equations (BE) and (2.24) provides a significant dimension-reduction with respect to time as the infimand does not depend on the past states. This is often referred to as the "dynamic programming principle". The reduction is essentially due to the time-separable structure of (1.11) where, conditionally on the current state, the future is "independent" of the previous states. We will see a further dimension-reduction with respect to scenarios when the random elements in the problem exhibit certain form of independence; see Theorem 2.100 below. Under appropriate conditions, the functions J_t in (2.24) turn out to be deterministic.

Recall from Sect. 1.3.3 that (1.11) fits the general framework with $x = (X, U)$ and

$$h(x, \omega) = \sum_{t=0}^{T} L_t(X_t, U_t, \omega) + \sum_{t=1}^{T} \delta_{\{0\}}(\Delta X_t - A_t(\omega)X_{t-1} - B_t(\omega)U_{t-1} - W_t(\omega)).$$

When applied to (1.11), Theorem 2.70 yields the dynamic programming principle in Theorem 2.91 below. In addition to $p \in \mathcal{N}^{\perp}$, the assumptions involve "co-state" variables $y_t \in L^1(\mathbb{R}^N)$ which will feature in the duality theory in Sect. 3.4.3. To simplify the notation, we define $y_0 := y_{T+1} := 0$ and $X_{T+1} := 0$.

Assumption 2.90 *Problem* (1.11) *is feasible and there exist $p \in \mathcal{N}^\perp$, $y_t \in L^1$ and $m_t \in L^1$ such that $A_t^* y_t$, $B_t^* y_t$ and $W_t \cdot y_t$ are integrable for all t and that*

1. *for almost every $\omega \in \Omega$ and every $(X, U) \in \mathbb{R}^n$ satisfying*

$$\Delta X_t = A_t(\omega)X_{t-1} + B_t(\omega)U_{t-1} + W_t(\omega) \quad \forall t = 1, \dots, T,$$

we have

$$L_t(X_t, U_t, \omega) \geq (X_t, U_t) \cdot p_t(\omega) - X_t \cdot \Delta y_{t+1}(\omega) - \Delta X_{t+1} \cdot y_{t+1}(\omega) - m_t(\omega)$$

for all $t = 0, \dots, T$,

2. *for all $(X, U) \in \mathcal{N}$ satisfying the system equations,*

$$E\left[\sum_{t=0}^{T} L_t(X_t, U_t)\right] = E\left[\sum_{t=0}^{T} [L_t(X_t, U_t) - (X_t, U_t) \cdot p_t]\right].$$

If the cost functions L_t are lower bounded, then Assumption 2.90 holds with $p = 0$ and $y = 0$ as soon as the problem is feasible.

Theorem 2.91 *Let $(I_t, J_t)_{t=0}^T$ be an L-bounded solution of* (2.24) *and let Assumption 2.90 hold. Then the optimum value of the optimal control problem coincides with that of*

$$\begin{aligned} &\text{minimize} && E\left[\sum_{s=0}^{t-1}(E_t L_s)(X_s, U_s) + J_t(X_t)\right] \quad \text{over } (X^t, U^t) \in \mathcal{N}^t, \\ &\text{subject to} && \Delta X_s = A_s X_{s-1} + B_s U_{s-1} + W_s \quad s = 1, \dots, t \text{ a.s} \end{aligned}$$

for all $t = 0, \dots, T$ and, moreover, an $(\bar{X}, \bar{U}) \in \mathcal{N}$ solves (1.11) *if and only if it satisfies the system equations and*

$$\bar{U}_t \in \operatorname*{argmin}_{U_t \in \mathbb{R}^M} E_t(L_t + I_t)(\bar{X}_t, U_t) \text{ a.s.}$$

for all $t = 0, \dots, T$. If the measurable mappings

$$M_t(\omega) := \{U_t \in \mathbb{R}^M \mid (E_t(L_t + I_t))^\infty(0, U_t, \omega) \leq 0\}$$

are linear-valued for all $t = 0, \dots, T$, then there exists an optimal control U with $U_t \in M_t^\perp$ almost surely.

Proof Let $k(x, \omega) := h(x, \omega) - x \cdot p(\omega)$. Assumption 2.90.2 means that $Ek = Eh$ on $\mathcal{N}$. Summing up the lower bounds in Assumption 2.90.1 shows that k is lower bounded. All the claims follow from Theorem 2.70 once we show that the sequence of normal integrands $(h_t)_{t=0}^T$ given by

$$
\begin{aligned}
h_t(x^t) := &\sum_{s=0}^{t-1}(E_t L_s)(X_s, U_s) + E_t(L_t + I_t)(X_t, U_t) \\
&+ \sum_{s=1}^{t} \delta_{\{0\}}(-\Delta X_s + A_s X_{s-1} + B_s U_{s-1} + W_s)
\end{aligned}
\tag{2.25}
$$

is an L-bounded solution of (BE). (For brevity, we omit the ω from the notation.) Indeed, by Theorem 2.70, the optimum value of (1.11) can then be expressed as

$$
\begin{aligned}
\inf_{x^t \in \mathcal{N}^t} Eh_t(x^t) &= \inf_{x^t \in \mathcal{N}^t} \{E[\sum_{s=0}^{t-1}(E_t L_s)(X_s, U_s, \omega) + E_t(L_t + I_t)(X_t, U_t)] \mid \\
&\qquad \Delta X_s = A_s X_{s-1} + B_s U_{s-1} + W_s \; s = 1, \dots, t\} \\
&= \inf_{x^t \in \mathcal{N}^t} \{E[\sum_{s=0}^{t-1}(E_t L_s)(X_s, U_s, \omega) + J_t(X_t)] \mid \\
&\qquad \Delta X_s = A_s X_{s-1} + B_s U_{s-1} + W_s \; s = 1, \dots, t\},
\end{aligned}
$$

where the second equality holds by the interchange rule in Theorem 1.64, Also, if (2.25) holds, then by Theorem 2.70, an $(\bar{X}, \bar{U}) \in \mathcal{N}$ solves (1.11) if and only if $(\bar{X}_t, \bar{U}_t)$ minimizes (2.25) almost surely. By Theorem A.17, (2.25) implies

$$
\begin{aligned}
h_t^\infty(x^t) := &\sum_{s=0}^{t-1}(E_t L_s)^\infty(X_s, U_s) + E_t(L_t + I_t)^\infty(X_t, U_t) \\
&+ \sum_{s=1}^{t} \delta_{\{0\}}(-\Delta X_s + A_s X_{s-1} + B_s U_{s-1})
\end{aligned}
$$

so

$$
N_t(\omega) = \{x_t \in \mathbb{R}^{N+M} \mid h_t^\infty(0, x_t, \omega) \le 0\} = \{(X_t, U_t) \mid X_t = 0, \; U_t \in M_t(\omega)\}
$$

and the last claim thus follows from Theorem 2.70 as well.

Assume that (2.25) solve (BE) from t onward. Then

$$\begin{aligned}
\tilde{h}_{t-1}(x^{t-1}) &= \inf_{x_t \in \mathbb{R}^{n_t}} h_t(x^{t-1}, x_t) \\
&= \sum_{s=0}^{t-1} (E_t L_s)(X_s, U_s) + \inf_{(X_t, U_t)} \{E_t(L_t + I_t)(X_t, U_t) \mid \\
&\qquad\qquad X_t = X_{t-1} + A_t X_{t-1} + B_t U_{t-1} + W_t\} \\
&\quad + \sum_{s=1}^{t-1} \delta_{\{0\}}(-\Delta X_s + A_s X_{s-1} + B_s U_{s-1} + W_s) \\
&= \sum_{s=0}^{t-1} (E_t L_s)(X_s, U_s) + J_t(X_{t-1} + A_t X_{t-1} + B_t U_{t-1} + W_t) \\
&\quad + \sum_{s=1}^{t-1} \delta_{\{0\}}(-\Delta X_s + A_s X_{s-1} + B_s U_{s-1} + W_s).
\end{aligned}$$

Thus, by the definitions of I and J,

$$\begin{aligned}
\tilde{h}_{t-1}(x^{t-1}) &= \sum_{s=0}^{t-1} (E_t L_s)(X_s, U_s) + I_{t-1}(X_{t-1}, U_{t-1}) \\
&\quad + \sum_{s=1}^{t-1} \delta_{\{0\}}(-\Delta X_s + A_s X_{s-1} + B_s U_{s-1} + W_s). \qquad (2.26)
\end{aligned}$$

Since I_{t-1} and L_{t-1} are L-bounded by assumption, Theorems 2.15 and 2.13 imply that

$$\begin{aligned}
h_{t-1}(x^{t-1}) &= E_{t-1}\tilde{h}_{t-1}(x^{t-1}) \\
&= \sum_{s=0}^{t-2} (E_{t-1} L_s)(X_s, U_s) + E_{t-1}(L_{t-1} + I_{t-1})(X_{t-1}, U_{t-1}) \\
&\quad + \sum_{s=1}^{t-1} \delta_{\{0\}}(-\Delta X_s + A_s X_{s-1} + B_s U_{s-1} + W_s)
\end{aligned}$$

and that h_{t-1} is L-bounded. Thus, the functions given by (2.25) are L-bounded and solve (BE) from $t - 1$ onward. It now suffices to note that, by Theorems 2.15 and 2.13 again, the function h_T given by (2.25) is L-bounded and solves (BE) from $t = T$ onward. □

The following gives a dual formulation of the lower bounds in Assumption 2.90.1 in terms of the scenariowise conjugates of L_t.

Remark 2.92 Assumption 2.90.1 hold if and only if

$$E[L_t^*(p_t - (\Delta y_{t+1} + A_{t+1}^* y_{t+1}, B_{t+1}^* y_{t+1})) - W_{t+1} \cdot y_{t+1}] < \infty,$$

for all t. Here $y_0 := y_{T+1} := 0$, $A_{T+1} := 0$ and $B_{T+1} := 0$. The assumption will be analyzed further in Sect. 3.4, where p and y appear in a problem dual to (1.11).

Proof Given a feasible (X, U), Fenchel's inequality gives

$$\begin{aligned}
&L_t(X_t, U_t) + L_t^*(p_t - (\Delta y_{t+1} + A_{t+1}^* y_{t+1}, B_{t+1}^* y_{t+1})) - W_{t+1} \cdot y_{t+1} \\
&\geq (X_t, U_t) \cdot p_t - X_t \cdot (\Delta y_{t+1} + A_{t+1}^* y_{t+1}) - U_t \cdot (B_{t+1}^* y_{t+1}) - W_{t+1} \cdot y_{t+1} \\
&= (X_t, U_t) \cdot p_t + X_t \cdot y_t - (X_t + A_{t+1} X_t + B_{t+1} U_t + W_{t+1}) \cdot y_{t+1} \\
&= (X_t, U_t) \cdot p_t + X_t \cdot y_t - (X_{t+1}) \cdot y_{t+1} \\
&= (X_t, U_t) \cdot p_t - X_t \cdot \Delta y_{t+1} - \Delta X_{t+1} \cdot y_{t+1},
\end{aligned}$$

which gives the equivalence. □

We will establish the existence of solutions to the Bellman equations (2.24) under the following.

Assumption 2.93 *There exist $p \in \mathcal{N}^\perp$, $y_t \in L^1$ and $m_t \in L^1$ such that $A_t^* y_t$, $B_t^* y_t$ and $W_t \cdot y_t$ are integrable,*

$$E_t[A_t^* y_t] = A_t^* E_t y_t, \quad E_t[B_t^* E_t y_t] = B_t^* E_t y_t, \quad E_t[W_t \cdot y] = W_t \cdot E_t y_t$$

for all t and

1. *for almost every $\omega \in \Omega$ and every $(X, U) \in \mathbb{R}^n$ satisfying*

   $$\Delta X_t = A_t(\omega) X_{t-1} + B_t(\omega) U_{t-1} + W_t(\omega) \quad \forall t = 1, \dots, T,$$

   we have

   $$L_t(X_t, U_t, \omega) \geq (X_t, U_t) \cdot p_t(\omega) - X_t \cdot \Delta y_{t+1}(\omega) - \Delta X_{t+1} \cdot y_{t+1}(\omega) - m_t(\omega)$$

   for all $t = 0, \dots, T$,
2. *the set*

   $$\{(X, U) \in \mathcal{N} \mid \sum_{t=0}^{T} [L_t^\infty(X_t, U_t) - p_t \cdot (X_t, U_t)] \leq 0, \qquad \Delta X_t = A_t X_{t-1} + B_t U_{t-1} \ \forall t \ a.s.\}$$

   is linear.

Condition 1 in Assumption 2.93 is the same as that in Assumption 2.90. The integrability conditions holds automatically under natural conditions to be given in Lemma 3.7. Theorem 2.71 yields the following existence result for the dynamic programming equations (2.24).

Theorem 2.94 *Under Assumption 2.93, the recursion* (2.24) *has a unique solution* $(J_t, I_t)_{t=0}^T$ *of L-bounded convex normal integrands, and*

$$M_t(\omega) := \{U_t \in \mathbb{R}^M \mid (E_t(L_t + I_t))^\infty(0, U_t, \omega) \leq 0\}$$

is linear-valued for all t.

Proof Summing up the lower bounds in Assumption 2.93.1 shows that $k(x, \omega) := h(x, \omega) - x \cdot p(\omega)$ is lower bounded. By Theorem A.17

$$h^\infty(x, \omega) = \sum_{t=0}^T L_t^\infty(X_t, U_t, \omega) + \sum_{t=1}^T \delta_{\{0\}}(\Delta X_t - A_t(\omega)X_{t-1} - B_t(\omega)U_{t-1}).$$

Thus, Assumption 2.93.2 means that $\{x \in \mathcal{N} \mid k^\infty(x) \leq 0\ a.s.\}$ is linear. By Theorem 2.71, (BE) has a unique solution $(h_t)_{t=0}^T$ for h and

$$N_t(\omega) := \{x_t \in \mathbb{R}^{n_t} \mid h_t^\infty(0, x^t, \omega) \leq 0\}$$

is linear-valued for all t.

Assume that $(I_{t'}, J_{t'})_{t'=t+1}^T$ are normal integrands satisfying (2.24) from time $t+1$ onward, and that (for brevity, we omit ω from the notation)

$$\tilde{h}_t(x^t) = \sum_{s=0}^t (E_{t+1}L_s)(X_s, U_s) + I_t(X_t, U_t)$$

$$+ \sum_{s=1}^t \delta_{\{0\}}(\Delta X_s - A_s X_{s-1} + B_s U_{s-1} + W_s),$$

$$J_{t+1}(X_{t+1}) \geq X_{t+1} \cdot E_{t+1} y_{t+1} - \sum_{t'=t+1}^T E_{t+1} m_{t'}.$$

Here E_{T+1} is defined as the identity mapping so the above hold for $t = T$. The lower bound gives

$$I_t(X_t, U_t) = J_{t+1}(X_t + A_{t+1}X_t + B_{t+1}U_t + W_{t+1})$$

$$\geq (X_t + A_{t+1}X_t + B_{t+1}U_t + W_{t+1}) \cdot E_{t+1} y_{t+1} - \sum_{t'=t+1}^T E_{t+1} m_{t'}$$

$$= X_t \cdot E_{t+1} y_{t+1} + X_t \cdot A^*_{t+1} E_{t+1} y_{t+1} + U_t \cdot B^*_{t+1} E_{t+1} y_{t+1}$$
$$+ W_{t+1} \cdot E_{t+1} y_{t+1} - \sum_{t'=t+1}^{T} E_{t+1} m_{t'}. \tag{2.27}$$

In particular, I_t is L-bounded as well. Under Assumption 2.93.1, all the terms in the expression of $\tilde{h}_t$ above are L-bounded, by Theorem 2.13. Thus, by Theorem 2.15 and Theorem 2.19,

$$h_t(x^t) = (E_t \tilde{h}_t)(x^t)$$
$$= \sum_{s=0}^{t-1} (E_t L_s)(X_s, U_s) + E_t(L_t + I_t)(X_t, U_t)$$
$$+ \sum_{s=1}^{t} \delta_{\{0\}}(\Delta X_s - A_s X_{s-1} + B_s U_{s-1} + W_s).$$

In particular,

$$N_t(\omega) = \{x_t \in \mathbb{R}^{n_t} \mid h_t^\infty(0, x_t, \omega) \le 0\}$$
$$= \{(X_t, U_t) \in \mathbb{R}^{n_t} \mid X_t = 0,\ (E_t(L_t + I_t))^\infty(0, U_t, \omega) \le 0\},$$

which is linear by Theorem 2.71. Thus, by Theorems 1.40, 1.42, and 1.45, the functions

$$J_t(X_t) := \inf_{U_t \in \mathbb{R}^M} E_t(L_t + I_t)(X_t, U_t)$$

and

$$I_{t-1}(X_{t-1}, U_{t-1}) := J_t(X_{t-1} + A_t X_{t-1} + B_t U_{t-1} + W_t)$$

are convex normal integrands. As in (2.26),

$$\tilde{h}_{t-1}(x^{t-1}) = \sum_{s=0}^{t-1} (E_t L_s)(X_s, U_s) + I_{t-1}(X_{t-1}, U_{t-1})$$
$$+ \sum_{s=1}^{t-1} \delta_{\{0\}}(-\Delta X_s + A_s X_{s-1} + B_s U_{s-1} + W_s).$$

By Lemma 2.11 and Example 2.7, Assumption 2.93 implies

$$\begin{aligned}(E_{t+1}L_t)(X_t,U_t) \geq (X_t,U_t)\cdot E_{t+1}p_t - X_t\cdot E_{t+1}[\Delta y_{t+1}]\\ - X_t\cdot A^*_{t+1}E_{t+1}y_{t+1} - U_t\cdot B^*_{t+1}E_{t+1}y_{t+1}\\ - W_{t+1}\cdot E_{t+1}y_{t+1} - E_{t+1}m_t.\end{aligned}$$

Combining this with (2.27) gives

$$(E_{t+1}L_t + I_t)(X_t,U_t) \geq (X_t,U_t)\cdot E_{t+1}p_t + X_t\cdot E_{t+1}y_t - \sum_{t'=t}^{T} E_{t+1}m_{t'}.$$

By Lemma 2.11, Theorems 2.15, 2.19 and Example 2.7,

$$E_t(L_t + I_t)(X_t,U_t) \geq X_t\cdot E_t y_t - \sum_{t'=t}^{T} E_t m_{t'}$$

and thus,

$$J_t(X_t) \geq X_t\cdot E_t y_t - \sum_{t'=t}^{T} E_t m_{t'}.$$

The claim thus follows by induction on t. □

The following two examples give sufficient conditions for Assumptions 2.90 and 2.93. The first one corresponds to Lemma 2.74 and the second one to Lemma 2.75. As in Assumptions 2.90 and 2.93, we define $y_0 := y_{T+1} := 0$ and $X_{T+1} := 0$.

Example 2.95 Assumptions 2.90 and 2.93 hold if (1.11) is feasible and

1. the set

$$\{(X,U)\in\mathcal{N} \mid \sum_{t=0}^{T} L_t^\infty(X_t,U_t) \leq 0,\ \Delta X_t = A_t X_{t-1} + B_t U_{t-1}\ \forall t\ a.s.\}$$

   is linear,
2. there exist $p\in\mathcal{N}^\perp$ and $\epsilon>0$ such that, for every $\lambda\in[1+\epsilon,1-\epsilon]$ and $t=0,\ldots,T$, there exist $y_t\in L^1$ and $m_t\in L^1$ such that $A_t^* y_t$, $B_t^* y_t$ and $W_t\cdot y_t$ are integrable,

$$E_t[A_t^* y_t] = A_t^* E_t y_t,\quad E_t[B_t^* E_t y_t] = B_t^* E_t y_t,\quad E_t[W_t\cdot y] = W_t\cdot E_t y_t$$

and, for almost every $\omega \in \Omega$ and every $(X, U) \in \mathbb{R}^n$ satisfying

$$\Delta X_t = A_t(\omega)X_{t-1} + B_t(\omega)U_{t-1} + W_t(\omega) \quad \forall t = 1, \dots, T,$$

we have

$$L_t(X_t, U_t, \omega) \geq \lambda(X_t, U_t) \cdot p_t(\omega) - X_t \cdot \Delta y_{t+1}(\omega) - \Delta X_{t+1} \cdot y_{t+1}(\omega) - m_t(\omega)$$

for every $t = 0, \dots, T$.

Proof Choosing $\lambda = 1$ in condition 2 here implies Assumptions 2.90.1 and 2.93.1. Summing up the lower bounds in condition 2, we see that assumptions of Lemma 2.74 are satisfied. Thus, Assumption 2.72 holds. Clearly, Assumption 2.72.3 is the same as Assumption 2.90.2 while Assumption 2.72.2 is the same as Assumption 2.93.2. □

Example 2.96 Assumptions 2.90 and 2.93 hold if (1.11) is feasible and there exist $p \in \mathcal{N}^\perp$, $y_t \in L^1$ and $m_t \in L^1$ such that $A_t^* y_t$, $B_t^* y_t$ and $W_t \cdot y_t$ are integrable,

$$E_t[A_t^* y_t] = A_t^* E_t y_t, \quad E_t[B_t^* E_t y_t] = B_t^* E_t y_t, \quad E_t[W_t \cdot y] = W_t \cdot E_t y_t$$

and

1. for almost every $\omega \in \Omega$ and every $(X, U) \in \mathbb{R}^n$ satisfying

$$\Delta X_t = A_t(\omega)X_{t-1} + B_t(\omega)U_{t-1} + W_t(\omega) \quad \forall t = 1, \dots, T,$$

we have

$$L_t(X_t, U_t, \omega) \geq (X_t, U_t) \cdot p_t(\omega) - X_t \cdot \Delta y_{t+1}(\omega) - \Delta X_{t+1} \cdot y_{t+1}(\omega) - m_t(\omega)$$

for all $t = 0, \dots, T$,

2. the set

$$\{(X, U) \in \mathcal{N} \mid \sum_{t=0}^{T} [L_t^\infty(X_t, U_t) - (X_t, U_t) \cdot p_t] \leq 0,$$
$$\Delta X_t = A_t X_{t-1} + B_t U_{t-1} \; \forall t \; a.s.\}$$

is linear,

3. $\{(X, U) \in \mathcal{N} \mid (X_t, U_t) \in \operatorname{dom} L_t \; a.s.\} \subseteq \operatorname{dom} EL_t$.

Proof It suffices to show that Assumption 2.90.2 is satisfied. Summing up the lower bounds in 1, we get

$$\sum_{t=0}^{T} L_t(X_t, U_t) \geq (X \cdot U) \cdot p - \sum_{t=0}^{T} m_t \tag{2.28}$$

for every $(X, U) \in \mathcal{N}$ satisfying the system equations. If either

$$E[\sum_{t=0}^{T} L_t(X_t, U_t)] < \infty \quad \text{or} \quad E[\sum_{t=0}^{T} L_t(X_t, U_t) - (X, U) \cdot p] < \infty,$$

then $(X_t, U_t) \in \operatorname{dom} L_t$ almost surely, so condition 3 implies

$$E[\sum_{t=0}^{T} L_t(X_t, U_t)] < \infty.$$

Then (2.28) and Lemma 2.67 imply $E[(X, U) \cdot p] = 0$ so Assumption 2.90.2 holds, by Lemma B.3. □

The following characterizes induced constraints and gives sufficient conditions for problem (1.11) to have relatively complete recourse; see Sect. 2.2.3. We will denote the $\mathcal{F}_t$-conditional essential infimum and supremum of a measurable mapping S by $\operatorname{essinf}_t S$ and $\operatorname{esssup}_t S$, respectively; see Sect. 2.1.4.

Example 2.97 (Induced Constraints) Assume that

$$\begin{aligned} \operatorname{cl} \operatorname{dom} h(\cdot, \omega) = \{(X, U) \in \mathbb{R}^{(N+M)(T+1)} \mid & (X_t, U_t) \in \operatorname{cl} \operatorname{dom} L_t(\cdot, \cdot, \omega), \\ & \Delta X_t = A_t(\omega) X_{t-1} + B_t(\omega) U_{t-1} + W_t(\omega) \ \forall t = 1, \ldots, T\}, \end{aligned}$$

$\operatorname{dom} L_t$ is $\mathcal{F}_t$-measurable for all t and that the recursion

$$\begin{aligned} \hat{S}_T &:= \operatorname{cl} \operatorname{dom} L_T, \\ \tilde{S}_t(\omega) &:= \{(X_t, U_t) \in \operatorname{cl} \operatorname{dom} L_t(\cdot, \cdot, \omega) \mid \exists U_{t+1} \in \mathbb{R}^M : \\ &\qquad (X_t + A_{t+1}(\omega) X_t + B_{t+1}(\omega) U_t + W_{t+1}(\omega), U_{t+1}) \in \hat{S}_{t+1}(\omega)\}, \\ \hat{S}_t &:= \operatorname{essinf}_t \tilde{S}_t \end{aligned}$$

is well-defined and each $\tilde{S}_t$ is closed-valued. Then the induced constraints are given by

$$\begin{aligned} D_t(\omega) := \{(X^t, U^t) \in \mathbb{R}^{(N+M)(t+1)} \mid & \\ & (X_{t'}, U_{t'}) \in \operatorname{cl} \operatorname{dom} L_{t'}(\cdot, \cdot, \omega) \ \forall t' < t, \ (X_t, U_t) \in \hat{S}_t(\omega), \\ & \Delta X_{t'} = A_{t'}(\omega) X_{t'-1} + B_{t'}(\omega) U_{t'-1} + W_{t'}(\omega) \ \forall t' = 1, \ldots, t\}. \end{aligned}$$

Problem (1.11) has relatively complete recourse if and only if $\tilde{S}_t$ is $\mathcal{F}_t$-measurable for every t. This happens, in particular, when the random set

$$\mathrm{dom}_{X_t} L_t(\omega) := \{X_t \in \mathbb{R}^N \mid \exists U_t \in \mathbb{R}^M : (X_t, U_t) \in (\mathrm{dom}\, L_t)(\omega)\}$$

equals $\mathbb{R}^N$ for almost every ω and every t.

Proof The first claim follows from Theorem 2.47 and induction on t. We have that $\tilde{S}_t$ is $\mathcal{F}_t$-measurable if and only if $D_t = D^t_{t+1}$. The latter holds for every t if and only $D_t = (\mathrm{cl}\,\mathrm{dom}\, h)^t$ for every t. Since each D_t is closed-valued, this means that the problem has relatively complete recourse. When $\mathrm{dom}_{X_t} L_t = \mathbb{R}^N$ almost surely for every t, an induction on t shows that $\tilde{S}_t = \mathrm{cl}\,\mathrm{dom}\, L_t$ almost surely for every t. □

The dynamic programming equations (2.24) of (1.11) can be formulated in many equivalent ways. If $(J_t, I_t)_{t=0}^T$ satisfy (2.24), then the functions

$$Q_t := E_t(L_t + I_t)$$

satisfy

$$\begin{aligned} I_T &= 0, \\ Q_t &= E_t(L_t + I_t), \\ I_{t-1}(X_{t-1}, U_{t-1}) &= \inf_{U_t \in \mathbb{R}^M} Q_t(X_{t-1} + A_t X_{t-1} + B_t U_{t-1} + W_t, U_t) \end{aligned} \tag{2.29}$$

and the optimality condition can be written as

$$U_t \in \operatorname*{argmin}_{U_t \in \mathbb{R}^M} Q_t(X_t, U_t).$$

Conversely, if Q_t and I_t satisfy (2.29) and the functions

$$J_t(X_t) := \inf_{U_t \in \mathbb{R}^M} Q_t(X_t, U_t)$$

are normal integrands, then J_t and I_t satisfy (2.24).

The mapping $S_t : \mathbb{R}^N \times \Omega \rightrightarrows \mathbb{R}^M$ defined by

$$S_t(X_t, \omega) := \operatorname*{argmin}_{U_t \in \mathbb{R}^M} Q_t(X_t, U_t, \omega)$$

is called the *optimal feedback mapping*. Under the assumptions of Theorem 2.91, the optimal controls $U = (U_t)_{t=0}^T$ and the corresponding states $X = (X_t)_{t=0}^T$ are the adapted processes with $U_t \in S_t(X_t)$ almost surely. The values of $S_t(X_t)$ depend on the past states and controls only through the current state.

Under the assumptions of Theorem 2.94, (2.29) admits solutions $(Q_t, I_t)_{t=0}^T$ where the functions Q_t are L-bounded and convex and, in particular, S_t are closed convex-valued. By Theorem 1.20 and Corollary 1.23, $\omega \mapsto S_t(X_t(\omega), \omega)$ is closed-valued and $\mathcal{F}_t$-measurable for any $\mathcal{F}_t$-measurable X_t. Since real-valued convex functions are continuous, Theorem 1.21 gives the following.

Remark 2.98 If J_t is real-valued, then $\omega \mapsto \operatorname{gph} S_t(\cdot, \omega)$ is closed-valued and $\mathcal{F}_t$-measurable. In particular, if $J_t(\cdot, \omega)$ is continuous and $S_t(\cdot, \omega)$ is single-valued and locally bounded, then $S_t(\cdot, \omega)$ is continuous.

Under p-homogeneity (see Lemma 2.22), the feedback mappings are homogeneous.

Remark 2.99 Assume that $W = 0$, each L_t is L-bounded and p-homogeneous with integrable C and that $(J_t, I_t)_{t=0}^T$ is an L-bounded solution of (2.24). Then J_t, I_t and Q_t are p-homogeneous and

$$S_t(\alpha X_t) = \alpha S_t(X_t) \quad \forall \alpha > 0.$$

If L_t are p-homogeneous merely for $t \geq t'$, then Q_t are p-homogeneous for $t \geq t'$.

Proof The functions J_{T+1} and I_{T+1} are trivially p-homogeneous. If J_{t+1} is p-homogeneous, then by Lemma 2.22 and Lemma A.86, J_t, I_t and Q_t are p-homogeneous and S_t is homogeneous. The claim thus follows by induction. □

If the problem data exhibits an appropriate form of independence, the cost-to-go functions become deterministic. The following gives more general conditions which imply specific measurability properties on the cost-to-go functions J_t.

Theorem 2.100 (Conditional Independence) *Assume that each L_t is L-bounded, $(J_t, I_t)_{t=0}^T$ is an L-bounded solution of (2.24) and that there are σ-algebras $\mathcal{H}_t \subseteq \mathcal{F}_t$ such that, for all t,*

1. *L_t is $\mathcal{H}_t$-conditionally independent of $\mathcal{F}_t$,*
2. *A_{t+1}, B_{t+1} and W_{t+1} are $\mathcal{H}_{t+1} \vee \mathcal{H}_t$-conditionally independent of $\mathcal{F}_t$,*
3. *$\mathcal{H}_{t+1}$ is $\mathcal{H}_t$-conditionally independent of $\mathcal{F}_t$.*

Then Q_t and S_t are $\mathcal{H}_t$-measurable.

Proof Assume that J_{t+1} is $\mathcal{H}_{t+1}$-measurable. Then

$$I_t(X_t, U_t) = J_{t+1}(X_t + A_{t+1}X_t + B_{t+1}U_t + W_{t+1})$$

is $\sigma(A_{t+1}, B_{t+1}, W_{t+1}) \vee \mathcal{H}_{t+1}$-measurable. Applying Lemma B.17 with $\mathcal{G}_0 = \mathcal{H}_t$, $\mathcal{G}_1 = \mathcal{H}_{t+1}$, $\mathcal{G}_2 = \sigma(A_{t+1}, B_{t+1}, W_{t+1})$ and $\mathcal{G}' = \mathcal{F}_t$, we find that the assumptions imply that I_t is $\mathcal{H}_t$-conditionally independent of $\mathcal{F}_t$. By Theorem 2.21,

$$E_t(L_t + I_t) = E^{\mathcal{H}_t}(L_t + I_t),$$

so J_t is $\mathcal{H}_t$ measurable, by Theorem 1.40. Similarly, by Theorems 2.21 and 1.40, condition 1 implies that J_T is $\mathcal{H}_T$-measurable, so the claims follow by induction on t. □

When $\mathcal{H}_t = \{\Omega, \emptyset\}$, Theorem 2.100 gives the following.

Corollary 2.101 *Assume that L_t are L-bounded, that $(J_t, I_t)_{t=0}^T$ is an L-bounded solution of* (2.24) *and that, for all t, L_t, A_{t+1}, B_{t+1} and W_{t+1} are independent of $\mathcal{F}_t$. Then Q_t and S_t are deterministic.*

If the uncertainty is represented by a Markov process ξ and the problem data depends appropriately on ξ, then the cost-to-go functions only depend on the current values of the system state and the Markov process. Recall that a *Markov process* is an adapted process $\xi = (\xi_t)_{t=0}^T$ such that ξ_{t+1} is ξ_t-conditionally independent of $\mathcal{F}_t$.

Example 2.102 (Canonical Representation) Let $\xi = (\xi_t)_{t=0}^{T+1}$ be a stochastic process as in Remark 1.76 and consider the optimal control problem with

$$A_t(\omega) = \hat{A}_t(\xi^t(\omega)),\ \ B_t(\omega) = \hat{B}_t(\xi^t(\omega)),\ \ W_t(\omega) = \hat{W}_t(\xi^t(\omega))$$

and

$$L_t(X_t, U_t, \omega) = \hat{L}_t(X_t, U_t, \xi^{t+1}(\omega)),$$

where $\hat{L}_t$ is an $\mathcal{A}^{t+1}$-measurable normal integrand and $\hat{A}_t$, $\hat{B}_t$ and $\hat{W}$ are $\mathcal{A}^t$-measurable. Assume that each L_t is L-bounded and that ξ_{t+1} has regular ξ^t-conditional distribution ν_t. Then sequences $(J_t, I_t)_{t=0}^T$ of L-bounded normal integrands satisfy (2.24) if and only if there exist $\mathcal{A}^t$-measurable normal integrands $\hat{J}_t : \mathbb{R}^N \times \Xi^t \to \overline{\mathbb{R}}$ and $\mathcal{A}^{t+1}$-measurable normal integrands $\hat{I}_t : \mathbb{R}^N \times \mathbb{R}^M \times \Xi^{t+1} \to \overline{\mathbb{R}}$ such that

$$J_t(X_t, \omega) = \hat{J}_t(X_t, \xi^t(\omega)), \quad I_t(X_t, U_t, \omega) = \hat{I}_t(X_t, U_t, \xi^{t+1}(\omega)),$$

and

$$\begin{aligned} \hat{I}_T &= 0, \\ \hat{J}_t(X_t, s^t) &= \inf_{U_t \in \mathbb{R}^M} E^{\nu_t}(\hat{L}_t + \hat{I}_t)(X_t, U_t, s^t), \\ \hat{I}_{t-1}(X_{t-1}, U_{t-1}, s^t) &= \hat{J}_t(X_{t-1} + \hat{A}_t(s^t)X_{t-1} + \hat{B}_t(s^t)U_{t-1} + \hat{W}_t(s^t), s^t) \end{aligned} \tag{2.30}$$

P_{ξ^t}-almost surely for all $t = T, \ldots, 0$.

If $(\xi_t)_{t=0}^T$ is Markov and $\hat{A}_{t+1}$, $\hat{B}_{t+1}$, $\hat{W}_{t+1}$ and $\hat{L}_t$ are independent of s^{t-1}, then Q_t and S_t are $\sigma(\xi_t)$-measurable. If $(\xi_t)_{t=0}^T$ is a sequence of independent random variables and $\hat{A}_{t+1}$, $\hat{B}_{t+1}$, $\hat{W}_{t+1}$ and $\hat{L}_t$ are independent of s^t, then Q_t and S_t are deterministic.

Proof Assume first that $(J_t, I_t)_{t=0}^T$ satisfies (2.24). By Corollary 1.34, there exist, $\mathcal{A}^t$-measurable normal integrands $\hat{J}_t : \mathbb{R}^N \times \Xi^t \to \overline{\mathbb{R}}$ and $\mathcal{A}^{t+1}$-measurable normal integrands $\hat{I}_t : \mathbb{R}^N \times \mathbb{R}^M \times \Xi^{t+1} \to \overline{\mathbb{R}}$ such that

$$J_t(X_t, \omega) = \hat{J}_t(X_t, \xi^t(\omega)), \quad I_t(X_t, U_t, \omega) = \hat{I}_t(X_t, U_t, \xi^{t+1}(\omega)).$$

By assumption,

$$(L_t + I_t)(X_t, U_t, \omega) = (\hat{L}_t + \hat{I}_t)(X_t, U_t, \xi^{t+1}(\omega)).$$

Applying Corollary 2.60, with $\xi_1 = \xi^t$ and $\xi_2 = \xi_{t+1}$ gives

$$(E^{\sigma(\xi^t)}(L_t + I_t))(X_t, U_t, \omega) = E^{\nu_t}(\hat{L}_t + \hat{I}_t)(X_t, U_t, \xi^t(\omega))$$

almost surely everywhere, which gives (2.30). Reversing the argument gives sufficiency.

The second last claim follows from Theorem 2.100 with $\mathcal{H}_t = \sigma(\xi_t)$. Indeed, condition 2 in Theorem 2.100 is implied by the Markov property of ξ while condition 1 follows from the fact that conditional independence is preserved in compositions; see Example 2.20. The last claim follows similarly with $\mathcal{H}_t = \{\Omega, \emptyset\}$. □

Remark 2.103 (Markov Decision Processes) In addition to the assumptions of Example 2.102, assume that $\hat{L}_t$ is independent of s_{t+1}. Problem (1.11) can then be formulated as a *'Markov decision process'* $(\mathcal{S}_t, \mathcal{A}, D_t, Q_t, R_t)$, where $\mathcal{S}_t := (\mathbb{R}^N, \Xi^t)$ is the 'state space', $\mathcal{A} := \mathbb{R}^M$ is the 'action space',

$$R_t((X_t, s^t), U_t) := \hat{L}_t(X_t, U_t, s^t)$$

is the 'running cost', $D_t := \operatorname{dom} R_t$ is the set of allowed state-action pairs and Q_t is the 'transition kernel' given by

$$\begin{aligned} &Q_t((X_{t+1}, s^{t+1}) \in A \mid (X_t, s^t), U_t) \\ &\qquad := \int 1_A(X_t + A_{t+1}(s^{t+1})X_t + B_{t+1}(s^{t+1})U_t, s^{t+1})\nu_t(s^t, ds_{t+1}), \end{aligned}$$

where ν_t is ξ^t-regular conditional distribution of ξ_{t+1}; see Sect. 2.1.5.

Under the above structural assumptions, Theorem 2.94 yields sufficient conditions for the existence of solutions to the Bellman equation while Theorem 2.91 serves as a "verification theorem". In general, however, Markov decision processes lead to nontrivial measurability issues in the context dynamic programming. The above assumptions are satisfied in many applications in practice; see, e.g., Example 2.104 and Sect. 2.3.5 below.

If, in addition to the above assumptions, $(\xi_t)_{t=0}^T$ is Markov and $\hat{A}_{t+1}$, $\hat{B}_{t+1}$, $\hat{W}_{t+1}$ and $\hat{L}_t$ are independent of s^{t-1}, then the problem can be formulated as a Markov decision process with $(\mathcal{S}_t, \mathcal{A}, D_t, Q_t, R_t)$, where $\mathcal{S}_t := (\mathbb{R}^N, \Xi_t)$, $\mathcal{A} := \mathbb{R}^M$,

$$R_t((X_t, s_t), U_t) := \hat{L}_t(X_t, U_t, s_t)$$

and

$$\begin{aligned} &Q_t((X_{t+1}, s_{t+1}) \in A \mid (X_t, s_t), U_t) \\ &\quad := \int 1_A(X_t + A_{t+1}(s_t, s_{t+1})X_t + B_{t+1}(s_t, s_{t+1})U_t, s_{t+1})\nu_t(s_t, ds_{t+1}), \end{aligned}$$

where ν_t is ξ_t-regular conditional distribution of ξ_{t+1}.

Example 2.104 (Linear-Quadratic Control) Consider problem (1.11) in the case

$$L_t(X_t, U_t) = \frac{1}{2}X_t \cdot Q_t X_t + \frac{1}{2}U_t \cdot R_t U_t,$$

where $Q_t \in L^1(\mathbb{R}^{N\times N})$ and $R_t \in L^1(\mathbb{R}^{M\times M})$ are random symmetric matrices such that Q_t is positive semi-definite and R_t is positive definite. Then both (2.24) and (1.11) have unique solutions.

Assume, in addition, that the recursion

$$\begin{aligned} K_T &:= E_T[Q_T], \\ K_{t-1} &:= E_{t-1}[Q_{t-1} + (I + A_t)^* K_t (I + A_t)] \\ &\quad - \frac{1}{2}E_{t-1}[(I + A_t)^* K_t B_t](E_{t-1}[R_{t-1} + B_t^* K_t B_t])^{-1} E_{t-1}[B_t^* K_t (I + A_t)] \end{aligned}$$

is well-defined, $W_t \cdot K_t A_t$ and $W_t \cdot K_t B_t$ are integrable, W_t has zero mean and is independent of $\mathcal{F}_{t-1}$ and of $A_{t'}$, $B_{t'}$, $Q_{t'}$ and $R_{t'}$ for $t' \geq t$. Then

$$J_t(X_t) = \frac{1}{2}X_t \cdot K_t X_t + \frac{1}{2}\sum_{t'=t+1}^{T} E_t[W_{t'} \cdot K_{t'} W_{t'}]$$

and the optimal feedback mapping is given by $S_t(X_t) = -\Lambda_t X_t$, where

$$\Lambda_t = (E_t[R_t + B_{t+1}^* K_{t+1} B_{t+1}])^{-1} E_t[B_{t+1}^* K_{t+1}(I + A_{t+1})].$$

If Q_t, R_t, A_{t+1} and B_{t+1} are independent of $\mathcal{F}_t$ and $A_t \in L^2(\mathbb{R}^{N\times N})$ and $B_t \in L^2(\mathbb{R}^{N\times M})$, then the matrices K_t and Λ_t are well-defined, deterministic and given by

$$\begin{aligned}K_T &= E[Q_T],\\ K_{t-1} &= E[Q_{t-1} + (I + A_t)^* K_t (I + A_t)]\\ &\quad - \frac{1}{2}E[(I + A_t)^* K_t B_t](E[R_{t-1} + B_t^* K_t B_t])^{-1} E[B_t^* K_t (I + A_t)]\end{aligned}$$

and

$$\Lambda_t = (E[R_t + B_{t+1}^* K_{t+1} B_{t+1}])^{-1} E[B_{t+1}^* K_{t+1}(I + A_{t+1})].$$

Proof Assumptions of Theorems 2.91 and 2.94 are satisfied with $p = 0$, $y = 0$ and $m = 0$, so (2.24) and (1.11) have unique solutions. By Example 2.7,

$$(E_t L_t)(X_t, U_t) = \frac{1}{2}X_t \cdot E_t[Q_t]X_t + \frac{1}{2}U_t \cdot E_t[R_t]U_t,$$

for all t, so

$$J_T(X_T) = \frac{1}{2}X_T \cdot E_T[Q_T]X_T.$$

Assume that the expression for $J_{t'}$ in the claim is valid for $t' = t, \dots, T$. We get

$$\begin{aligned}&I_{t-1}(X_{t-1}, U_{t-1})\\ =&J_t((I + A_t)X_{t-1} + B_t U_{t-1} + W_t)\\ =&\frac{1}{2}X_{t-1} \cdot (I + A_t)^* K_t (I + A_t) X_{t-1} + \frac{1}{2}U_{t-1} \cdot B_t^* K_t B_t U_{t-1}\\ &+ \frac{1}{2}W_t \cdot K_t W_t + X_{t-1} \cdot (I + A_t)^* K_t B_t U_{t-1}\\ &+ W_t \cdot K_t (X_{t-1} + A_t X_{t-1} + B_t U_{t-1}) + \frac{1}{2}\sum_{t'=t+1}^{T} E_t[W_{t'} \cdot K_{t'} W_{t'}].\end{aligned}$$

When W_t has zero mean and is independent of $\mathcal{F}_{t-1}$, $Q_{t'}$, $R_{t'}$, $A_{t'}$ and $B_{t'}$ for $t' \geq t$ and W_t is independent of $\mathcal{F}_{t-1}$, Example 2.7 gives

$$\begin{aligned}&E_{t-1}(L_{t-1} + I_{t-1})(X_{t-1}, U_{t-1})\\ =&\frac{1}{2}X_{t-1} \cdot E_{t-1}[Q_{t-1} + (I + A_t)^* K_t (I + A_t)]X_{t-1}\end{aligned}$$

$$+\frac{1}{2}U_{t-1}\cdot E_{t-1}[R_{t-1}+B_t^*K_tB_t]U_{t-1}$$

$$+X_{t-1}\cdot E_{t-1}[(I+A_t)^*K_tB_t]U_{t-1}+\frac{1}{2}\sum_{t'=t}^{T}E_{t-1}[W_{t'}\cdot K_{t'}W_{t'}].$$

Thus,

$$\begin{aligned}J_{t-1}(X_{t-1}) &= \inf_{U_{t-1}\in\mathbb{R}^M} E_{t-1}(L_{t-1}+I_{t-1})(X_{t-1},U_{t-1})\\ &= \frac{1}{2}X_{t-1}\cdot K_{t-1}X_{t-1}+\frac{1}{2}\sum_{t'=t}^{T}E_t[W_{t'}\cdot K_{t'}W_{t'}],\end{aligned}$$

where the infimum is attained at the unique point $S_t(X_t) := -\Lambda_t X_t$. The second claim thus follows by induction on t.

Under the additional independence condition, the matrices K_t and Λ_t are deterministic provided they are well-defined. The integrability of Q_t and R_t and the square integrability of A_t and B_t imply that all the expectations are finite, while the positive definiteness of R_t implies that the matrix $E[R_{t-1}+B_t^*K_tB_t]$ is positive definite and, in particular, invertible. It follows that the recursion is indeed well-defined. □

The recursive equations for $(K_t)_{t=0}^T$ in Example 2.104 are known as *Riccati equations*.

2.3.4 Problems of Lagrange

Consider again problem (1.13)

$$\text{minimize}\quad E\left[\sum_{t=0}^{T}K_t(x_t,\Delta x_t)\right]\quad\text{over } x\in\mathcal{N}$$

from Sect. 1.3.4. This fits the general format (P) with

$$h(x,\omega)=\sum_{t=0}^{T}K_t(x_t,\Delta x_t,\omega).$$

Much like in the optimal control problem (1.11), the time-separable structure here allows us to express the solutions of the generalized Bellman equations (BE) in terms of normal integrands V_t and $\tilde{V}_t$ on $\mathbb{R}^d \times \Omega$ that solve the following dynamic programming equations

$$\begin{aligned}\tilde{V}_T &= 0,\\ V_t &= E_t\tilde{V}_t,\\ \tilde{V}_{t-1}(x_{t-1},\omega) &= \inf_{x_t\in\mathbb{R}^d}\{(E_tK_t)(x_t,\Delta x_t,\omega)+V_t(x_t,\omega)\}.\end{aligned} \tag{2.31}$$

Note that V_t is a function only of x_t and ω while the functions h_t in the general Bellman equations (BE) may depend on the whole path of x up to time t. The Eq. (2.31) are also known as *Bellman equations* and the functions V_t are often called 'value functions', 'cost-to-go functions' or 'Bellman functions'.

When applied to (1.13), Theorem 2.70 gives Theorem 2.106 below. In addition to $p \in \mathcal{N}^\perp$, the assumptions involve variables $y_t \in L^1(\mathbb{R}^d)$ which will feature in the duality theory in Sect. 3.4.4. Throughout, we set $y_{T+1} := 0$.

Assumption 2.105 *Problem* (1.13) *is feasible and there exist* $p \in \mathcal{N}^\perp$, $y_t \in L^1$ *and* $m_t \in L^1$ *such that*

1. *for almost every* $\omega \in \Omega$,

$$K_t(x_t,\Delta x_t,\omega) \geq x_t\cdot(p_t(\omega)+\Delta y_{t+1}(\omega)) + \Delta x_t\cdot y_t(\omega) - m_t(\omega) \quad \forall t$$

for every $x \in \mathbb{R}^n$,

2. $E[\sum_{t=0}^T K_t(x_t,\Delta x_t)] = E[\sum_{t=0}^T (K_t(x_t,\Delta x_t) - x_t\cdot p_t)]$ *for all* $x \in \mathcal{N}$.

Theorem 2.106 *Let* $(V_t)_{t=0}^T$ *be an L-bounded convex solution of* (1.13) *and let Assumption 2.105 hold. Then*

$$\inf(1.13) = \inf_{x^t\in\mathcal{N}^t} E\left[\sum_{s=0}^t (E_tK_s)(x_s,\Delta x_s) + V_t(x_t)\right] \quad t=0,\dots,T,$$

for all $t = 0,\dots,T$ *and, moreover, an* $\bar{x} \in \mathcal{N}$ *solves* (1.13) *if and only if*

$$x_t \in \operatorname*{argmin}_{x_t\in\mathbb{R}^d}\{(E_tK_t)(x_t,\Delta x_t)+V_t(x_t)\} \quad a.s.$$

for all $t = 0,\dots,T$. *If the measurable mappings*

$$N_t(\omega) := \{x_t \in \mathbb{R}^d \mid (E_tK_t)^\infty(x_t,x_t,\omega) + V_t^\infty(x_t,\omega) \leq 0\}$$

are linear-valued for all $t = 0,\dots,T$, *then there exists an optimal* $x \in \mathcal{N}$ *with* $x_t \perp N_t$ *almost surely.*

Proof Summing up the lower bounds in Assumption 2.105.1 shows that $k(x,\omega) := h(x,\omega) - x \cdot p(\omega)$ is lower bounded while Assumption 2.105.2 means that $Ek = Eh$ on $\mathcal{N}$. Assume that $(V_t)_{t=0}^T$ is an L-bounded convex solution of (2.31). All the claims follow from Theorem 2.70 once we show that $(h_t)_{t=0}^T$ given by

$$h_t(x^t,\omega) := \sum_{s=0}^{t} (E_t K_s)(x_s, \Delta x_s, \omega) + V_t(x_t,\omega)$$

is an L-bounded solution of (BE). Assume that $(h_{t'})_{t'=t}^T$ satisfies (BE) from time t onward. We get

$$\begin{aligned}
&\tilde{h}_{t-1}(x^{t-1},\omega) \\
:=& \inf_{x_t \in \mathbb{R}^d} h_t(x^{t-1}, x_t, \omega) \\
=& \sum_{s=0}^{t-1} (E_t K_s)(x_s, \Delta x_s, \omega) + \inf_{x_t \in \mathbb{R}^d} \{(E_t K_t)(x_t, \Delta x_t, \omega) + V_t(x_t,\omega)\} \\
=& \sum_{s=0}^{t-1} (E_t K_s)(x_s, \Delta x_s, \omega) + \tilde{V}_{t-1}(x_{t-1},\omega).
\end{aligned}$$

By Theorem 2.15 again,

$$h_{t-1} = E_{t-1}\tilde{h}_{t-1}.$$

Since $V_T = 0$, we have $h_T = E_T h$, by Theorem 2.15. Thus, by induction, $(h_t)_{t=0}^T$ solves (BE). □

We will establish the existence of solutions to the Bellman equations (2.31) under the following.

Assumption 2.107 *There exist $p \in \mathcal{N}^\perp$, $y_t \in L^1$ and $m_t \in L^1$ such that*

1. *for almost every $\omega \in \Omega$,*

$$K_t(x_t, \Delta x_t, \omega) \ge x_t \cdot (p_t(\omega) + \Delta y_{t+1}(\omega)) + \Delta x_t \cdot y_t(\omega) - m_t(\omega) \quad \forall t$$

   for every $x \in \mathbb{R}^n$,
2. *$\{x \in \mathcal{N} \mid \sum_{t=0}^T (K_t^\infty(x_t, \Delta x_t) - x_t \cdot p_t) \le 0\ a.s.\}$ is linear.*

Note that Assumption 2.107.1 is the same as Assumption 2.105.1. Theorem 2.71 gives the following existence result for the dynamic programming equations (2.31).

Theorem 2.108 *Under Assumption 2.107, the recursion* (2.31) *has a unique solution* $(V_t)_{t=0}^T$ *of L-bounded convex normal integrands, and the measurable mappings*

$$N_t(\omega) := \{x_t \in \mathbb{R}^d \mid (E_t K_t)^\infty(x_t, x_t, \omega) + V_t^\infty(x_t, \omega) \le 0\} \tag{2.32}$$

are almost surely linear-valued.

Proof Summing up the lower bounds in Assumption 2.107.1 shows that $k(x, \omega) := h(x, \omega) - x \cdot p(\omega)$ is lower bounded. By Theorem A.17,

$$h^\infty(x, \omega) = \sum_{t=0}^{T} K_t^\infty(x_t, \Delta x_t, \omega).$$

Thus, Assumption 2.107.2 means that $\{x \in \mathcal{N} \mid k^\infty(x) \le 0 \; a.s.\}$ is linear. By Theorem 2.71, (BE) has a unique solution $(h_t)_{t=0}^T$ for h and

$$N_t(\omega) := \{x_t \in \mathbb{R}^{n_t} \mid h_t^\infty(0, x_t, \omega) \le 0\} \tag{2.33}$$

is linear-valued for all t.

Assume that $(V_{t'})_{t'=t}^T$ are normal integrands satisfying (2.31) from time t onward, and that

$$\begin{aligned} h_t(x^t) &= \sum_{s=0}^{t} (E_t K_s)(x_s, \Delta x_s) + V_t(x_t) \\ V_t(x_t) &\ge -x_t \cdot (E_t y_{t+1}) - E_t[\sum_{t'=t+1}^{T} m_{t'}] \end{aligned} \tag{2.34}$$

almost surely. For brevity, we omit the ω from the notation. We get

$$\begin{aligned} \tilde{h}_{t-1}(x^{t-1}) &= \inf_{x_t \in \mathbb{R}^d} h_t(x^{t-1}, x_t) \\ &= \sum_{s=0}^{t-1} (E_t K_s)(x_s, \Delta x_s) + \tilde{V}_{t-1}(x_{t-1}), \end{aligned} \tag{2.35}$$

where

$$\tilde{V}_{t-1}(x_{t-1}) := \inf_{x_t \in \mathbb{R}^d} \{(E_t K_t)(x_t, \Delta x_t) + V_t(x_t)\}.$$

The equation in (2.34) and the linearity of (2.33) imply that (2.32) is linear. Thus, by Theorem 1.40, $\tilde{V}_{t-1}$ is a normal integrand.

By Lemma 2.11 and Example 2.7, Assumption 2.107.1 implies

$$(E_t K_t)(x_t, \Delta x_t) \geq x_t \cdot E_t \Delta y_{t+1} + \Delta x_t \cdot E_t y_t - E_t m_t.$$

Combining this with the inequality in (2.34) gives

$$\tilde{V}_{t-1}(x_{t-1}) \geq -x_{t-1} \cdot E_t y_t - E_t[\sum_{t'=t}^{T} m_{t'}].$$

By Theorem 2.13, there is a unique normal integrand V_{t-1} such that $V_{t-1} = E_{t-1}\tilde{V}_{t-1}$. By Lemma 2.11 and Example 2.7, V_{t-1} satisfies the inequality in (2.34). Taking conditional expectations on both sides of (2.35) and using Theorem 2.15 and Theorem 2.19, we get

$$h_{t-1}(x^{t-1}) = \sum_{s=0}^{t-1}(E_{t-1}K_s)(x_s, \Delta x_s) + V_{t-1}(x_{t-1}).$$

For $t = T$, $V_T = 0$ so (2.34) holds by Theorem 2.15. The claim thus holds by induction on t. □

The following analogue of Example 2.95 gives sufficient conditions for both Assumptions 2.105 and 2.107.

Example 2.109 Assumptions 2.105 and 2.107 hold if (1.13) is feasible,

1. $\{x \in \mathcal{N} \mid \sum_{t=0}^{T} K_t^\infty(x_t, \Delta x_t) \leq 0\, a.s.\}$ is linear,
2. there exist $p \in \mathcal{N}^\perp$ and $\epsilon > 0$ such that, for every $\lambda \in [1-\epsilon, 1+\epsilon]$ and $t = 0, \ldots, T$, there exist $y_t \in L^1$ and $m_t \in L^1$ with

$$K_t(x_t, \Delta x_t, \omega) \geq x_t \cdot (\lambda p_t(\omega) + \Delta y_{t+1}(\omega)) + \Delta x_t \cdot y_t(\omega) - m_t(\omega) \quad \forall t$$

for every $x \in \mathbb{R}^n$ and almost every $\omega \in \Omega$.

Proof Choosing $\lambda = 1$ in condition 2 here implies Assumptions 2.105.1 and 2.107.1. Summing up the lower bounds in condition 2, we see that assumptions of Lemma 2.74 are satisfied. Thus, Assumption 2.72 holds. Clearly, Assumption 2.72.3 is the same as Assumption 2.105.2 while Assumption 2.72.2 is the same as Assumption 2.107.2. □

The linearity condition in Example 2.109 holds, in particular, if $K_t^\infty \geq 0$ and $K_t^\infty(x_t, x_t) > 0$ for every $x_t \neq 0$ and $t = 0, \ldots, T$ almost surely. Indeed, in this case,

$$\{x \in \mathbb{R}^n \mid \sum_{t=0}^{T} K_t^\infty(x_t, \Delta x_t, \omega) \leq 0\} = \{0\}$$

for almost every $\omega \in \Omega$. The last condition in Example 2.109 will be discussed in Sect. 3.4.

Remark 2.110 Even under Assumptions 2.105 and 2.107, the terms $K_t(x_t, \Delta x_t)$ need not be integrable for an optimal solution x of (1.13). Indeed, defining

$$K_t(x_t, \Delta x_t, \omega) := x_t \cdot \Delta y_{t+1}(\omega) + y_t(\omega) \cdot \Delta x_t,$$

for a process $y \in L^1$ with $y_{T+1} := 0$, the assumptions of Example 2.109 are satisfied with $p = 0$. Since $\sum_{t=0}^{T} K_t(x_t, \Delta x_t, \omega) = 0$, any $x \in \mathcal{N}$ is optimal, but $K_t(x_t, \Delta x_t)$ need not be integrable.

The following characterizes induced constraints and gives sufficient conditions for the problem (1.13) to have relatively complete recourse; see Sect. 2.2.3. We will denote the $\mathcal{F}_t$-conditional essential infimum and supremum of a measurable mapping S by $\operatorname{essinf}_t S$ and $\operatorname{esssup}_t S$, respectively; see Sect. 2.1.4.

Example 2.111 (Induced Constraints) Assume that

$$\operatorname{cl}\operatorname{dom} h(\cdot, \omega) = \{x \in \mathbb{R}^n \mid (x_t, \Delta x_t) \in \operatorname{cl}\operatorname{dom} K_t(\cdot, \cdot, \omega)\},$$

$\operatorname{dom} K_t$ is $\mathcal{F}_t$-measurable for all t and that the recursion

$$\begin{aligned} S_T(\omega) &:= \mathbb{R}^d, \\ \tilde{S}_{t-1}(\omega) &:= \{x_{t-1} \in \mathbb{R}^d \mid \exists x_t \in S_t(\omega) : \ (x_t, \Delta x_t) \in \operatorname{cl}\operatorname{dom} K_t(\cdot, \cdot, \omega)\}, \\ S_{t-1} &:= \operatorname{essinf}_{t-1} \tilde{S}_{t-1} \end{aligned}$$

is well-defined and each $\tilde{S}_t$ is closed-valued. Then the induced constraints are given by

$$D_t(\omega) := \{x^t \in \mathbb{R}^{n^t} \mid (x_{t'}, \Delta x_{t'}) \in \operatorname{cl}\operatorname{dom} K_{t'}(\cdot, \cdot, \omega)\ \forall t' \le t,\ x_t \in S_t(\omega)\}.$$

Problem (1.13) has relatively complete recourse if and only if $\tilde{S}_t$ is $\mathcal{F}_t$-measurable for every t. This happens, in particular, if, for every t,

$$\{x_{t-1} \in \mathbb{R}^d \mid \exists x_t : (x_t, \Delta x_t) \in \operatorname{cl}\operatorname{dom} K_t(\cdot, \cdot, \omega)\} = \mathbb{R}^d$$

for almost every ω.

Proof The first claim follows from Theorem 2.47 and induction on t. We have that $\tilde{S}_t$ is $\mathcal{F}_t$-measurable if and only if $D_t = D_{t+1}^t$. The latter holds for every t if and only if $D_t = (\operatorname{cl}\operatorname{dom} h)^t$ almost surely for every t. Since each D_t is closed-valued, this means that the problem has relatively complete recourse. As to the last claim, an induction on t shows that $S_t = \mathbb{R}^d$ for all t. □

In general, the linear programming format in Example 1.83 and Example 2.87 does not allow for more explicit Bellman equations but when the matrix A has block-diagonal structure, we obtain the following.

Example 2.112 (Block-Diagonal Stochastic LP) Consider the stochastic linear programming problem from Example 1.83 and assume that the constraint $Ax - b \in K$ can be written as

$$T_t \Delta x_t + W_t x_t - b_t \in C_t \quad t = 0, \dots, T.$$

This is an instance of (1.13) with

$$K_t(\Delta x_t, x_t, \omega) = \begin{cases} c_t(\omega) \cdot x_t & \text{if } T_t(\omega)\Delta x_t + W_t(\omega)x_t - b_t(\omega) \in C_t(\omega), \\ +\infty & \text{otherwise.} \end{cases}$$

If T_t, W_t, b_t and C_t are $\mathcal{F}_t$-measurable, then K_t is an $\mathcal{F}_t$-measurable convex normal integrand and the dynamic programming recursion (2.31) can be written as

$$\begin{aligned} \tilde{V}_T &= 0, \\ V_t &= E_t \tilde{V}_t, \\ \tilde{V}_{t-1}(x_{t-1}, \omega) &= \inf_{x_t \in \mathbb{R}^d} \{c_t(\omega) \cdot x_t + V_t(x_t, \omega) \mid \\ &\qquad\qquad T_t(\omega)\Delta x_t + W_t(\omega)x_t - b_t(\omega) \in C_t(\omega)\}. \end{aligned}$$

Theorem 2.108 gives sufficient conditions for this to be well-defined.

The dynamic programming equations (2.31) can be written in many equivalent forms. If $(V_t, \tilde{V}_t)_{t=0}^{T+1}$ satisfy (2.31), then the functions

$$Q_t(x_{t-1}, x_t) = (E_t K_t)(\Delta x_t, x_t) + V_t(x_t)$$

satisfy the recursion

$$\begin{aligned} \tilde{V}_T &= 0, \\ Q_t(x_{t-1}, x_t) &= (E_t K_t)(\Delta x_t, x_t) + (E_t \tilde{V}_t)(x_t), \\ \tilde{V}_{t-1}(x_{t-1}) &= \inf_{x_t \in \mathbb{R}^d} Q_t(x_{t-1}, x_t) \end{aligned} \tag{2.36}$$

(for brevity, we omit ω from the notation) and the optimality conditions can be written as

$$x_t \in \operatorname*{argmin}_{x_t \in \mathbb{R}^d} Q_t(x_{t-1}, x_t) \; a.s.$$

Conversely, if Q_t and $\tilde{V}_t$ satisfy (2.36) then $\tilde{V}_t$ and $V_t := E_t \tilde{V}_t$ satisfy (2.31).

The mapping $S_t : \mathbb{R}^d \times \Omega \rightrightarrows \mathbb{R}^d$ defined by

$$S_t(x_{t-1}, \omega) := \operatorname*{argmin}_{x_t \in \mathbb{R}^d} Q_t(x_{t-1}, x_t, \omega)$$

is called the *optimal feedback mapping*. Under the assumptions of Theorem 2.106, the optimal x_t are the $\mathcal{F}_t$-measurable selections of $S_t(x_{t-1})$. The values of $S_t(x_{t-1})$ depend on the past decisions only through x_{t-1}.

Under the assumptions of Theorem 2.108, recursion (2.36) admits solutions $(Q_t, \tilde{V}_t)_{t=0}^T$ where the functions Q_t are L-bounded and convex and, in particular, S_t are closed convex-valued. By Theorem 1.20 and Corollary 1.23, $\omega \mapsto S_t(x_{t-1}(\omega), \omega)$ is closed-valued and $\mathcal{F}_t$-measurable for any $\mathcal{F}_{t-1}$-measurable x_{t-1}. Theorem 1.21 gives the following.

Remark 2.113 If $\tilde{V}_t(\cdot, \omega)$ is continuous for every $\omega \in \Omega$, then

$$\omega \mapsto \operatorname{gph} S_t(\cdot, \omega)$$

is closed-valued and $\mathcal{F}_t$-measurable. In particular, if $\tilde{V}_t(\cdot, \omega)$ is continuous and $S_t(\cdot, \omega)$ is single-valued and locally bounded, then $S_t(\cdot, \omega)$ is continuous.

Under p-homogeneity (see Lemma 2.22), the feedback mappings are homogeneous.

Remark 2.114 Assume that each K_t is L-bounded and p-homogeneous with integrable C and that $(V_t, \tilde{V}_t)_{t=0}^T$ is an L-bounded solution of (2.31). Then $\tilde{V}_t$, V_t and Q_t are p-homogeneous and

$$S_t(\alpha x_{t-1}) = \alpha S_t(x_{t-1}) \quad \forall \alpha > 0.$$

If K_t are p-homogeneous merely for $t \geq t'$, then Q_t are p-homogeneous for $t \geq t'$.

Proof The function V_T is trivially p-homogeneous. If V_{t+1} is p-homogeneous, then by Lemma 2.22 and Theorem A.86, $\tilde{V}_t$, V_t and Q_t are p-homogeneous and S_t is homogeneous. The claim thus follows by induction. □

Theorem 2.115 (Conditional Independence) *Assume that K_t are L-bounded, that $(V_t)_{t=0}^T$ and $(\tilde{V}_t)_{t=0}^T$ are an L-bounded solution of* (2.31) *and that there is a sequence of σ-algebras $(\mathcal{H}_t)_{t=0}^T$ such that $\mathcal{H}_t \subseteq \mathcal{F}_t$ and*

1. *K_t is $\mathcal{H}_t$-conditionally independent of $\mathcal{F}_t$,*
2. *$\mathcal{H}_{t+1}$ is $\mathcal{H}_t$-conditionally independent of $\mathcal{F}_t$.*

Then Q_t and S_t are $\mathcal{H}_t$-measurable. In particular, if $\mathcal{H}_t = \{\Omega, \emptyset\}$, then Q_t and S_t are deterministic.

Proof By definition, $V_T = 0$ which is $\mathcal{H}_T$-measurable. Assume now that V_{t+1} is $\mathcal{H}_{t+1}$-measurable. By Theorem 2.21, condition 1 implies $E_{t+1}K_{t+1}$ is $\mathcal{H}_{t+1}$-measurable. By Theorem 1.21, $\tilde{V}_t$ is then $\mathcal{H}_{t+1}$-measurable as well. By Theorem 2.21 again, condition 2 implies $V_t = E^{\mathcal{H}_t}\tilde{V}_t$, so the claim follows by induction. □

Recall that a process $(\xi_t)_{t=0}^T$ is Markov if ξ_{t+1} is ξ_t-conditionally independent of $\mathcal{F}_t$.

Example 2.116 (Canonical Representation) Let $\xi = (\xi_t)_{t=0}^{T+1}$ be a stochastic process as in Remark 1.76 and consider problem (1.13) with

$$K_t(x_t, \Delta x_t, \omega) = \hat{K}_t(x_t, \Delta x_t, \xi^{t+1}(\omega))$$

where $\hat{K}_t$ is an $\mathcal{A}^{t+1}$-measurable normal integrand. Assume that each K_t is L-bounded and that ξ_{t+1} has regular ξ^t-conditional distribution ν_t. Then sequences $(V_t, \tilde{V}_t)_{t=0}^{T+1}$ of L-bounded normal integrands satisfy (2.31) if and only if there exist $\mathcal{A}^t$-measurable normal integrands $\hat{V}_t : \mathbb{R}^d \times \Xi^t \to \overline{\mathbb{R}}$ and $\mathcal{A}^{t+1}$-measurable normal integrands $\hat{\tilde{V}}_t : \mathbb{R}^d \times \Xi^{t+1} \to \overline{\mathbb{R}}$ such that

$$V_t(x_t, \omega) = \hat{V}_t(x_t, \xi^t(\omega)), \quad \tilde{V}_t(x_t, \omega) = \hat{\tilde{V}}_t(x_t, \xi^{t+1}(\omega)),$$

and

$$\begin{aligned} \hat{V}_T &= 0, \\ \hat{V}_t &= E^{\nu_t}\hat{\tilde{V}}_t, \\ \hat{\tilde{V}}_{t-1}(x_{t-1}, \xi^t) &= \inf_{x_t \in \mathbb{R}^d}\{(E^{\nu_t}K_t)(x_t, \Delta x_t, \xi^t) + \hat{V}_t(x_t, \xi^t)\} \end{aligned} \tag{2.37}$$

P_{ξ^t}-almost surely for all $t = T, \dots, 0$.

If $(\xi_t)_{t=0}^T$ is Markov and $\hat{K}_t$ is independent of s^{t-1}, then Q_t and S_t are $\sigma(\xi_t)$-measurable. If $(\xi_t)_{t=0}^T$ is a sequence of independent random variables and $\hat{K}_t$ is independent of s^t, then Q_t and S_t are deterministic.

Proof Assume first that $(V_t, \tilde{V}_t)_{t=0}^T$ satisfies (2.31). By Corollary 1.34, there exist, $\mathcal{A}^t$-measurable normal integrands $\hat{V}_t : \mathbb{R}^d \times \Xi^t \to \overline{\mathbb{R}}$ and $\mathcal{A}^{t+1}$-measurable normal integrands $\hat{\tilde{V}}_t : \mathbb{R}^d \times \Xi^{t+1} \to \overline{\mathbb{R}}$ such that

$$V_t(x_t, \omega) = \hat{V}_t(x_t, \xi^t(\omega)), \quad \tilde{V}_t(x_t, \omega) = \hat{\tilde{V}}_t(x_t, \xi^{t+1}(\omega)).$$

By assumption,

$$K_t(x_t, \Delta x_t, \omega) = \hat{K}_t(x_t, \Delta x_t, \xi^{t+1}(\omega))$$

Applying Corollary 2.60, with $\xi_1 = \xi^t$ and $\xi_2 = \xi_{t+1}$ gives

$$(E^{\sigma(\xi^t)} K_t)(x_t, \omega) = (E^{\nu_t} \hat{K}_t)(x_t, \xi^t(\omega))$$

and

$$\hat{V}_t(x_t, \xi^t) = (E^{\nu_t} \hat{\hat{V}}_t)(x_t, \xi^t(\omega))$$

almost surely everywhere, which gives (2.37). Reversing the argument gives sufficiency.

The second last claim follows from Theorem 2.115 with $\mathcal{H}_t = \sigma(\xi_t)$. Indeed, condition 2 in Theorem 2.115 is implied by the Markov property of ξ while condition 1 follows from the fact that conditional independence is preserved in compositions; see Example 2.20. The last claim follows similarly with $\mathcal{H}_t = \{\Omega, \emptyset\}$. □

2.3.5 *Financial Mathematics*

Consider again problem (1.15)

$$\begin{aligned} &\text{minimize} \quad EV\left(c - \sum_{t=0}^{T-1} x_t \cdot \Delta s_{t+1}\right) \quad \text{over} \quad x \in \mathcal{N}, \\ &\text{subject to} \quad x_t \in D_t \quad t = 0, \dots, T \ a.s., \end{aligned}$$

from Sect. 1.3.5. As observed in Example 1.91, this can be written in the optimal control format as

$$\begin{aligned} \text{minimize} \quad & EV(c - X_T) \quad \text{over} \quad (X, U) \in \mathcal{N}, \\ \text{subject to} \quad & \Delta X_t = U_{t-1} \cdot \Delta s_t \quad \forall t = 1, \dots, T, \\ & U_t \in D_t \quad \forall t = 0, \dots, T-1, \\ & X_0 = 0. \end{aligned}$$

This is an instance of problem (1.11) with $N = 1$, $M = |J|$, $A_t = 0$, $B_t = \Delta s_t$, $W_t = 0$ and

$$\begin{aligned} L_T(X_T, U_T, \omega) &= V(c(\omega) - X_T, \omega), \\ L_t(X_t, U_t, \omega) &= \delta_{D_t(\omega)}(U_t) \quad t = 1, \dots, T-1, \\ L_0(X_0, U_0, \omega) &= \delta_{\{0\} \times D_t(\omega)}(X_0, U_0). \end{aligned}$$

Assuming that the loss function V and the claim c are $\mathcal{F}_T$-measurable, the dynamic programming equations (2.24) can be written as

$$\begin{aligned} J_T(X_T,\omega) &= V(c(\omega)-X_T,\omega), \\ I_t(X_t,U_t,\omega) &= J_{t+1}(X_t+U_t\cdot\Delta s_{t+1}(\omega),\omega) \quad t=T-1,\dots,0, \\ J_t(X_t,\omega) &= \inf_{U_t\in D_t}(E_tI_t)(X_t,U_t,\omega) \quad t=T-1,\dots,1, \\ J_0(X_0,\omega) &= \inf_{U_0\in D_t}(E_0I_0)(0,U_0,\omega)+\delta_{\{0\}}(X_0). \end{aligned} \tag{2.38}$$

As before, we assume that V is a nondecreasing, nonconstant convex normal integrand. This implies that the recession function V^∞ of V is nondecreasing and strictly positive on strictly positive reals.

Assumption 2.117 *Problem* (1.15) *is feasible and*

1. *the set*

$$\{x\in\mathcal{N}\mid\sum_{t=0}^{T-1}x_t\cdot\Delta s_{t+1}\geq 0,\ x_t\in D_t^\infty\ \forall t\ a.s.\}$$

   is linear,
2. *there exist* $y\in L^1$ *and* $\epsilon>0$ *such that*

$$cy,\ y\Delta s_{t-1},\ \sigma_{D_t}(E_t[y\Delta s_{t+1}])\in L^1 \quad t=0,\dots,T-1$$

   and $EV^*(\lambda y)<\infty$ *for all* $\lambda\in[1-\epsilon,1+\epsilon]$.

If there are no portfolio constraints, condition 1 in Assumption 2.117 becomes the classical "no-arbitrage" condition; see Remark 2.120 below. Condition 2 in Assumption 2.117 holds, in particular, if V is lower bounded since then, $EV^*(0)<\infty$ so one can simply take $y=0$. More general conditions will be given in Remark 2.121 below.

Theorem 2.118 *Assume that* V *and* c *are* $\mathcal{F}_T$*-measurable and that Assumption 2.117 holds. Then* (2.38) *admits a unique solution* $(J_t,I_t)_{t=0}^T$ *of L-bounded convex normal integrands, problem* (1.15) *has optimal solutions* $\bar{x}\in\mathcal{N}$ *and they are characterized by*

$$\bar{x}_t\in\operatorname*{argmin}_{U_t\in D_t}(E_tI_t)(X_t,U_t) \quad t=T-1,\dots,1\ a.s.,$$

where

$$X_t=\sum_{t'=0}^{t-1}\bar{x}_{t'}\cdot\Delta s_{t'+1}.$$

Moreover, the optimum value of (1.15) *equals that of the problem*

$$\begin{array}{lll} \text{minimize} & EJ_t\left(\sum_{t'=0}^{t-1} x_{t'} \cdot \Delta s_{t'+1}\right) \quad \text{over} \quad x^t \in \mathcal{N}^t, \\ \text{subject to} & x_{t'} \in D_{t'} \quad t' = 0, \dots, t \ a.s. \end{array}$$

for all $t = 0, \dots, T$.

Proof By Theorems 2.91 and 2.94, it suffices to show that the conditions of Example 2.95 hold. By Theorem A.17, the set in condition 1 of Example 2.95 becomes

$$\begin{aligned} \{(X, U) \in \mathcal{N} \mid V^\infty(c - X_T) \le 0,\ U_t \in D_t^\infty\ t = 0, \dots, T-1, \\ X_0 = 0,\ \Delta X_t = U_{t-1} \cdot \Delta s_t\ t = 1, \dots, T\ a.s.\}. \end{aligned}$$

Since V is nondecreasing and nonconstant, we have $V^\infty(\alpha) \le 0$ if and only if $\alpha \le 0$. Condition 1 of Example 2.95 thus coincides with condition 1 in Assumption 2.117. By Fenchel's inequality,

$$\begin{aligned} L_T(X_T, U_T) &\ge -V^*(\lambda y) + \lambda(c - X_T)y, \\ L_t(X_t, U_t) &\ge -\sigma_{D_t}(\lambda E_t[y\Delta s_{t+1}]) + \lambda U_t \cdot E_t[y\Delta s_{t+1}] \quad t = 0, \dots, T-1. \end{aligned}$$

For brevity, we omit the ω from the notation. The second inequality can be written in terms of $\Delta X_{t+1} = U_t \cdot \Delta s_{t+1}$ as

$$\begin{aligned} L_t(X_t, U_t) &\ge -\lambda\sigma_{D_t}(E_t[y\Delta s_{t+1}]) + \lambda U_t \cdot (E_t[y\Delta s_{t+1}] - y\Delta s_{t+1}) \\ &\qquad + \lambda y U_t \cdot \Delta s_{t+1} \\ &\ge \lambda U_t \cdot (E_t[y\Delta s_{t+1}] - y\Delta s_{t+1}) - \lambda\sigma_{D_t}(E_t[y\Delta s_{t+1}]) + \lambda y \Delta X_{t+1} \end{aligned}$$

so condition 2 in Example 2.95 holds with $p_T := 0$, $p_t := (0, y\Delta s_{t+1} - E_t[y\Delta s_{t+1}])$ for $t < T$, $y_t := -\lambda y$ for $t = 1, \dots, T$, $m_T := V^*(\lambda y) - \lambda c y$ and $m_t := \lambda\sigma_{D_t}(E_t[y\Delta s_{t+1}])$ for $t < T$. □

Unless $\operatorname{dom} V = \mathbb{R}$, problem (1.15) may fail to have relatively complete recourse; see Sect. 2.2.3. We will denote the $\mathcal{F}_t$-conditional essential infimum and supremum of a measurable mapping S by $\operatorname{essinf}_t S$ and $\operatorname{esssup}_t S$, respectively; see Sect. 2.1.4.

Example 2.119 (Induced Constraints) Consider problem (1.15) and assume that c is $\mathcal{F}_0$-measurable and $\operatorname{dom} V = \mathbb{R}_-$ so that

$$(\operatorname{dom} h)(\omega) := \{x \in \mathbb{R}^n \mid c(\omega) - \sum_{t=0}^{T-1} x_t \cdot \Delta s_{t+1}(\omega) \le 0,\ x_t \in D_t(\omega)\ \forall t\}.$$

Let $C_T := \{0\}$,

$$C_t := \operatorname{esssup}_t\{-\Delta s_{t+1}\}\ a.s. \quad \forall t = 0, \dots, T-1$$

and assume that

$$\inf_{x \in D_t} \sigma_{C_t}(x) = 0 \quad a.s. \tag{2.39}$$

Then the induced constraints for problem (1.15) are given by

$$D_t(\omega) = \{x^t \in \mathbb{R}^{n^t} \mid c(\omega) - \sum_{t'=0}^{t-1} x_{t'} \cdot \Delta s_{t'+1}(\omega) + \sigma_{C_t}(x_t, \omega) \le 0,$$
$$x_{t'} \in D_{t'}(\omega)\ \forall t' \le t\}.$$

If, on the other hand, $\operatorname{dom} V = \mathbb{R}$, then

$$(\operatorname{dom} h)(\omega) := \{x \in \mathbb{R}^n \mid x_t \in D_t(\omega)\ \forall t\}$$

and problem (1.15) has relatively complete recourse.

Proof We make the induction assumption that $(S_{t'})_{t'=t}^T$ solves (2.17) from t onward. Since $0 \in D_t$, (2.39) implies

$$\begin{aligned}\tilde{S}_{t-1}(\omega) &:= \{x^{t-1} \in \mathbb{R}^{n^{t-1}} \mid x^t \in S_t(\omega)\} \\ &= \{x^{t-1} \in \mathbb{R}^{n^{t-1}} \mid c(\omega) - \sum_{t'=0}^{t-1} x_{t'} \cdot \Delta s_{t'+1}(\omega) \le 0,\ x_{t'} \in D_{t'}(\omega)\ \forall t' < t\}.\end{aligned}$$

By Theorems 2.47.2 and 2.42.1,

$$(\operatorname{essinf}_{t-1} \tilde{S}_{t-1})(\omega) = S_{t-1}(\omega).$$

Clearly, $S_T = S$, so the first claim follows by backward induction on t. The last claim follows from the assumption that each D_t is $\mathcal{F}_t$-measurable. □

In Example 2.119, the quantity

$$\sigma_{C_t}(x, \omega) := \sup_{s \in C_t(\omega)} x \cdot s$$

can be thought of as the worst case loss when choosing the portfolio x at time t. Since $0 \in D_t$, condition (2.39) means that, almost surely, one cannot guarantee strictly positive profit over period $(t, t+1]$. Condition (2.39) holds, in particular, if $0 \in \operatorname{cl} \operatorname{co} C_t$.

The following characterizes Assumption 2.117.1.

Remark 2.120 (No-Arbitrage Condition) Let

$$K_t(\omega) := \{x_t \in \mathbb{R}^J \mid x_t \cdot \Delta s_{t+1}(\omega) \geq 0\}.$$

The following are equivalent:

1. the set

$$\{x \in \mathcal{N} \mid \sum_{t=0}^{T-1} x_t \cdot \Delta s_{t+1} \geq 0,\ x_t \in D_t^\infty\ \forall t\ a.s.\}$$

   is linear;
2. the set

$$\{x_t \in L^0(\mathcal{F}_t) \mid x_t \in K_t \cap D_t^\infty\ a.s.\}$$

   is linear for all $t = 0, \dots, T-1$;
3. the set

$$(\text{essinf}_t K_t) \cap D_t^\infty$$

   is almost surely linear-valued for all $t = 0, \dots, T-1$;
4. we have

$$\text{rint}(\text{co}\,\text{esssup}_t\{\Delta s_{t+1}\}) \cap \text{rint}(D_t^\infty)^\circ \neq \emptyset$$

   almost surely for all $t = 0, \dots, T-1$.

Note that, if there are no portfolio constraints, condition 1 means that the price process s satisfies the *no-arbitrage condition*:

$$x \in \mathcal{N},\ \sum_{t=0}^{T-1} x_t \cdot \Delta s_{t+1} \geq 0 \quad a.s. \implies \sum_{t=0}^{T-1} x_t \cdot \Delta s_{t+1} = 0 \quad a.s.$$

Proof Clearly, 1 implies 2. By Theorem 2.14,

$$L^0(\mathcal{F}_t; K_t \cap D_t^\infty) = L^0(\mathcal{F}_t; \text{essinf}_t(K_t \cap D_t^\infty)),$$

where $\text{essinf}_t(K_t \cap D_t^\infty) = (\text{essinf}_t K_t) \cap D_t^\infty$, by Theorem 2.42. Thus, by Corollary 1.74, 2 and 3 are equivalent. Since the polar of the sum of two cones is the intersection of the polar cones, Corollary A.58 gives

$$[(\text{essinf}_t K_t) \cap D_t^\infty]^\circ = \text{cl}[(\text{essinf}_t K_t)^\circ + (D_t^\infty)^\circ],$$

so condition 3 means that $\mathrm{cl}[(\mathrm{essinf}_t K_t)^\circ + (D_t^\infty)^\circ]$ is linear. The linearity means that $0 \in \mathrm{rint}\,\mathrm{cl}[(\mathrm{essinf}_t K_t)^\circ + (D_t^\infty)^\circ]$. By Theorem A.28 and Lemma A.5,

$$\mathrm{rint}\,\mathrm{cl}[(\mathrm{essinf}_t K_t)^\circ + (D_t^\infty)^\circ] = \mathrm{rint}[(\mathrm{essinf}_t K_t)^\circ] + \mathrm{rint}[(D_t^\infty)^\circ],$$

so condition 3 can be written as

$$\mathrm{rint}[-(\mathrm{essinf}_t K_t)^\circ] \cap \mathrm{rint}[(D_t^\infty)^\circ] \neq \emptyset. \tag{2.40}$$

By Corollaries 2.51 and 2.52,

$$(\mathrm{essinf}_t K_t)^\circ = \mathrm{cl}\,\mathrm{co}\,\mathrm{esssup}_t\,\mathrm{pos}\{-\Delta s_{t+1}\} = \mathrm{cl}\,\mathrm{co}\,\mathrm{pos}\,\mathrm{esssup}_t\{-\Delta s_{t+1}\}.$$

Since positive hull and convex hull commute, Theorem A.28 and Lemma A.7 imply

$$\mathrm{rint}[-(\mathrm{essinf}_t K_t)^\circ] = \mathrm{pos}\,\mathrm{rint}\,\mathrm{co}\,\mathrm{esssup}_t\{\Delta s_{t+1}\}.$$

Since a set intersects a cone if and only if its positive hull does so too, (2.40) is equivalent to condition 4.

To prove that 3 implies 1, we denote $C_t := \mathrm{esssup}_t\{-\Delta s_{t+1}\}$ and

$$h(x,\omega) := \delta_{\mathbb{R}_+}\Big(\sum_{t=0}^{T-1} x_t \cdot \Delta s_{t+1}(\omega)\Big) + \sum_{t=0}^{T} \delta_{D_t^\infty}(x_t,\omega).$$

Condition 1 means that $\mathcal{L} := \{x \in \mathcal{N} \mid h^\infty(x) \leq 0\ a.s.\}$ is linear. By Theorem 2.62, $\mathcal{L}$ is linear if and only if the Bellman equations (BE) associated h have a solution $(h_t)_{t=0}^T$ and $N_t(\omega) := \{x_t \in \mathbb{R}^n \mid h_t^\infty(0, x_t, \omega) \leq 0\}$ are linear-valued. By Example 2.53, $\mathrm{lev}_0\,\sigma_{C_t} = \mathrm{essinf}_t K_t$. Under condition 3, $\mathrm{lev}_0\,\sigma_{C_t} \cap D_t^\infty$ is linear-valued, so Remark A.18 implies

$$\inf_{x \in D_t^\infty(\omega)} \sigma_{C_t}(x,\omega) = 0$$

almost surely. Thus condition 3 implies (2.39) with $D_t = D_t^\infty$. By Example 2.119, $h_t = \delta_{S_t}$, where

$$S_t(\omega) := \{x^t \in \mathbb{R}^{n^t} \mid -\sum_{t'=0}^{t-1} x_{t'} \cdot \Delta s_{t'+1}(\omega) + \sigma_{C_t}(x_t,\omega) \leq 0,$$
$$x_{t'} \in D_{t'}^\infty(\omega)\ \forall t' \leq t\}.$$

Thus,

$$N_t(\omega) = \{x_t \in \mathbb{R}^J \mid \sigma_{C_t}(x_t, \omega) \le 0,\ x_t \in D_t^\infty(\omega)\} = \text{essinf}_t K_t \cap D_t^\infty,$$

so 3 and 1 are equivalent. □

Given a loss function V, its *asymptotic elasticities* are defined by

$$AE_-(V) := \limsup_{u\to-\infty} \frac{uV'(u)}{V(u)} \quad \text{and} \quad AE_+(V) := \liminf_{u\to+\infty} \frac{uV'(u)}{V(u)};$$

see Sect. A.11.

Remark 2.121 (Asymptotic Elasticity Conditions) Condition 2 of Assumption 2.117 holds if there exists $\hat{y} \in L^1$ such that

$$c\hat{y},\ \sigma_{D_t}(E_t[\hat{y}\Delta s_{t+1}]) \in L^1 \quad \forall t,$$

$EV^*(\hat{y}) < \infty$ and one of the following conditions hold:

1. there exist $\lambda \in (0, 1)$, $\bar{y} \in \operatorname{dom} EV^*$ and $C > 0$ such that

$$V^*(\lambda y, \omega) \le CV^*(y, \omega) \quad \forall y \in [0, \bar{y}(\omega)];$$

2. there exist $\lambda > 1$, $\bar{y} \in \operatorname{dom} EV^*$ and $C > 0$ such that

$$V^*(\lambda y, \omega) \le CV^*(y, \omega) \quad \forall y \ge \bar{y}(\omega).$$

If V is deterministic, then by Theorem A.87, condition 1 holds if $AE_-(V) < 1$ while condition 2 holds if $AE_+(V) > 1$. Most familiar loss functions, such as the exponential, logarithmic and power functions, satisfy these conditions; see Example A.88.

Proof Under condition 1,

$$\begin{aligned} V^*(\lambda\hat{y}) &= 1_{\{\hat{y}\le\bar{y}\}}V^*(\lambda\hat{y}) + 1_{\{\hat{y}>\bar{y}\}}V^*(\lambda\hat{y}) \\ &\le 1_{\{\hat{y}\le\bar{y}\}}V^*(\lambda\hat{y}) + 1_{\{\hat{y}>\bar{y}\}}\max\{V^*(\lambda\bar{y}), V^*(\hat{y})\} \\ &\le 1_{\{\hat{y}\le\bar{y}\}}CV^*(\hat{y}) + 1_{\{\hat{y}>\bar{y}\}}\max\{CV^*(\bar{y}), V^*(\hat{y})\}, \end{aligned}$$

where the first inequality comes from the convexity of V^*. Since $\bar{y}, \hat{y} \in \operatorname{dom} EV^*$ and $V^* \ge 0$ (by the assumption $V(0) = 0$), the last expression is integrable so $V^*(\lambda\hat{y})$ is integrable as well. By convexity, condition 2 of Assumption 2.117 thus holds with $\epsilon = (1-\lambda)/2$ and $y = (1-\epsilon)\hat{y}$.

Similarly, condition 2 gives

$$\begin{aligned}V^*(\lambda\hat{y}) &= 1_{\{\hat{y}\le\bar{y}\}}V^*(\lambda\hat{y}) + 1_{\{\hat{y}>\bar{y}\}}V^*(\lambda\hat{y})\\&\le 1_{\{\hat{y}\le\bar{y}\}}\max\{V^*(\hat{y}),\, V^*(\lambda\bar{y})\} + 1_{\{\hat{y}>\bar{y}\}}V^*(\lambda\hat{y})\\&\le 1_{\{\hat{y}\le\bar{y}\}}\max\{V^*(\hat{y}),\, CV^*(\bar{y})\} + 1_{\{\hat{y}>\bar{y}\}}CV^*(\hat{y})\end{aligned}$$

so $V(\lambda\hat{y})$ is again integrable. By convexity, condition 2 of Assumption 2.117 thus holds with $\epsilon = (\lambda - 1)/2$ and $y = (1+\epsilon)\hat{y}$. □

Note that if there are no portfolio constraints, then $\sigma_{D_t} = \delta_{\{0\}}$ so the condition $\sigma_{D_t}(E_t[y\Delta s_{t+1}]) \in L^1$ in Assumption 2.117 means that $E_t[y\Delta s_{t+1}] = 0$ almost surely. If Q is a probability measure with $dQ/dP = y$, then by Theorem B.15, this means that s is a martingale with respect to Q. We will see in Theorem 4.29 that, in the absence of portfolio constraints, the conditions in Remark 2.120 are equivalent to the existence of a martingale measure Q equivalent to P. By Remark 2.121, the lower bound in Lemma 2.74 then holds if

$$c\frac{dQ}{dP},\ V^*\left(\frac{dQ}{dP}\right) \in L^1$$

and V is deterministic with $AE_-(V) < 1$ or $AE_+(V) > 1$.

Example 2.122 (Superhedging) Problem (1.16) fits the general format of (1.11) with time running from $t = -1, \ldots T$, $\mathcal{F}_{-1} = \{\Omega, \emptyset\}$, $N = 1$, $M = |J|$, $A_t = 0$ for all $t = 0\ldots, T$, $B_0 = 0$, $B_t = \Delta s_t$ for $t = 1, \ldots T-1$, $W_t = 0$ for all t, $X_{-1} = \alpha$ and

$$\begin{aligned}L_T(X_T, U_T, \omega) &= \delta_{\mathbb{R}_-}(c(\omega) - X_T),\\L_t(X_t, U_t, \omega) &= \delta_{D_t}(U_t, \omega) \quad t = 0, \ldots, T-1,\\L_{-1}(X_{-1}, U_{-1}, \omega) &= X_{-1}.\end{aligned}$$

Assume that the problem is feasible and that

1. the set $\{x \in \mathcal{N} \mid \sum_{t=0}^{T-1} x_t \cdot \Delta s_{t+1} \ge 0,\ x_t \in D_t^\infty\ \forall t\ a.s.\}$ is linear,
2. there exists $y \in L^1_+$ such that $E[y] > 0$,

$$cy,\ y\Delta s_{t+1},\ \sigma_{D_t}(E_t[y\Delta s_{t+1}]) \in L^1 \quad t = 0, \ldots, T-1.$$

Then (2.24) has a unique solution given by

$$\begin{aligned}J_t(X_t) &:= \delta_{\mathbb{R}_-}(c_t - X_t) \quad t = T, \ldots, 0,\\J_{-1}(X_{-1}) &:= X_{-1} + \delta_{\mathbb{R}_-}(\operatorname{esssup} c_0 - X_{-1}),\end{aligned}$$

where $c_T = \text{esssup}_T c$ and

$$c_t(\omega) := \inf_{U_t \in D_t(\omega)} \sigma_{\text{esssup}_t\{(-\Delta s_{t+1}, c_{t+1})\}}(U_t, 1, \omega) \quad t = T-1, \ldots, 0.$$

Here each c_t is real-valued and $\mathcal{F}_t$-measurable. The process $(c_t)_{t=0}^T$ can be thought of as the superhedging cost process of the claim c. In particular, the optimum value of (1.16) equals $\text{esssup}\, c_0$ and the optimal trading strategies U are characterized by the conditions

$$\sigma_{\text{esssup}_t\{(-\Delta s_{t+1}, c_{t+1})\}}(U_t, 1, \omega) \le X_t, \quad U_t \in D_t(\omega),$$

where X is the wealth process corresponding to U and the initial wealth $\text{esssup}\, c_0$. The infimum in the definition of c_t is attained almost surely by an $\mathcal{F}_t$-measurable U_t, and such an U_t is optimal.

Proof We will verify the conditions of Example 2.96 so that both Theorems 2.91 and 2.94 are applicable. It is clear that the superhedging problem satisfies condition 2 of Example 2.96. Scaling if necessary, we may assume that $E[y] = 1$. Let $y_t := -y$ for all $t = 0, \ldots, T$ and

$$\begin{aligned} p_{-1} &:= (E[y] - y, 0), \\ p_t &:= (0, E_t[y\Delta s_{t+1}] - y\Delta s_{t+1}) \quad t = 0, \ldots T, \end{aligned}$$

where, again, $s_{T+1} := 0$. Clearly, $p \in \mathcal{N}^\perp$. We have

$$\begin{aligned} L_T^*(V_T, Y_T, \omega) &= \delta_{\mathbb{R}_- \times \{0\}}(V_T, Y_T) + V_T c(\omega), \\ L_t^*(V_t, Y_t, \omega) &= \delta_{\{0\}}(V_t) + \sigma_{D_t}(Y_t, \omega) \quad t = 0, \ldots, T-1, \\ L_{-1}^*(V_{-1}, Y_{-1}, \omega) &= \delta_{\{(1,0)\}}(V_{-1}, Y_{-1}). \end{aligned}$$

By Remark 2.92, the second assumption implies condition 1 in Example 2.96. By Theorem A.17,

$$\begin{aligned} L_T^\infty(X_T, U_T, \omega) &= \delta_{\mathbb{R}_-}(-X_T), \\ L_t^\infty(X_t, U_t, \omega) &= \delta_{D_t^\infty}(U_t, \omega) \quad t = 0, \ldots, T-1, \\ L_{-1}^\infty(X_{-1}, U_{-1}, \omega) &= X_{-1}, \end{aligned}$$

so condition 3 in Example 2.96 means that the set

$$\mathcal{L} := \{(X, U) \in \mathcal{N} \mid U_t \in D_t^\infty,\ \alpha - (X, U) \cdot p \le 0,\ X_T \ge 0,$$
$$\Delta X_t = U_t \cdot \Delta s_{t+1} \quad \forall t\ a.s.\}$$

is linear. Let $(X, U) \in \mathcal{L}$. Since $\mathcal{F}_{-1}$ is the trivial σ-algebra, taking the expectation of the constraint $\alpha - (X, U)\cdot p \leq 0$ implies, by Lemma 2.67, that $E[(X, U)\cdot p] = 0$ and then that $\alpha \leq 0$. Combining the system equations with the constraint $X_T \geq 0$, we get $\sum U_t \cdot \Delta s_{t+1} \geq -\alpha \geq 0$, so $\sum U_t \cdot \Delta s_{t+1} = 0$ and thus, $\alpha = 0$, by assumption 1. Since $E[(X, U)\cdot p] = 0$, the second constraint now gives $(X, U)\cdot p = 0$ almost surely. Thus $-(X, U) \in \mathcal{L}$, so $\mathcal{L}$ is linear. In summary, all the conditions of Example 2.96 hold.

By Theorem 2.94, (2.24) has a unique solution $(J_t, I_t)_{t=-1}^T$ and the mappings

$$M_t(\omega) := \{U_t \in \mathbb{R}^M \mid (E_t(L_t + I_t))^\infty(0, U_t, \omega) \leq 0\} \tag{2.41}$$

are linear-valued for every t. Assume that $(J_{t'})_{t'=t+1}^T$ in the statement solves (2.24) from $t + 1$ onward for $t \geq 0$. Since the problem is feasible, Theorem 2.91 implies that J_{t+1} is proper and thus, that $c_{t+1} < \infty$ almost surely. We have

$$(L_t + I_t)(X_t, U_t) = \delta_{\mathbb{R}_-}(c_{t+1} - X_t - U_t \cdot \Delta s_{t+1}) + \delta_{D_t}(U_t) = \delta_{\tilde{S}_t}(X_t, U_t),$$

where

$$\tilde{S}_t(\omega) = \{(X_t, U_t) \in \mathbb{R} \times \mathbb{R}^J \mid c_{t+1}(\omega) \leq X_t - U_t \cdot \Delta s_{t+1}(\omega),\ U_t \in D_t(\omega)\}.$$

Theorem 2.14 gives $E_t(L_t + I_t)(X_t, U_t) = \delta_{S_t}(X_t, U_t)$, where $S_t := \operatorname{essinf}_t \tilde{S}_t$. By Example 2.53,

$$S_t(\omega) = \{(X_t, U_t) \in \mathbb{R} \times \mathbb{R}^J \mid \sigma_{\operatorname{esssup}_t\{(-\Delta s_{t+1}, c_{t+1})\}}(U_t, 1, \omega) \leq X_t,\ U_t \in D_t(\omega)\},$$

so, by Theorem A.17,

$$S_t^\infty(\omega) = \{(X_t, U_t) \in \mathbb{R} \times \mathbb{R}^J \mid \sigma_{\operatorname{esssup}_t\{(-\Delta s_{t+1}, c_{t+1})\}}(U_t, 0, \omega) \leq X_t,\ U_t \in D_t^\infty(\omega)\}.$$

Thus

$$M_t(\omega) = \{U_t \in D_t^\infty(\omega) \mid \sigma_{\operatorname{esssup}_t\{(-\Delta s_{t+1}, c_{t+1})\}}(U_t, 0, \omega) \leq 0\}.$$

Since M_t is linear-valued, Theorem 1.40 implies that the infimum in the definition of c_t is attained, c_t is real-valued and

$$J_t(X_t) := \inf_{U_t \in \mathbb{R}^M} E_t(L_t + I_t)(X_t, U_t) = \delta_{\mathbb{R}_-}(c_t - X_t).$$

By induction on t, J_t in statement solve the Bellman equations (2.24) from $t = 0$ onward. It follows that $I_{-1}(X_{-1}, U_{-1}) = \delta_{\mathbb{R}_-}(c_0 - \alpha)$ so

$$(L_{-1} + I_{-1})(X_{-1}, U_{-1}) = \alpha + \delta_{[c_0,\infty)}(\alpha).$$

By Theorems 2.15 and 2.14,

$$J_{-1}(X_{-1}) = \inf_{U_{-1}\in\mathbb{R}^M} E_{-1}(L_{-1} + I_{-1})(X_{-1}, U_{-1}) = \alpha + \delta_{[\operatorname{esssup}_{-1} c_0,\infty)}(\alpha),$$

where $\operatorname{esssup}_{-1} c_0 = \operatorname{esssup} c_0$, since $\mathcal{F}_{-1} = \{\Omega, \emptyset\}$. By Theorem 2.91, the optimum value of (1.16) equals

$$\inf_{X_{-1}} J_{-1}(X_{-1}) = \operatorname{esssup} c_0$$

and the minimizers of $E_t(L_t + I_t)(X_t, U_t) = \delta_{S_t}(X_t, U_t)$ with respect to U_t are the optimal controls. □

Note that the optimality condition in Example 2.122 involves the value process X which depends on the optimum value of (1.16). However, the $\mathcal{F}_t$-measurable minimizers U_t in the definition of c_t are optimal but independent of X.

Remark 2.123 (Delta Hedging) We continue Example 2.122 and assume that the claim is of the form $c = -g(-s_T)$ for a continuous function $g : \mathbb{R}^J \to \overline{\mathbb{R}}$ and that the sets $C_t := \operatorname{esssup}_t\{-s_{t+1}\}$ form a deterministic increasing sequence of polyhedral convex sets (e.g. $C_t = \mathbb{R}^J_+$ for all t). If the portfolio constraints D_t are deterministic, then

$$c_t = -g_t(-s_t) \quad t = 0, \dots, T,$$

where $g_T = g$ and

$$g_t = [(g_{t+1} + \delta_{C_t})^* + \delta_{D_t}]^*.$$

In particular, if s_0 is deterministic, the superhedging cost of c is $-g_0(-s_0)$. The $\mathcal{F}_t$-measurable selections of $\partial g_t(-s_t)$ are optimal superhedging strategies. Such strategies are known as *delta hedging strategies*. They make the portfolio value and the superhedging cost c_t roughly equally sensitive to small movements of the underlying prices s.

Proof Assume that $c_{t+1} = -g_{t+1}(-s_{t+1})$ and that g_{t+1} is continuous on C_t. By Theorem 2.43,

$$\begin{aligned}
\operatorname{esssup}_t\{(-\Delta s_{t+1}, c_{t+1})\} &= (s_t, 0) + \operatorname{esssup}_t\{(-s_{t+1}, -g_{t+1}(-s_{t+1}))\} \\
&= (s_t, 0) + \operatorname{cl}\{(-\xi, -g_{t+1}(-\xi)) \mid \xi \in \operatorname{esssup}_t\{s_{t+1}\}\} \\
&= (s_t, 0) + \operatorname{cl}\{(\xi, -g_{t+1}(\xi)) \mid \xi \in C_t\}.
\end{aligned}$$

It follows that

$$\sigma_{\mathrm{esssup}_t\{(-\Delta s_{t+1},c_{t+1})\}}(U_t,1) = U_t\cdot s_t + \sup_\xi\{\xi\cdot U_t - g_{t+1}(\xi)\mid \xi\in C_t\}\}$$
$$= U_t\cdot s_t + (g_{t+1}+\delta_{C_t})^*(U_t)$$

so

$$\begin{aligned} c_t &= \inf_{U_t\in D_t}\{U_t\cdot s_t + (g_{t+1}+\delta_{C_t})^*(U_t)\}\\ &= -[(g_{t+1}+\delta_{C_t})^* + \delta_{D_t}]^*(-s_t)\\ &= -g_t(-s_t). \end{aligned}$$

Since $(g_{t+1}+\delta_{C_t})^*+\delta_{D_t}\ge (g_{t+1}+\delta_{C_t})^*$, we have

$$g_t \le (g_{t+1}+\delta_{C_t})^{**} \le g_{t+1}+\delta_{C_t}.$$

Thus, since C_t are increasing in t, the induction assumption gives $C_t\subseteq \operatorname{dom} g_t$. On the other hand, since g_t is closed and since c_t is almost surely real-valued (see Example 2.122), g_t has to be real-valued on C_t. It is not difficult to show that a real-valued lsc convex function on a polyhedral convex set is continuous. The first claim thus follows by induction.

As observed in Example 2.122, the $\mathcal{F}_t$-measurable scenariowise minimizers U_t in the definition of c_t are optimal superhedging strategies. By the above, they are the $\mathcal{F}_t$-measurable selections of

$$\operatorname*{argmin}_{U_t}\{U_t\cdot s_t + (g_{t+1}+\delta_{C_t})^*(U_t)+\delta_{D_t}(U_t)\},$$

or equivalently, of $\partial g_t(-s_t)$; see Sect. A.6. □

The following gives a simple existence result for the hedging problem when the hedging error is measured by the variance. Variance is a popular criterion in risk management and portfolio optimization but one should note that it is not quite rational as it does not distinguish between profits and losses.

Example 2.124 (Variance-Optimal Hedging) The problem

$$\text{minimize}\quad E\left(\alpha+\sum_{t=0}^{T-1}x_t\cdot\Delta s_{t+1}-c\right)^2\quad \text{over } \alpha\in\mathbb{R},\ x\in\mathcal{N}$$

has a solution. The variable α can be interpreted as an initial value of a self-financing trading strategy x. The problem fits the format of linear-quadratic control in Example 2.104 so the dynamic programming equations have a particularly simple form.

Proof This fits the general format (P) with t running from -1 to $T-1$, $\mathcal{F}_{-1} = \{\Omega, \emptyset\}$, $x_{-1} = \alpha$ and

$$h(x,\omega) = (\alpha + \sum_{t=0}^{T-1} x_t \cdot \Delta s_{t+1}(\omega) - c(\omega))^2.$$

By Theorem A.17,

$$h^\infty(x,\omega) = \begin{cases} 0 & \text{if } \alpha + \sum_{t=0}^{T-1} x_t \cdot \Delta s_{t+1}(\omega) = 0, \\ +\infty & \text{otherwise,} \end{cases}$$

so Assumption 2.64 holds and the claim follows from Theorem 2.65. □

Example 2.125 (Rates of Return) Consider the formulation

$$\begin{aligned} &\text{minimize} && EV(c - X_T) \quad \text{over} \quad (X,U) \in \mathcal{N}, \\ &\text{subject to} && \Delta X_t = R_t \cdot U_{t-1} \quad \forall t = 1,\dots,T, \\ &&& (X_t, U_t) \in C_t \quad \forall t = 0,\dots,T-1 \end{aligned}$$

from Example 1.91. This fits the control format with $N = 1$, $M = |J|$, $A_t = 0$, $B_t = R_t$, $W_t = 0$ and

$$\begin{aligned} L_T(X_T, U_T, \omega) &= V(c(\omega) - X_T, \omega), \\ L_t(X_t, U_t, \omega) &= \delta_{C_t(\omega)}(X_t, U_t) \quad t = 0,\dots,T-1. \end{aligned}$$

Assuming that V and c are $\mathcal{F}_T$-measurable, the dynamic programming equations (2.24) can be written as

$$\begin{aligned} J_T(X_T,\omega) &= V(c(\omega) - X_T), \\ I_t(X_t, U_t, \omega) &= J_{t+1}(X_t + R_{t+1}(\omega) \cdot U_t, \omega), \\ J_t(X_t,\omega) &= \inf_{U_t \in \mathbb{R}^J} \{(E_t I_t)(X_t, U_t, \omega) \mid (U_t, X_t) \in C_t(\omega)\}. \end{aligned} \tag{2.42}$$

Under the assumptions of Theorem 2.91, the optimal controls U_t are the $\mathcal{F}_t$-measurable selections of the *optimal feedback mapping* $S_t : \mathbb{R}^N \times \Omega \rightrightarrows \mathbb{R}^M$ defined by

$$S_t(X_t,\omega) := \operatorname*{argmin}_{U_t \in \mathbb{R}^M} Q_t(X_t, U_t, \omega),$$

where

$$Q_t(X_t, U_t, \omega) := (E_t I_t)(X_t, U_t, \omega) + \delta_{C_t(\omega)}(U_t, X_t).$$

Example 2.126 (Exponential Utility) Consider the model of Example 2.125 in the case where $C_t(\omega) = \mathbb{R} \times \tilde{D}_t(\omega)$ for $t > 0$ and $V(u) = \exp(\rho u)/\rho$ for a given initial wealth $w \in \mathbb{R}$ and risk aversion parameter $\rho > 0$. A simple induction argument in (2.42) gives

$$J_t(X_t) = \alpha_t V(-X_t) \quad t = T, \dots, 1,$$
$$S_t(X_t) = \operatorname*{argmin}_{U_t \in \tilde{D}_t} E_t[\alpha_{t+1} \exp(-\rho R_{t+1} \cdot U_t)],$$

where $\alpha_T = \exp(\rho c)$ and, for $t = T - 1, \dots, 1$,

$$\alpha_t = \inf_{U_t \in \tilde{D}_t} E_t[\alpha_{t+1} \exp(-\rho R_{t+1} \cdot U_t)].$$

Note that S_t does not depend on the level of wealth X_t.

Assuming further that V, J_t and I_t are L-bounded and each R_{t+1} is independent of $\mathcal{F}_t$ and that $\tilde{D}_t$ and c are deterministic, it follows by induction on t and by Corollary 2.101 that α_t, Q_t and S_t are deterministic. If, in addition, each R_t is normally distributed with mean μ_t and covariance Σ_t, then

$$\begin{aligned}
&E_t[\alpha_{t+1} e^{-\rho R_{t+1} \cdot U_t}]\\
&= \alpha_{t+1} E[e^{-\rho R_{t+1} \cdot U_t}]\\
&= \alpha_{t+1} \int_{\mathbb{R}^J} e^{-\rho R \cdot U_t} \frac{1}{\sqrt{(2\pi)^{|J|} |\Sigma_{t+1}|}} e^{-\frac{1}{2}(R-\mu_{t+1}) \cdot \Sigma_{t+1}^{-1}(R-\mu_{t+1})} dR\\
&= \alpha_{t+1} \int_{\mathbb{R}^J} e^{-\rho U_t \cdot \mu_{t+1} + \frac{\rho^2}{2} U_t \cdot \Sigma_{t+1} U_t} \frac{1}{\sqrt{(2\pi)^{|J|} |\Sigma_{t+1}|}}\\
&\qquad e^{-\frac{1}{2}(R-\rho\Sigma_{t+1}U_t-\mu_{t+1}) \cdot \Sigma_{t+1}^{-1}(R-\rho\Sigma_{t+1}U_t-\mu_{t+1})} dR\\
&= \alpha_{t+1} e^{-\rho U_t \cdot \mu_{t+1} + \frac{\rho^2}{2} U_t \cdot \Sigma_{t+1} U_t}.
\end{aligned}$$

When there are no portfolio constraints, the last expression is minimized by $S_t(X_t) = \Sigma_t^{-1} \mu_t / \rho$. This means that, if Σ_t and μ_t are constant, it is optimal to hold a constant amount of wealth in the risky assets. Note, however, that the assumption that R_t be normally distributed is questionable. For example, when the prices are strictly positive, each R_t is componentwise larger than -1.

The following example covers utility functions with constant relative risk aversion (CRRA); see Remark A.85.

Example 2.127 (CRRA Utility) Consider the model of Example 2.125 and assume that $c = 0$, C_t are conical, V is p-homogeneous with integrable C (see Lemma 2.22) and that $(I_t, J_t)_{t=0}^T$ is an L-bounded solution of (2.42). Since indicator functions of convex cones are p-homogeneous, we get, as in Remark 2.99, that I_t, J_t and Q_t are p-homogeneous and

$$S_t(\alpha X_t) = \alpha S_t(X_t) \quad \forall \alpha > 0.$$

If $X_t > 0$, then

$$\frac{1}{X_t} S_t(X_t) = S_t(1)$$

which means that the elements U_t of the optimal feedback mapping are such that the proportions U_t/X_t of wealth allocated in the risky assets do not depend on the level of wealth.

Example 2.128 (Canonical Representation) Let $\xi = (\xi_t)_{t=0}^{T+1}$ be a stochastic process and consider the problem in Example 2.125 in the canonical form

$$\begin{aligned} &\text{minimize} && EV(\hat{c}(\xi^T) - X_T) \quad \text{over} \quad (X, U) \in \mathcal{N}, \\ &\text{subject to} && X_0 = w, \\ &&& \Delta X_t = \hat{R}_t(\xi^t) \cdot U_{t-1} \quad \forall t = 1, \dots, T, \\ &&& (X_t, U_t) \in \hat{C}_t(\xi^t) \quad \forall t = 0, \dots, T-1, \end{aligned}$$

where $\hat{c}$ is $\mathcal{A}^T$-measurable, and $\hat{R}_t$ and $\hat{C}_t$ are $\mathcal{A}^t$-measurable. The process ξ could simply be the price process s but it could also involve other risk factors. The problem fits the format of Example 2.102 with

$$\begin{aligned} \hat{L}_T(X_T, U_T, \xi^{T+1}) &= V(\hat{c}(\xi^T) - X_T), \\ \hat{L}_t(X_t, U_t, \xi^{t+1}) &= \delta_{\hat{C}_t(\xi^t)}(X_t, U_t). \end{aligned}$$

Assume that each L_T is L-bounded and that ξ_{t+1} has regular ξ^t-conditional distribution ν_t. Then sequences $(J_t, I_t)_{t=0}^T$ of L-bounded normal integrands satisfy (2.42) if and only if there exist $\mathcal{A}^t$-measurable normal integrands $\hat{J}_t : \mathbb{R} \times \Xi^t \to \overline{\mathbb{R}}$ and $\mathcal{A}^{t+1}$-measurable normal integrands $\hat{I}_t : \mathbb{R} \times \mathbb{R}^{|J|} \times \Xi^{t+1} \to \overline{\mathbb{R}}$ such that

$$J_t(X_t, \omega) = \hat{J}_t(X_t, \xi^t(\omega)), \quad I_t(X_t, U_t, \omega) = \hat{I}_t(X_t, U_t, \xi^{t+1}(\omega)),$$

and

$$\hat{I}_T = 0,$$
$$\hat{J}_t(X_t, s^t) = \inf_{U_t \in \mathbb{R}^M} E^{\nu_t}(\hat{L}_t + \hat{I}_t)(X_t, U_t, s^t),$$
$$\hat{I}_{t-1}(X_{t-1}, U_{t-1}, s^t) = \hat{J}_t(X_{t-1} + \hat{R}_t(s^t) \cdot U_{t-1}, s^t)$$

P_{ξ^t}-almost surely for all $t = T, \dots, 0$.

If $(\xi_t)_{t=0}^T$ is Markov and $\hat{R}_{t+1}$ and $\hat{C}_t$ are independent of s^{t-1} and $\hat{c}$ is independent of s^{T-1}, then Q_t and S_t are $\sigma(\xi_t)$-measurable. If $(\xi_t)_{t=0}^T$ is a sequence of independent random variables and $\hat{R}_{t+1}$ and $\hat{C}_t$ are independent of s^t and $\hat{c}$ is deterministic, then Q_t and S_t are deterministic.

Note that claims of the form $c(\omega) = \hat{c}(\xi_T(\omega))$ cover many path-dependent options. For example, Asian options whose payouts are functions of the running average of a price process s^k for some $k \in J$ can be modelled by adding to ξ the scalar process $A_t = \sum_{t'=0}^t s_{t'}^k$ that satisfies $A_{t+1} = A_t + (1 + R_{t+1}^k)s_t^k$. Similarly, options whose payouts depend on the running maximum $M_t = \sup_{t' \le t} s_{t'}^k$ can be modelled with $M_{t+1} = \max(M_t, (1 + R_{t+1}^k)s_t^k)$; similarly with the running minimum. If the price process s is Markov, the process $\xi := (s, A, M, R)$ is Markov as well.

2.4 Bibliographical Notes

The generalized Bellman equations for problem (P) were first studied in [131] and [53]. They extend many earlier formulations of the stochastic dynamic programming principle such as those in [7, 9, 36, 47]. Like the present chapter, [131] studied the convex case where decisions are described by finite dimensional vectors. The results were extended in [53] to nonconvex problems where decisions are described by elements of Polish spaces. Both assumed that the objective is lower bounded and that the decisions are taken from a compact set, uniformly compact in [131]. The compactness assumptions were relaxed in [101] in the convex case and in [109] in the nonconvex case. The general results on L-bounded objectives presented in Sect. 2.2 are from [107]. Dynamic programming formulations in terms of the essential infimums of random variables as in Remark 2.3 and Remark 2.4 have been studied e.g. in [150] and in [120] in the context of problem (1.15).

Conditional expectations of convex normal integrands were introduced in [19]; see also [21, 22]. The definition was extended to general (possibly nonconvex) lower-bounded $\mathcal{B} \otimes \mathcal{F}$-measurable integrands in [48]. More general conditions for the existence of a conditional expectation of a normal integrand have been given in [32, 152, 153]; see [153] for a more detailed survey. Our approach and the conditions for existence in Sect. 2.1.1 are close to those in [22]. Indeed, by Lemma 2.5, a convex normal integrand is L-bounded if and only if dom $Eh^* \cap L^1 \neq \emptyset$ which is the condition used in [22, Theorem 2]. The essential infimum of a random set in Theorem 2.14 was introduced in [153] under the name of 'conditional core'.

Most results in Sect. 2.1.2 can be found in the above references. Theorem 2.21, Lemma 2.22 and Theorem 2.24 and parts 2 and 3 of Theorem 2.15 seem to be new. Theorems 2.23 and 2.26 can be found in [22]. The first part of Corollary 2.29 is from [63, Theorem 5.2] while the last two seem new. Corollary 2.28 and Theorem 2.31 can be found in [153]. Theorem 2.34 extends Corollary 1 of [22, Theorem 3] by slightly relaxing the assumptions on the domains of the integral functionals. The Jensen's inequality in Theorem 2.35 extends Corollary 1 in Section 2.1.2 of [153] by relaxing the integrability requirements. Theorem 2.36 and Corollary 2.37 on the necessity of the measurability condition seem to be new.

The closed convex hull of the essential supremum in Theorem 2.39 coincides with the 'conditional convex hull' introduced in [82]. In complete probability spaces, our definition coincides with the definition of 'conditional closure' in [51]. The argument in the proof of Theorem 2.39 is close to that of [51, Theorem 3.1]. In [51, 82], the mappings were assumed graph-measurable and the underlying probability space was complete. In this case, graph-measurability is equivalent to measurability; see Remark 1.4.

The additivity property of the conditional essential supremum in Theorem 2.43.5 can be found in [82, Proposition 5.4]. Theorem 2.43.1 is essentially [51, Lemma 3.1]. By the last claim of Theorem 2.44, the first formula in Corollary 2.50 is equivalent to the equality in [82, Theorem 5.2]. The first equality of Corollary 2.51 was given in [82, Proposition 5.5]. Corollary 2.50 was essentially stated after [82, Proposition 5.5] without proof. The remaining statements in Sect. 2.1.4 seem to be new. In particular, the notions of conditional essential supremum and infimum of a normal integrand seem to appear here for the first time.

In [131], the conditional expectation of normal integrand is defined via regular conditional distributions of random variables as in Sect. 2.1.5. By Theorem 2.57 such a definition coincides with Definition 2.1 as soon as the regular conditional distribution exists and the integrand is L-bounded. Lemma 2.55 generalizes [131, Lemma 2] by relaxing the assumptions that the integrand be inf-compact with nonanticipative domain. Lemma 2.55 is close to [153, Theorem 1.7] which is concerned with jointly measurable integrands that are lsc in the x-argument. Other than that, the results in Sect. 2.1.5 seem new.

The optimality conditions in Theorem 2.2 and the existence results for the generalized Bellman equations for lower bounded integrands in Theorem 2.62 are from [101]. They extend those of [131], and of [53] in the convex case, by relaxing the compactness assumption on the set of feasible strategies. The extensions to L-bounded integrands in Sect. 2.2.2 are from [107]. The extensions are largely based on Lemma 2.67 which first appeared in [110] where it was used to extend the main result of [101] on the lower semicontinuity of the optimum value function of (P); see Chap. 4 below. In the literature of stochastic control, results such as Theorem 2.2 and Theorem 2.70 relating the solutions of the Bellman equations to the solutions of the optimization problem are often called "verification theorems"; see e.g. [56]. Example 2.68 is a classic result in stochastic analysis; see e.g. Theorems 1 and 2 of [66].

The notions of induced constraints and relatively complete recourse studied in Sect. 2.2.3 were introduced in the case of two-stage models in [133, 157]. In the case of multistage models, relatively complete recourse was used in [131, 138] to prove the existence of dual solutions, a topic that will be studied in the following chapters. In [131], the domain dom h was assumed closed and uniformly bounded so that the closures in Corollary 2.79 would be superfluous. The boundedness was relaxed in the "bounded recourse condition" of [138] in the context of stochastic problems of Bolza. This condition will be extended to general convex stochastic optimization problems in Example 5.39. Related concepts are analyzed in [45, Chapter 4.4] but one should note that the definition of "relatively complete recourse" in [45, Chapter 4.4] differs from that in [131] and Sect. 2.2.3. The general formulations of induced constraints in terms of conditional essential infimums of random sets in Sect. 2.2.3 seem to be new. The canonical representation of the Bellman equations in Sect. 2.2.4 goes back to [131]. Theorem 2.84 extends [131, Theorem 1] by relaxing many of its assumptions.

Unlike the other applications studied in Sect. 2.3, the mathematical programming models in Sect. 2.3.1 lack the temporal structure that allows for the dimension reductions in the dynamic programming recursions found e.g. in optimal control. The existence results in Sect. 2.3.1 extend those of [136] by relaxing the compactness assumption on the set of feasible solutions. Existence of solutions to Bellman equations for linearly constrained problems have been established also in [92]. The results on optimal stopping problems in Sect. 2.3.2 are classic in stochastic analysis (see e.g. [111]) but our proof via the convex relaxation (1.10) seems new.

The dynamic programming equations (2.24) for optimal control (1.11) are essentially the same as those in e.g. [12, Section 2.2], [14, Section 1.2], [2, page 100] and [29, Proposition 4.12] when applied to the convex case. Similar equations have been extensively studied in the vast literature on Markov decision processes; see e.g. [6, 116, 118]. We have introduced the functions I_t to clarify that the conditional expectations are taken in the sense of normal integrands. This resolves many of the measurability problems that arise in earlier formulations. Essentially, our formulation of (1.11) in terms of normal integrands is what [14, Section 1.2.II] calls "semicontinuous models". In [14], these models were, however, interpreted quite narrowly with the exclusion of even the classical linear quadratic model in Example 2.104. The discrepancy seems to have come from the belief that "in the usual stochastic programming model, the controls cannot influence the distribution of future states"; see page 12 of [14]. In the models of Sect. 2.3.3, however, the controls do influence the distribution of future states. What is crucial for convexity is that the state equations are affine. In nonconvex dynamic programming recursions of [53, 109], even this assumption can be relaxed. The literature on optimal stochastic control is vast but the optimality conditions and the existence results for the convex control problems in Theorem 2.91 and Theorem 2.94 seem new in the presented generality. The formulation of the dynamic programming recursion in terms of the Q-functions in (2.29) is used in the context of reinforcement learning e.g. in [13]. Example 2.99 on p-homogeneous objectives seems new. It is inspired by the more specific results on financial models in Example 2.127, the idea of which goes back to Merton [85] and Samuelson [143]. The results concerning problems of Lagrange in Sect. 2.3.4 are new but a closely related model was treated in [101, Corollary 4].

The existence result for portfolio optimization in Theorem 2.118 extends those in Theorems 2.7 and 2.10 of [120] by allowing for portfolio constraints and more general utility functions. In models without portfolio constraints, the equivalence of 1 and 4 in Remark 2.120 was stated in [66, Theorem 3] (where the statement has a typo as one should have the discounted price process in condition (g)). Example 2.122 and Remark 2.123 are closely related to [51, Section 4.2] which studied 'robust superhedging' of contingent claims; see also [52]. The existence result for variance-optimal hedging in Example 2.124 extends the existence result in [57, Theorem 10.39]. The properties of the optimal trading strategies in Example 2.126 and Example 2.127 are classic in the literature of financial mathematics; see e.g. [85]. The implications of Markovianity in Example 2.128 in the presence of a random claim C seem new.

Chapter 3
Duality

In the remaining three chapters, we will study problem (P) within the functional analytic duality framework of Sect. A.9. This will yield a dual problem whose optimum value coincides with that of (P) and whose optimal solutions can be used to characterize those of (P). The purpose of the present chapter is to lay out the duality framework while the remaining two chapters will give sufficient conditions for the absence of a duality gap and the existence of primal and dual solutions, respectively.

Problem (P) does not directly fit into the general duality theory of Sect. A.9 as the spaces L^0 and $\mathcal{N}$ fail to be locally convex, in general. We will therefore first, in Sect. 3.2, develop a duality theory for the optimization problem

$$\text{minimize } Eh(x) := \int h(x(\omega), \omega) dP(\omega) \text{ over } x \in \mathcal{X} \cap \mathcal{N} \tag{$P_{\mathcal{X}}$}$$

obtained from (P) by posing the additional restriction that the strategy x belongs to a given locally convex space $\mathcal{X} \subseteq L^0(\Omega, \mathcal{F}, P; \mathbb{R}^n)$. Section 3.3 will then extend the duality theory to the original problem (P) where we optimize over general adapted strategies in L^0.

We will assume that the normal integrand h is of the form

$$h(x, \omega) = f(x, \bar{u}(\omega), \omega)$$

for a convex normal integrand f on $\mathbb{R}^n \times \mathbb{R}^m \times \Omega$ and an element $\bar{u}$ of a locally convex vector space $\mathcal{U}$ of $\mathbb{R}^m$-valued random variables. Here m is a given integer. In some applications, the random vector $\bar{u}$ is introduced only in order to dualize (P) but in others, it has practical significance. For example, in problems of financial mathematics, $\bar{u}$ can be used to describe the cash-flows of assets or liabilities. In stochastic control, $\bar{u}$ may be taken to be the additive noise in the system equations. The details and further applications will be given in Sect. 3.4 below.

T. Pennanen, A.-P. Perkkiö, *Convex Stochastic Optimization*, Probability Theory and Stochastic Modelling 107, https://doi.org/10.1007/978-3-031-76432-5_3

We will apply the duality theory of Sect. A.9 with the Rockafellian $F : \mathcal{X} \times \mathcal{X} \times \mathcal{U} \to \overline{\mathbb{R}}$ defined by

$$F(x, z, u) := Ef(x, u) + \delta_{\mathcal{N}}(x - z).$$

Clearly, $F(x, 0, \bar{u}) = Eh(x) + \delta_{\mathcal{N}}(x)$ so problem $(P_{\mathcal{X}})$ is the same as that of minimizing $F(x, 0, \bar{u})$ over $x \in \mathcal{X}$. We are thus in the setting of Sect. A.9 with the dualizing parameter $(z, u) \in \mathcal{X} \times \mathcal{U}$ and $\bar{v} = 0$. One should note the slight conflict of notation with the appendix as (z, u) and $(0, \bar{u})$ here correspond to u and $\bar{u}$, respectively, in Sect. A.9. According to the general theory in Sect. A.9, the dual problem associated with the above parameterization of $(P_{\mathcal{X}})$ is the maximization problem

$$\text{maximize } \langle \bar{u}, y \rangle - F^*(0, p, y) \text{ over } (p, y) \in \mathcal{V} \times \mathcal{Y}, \tag{D}$$

where $\mathcal{V}$ and $\mathcal{Y}$ are linear spaces in separating duality with $\mathcal{X}$ and $\mathcal{U}$, respectively, and F^* is the corresponding convex conjugate of F.

In order to write the dual problem more explicitly, we will assume that the spaces $\mathcal{X}, \mathcal{U}, \mathcal{V}$ and $\mathcal{Y}$ are decomposable spaces of random variables. We will also assume that they are "solid" in the sense defined in Sect. 3.1.1 below. This allows us to apply the duality theory of integral functionals reviewed in Sect. 3.1. It turns out that the dual variable associated with the nonadapted perturbation z of x is a random variable p that becomes a "shadow price of information" in the optimality conditions. The optimality conditions turn out to be scenariowise conditions that reduce to the classical Karush-Kuhn-Tucker-Rockafellar conditions in the deterministic case; see Sect. A.9.

After deriving the duality framework for problem $(P_{\mathcal{X}})$ in Sect. 3.2, we will move back to the original problem (P) in Sect. 3.3 and find that its optimum value as a function of the parameters (z, u) has the same lower semicontinuous hull as that of problem $(P_{\mathcal{X}})$. It follows that their dual problems coincide and, by a scenariowise application of Fenchel inequality and Lemma 2.67, we find that the two problems have the same scenariowise optimality conditions. Being able to treat the original problem (P) instead of its restriction $(P_{\mathcal{X}})$ to a locally convex space $\mathcal{X}$ is important in applications where primal solutions cannot be found in $\mathcal{X}$ but only in the space $\mathcal{N}$ of all adapted processes. This is typical e.g. in problems of financial mathematics.

In many stochastic optimization duality results in the literature, the dual problems and optimality conditions are given in terms of the dual variable y only, without the shadow price of information p. Under natural measurability conditions, such dual problems can be recovered from our general duality framework by considering a reduced dual problem obtained by optimizing the dual objective over the shadow price of information for a given y. Moreover, the shadow price of information can often be expressed explicitly in terms of y.

3.1 Integral Functionals in Duality

Convex duality is based on the theory of conjugate functions on dual pairs of locally convex topological vector spaces; see Sects. A.6 and A.9. The first part of this section reviews dual pairs of spaces of random variables while the second part reviews conjugation of integral functionals on such spaces. This forms the functional analytic setting for the duality theory of stochastic optimization developed in the following sections. For full generality, we make minimal assumptions on the spaces of random variables.

3.1.1 Dual Spaces of Random Variables

Let $\mathcal{U}$ and $\mathcal{Y}$ be decomposable linear spaces (see Sect. 1.1.7) of $\mathbb{R}^m$-valued random variables such that $u \cdot y \in L^1$ for all $u \in \mathcal{U}$ and $y \in \mathcal{Y}$. We will assume that $\mathcal{U}$ and $\mathcal{Y}$ are in *separating duality* under the bilinear form

$$\langle u, y\rangle := E[u \cdot y]$$

in the sense that for every nonzero $u \in \mathcal{U}$, there exists a $y \in \mathcal{Y}$ such that $\langle u, y\rangle \neq 0$ and vice versa. As usual, we identify random variables that coincide almost surely so the elements of $\mathcal{U}$ and $\mathcal{Y}$ are actually *equivalence classes* of random variables.

We will also assume that the spaces are *solid* in the sense that if $\bar{u} \in \mathcal{U}$ and $u \in L^0$ are such that $|u^i| \leq |\bar{u}^i|$ almost surely for every $i = 1, \dots, m$, then $u \in \mathcal{U}$; similarly for $\mathcal{Y}$. Solidity implies that

$$\mathcal{U} = \mathcal{U}_1 \times \cdots \times \mathcal{U}_m \quad \text{and} \quad \mathcal{Y} = \mathcal{Y}_1 \times \cdots \times \mathcal{Y}_m, \tag{3.1}$$

where $\mathcal{U}_i$ and $\mathcal{Y}_i$ are solid decomposable linear spaces of real-valued random variables in separating duality under the bilinear form $(u_i, y_i) \mapsto E[u_i y_i]$. In particular,

$$u_i y_i \in L^1 \quad \text{and} \quad \langle u, y\rangle = \sum_{i=1}^m E[u_i y_i] \quad \forall u \in \mathcal{U},\ y \in \mathcal{Y}. \tag{3.2}$$

Remark 3.1 A linear space $\mathcal{U}$ of random variables is solid, in particular, if

$$u \in L^0,\ \bar{u} \in \mathcal{U},\ |u| \leq |\bar{u}| \Rightarrow u \in \mathcal{U}$$

where, as usual, $|\cdot|$ denotes the Euclidean norm on $\mathbb{R}^m$. This stronger property means that there exists a solid linear space $\mathcal{U}_0$ of real-valued random variables such that

$$\mathcal{U} = \{u \in L^0 \mid |u| \in \mathcal{U}_0\} \tag{3.3}$$

or, equivalently, that

$$\mathcal{U} = \mathcal{U}_0^m.$$

A benefit of our general definition of solidity is that it does not require all components of u to belong to the same space.

Proof The stronger solidity property means that

$$\mathcal{U} = \{u \in L^0 \mid \exists \bar{u} \in \mathcal{U} : |u| \leq |\bar{u}| \ a.s.\}$$

which means that (3.3) holds with

$$\mathcal{U}_0 := \{\xi \in L^0 \mid \exists u \in \mathcal{U} : |\xi| \leq |u| \ a.s.\}.$$

Linearity and solidity together with (3.3) imply $\mathcal{U} = \mathcal{U}_0^m$. Assume now that $\mathcal{U} = \mathcal{U}_0^m$ for a linear solid $\mathcal{U}_0$. Let $u \in L^0$ and $\bar{u} \in \mathcal{U}$ with $|u| \leq |\bar{u}|$. There is a constant $c > 0$ such that

$$c \sum_{i=1}^m |u^i| \leq |u| \leq |\bar{u}| \leq \sum_{i=1}^m |\bar{u}^i|,$$

where $\sum_i |\bar{u}^i| \in \mathcal{U}_0$, by linearity. Thus, for each i, $|u^i| \leq \sum_i |\bar{u}^i|/c$ so $u^i \in \mathcal{U}_0$, by solidity. □

Most spaces of random variables encountered in applications are solid and decomposable. Examples include the classical Lebesgue, Orlicz and Lorentz spaces as well as spaces of finite moments; see Sect. 5.1.1. Cartesian products of solid and decomposable spaces are solid and decomposable. If $\mathcal{U}_i$ is in separating duality with $\mathcal{Y}_i$, then $\mathcal{U}_1 \times \cdots \times \mathcal{U}_m$ is in separating duality with $\mathcal{Y}_1 \times \cdots \times \mathcal{Y}_m$. Spaces that do not satisfy our assumptions include the spaces of continuous functions or various Sobolev spaces of functions on $\mathbb{R}^n$ as they are neither decomposable nor solid. The space L^0 of all random variables is decomposable and solid, but if $(\Omega, \mathcal{F}, P)$ is atomless, it cannot be paired with a nontrivial space of random variables. Indeed, if $y \in L^0$ is nonzero, then, by Remark 1.56, $u \mapsto E[u \cdot y]$ is improper on L^0.

In applications, one is often given a decomposable space $\mathcal{U} \subseteq L^1$ of random variables and then needs to find an appropriate dual space $\mathcal{Y}$. In general, there are several possibilities. The smallest decomposable space that is in separating duality with $\mathcal{U}$ with respect to the bilinear form $\langle u, y \rangle = E[u \cdot y]$ is the space L^∞ of essentially bounded functions. The largest one is the *Köthe dual*

$$\mathcal{U}' := \{y \in L^0 \mid u \cdot y \in L^1 \quad \forall u \in \mathcal{U}\}.$$

Köthe duals have simple characterizations for many familiar spaces of random variables; see Sect. 5.1.1. For example, $(L^p)' = L^q$ for any $p \in [1, \infty]$ and the usual conjugate exponent q of p. The following is easily verified.

Remark 3.2 Given a linear space $\mathcal{U}$ of random variables, we have

1. If $L^\infty \subseteq \mathcal{U}$, then $\mathcal{U}' \subseteq L^1$;
2. If $\mathcal{U} \subseteq L^1$, then $L^\infty \subseteq \mathcal{U}'$;
3. If $u1_A \in \mathcal{U}$ for all $u \in \mathcal{U}$ and $A \in \mathcal{F}$, then $y1_A \in \mathcal{U}'$ for all $y \in \mathcal{U}'$ and $A \in \mathcal{F}$;
4. If $\mathcal{U}$ is solid, then $\mathcal{U}'$ is solid.

In particular, if $\mathcal{U} \subseteq L^1$ is solid and decomposable, then $\mathcal{U}' \subseteq L^1$ is solid and decomposable.

A solid space containing all constant functions is decomposable. The following shows that the converse does not hold in general.

Counterexample 3.3 Let $u \geq 1$ be an unbounded real-valued random variable and let $\mathcal{U}$ be the sum of L^∞ and the linear span of the set $\{u1_A \mid A \in \mathcal{F}\}$. Then $\mathcal{U}$ is decomposable, by construction, but not solid, since it does not contain $\sqrt{u}$. Indeed, assume that

$$\sqrt{u} = \bar{u} + \sum_{\nu=1}^{N} \alpha^\nu u 1_{A^\nu}$$

for some $\bar{u} \in L^\infty$, a finite partition $(A^\nu)_{\nu=1}^N$ and $\alpha^\nu \in \mathbb{R}$. Since $\sqrt{u}$ is unbounded, $1_{A^\nu} u$ has to be unbounded for some ν. We have

$$\sqrt{u} - \alpha^\nu u = \bar{u}$$

on A^ν, which is impossible if the left side is unbounded.

Given a topology on $\mathcal{U}$, the corresponding *topological dual* of $\mathcal{U}$ is the linear space of all continuous linear functionals on $\mathcal{U}$. A topology is *compatible* with the bilinear form on $\mathcal{U} \times \mathcal{Y}$ if every continuous linear functional on $\mathcal{U}$ can be expressed in the form

$$u \mapsto \langle u, y \rangle$$

for some $y \in \mathcal{Y}$. Such topologies can be characterized in terms of the "weak" and "Mackey" topologies associated with the bilinear form. The *weak topology* $\sigma(\mathcal{U}, \mathcal{Y})$ on $\mathcal{U}$ is the topology generated by the linear functionals $u \mapsto \langle u, y \rangle$ where $y \in \mathcal{Y}$. Similarly for $\mathcal{Y}$. The *Mackey topology* $\tau(\mathcal{U}, \mathcal{Y})$ is the topology generated by the sublinear functionals

$$\sigma_D(u) := \sup_{y \in D} \langle u, y \rangle,$$

where $D \subset \mathcal{Y}$ is $\sigma(\mathcal{Y},\mathcal{U})$-compact. Similarly for $\mathcal{Y}$. Given a topology on $\mathcal{U}$, the corresponding topological dual can be identified with $\mathcal{Y}$ if and only if the topology is between $\sigma(\mathcal{U},\mathcal{Y})$ and $\tau(\mathcal{U},\mathcal{Y})$; see Sect. A.7. If $\mathcal{U}$ is a Fréchet (e.g. Banach) space under a given topology s and if $\mathcal{Y}$ the topological dual of $\mathcal{U}$, then $\tau(\mathcal{U},\mathcal{Y})$ coincides with s; see Remark A.48. In particular if $\mathcal{U} = L^p$ and $\mathcal{Y} = L^q$ with $p \in [1,\infty)$ and q the conjugate exponent of p, then the Mackey topology $\tau(\mathcal{U},\mathcal{Y})$ is just the usual L^p norm topology. However, $\tau(L^\infty, L^1)$ is, in general, strictly weaker than the norm topology of L^∞; see Sect. 5.1.1.

The following relates the weak and Mackey topologies on $\mathcal{U}$ to those on L^1 and L^∞ as well as to the metric topology of L^0 studied in Sect. 1.1.5.

Lemma 3.4 *We have $L^\infty \subseteq \mathcal{U} \subseteq L^1$ and $L^\infty \subseteq \mathcal{Y} \subseteq L^1$ and*

$$\sigma(L^1, L^\infty)|_{\mathcal{U}} \subseteq \sigma(\mathcal{U},\mathcal{Y}), \quad \sigma(\mathcal{U},\mathcal{Y})|_{L^\infty} \subseteq \sigma(L^\infty, L^1),$$

$$\tau(L^1, L^\infty)|_{\mathcal{U}} \subseteq \tau(\mathcal{U},\mathcal{Y}), \quad \tau(\mathcal{U},\mathcal{Y})|_{L^\infty} \subseteq \tau(L^\infty, L^1).$$

The L^0-topology on $\mathcal{U}$ is weaker than $\tau(\mathcal{U},\mathcal{Y})$.

Proof Since $\mathcal{U}$ and $\mathcal{Y}$ are decomposable, $L^\infty \subseteq \mathcal{U}$ and $L^\infty \subseteq \mathcal{Y}$. Let $u \in \mathcal{U}$ and define $y \in L^\infty$ by $y^i = \operatorname{sign}(u^i)$. We have $\|u\|_{L^1} = E[u \cdot y] \in \mathbb{R}$. Thus, $\mathcal{U} \subseteq L^1$, and, by symmetry, $\mathcal{Y} \subseteq L^1$. The inclusions $L^\infty \subseteq \mathcal{U} \subseteq L^1$ and $L^\infty \subseteq \mathcal{Y} \subseteq L^1$ give the relations for the σ-topologies. Since, by symmetry, analogous relations are valid for the σ-topologies on $\mathcal{Y}$, $\sigma(L^\infty, L^1)$-compact subsets of L^∞ are $\sigma(\mathcal{Y},\mathcal{U})$-compact. Since $\tau(\mathcal{U},\mathcal{Y})$ is generated by the support functions of $\sigma(\mathcal{Y},\mathcal{U})$-compact sets, we get $\tau(L^1, L^\infty)|_{\mathcal{U}} \subseteq \tau(\mathcal{U},\mathcal{Y})$. The remaining inclusion is verified similarly. As noted above, $\tau(L^1, L^\infty)$-topology is the L^1-norm topology on L^1. Since the L^0-topology on L^1 is weaker than the norm topology, the last claim follows from $\tau(L^1, L^\infty)|_{\mathcal{U}} \subseteq \tau(\mathcal{U},\mathcal{Y})$. □

Let $\mathcal{X}$ and $\mathcal{V}$ be decomposable solid spaces of $\mathbb{R}^n$-valued random variables in separating duality under the bilinear form

$$\langle x, v\rangle := E[x \cdot v].$$

A linear mapping $\mathcal{A} : \mathcal{X} \to \mathcal{U}$ is *weakly continuous* if it is continuous with respect to the weak topologies on $\mathcal{X}$ and $\mathcal{U}$. By Lemma A.53, this means that $x \mapsto \langle \mathcal{A}x, y\rangle$ is $\sigma(\mathcal{X},\mathcal{V})$-continuous for all $y \in \mathcal{Y}$, or equivalently, there exists a linear mapping $\mathcal{A}^* : \mathcal{Y} \to \mathcal{V}$ such that

$$\langle \mathcal{A}x, y\rangle = \langle x, \mathcal{A}^* y\rangle \quad \forall x \in \mathcal{X},\ y \in \mathcal{Y}.$$

The mapping $\mathcal{A}^*$ is known as the *adjoint* of $\mathcal{A}$.

Recall that the scenariowise Moore-Penrose inverse $A^\dagger$ of a random matrix A is measurable; see Example 1.25.

Lemma 3.5 *Let $A \in L^0(\mathbb{R}^{m\times n})$ be a random matrix such that $Ax \in \mathcal{U}$ for all $x \in \mathcal{X}$. The linear mapping $\mathcal{A} : \mathcal{X} \to \mathcal{U}$ defined pointwise by*

$$\mathcal{A}x = Ax \quad a.s.$$

is weakly continuous if and only if $A^*y \in \mathcal{V}$ for all $y \in \mathcal{Y}$, and in this case its adjoint is given pointwise by

$$\mathcal{A}^*y = A^*y \quad a.s.,$$

the weak closure of $\operatorname{rge}\mathcal{A}$ *is* $\mathcal{U}(\operatorname{rge} A)$ *and if $A^\dagger u \in \mathcal{X}$ for all $u \in \mathcal{U}$, then* $\operatorname{rge}\mathcal{A}$ *is weakly closed in $\mathcal{U}$. If $\mathcal{V}$ is the Köthe dual of $\mathcal{X}$, then $A^*y \in \mathcal{V}$ for all $y \in \mathcal{Y}$.*

Proof For any $x \in \mathcal{X}$ and $y \in \mathcal{Y}$,

$$\langle \mathcal{A}x, y\rangle = E[(Ax)\cdot y] = E[x \cdot A^*y],$$

which proves the equivalence and the adjoint formula. When $\mathcal{V}$ is the Köthe dual of $\mathcal{X}$, the above equation implies that $x \cdot A^*y \in L^1$ for all $x \in \mathcal{X}$, so $A^*y \in \mathcal{V}$.

It is clear that $\operatorname{rge}\mathcal{A} \subseteq \mathcal{U}(\operatorname{rge} A)$. By Corollary 1.57, the set $\mathcal{U}(\operatorname{rge} A)$ is L^0-closed so, by Lemma 3.4, it is also weakly closed. It follows that $\operatorname{cl}\operatorname{rge}\mathcal{A} \subseteq \mathcal{U}(\operatorname{rge} A)$. Given $u \in \mathcal{U}(\operatorname{rge} A)$, there exists, by Corollary 1.28, an $x \in L^0$ with $u = Ax$. Defining $x^\nu := x1_{\{|x|\le\nu\}}$, we have $x^\nu \in \mathcal{X}$ and $Ax^\nu = u1_{\{|x|\le\nu\}}$, so

$$E[Ax^\nu \cdot y] \to E[u\cdot y] \quad \forall y \in \mathcal{Y},$$

by dominated convergence. Thus, $Ax^\nu \to u$ weakly, so $\operatorname{cl}\operatorname{rge}\mathcal{A} = \mathcal{U}(\operatorname{rge} A)$.

If $A^\dagger u \in \mathcal{X}$ for all $u \in \mathcal{U}$, then any $u \in \operatorname{rge}\mathcal{A}$ can be expressed as $u = \mathcal{A}(\mathcal{A}^\dagger u)$, where $\mathcal{A}^\dagger : \mathcal{U} \to \mathcal{X}$ is defined pointwise by $(\mathcal{A}^\dagger u)(\omega) := A^\dagger(\omega)u(\omega)$. By the first claim, $\mathcal{A}$ and $\mathcal{A}^\dagger$ are both weakly continuous. The set $\operatorname{rge}\mathcal{A}$ is thus closed since it is the kernel of the continuous mapping $u \mapsto \mathcal{A}\mathcal{A}^\dagger u - u$. □

The following characterizes the adjoint of the conditional expectation operator with respect to a σ-algebra $\mathcal{G} \subseteq \mathcal{F}$; see Sect. B.4.

Lemma 3.6 *Let $\mathcal{G} \subset \mathcal{F}$ be a σ-algebra such that $E^{\mathcal{G}}u \in \mathcal{U}$ for all $u \in \mathcal{U}$. The mapping $E^{\mathcal{G}} : \mathcal{U} \to \mathcal{U}$ is weakly continuous if and only if $E^{\mathcal{G}}y \in \mathcal{Y}$ for all $y \in \mathcal{Y}$ and, in this case, its adjoint is given by*

$$(E^{\mathcal{G}})^*y = E^{\mathcal{G}}y \quad a.s.$$

If $\mathcal{Y}$ is the Köthe dual of $\mathcal{U}$, then $E^{\mathcal{G}}y \in \mathcal{Y}$ for all $y \in \mathcal{Y}$.

Proof If u^i, y^i, $(E^{\mathcal{G}}u)^iy^i$ and $u^i(E^{\mathcal{G}}y)^i$ are integrable, Lemma B.13 gives

$$E[E^{\mathcal{G}}u \cdot y] = E[(E^{\mathcal{G}}u)\cdot E^{\mathcal{G}}y] = E[u \cdot E^{\mathcal{G}}y]. \tag{3.4}$$

Thus, if $E^{\mathcal{G}}\mathcal{U} \subset \mathcal{U}$ and $E^{\mathcal{G}}\mathcal{Y} \subset \mathcal{Y}$, then, by (3.2), the function $u \mapsto E^{\mathcal{G}}u$ is weakly continuous. On the other hand, if $E^{\mathcal{G}} : \mathcal{U} \to \mathcal{U}$ is weakly continuous, then $u \mapsto E[E^{\mathcal{G}}u \cdot y]$ is $\sigma(\mathcal{U}, \mathcal{Y})$-continuous for $y \in \mathcal{Y}$. Thus, there exists a $y' \in \mathcal{Y}$ such that $E[E^{\mathcal{G}}u \cdot y] = E[u \cdot y']$ for all $u \in \mathcal{U}$. Since $y \in L^1$, (3.4) gives

$$E[E^{\mathcal{G}}u \cdot y] = E[u \cdot E^{\mathcal{G}}y] \quad \forall u \in L^{\infty}.$$

Thus, $y' = E^{\mathcal{G}}y$ almost surely so $E^{\mathcal{G}}\mathcal{Y} \subset \mathcal{Y}$.

Assume now that $\mathcal{Y}$ is the Köthe dual of $\mathcal{U}$ and let $y \in \mathcal{Y}$. It suffices to show that $E^{\mathcal{G}}y \in \mathcal{Y}$. By (3.2), it suffices to treat the case where at most one component y^i of y is nonzero. Without loss of generality, we may assume that y^i is nonnegative so that $E^{\mathcal{G}}y^i$ is nonnegative as well. If $\mathcal{Y}$ is the Köthe dual, it suffices to show that $E[u^i(E^{\mathcal{G}}y^i)] < \infty$ for every nonnegative $u \in \mathcal{U}$. By Lemma B.13, $E[u^i(E^{\mathcal{G}}y^i)] = E[E^{\mathcal{G}}(u^i)y^i]$, where the right side is finite, since $E^{\mathcal{G}}\mathcal{U} \subset \mathcal{U}$. □

Many familiar spaces of random variables satisfy the condition $E^{\mathcal{G}}\mathcal{U} \subset \mathcal{U}$ in Lemma 3.6 for every σ-algebra $\mathcal{G} \subseteq \mathcal{F}$; see Remark 5.9. The condition may fail, e.g., in Musielak-Orlicz spaces with random Young functions.

When the matrix A in Lemma 3.5 is $\mathcal{G}$-measurable, then the corresponding mapping $\mathcal{A} : \mathcal{X} \to \mathcal{U}$ commutes with the $\mathcal{G}$-conditional expectation.

Lemma 3.7 *Let $\mathcal{G} \subseteq \mathcal{F}$ be a σ-algebra such that $E^{\mathcal{G}}u \in \mathcal{U}$ for all $u \in \mathcal{U}$ and let $A \in L^0(\mathbb{R}^{m\times n})$ be a $\mathcal{G}$-measurable random matrix such that $Ax \in \mathcal{U}$ for all $x \in \mathcal{X}$. Then*

$$E^{\mathcal{G}}[Ax] = AE^{\mathcal{G}}x$$

for all $x \in \mathcal{X}$.

Proof Let $x \in \mathcal{X}$. Changing x by setting all but its jth component to zero, we still have $x \in \mathcal{X}$, by solidity of $\mathcal{X}$. Thus, $u := (A_{ij}x^j)_{i=1}^m \in \mathcal{U}$ since $Ax \in \mathcal{U}$ for all $x \in \mathcal{X}$. Changing u by setting all but its ith component to zero, we still have $u \in \mathcal{U}$, by solidity of $\mathcal{U}$. Thus, by Lemma 3.4, $A_{i,j}x^j$ is integrable for every i and j. By Lemma B.13.2, $E^{\mathcal{G}}[A_{ij}x^j] = A_{ij}E^{\mathcal{G}}x^j$. Applying Lemma B.13.1 to each component of Ax then gives $E^{\mathcal{G}}[Ax] = AE^{\mathcal{G}}x$. □

3.1.2 Conjugates of Integral Functionals

This section studies convex integral functionals on paired decomposable spaces $\mathcal{U}$ and $\mathcal{Y}$ of random variables. More precisely, we fix a normal integrand h and study the integral functionals $Eh : \mathcal{U} \to \overline{\mathbb{R}}$ and $Eh^* : \mathcal{Y} \to \overline{\mathbb{R}}$ defined by

$$Eh(u) := \int_{\Omega} h(u(\omega), \omega)dP(\omega)$$

and

$$Eh^*(y) := \int_\Omega h^*(y(\omega), \omega) dP(\omega),$$

where

$$h^*(v, \omega) := \sup_{x \in \mathbb{R}^m} \{x \cdot v - h(x, \omega)\}.$$

By Theorem 1.46, h^* is a convex normal integrand. The next result characterizes the conjugate and the subdifferential of Eh with respect to the pairing of $\mathcal{U}$ with $\mathcal{Y}$; see Sect. A.6. Recall that the conjugate of f is defined for each $y \in \mathcal{Y}$ by

$$(Eh)^*(y) := \sup_{u \in \mathcal{U}} \{\langle u, y\rangle - Eh(u)\}$$

while the subdifferential $\partial Eh(u)$ of Eh at a $u \in \mathcal{U}$ is the closed convex set

$$\partial Eh(u) := \{y \in \mathcal{Y} \mid Eh(u') \geq Eh(u) + \langle u' - u, y\rangle \quad \forall u' \in \mathcal{U}\}.$$

Given $u \in \mathcal{U}$, the mapping

$$\omega \mapsto \partial h(u)(\omega) := \partial h(u(\omega), \omega)$$

is measurable, by Theorem 1.51.

Theorem 3.8 *If h is a convex normal integrand with* dom $Eh \neq \emptyset$, *then*

$$(Eh)^* = Eh^*$$

and

$$\partial Eh(u) = \{y \in \mathcal{Y} \mid y \in \partial h(u) \ a.s.\}$$

for any $u \in \mathcal{U}$ such that $Eh(u)$ is finite.

Proof By Lemma B.3,

$$\langle u, y\rangle - Eh(u) = E[u \cdot y - h(u)]$$

for every $u \in \mathcal{U}$ and $y \in \mathcal{Y}$. The first claim thus follows by applying Theorem 1.64 to the normal integrand $h_y(u, \omega) := h(u, \omega) - u \cdot y(\omega)$. As to the second, we have, by definition, $y \in \partial Eh(u)$ if and only if $Eh(u) + (Eh)^*(y) = \langle u, y\rangle$. By the first claim, this is equivalent to

$$Eh(u) + Eh^*(y) = \langle u, y\rangle.$$

By Fenchel's inequality, $h(u) + h^*(y) \geq u \cdot y$ almost surely so, by Lemma B.3, $y \in \partial Eh(u)$ if and only if

$$E[h(u) + h^*(y) - u \cdot y] = 0.$$

By Fenchel's inequality again, this holds if and only if

$$h(u) + h^*(y) = u \cdot y$$

almost surely. This means that $y \in \partial h(u)$ almost surely. □

Recall that a convex function g in a locally convex vector space is lsc with respect to the weak topology if it is lsc merely with respect to the Mackey topology; see Corollary A.52. The converse is immediate. From now on, we will simply say that a convex function g on a locally convex vector space is *lsc* if it is lower semicontinuous with respect to the Mackey topology. Accordingly, we say that a convex set is *closed* if it is closed with respect to the Mackey topology. The *closure* of a function g is defined by

$$\operatorname{cl} g = \begin{cases} \operatorname{lsc} g & \text{if } \operatorname{lsc} g(u) > -\infty \text{ for all } u \in U, \\ -\infty & \text{otherwise;} \end{cases}$$

see Sect. A.8. By Theorem A.54, $\operatorname{cl} g = g^{**}$. The function g is said to be *closed* at a point u if $g(u) = (\operatorname{cl} g)(u)$. A function which is closed at every point is said to be *closed*. Clearly, a convex set is closed if and only if its indicator closed.

Corollary 3.9 *Let h be a convex normal integrand. The following are equivalent:*

1. $\operatorname{dom} Eh \neq \emptyset$ *and* $\operatorname{dom} Eh^* \neq \emptyset$;
2. *Eh is closed and proper;*
3. *Eh^* is closed and proper;*
4. $\operatorname{dom} Eh \neq \emptyset$ *and there exist $y \in \mathcal{Y}$ and $\alpha \in L^1$ such that*

$$h(u, \omega) \geq u \cdot y(\omega) - \alpha(\omega) \quad \forall u \in \mathbb{R}^m;$$

5. $\operatorname{dom} Eh^* \neq \emptyset$ *and there exist $u \in \mathcal{U}$ and $\alpha \in L^1$ such that*

$$h^*(y, \omega) \geq u(\omega) \cdot y - \alpha(\omega) \quad \forall u \in \mathbb{R}^m$$

and imply that Eh and Eh^* are conjugates of each other and that $y \in \partial Eh(u)$ if and only if $y \in \partial h(u)$ almost surely.

Proof We prove the equivalence of 1, 2 and 4. The equivalence with 3 and 5 then follow by symmetry. By Theorem 3.8, 1 implies that Eh and Eh^* are conjugates of each other so both are closed and proper and thus, 2 and 3 hold. Assuming 2, the biconjugate theorem gives the existence of $y \in \mathcal{Y}$ and $a \in \mathbb{R}$ such that

$$Eh(u) \geq \langle u, y \rangle - a \quad \forall u \in \mathcal{U}.$$

Thus, by Theorem 3.8

$$a \geq (Eh)^*(y) = Eh^*(y).$$

By Fenchel's inequality,

$$h(u, \omega) + h^*(y, \omega) \geq u \cdot y,$$

so 4 holds with $\alpha(\omega) = h^*(y(\omega), \omega)$. If 4 holds, $Eh^*(y) \leq E\alpha$, so 1 holds.

Assume 1. By Theorem 3.8, $y \in \partial Eh(u)$ implies $y \in \partial h(u)$ almost surely. On the other hand, if $u \in \mathcal{U}$ and $y \in \mathcal{Y}$ are such that $y \in \partial h(u)$ almost surely, then $h(u) + h^*(y) = u \cdot y$ almost surely. The properness of Eh and Eh^* implies that the negative parts of $h(u)$ and $h^*(y)$ are integrable so, by Lemma B.3, $Eh(u) + Eh^*(y) = \langle u, y \rangle$, which means that $y \in \partial Eh(u)$. □

Applying Corollary 3.9 to the indicator function of a closed-valued measurable mapping, gives the following analogue of Corollary 1.57.

Corollary 3.10 *Given a closed convex-valued measurable mapping* $S : \Omega \rightrightarrows \mathbb{R}^m$, *the set*

$$\mathcal{U}(S) := \{u \in \mathcal{U} \mid u \in S \ a.s.\}$$

is closed and convex.

Proof If $\mathcal{U}(S) = \emptyset$ the claim holds trivially. If $\mathcal{U}(S) \neq \emptyset$ it follows by applying Corollary 3.9 to $h(u, \omega) := \delta_S(u, \omega)$. Indeed, we have $0 \in \operatorname{dom} Eh^*$ so condition 1 of Corollary 3.9 holds. Thus, the function $Eh = \delta_{\mathcal{U}(S)}$ is closed. □

Given a convex set $C \subset \mathcal{U}$, its *core* is the set $\operatorname{core} C$ of points $u \in C$ such that the *positive hull*

$$\operatorname{pos}(C - u) := \bigcup_{\lambda > 0} \lambda(C - u)$$

is the whole space $\mathcal{U}$; see Sect. A.2. Recall that the Köthe dual of $\mathcal{U}$ is the linear space

$$\mathcal{U}' := \{y \in L^0 \mid u \cdot y \in L^1 \quad \forall u \in \mathcal{U}\}.$$

Theorem 3.11 *If Eh is proper and $\mathcal{Y} = \mathcal{U}'$, then*

$$\partial Eh(u) = \{y \in L^0 \mid y \in \partial h(u) \text{ a.s.}\} \neq \emptyset$$

for all $u \in \operatorname{core}\operatorname{dom} Eh$ and Eh is closed as soon as $\operatorname{core}\operatorname{dom} Eh \neq \emptyset$.

Proof Let $u \in \operatorname{core}\operatorname{dom} Eh$. We have $u \in \operatorname{int}\operatorname{dom} h$ almost surely. Indeed, given a finite set $\{w_i\}_{i \in I} \subset \mathbb{R}^m$ whose convex hull is a neighborhood of the origin, there is an $\epsilon > 0$ such that $u + \epsilon w_i \in \operatorname{dom} Eh$ for all $i \in I$. Thus, $u + \epsilon w_i \in \operatorname{dom} h$ for all $i \in I$ almost surely so $u \in \operatorname{int}\operatorname{dom} h$ almost surely.

By Theorem A.29 and Corollary A.62, $\partial h(u) \neq \emptyset$ almost surely. By the measurable selection theorem, there exists a $y \in L^0$ with $y \in \partial h(u)$ almost surely, i.e.

$$h(u(\omega) + u', \omega) \geq h(u(\omega), \omega) + u' \cdot y(\omega) \quad \forall u' \in \mathbb{R}^m.$$

Given any $u' \in \mathcal{U}$ and $\beta > 0$, this implies,

$$E[u' \cdot y] \leq \frac{Eh(u + \beta u') - Eh(u)}{\beta}.$$

Since $u \in \operatorname{core}\operatorname{dom} Eh$, there is a $\beta > 0$ such that the right side is finite. Thus, $y \in \mathcal{U}'$ so $y \in \mathcal{Y}$, by assumption. Together with the above inequalities, this proves the first claim. The condition $y \in \partial h(u)$ means that

$$h^*(y) = u \cdot y - h(u)$$

so we also get that both Eh and Eh^* have nonempty domains. The last claim thus follows from Corollary 3.9. □

The *relative core* of a set $C \subset \mathcal{U}$ is the core of C relative to its affine hull; see Sect. A.2. By Lemma A.4, $\operatorname{rcore} C$ is set of points $u \in C$ such that $\operatorname{pos}(C - u)$ is linear.

Remark 3.12 Assume that Eh is proper, $\mathcal{Y} = \mathcal{U}'$,

$$\mathcal{U}(\operatorname{aff}\operatorname{dom} h) \subseteq \operatorname{aff}\operatorname{dom} Eh,$$

and that $\mathcal{U}$ satisfies the stronger solidity property in Remark 3.1. Then

$$\partial Eh(u) \neq \emptyset$$

for all $u \in \operatorname{rcore}\operatorname{dom} Eh$ and Eh is closed as soon as $\operatorname{rcore}\operatorname{dom} Eh \neq \emptyset$.

Proof Let $u \in \operatorname{rcore}\operatorname{dom} Eh$ and let π be the scenariowise projection to $\operatorname{aff}\operatorname{dom} h - u$. We have $u \in \operatorname{rint}\operatorname{dom} h$ almost surely. Indeed, let $\{w_i\}_{i \in I} \subset \mathbb{R}^m$ be a finite set whose convex hull is a neighborhood of the origin in $\mathbb{R}^m$. Since $\mathcal{U}(\operatorname{aff}\operatorname{dom} h) \subset$

aff dom Eh by assumption, there is an $\epsilon > 0$ such that $u + \epsilon\pi w_i \in \operatorname{dom} Eh$ for all $i \in I$. It follows that $u + \epsilon\pi w_i \in \operatorname{dom} h$ for all $i \in I$ almost surely so $u \in \operatorname{rint}\operatorname{dom} h$ almost surely. By Corollaries A.36 and A.63, $\partial h(u) \neq \emptyset$ almost surely. By the measurable selection theorem, there exists a $y \in L^0$ with $y \in \partial h(u)$ almost surely, i.e.

$$h(u(\omega) + u', \omega) \geq h(u(\omega), \omega) + u' \cdot y(\omega) \quad \forall u' \in \mathbb{R}^m. \tag{3.5}$$

Clearly, $\pi y \in \partial h(u)$ as well.

Let $u' \in \mathcal{U}$. Since, by Corollary A.82, $|\pi u'| \leq |u'|$ almost surely, we have $\pi u' \in \mathcal{U}$ since $\mathcal{U}$ satisfies the stronger solidity property in Remark 3.1. By assumption, $\mathcal{U}(\operatorname{aff}\operatorname{dom} h) \subset \operatorname{aff}\operatorname{dom} Eh$, so there exists $\lambda > 0$ such that $u + \lambda\pi u' \in \operatorname{dom} Eh$. Combining this with (3.5) gives

$$E[\lambda u' \cdot \pi y] = E[\lambda\pi u' \cdot y] \leq Eh(u + \lambda\pi u') - Eh(u) < \infty.$$

Since $u' \in \mathcal{U}$ was arbitrary, this implies that πy is in the Köthe dual of $\mathcal{U}$ and thus, by assumption, in $\mathcal{Y}$. This proves the first claim. The last claim follows from the conditions $\pi y \in \partial h(u)$ and $\pi y \in \mathcal{Y}$ just like in the proof of Theorem 3.11. □

Note that the last assumption in Remark 3.12 is slightly stronger than mere solidity of $\mathcal{U}$ which means that $u \in \mathcal{U}$ for all $u \in L^0$ such that $|u^i| \leq |\bar{u}^i|$ for some $\bar{u} \in \mathcal{U}$.

The following is a corollary of Theorem 1.67.

Theorem 3.13 (Jensen's Inequality) *Let $\mathcal{G} \subset \mathcal{F}$ be a σ-algebra such that $E^{\mathcal{G}}\mathcal{U} \subset \mathcal{U}$ and $E^{\mathcal{G}}\mathcal{Y} \subset \mathcal{Y}$ and let h be a $\mathcal{G}$-measurable convex normal integrand such that $Eh^*(y) < \infty$ for some $y \in \mathcal{Y}$. Then*

$$Eh(E^{\mathcal{G}}u) \leq Eh(u)$$

for every $u \in \mathcal{U}$.

Proof By Fenchel's inequality,

$$Eh(u) \geq E[u \cdot y] - Eh^*(y)$$

for all $u \in \mathcal{U}$, so the claim follows from Theorem 1.67. □

The following extends Theorem 2.36 from L^1 to more general $\mathcal{U}$.

Theorem 3.14 *Let $\mathcal{G} \subset \mathcal{F}$ be a σ-algebra such that $E^{\mathcal{G}}\mathcal{U} \subset \mathcal{U}$ and $E^{\mathcal{G}}\mathcal{Y} \subset \mathcal{Y}$. Given a closed convex-valued random set S with $\mathcal{U}(S) \neq \emptyset$, the following are equivalent:*

1. *S is $\mathcal{G}$-measurable;*
2. *$E^{\mathcal{G}}u \in \mathcal{U}(S)$ for every $u \in \mathcal{U}(S)$;*
3. *$E^{\mathcal{G}}S \subseteq S$ almost surely.*

Proof By Theorem 2.36, it suffices to show that condition 2 here implies that of Theorem 2.36. Condition 2 means that $E\delta_S(E^{\mathcal{G}}u) \leq E\delta_S(u)$ for all $u \in \mathcal{U}$. By Corollary 3.9, Lemmas A.47 and 3.6, this means that $E\sigma_S(E^{\mathcal{G}}y) \leq E\sigma_S(y)$ for all $y \in \mathcal{Y}$. In particular,

$$E\sigma_S(E^{\mathcal{G}}y) \leq E\sigma_S(y) \quad \forall y \in L^\infty.$$

Applying the same argument in the pairing of L^1 with L^∞ now gives

$$E\delta_S(E^{\mathcal{G}}u) \leq E\delta_S(u) \quad \forall u \in L^1,$$

which is condition 2 in Theorem 2.36. □

The following is a corollary of Theorem 1.58.

Theorem 3.15 *Let h be a convex normal integrand such that $Eh : \mathcal{U} \to \overline{\mathbb{R}}$ is proper and closed. Then*

$$(Eh)^\infty = Eh^\infty,$$
$$\sigma_{\operatorname{dom} Eh} = E\sigma_{\operatorname{dom} h}$$

and

$$\operatorname{cl}\operatorname{dom} Eh = \mathcal{U}(\operatorname{cl}\operatorname{dom} h).$$

Proof By Corollary 3.9, there exists $y \in \mathcal{Y}$ such that $h^*(y)$ is integrable. By Fenchel's inequality,

$$\bar{h}(u, \omega) := h(u, \omega) - u \cdot \bar{y}(\omega) \geq -h^*(y(\omega), \omega),$$

so, by Theorem 1.55, $E\bar{h}$ is proper and lsc on L^0. By Theorem 1.58,

$$(E\bar{h})^\infty = E\bar{h}^\infty.$$

Since $\bar{h}^\infty(u, \omega) = h^\infty(u, \omega) - u \cdot y(\omega)$ and $(E\bar{h})^\infty(u) = (Eh)^\infty - E[u \cdot y]$ on $\mathcal{U}$, we get $(Eh)^\infty = Eh^\infty$ on $\mathcal{U}$. The second expression follows from the first one and Lemma A.45. Applying Corollary 3.9 and Theorem A.54 to the second expression, we get $\operatorname{cl}\delta_{\operatorname{dom} Eh} = E\delta_{\operatorname{cl}\operatorname{dom} h}$, which is the last expression. □

3.2 Duality for Integrable Strategies

We now return to the problem

$$\text{minimize } Eh(x) := \int h(x(\omega), \omega) dP(\omega) \text{ over } x \in \mathcal{X} \cap \mathcal{N} \qquad (P_{\mathcal{X}})$$

from the introduction of this chapter. Again, we assume that $h(x, \omega) = f(x, \bar{u}(\omega), \omega)$ for a convex normal integrand f and random vector $\bar{u} \in \mathcal{U}$. By Theorem 1.20.5, such an h is a normal integrand. As observed in the introduction, $(P_{\mathcal{X}})$ fits the duality framework of Sect. A.9 with the Rockafellian $F : \mathcal{X} \times \mathcal{X} \times \mathcal{U} \to \overline{\mathbb{R}}$ defined by

$$F(x, z, u) := Ef(x, u) + \delta_{\mathcal{N}}(x - z)$$

and the dualizing parameter $(z, u) \in \mathcal{X} \times \mathcal{U}$. Clearly,

$$F(x, z, u) = Ef(x, u) + \delta_{\mathcal{X}_a}(x - z) \quad \forall (x, z, u) \in \mathcal{X} \times \mathcal{X} \times \mathcal{U},$$

where

$$\mathcal{X}_a := \mathcal{X} \cap \mathcal{N}$$

is the linear space of the adapted strategies in $\mathcal{X}$.

In order to apply the results of Sect. 3.1, we assume that $\mathcal{X}$ and $\mathcal{U}$ are solid decomposable spaces in separating duality with solid decomposable spaces $\mathcal{V} \subseteq L^0(\Omega, \mathcal{F}, P; \mathbb{R}^n)$ and $\mathcal{Y} \subseteq L^0(\Omega, \mathcal{F}, P; \mathbb{R}^m)$, respectively, under the bilinear forms

$$\langle x, v \rangle := E[x \cdot v] \quad \text{and} \quad \langle u, y \rangle := E[u \cdot y].$$

Solidity implies that

$$\mathcal{X} = \mathcal{X}_0 \times \cdots \times \mathcal{X}_T \quad \text{and} \quad \mathcal{V} = \mathcal{V}_0 \times \cdots \times \mathcal{V}_T,$$

where $\mathcal{X}_t$ and $\mathcal{V}_t$ are solid decomposable spaces of $\mathbb{R}^{n_t}$-valued random variables in separating duality under the bilinear form $(x_t, v_t) \mapsto E[x_t \cdot v_t]$. It follows that

$$\langle x, v \rangle = \sum_{t=0}^{T} E[x_t \cdot v_t] \quad \forall x \in \mathcal{X},\ v \in \mathcal{V}$$

and

$$\mathcal{X}_a = \mathcal{X}_0(\mathcal{F}_0) \times \cdots \times \mathcal{X}_T(\mathcal{F}_T).$$

We will denote the orthogonal complement of $\mathcal{X}_a$ by

$$\mathcal{X}_a^\perp := \{v \in \mathcal{V} \mid \langle x, v\rangle = 0 \quad \forall x \in \mathcal{X}_a\}.$$

The following generalizes Lemma 2.66 which characterizes the set

$$\mathcal{N}^\perp := \{v \in L^1 \mid \langle x, v\rangle = 0 \quad \forall x \in \mathcal{N} \cap L^\infty\}.$$

Lemma 3.16 *The set $\mathcal{X}_a$ is closed and*

$$\mathcal{X}_a^\perp = \mathcal{N}^\perp \cap \mathcal{V} = \{v \in \mathcal{V} \mid E_t v_t = 0 \quad t = 0, \dots, T\}.$$

Proof By Lemma 1.80, $\mathcal{N}$ is closed in L^0 so the first claim follows from Lemma 3.4. Since

$$\mathcal{X}_a = \mathcal{X}_0(\mathcal{F}_0) \times \cdots \times \mathcal{X}_T(\mathcal{F}_T),$$

we have $v \in \mathcal{X}_a^\perp$ if and only if $E[x_t \cdot v_t] = 0$ for every $x_t \in \mathcal{X}_t(\mathcal{F}_t)$. Here, $E[x_t \cdot v_t] = E[x_t \cdot (E_t v_t)]$, by Lemma B.13. The claim now follows from the fact that $\mathcal{X}_t(\mathcal{F}_t)$ separates points in $\mathcal{V}_t(\mathcal{F}_t)$. Indeed, since the spaces are decomposable, we have $L^\infty(\mathcal{F}_t) \subseteq \mathcal{X}_t(\mathcal{F}_t)$ and $\mathcal{V}_t(\mathcal{F}_t) \subseteq L^1(\mathcal{F}_t)$. □

According to the general duality framework of Sect. A.9, the *dual problem* associated with the above specifications is the concave maximization problem

$$\text{maximize} \quad \langle \bar{u}, y\rangle - F^*(0, p, y) \quad \text{over } (p, y) \in \mathcal{V} \times \mathcal{Y}, \qquad (D)$$

where $F^* : \mathcal{V} \times \mathcal{V} \times \mathcal{Y} \to \overline{\mathbb{R}}$ is the conjugate of F, i.e.

$$F^*(v, p, y) := \sup_{x,z,u} \{\langle x, v\rangle + \langle z, p\rangle + \langle u, y\rangle - F(x, z, u)\}.$$

An explicit expression for F^* will be given in Theorem 3.20 below. By Fenchel's inequality,

$$F(x, 0, u) \geq \langle u, y\rangle - F^*(0, p, y) \quad \forall x \in \mathcal{X},\ u \in \mathcal{U},\ p \in \mathcal{V},\ y \in \mathcal{Y},$$

so

$$\inf (P_{\mathcal{X}}) \geq \sup (D),$$

where $\inf(P_{\mathcal{X}})$ and $\sup(D)$ denote the optimum values of $(P_{\mathcal{X}})$ and (D), respectively. A *duality gap* is said to exist if the inequality is strict. Conversely, we say that there is no duality gap if $\inf (P_{\mathcal{X}}) = \sup (D)$.

The *Lagrangian* associated with the function F is the convex-concave function L on $\mathcal{X} \times \mathcal{V} \times \mathcal{Y}$ given by

$$L(x, p, y) := \inf_{(z,u)\in\mathcal{X}\times\mathcal{U}} \{F(x, z, u) - \langle z, p\rangle - \langle u, y\rangle\}.$$

The associated minimax problem is to find a saddle value and/or a saddle point of the concave-convex function

$$L_{\bar{u}}(x, p, y) := L(x, p, y) + \langle \bar{u}, y\rangle$$

when minimizing over x and maximizing over (p, y). If

$$\inf_x \sup_{p,y} L_{\bar{u}}(x, p, y) = \sup_{p,y} \inf_x L_{\bar{u}}(x, p, y),$$

the common value is called the *saddle value* of $L_{\bar{u}}$. A point (x, p, y) is a *saddle point* of $L_{\bar{u}}$ if

$$L_{\bar{u}}(x, p', y') \leq L_{\bar{u}}(x, p, y) \leq L_{\bar{u}}(x', p, y) \quad \forall x' \in \mathcal{X},\ p' \in \mathcal{V},\ y' \in \mathcal{Y}.$$

Clearly, the existence of a saddle point implies the existence of a saddle value. By definition, the conjugate of F can be expressed in term of the Lagrangian as

$$F^*(v, p, y) = \sup_{x\in\mathcal{X}} \{\langle x, v\rangle - L(x, p, y)\}.$$

It follows that the dual problem coincides with the maximization half of the minimax problem. Similarly, if $F(x, z, u)$ is closed in (z, u), then by Theorem A.54, the primal problem coincides with the minimization half of the minimax problem.

The next three theorems are direct consequences of Theorems A.71, A.72, and A.73 in the appendix. They all involve the assumption that the integral functional Ef be closed in u. This means that $Ef(x, \cdot)$ is closed in $\mathcal{U}$ for each $x \in \mathcal{X}$. Combined with Lemma 3.16, this implies that the function F is closed in (z, u). Recall that, by the biconjugate theorem Theorem A.54, a convex function is closed if and only if it coincides with its biconjugate.

The *optimum value function* $\varphi : \mathcal{X} \times \mathcal{U} \to \overline{\mathbb{R}}$ associated with F will be denoted by

$$\varphi(z, u) := \inf_{x\in\mathcal{X}} F(x, z, u).$$

By definition of F,

$$\varphi(z, u) = \inf_{x\in\mathcal{X}} \{Ef(x, u) \mid x - z \in \mathcal{N}\} = \inf_{x\in\mathcal{X}} \{Ef(x, u) \mid x - z \in \mathcal{X}_a\}.$$

Clearly, $\varphi(0, \bar{u}) = \inf(P_{\mathcal{X}})$. Note that we deviate slightly from the notation of Sect. A.9 where the primal optimum value function has a subindex v. We omit the subindex here since we set $v = 0$ throughout. Clearly, $\varphi^*(p, y) = F^*(0, p, y)$ so the dual problem can be written also as

$$\text{maximize} \quad \langle \bar{u}, y\rangle - \varphi^*(p, y) \quad \text{over } (p, y) \in \mathcal{V} \times \mathcal{Y}.$$

The following is a direct consequence of Theorem A.71.

Theorem 3.17 *The implications* $1 \Leftrightarrow 2 \Rightarrow 3$ *hold among the following conditions:*

1. $\inf(P_{\mathcal{X}}) = \sup(D)$;
2. φ *is closed at* $(0, \bar{u})$;
3. *The function* $L_{\bar{u}}$ *has a saddle value.*

If $Ef(x, u)$ *is closed in* u, *then* $1 \Leftrightarrow 2 \Leftrightarrow 3$.

The integral functional $Ef(x, u)$ is closed in u in particular if it is jointly closed in (x, u). By Theorem 3.8, this happens if $\operatorname{dom} Ef^* \cap (\mathcal{V} \times \mathcal{Y}) \neq \emptyset$.

The following restatement of Theorem A.72 characterizes situations where there is no duality gap and, furthermore, the dual admits solutions. Recall that the subdifferential $\partial\varphi(z, u)$ of φ at a point $(z, u) \in \mathcal{X} \times \mathcal{U}$ is the closed convex set of points $(v, y) \in \mathcal{V} \times \mathcal{Y}$ such that

$$\varphi(z', u') \geq \varphi(z, u) + \langle z' - z, v\rangle + \langle u' - u, y\rangle \quad \forall (z', u') \in \mathcal{X} \times \mathcal{U}.$$

Theorem 3.18 *If* $\varphi(0, u) < \infty$, *then the implications* $1 \Leftrightarrow 2 \Rightarrow 3$ *hold among the following conditions:*

1. (p, y) *solves* (D) *and* $\inf(P_{\mathcal{X}}) = \sup(D)$;
2. *either* $(p, y) \in \partial\varphi(0, \bar{u})$ *or* $\varphi(0, \bar{u}) = -\infty$;
3. $\inf_x \sup_{p,y} L_{\bar{u}}(x, p, y) = \inf_x L_{\bar{u}}(x, p, y)$.

If, in addition, $Ef(x, u)$ *is closed in* u, *then* $1 \Leftrightarrow 2 \Leftrightarrow 3$.

The following restatement of Theorem A.73 characterizes the situations where both primal and dual solutions exist and there is no duality gap.

Theorem 3.19 *The implications* $1 \Leftrightarrow 2 \Rightarrow 3$ *hold among the following conditions:*

1. x *solves* $(P_{\mathcal{X}})$, (p, y) *solves* (D) *and* $\inf(P_{\mathcal{X}}) = \sup(D) \in \mathbb{R}$;
2. $(0, p, y) \in \partial F(x, 0, \bar{u})$;
3. $0 \in \partial_x L(x, p, y)$ *and* $(0, \bar{u}) \in \partial_{(p,y)}[-L](x, p, y)$.

If $Ef(x, u)$ *is closed in* u, *then* $1 \Leftrightarrow 2 \Leftrightarrow 3$.

The subdifferential conditions in part 3 of Theorem 3.19 are known as the (generalized) *Karush-Kuhn-Tucker-Rockafellar* (KKTR) conditions; see Sect. A.9.

In order to write the dual problem and the optimality conditions more explicitly in terms of the problem data, we will first derive explicit expressions for F^*. Section 3.2.1 will focus on the Lagrangian and the associated minimax problem.

Theorem 3.20 *If* $\operatorname{dom} Ef \cap (\mathcal{X} \times \mathcal{U}) \neq \emptyset$*, then* Ef^* *is closed,*

$$F^*(v, p, y) = Ef^*(v + p, y) + \delta_{\mathcal{X}_a^\perp}(p)$$

and, in particular,

$$\varphi^*(p, y) = Ef^*(p, y) + \delta_{\mathcal{X}_a^\perp}(p).$$

If, in addition, $\operatorname{dom} Ef^* \cap (\mathcal{V} \times \mathcal{Y}) \neq \emptyset$*, then the functions* Ef*,* Ef^**,* F *and* F^* *are all closed and proper.*

Proof Recall that $F(x, z, u) = Ef(x, u) + \delta_{\mathcal{X}_a}(x - z)$, where $\mathcal{X}_a$ is closed by Lemma 3.16. By Theorem 3.8, the first assumption implies the closedness of Ef^* and, by the interchange rule in Theorem 1.64, that

$$\begin{aligned}
F^*(v, p, y) &= \sup_{x \in \mathcal{X}, z \in \mathcal{X}, u \in \mathcal{U}} \{\langle x, v\rangle + \langle z, p\rangle + \langle u, y\rangle - Ef(x, u) \,|\, x - z \in \mathcal{X}_a\} \\
&= \sup_{x \in \mathcal{X}, z' \in \mathcal{X}, u \in \mathcal{U}} \{E[x \cdot (v + p) + u \cdot y - f(x, u) - z' \cdot p] \,|\, z' \in \mathcal{X}_a\} \\
&= Ef^*(v + p, y) + \delta_{\mathcal{X}_a^\perp}(p).
\end{aligned}$$

The expression for φ^* now follows from the fact that $\varphi(p, y) = F^*(0, p, y)$, by definition. When $\operatorname{dom} Ef \cap (\mathcal{X} \times \mathcal{U}) \neq \emptyset$ and $\operatorname{dom} Ef^* \cap (\mathcal{V} \times \mathcal{Y}) \neq \emptyset$, both Ef and Ef^* are closed and proper, by Corollary 3.9 and then, the functions F and F^* are closed as sums of closed functions. The properness of Ef and Ef^* clearly implies the properness of F and F^*. □

Corollary 3.21 *If* $\operatorname{dom} Ef \cap (\mathcal{X} \times \mathcal{U}) \neq \emptyset$*, the dual problem* (D) *can be written as*

$$\text{maximize} \quad \langle \bar{u}, y\rangle - Ef^*(p, y) \quad \text{over} \quad (p, y) \in \mathcal{X}_a^\perp \times \mathcal{Y}$$

as well as

$$\text{maximize} \quad E[\bar{u} \cdot y - f^*(p, y)] \quad \text{over} \quad (p, y) \in \mathcal{X}_a^\perp \times \mathcal{Y}$$

Proof The first claim follows directly from Theorem 3.20. As to the second, Fenchel's inequality gives

$$f^*(p, y) \geq u \cdot y - f(x, u)$$

so the assumption $\operatorname{dom} Ef \cap (\mathcal{X} \times \mathcal{U}) \neq \emptyset$ implies that the negative part of $f^*(p, y)$ is integrable for every $(p, y) \in \mathcal{V} \times \mathcal{Y}$. The claim thus follows from Lemma B.3. □

The first condition in Theorem 3.20 clearly holds if $(P_{\mathcal{X}})$ is feasible. If, in addition, the dual problem is feasible, then by Corollary 3.21, the second condition in Theorem 3.20 holds as well. Note that the dual is feasible e.g. if F is bounded from below since then $F^*(0,0)$ is finite.

In the deterministic setting, $\mathcal{X}_a^\perp = \{0\}$ so the dual problem becomes

$$\text{maximize} \quad \bar{u} \cdot y - f^*(0, y) \quad \text{over} \quad y \in \mathbb{R}^m$$

and we recover a finite-dimensional instance of the general conjugate duality framework; see Sect. A.9. In general, the dual objective can be written also as

$$\langle \bar{u}, y \rangle - Ef^*(p, y) = E \inf_{(x,u) \in \mathbb{R}^n \times \mathbb{R}^m} [f(x, u) - x \cdot p + (\bar{u} - u) \cdot y].$$

This is the optimum value in a relaxed version of the primal problem (P) where we are now allowed to optimize over both x and u and the information constraint $x \in \mathcal{N}$ has been removed so the minimization can be done scenariowise; see Theorem 1.64. The constraints $x \in \mathcal{X}_a$ and $u = \bar{u}$ have been replaced by linear penalties given by the dual variables p and y.

Recall that the optimum value of (D) is always less than or equal to that of $(P_{\mathcal{X}})$. If the value function φ is closed at $(0, \bar{u})$ then, by Theorem 3.17, the optimum values are equal. If $(p, y) \in \partial\varphi(0, \bar{u})$ then, by Theorem 3.18, there is no duality gap and (p, y) solves the dual. This implies, in particular, that p is a subgradient of φ with respect to the first argument at $(0, \bar{u})$, i.e.

$$Ef(x + z, \bar{u}) - \langle z, p \rangle \geq \varphi(0, \bar{u}) \quad \forall x \in \mathcal{X}_a, z \in \mathcal{X}.$$

In other words, p describes a linear penalty that would make it disadvantageous to use nonadapted strategies in $(P_{\mathcal{X}})$. Such a $p \in \mathcal{X}_a^\perp$ is known as a *shadow price of information*.

When the dimension n_t of x_t and p_t is independent of time, the elements of $\mathcal{X}_a^\perp$ can be seen as nonadapted *martingale increments*. Indeed, we then have $p \in \mathcal{X}_a^\perp$ if and only if $p_t = \Delta m_{t+1}$ for $m_t \in \mathcal{V}_t$ and $m_{T+1} \in \mathcal{V}_T$ such that

$$E_t[\Delta m_{t+1}] = 0.$$

This is the usual martingale condition, but here m need not be adapted.

Theorem 3.20 can be used to restate Theorems 3.17, 3.18, and 3.19 more explicitly. In particular, the first part of Theorem 3.19 can be written as follows.

Theorem 3.22 *If $(P_{\mathcal{X}})$ and (D) are feasible, then the following are equivalent:*

1. *x solves $(P_{\mathcal{X}})$, (p, y) solves (D) and $\inf(P_{\mathcal{X}}) = \sup(D)$;*
2. *$x \in \mathcal{X}_a$, $(p, y) \in \mathcal{X}_a^\perp \times \mathcal{Y}$ and*

$$(p, y) \in \partial f(x, \bar{u}) \quad a.s.$$

Proof By Theorem 3.19, 1 is equivalent to $(0, p, y) \in \partial F(x, 0, \bar{u})$ which means that $F(x, 0, \bar{u}) + F^*(0, p, y) = \langle \bar{u}, y \rangle$. By Lemma 3.20, this means that $x \in \mathcal{X}_a$, $p \in \mathcal{X}_a^\perp$ and

$$Ef(x, \bar{u}) + Ef^*(p, y) = E[x \cdot p] + E[\bar{u} \cdot y]. \tag{3.6}$$

Given $(x', u') \in \mathcal{X} \times \mathcal{U}$ and $(p', y') \in \mathcal{V} \times \mathcal{Y}$, we have

$$f(x', u') + f^*(p', y') \geq x' \cdot p' + u \cdot y', \tag{3.7}$$

by Fenchel's inequality, so the feasibility assumptions imply that the negative parts of $f(x', u')$ and $f^*(p', y')$ are integrable. Thus, by Lemma B.3,

$$Ef(x', u') + Ef^*(p', y') = E[f(x', u') + f^*(p', y')]$$

so (3.6) means that $(x, \bar{u})$ and (p, y) satisfy (3.7) as an equality, i.e. $(p, y) \in \partial f(x, \bar{u})$. □

If the subdifferential $\partial\varphi(0, \bar{u})$ is nonempty, then by Theorem 3.18, there is no duality gap and a dual has a solution. Theorem 3.22 thus implies the following.

Corollary 3.23 *If $\partial\varphi(0, \bar{u}) \neq \emptyset$, then $\inf(P_{\mathcal{X}}) = \sup(D)$, the dual optimum is attained and the following are equivalent for an $x \in \mathcal{X}_a$:*

1. *x solves $(P_{\mathcal{X}})$;*
2. *there exists $(p, y) \in \mathcal{X}_a^\perp \times \mathcal{Y}$ with*

$$(p, y) \in \partial f(x, \bar{u}) \quad a.s.$$

3.2.1 Lagrangian Integrands and KKTR-Conditions

This section focuses on the Lagrangian L and the associated minimax problem. The Lagrangian L itself has a somewhat cumbersome expression but it turns out that it is "equivalent" to a simpler function that has the same saddle value and saddle points. The expressions derived below, involve the *Lagrangian integrand* $l : \mathbb{R}^n \times \mathbb{R}^m \times \Omega \to \overline{\mathbb{R}}$ defined by

$$l(x, y, \omega) := \inf_{u \in \mathbb{R}^m} \{f(x, u, \omega) - u \cdot y\}.$$

For any (x, y, ω), the function $l(\cdot, y, \omega)$ is convex, by Theorem A.1, and $l(x, \cdot, \omega)$ is upper semicontinuous and concave. Clearly,

$$f^*(v, y, \omega) = \sup_{x \in \mathbb{R}^n} \{x \cdot v - l(x, y, \omega)\}$$

so, by Theorem A.54,

$$(\mathrm{cl}_x\, l)(x, y, \omega) = \sup_{v \in \mathbb{R}^n} \{x \cdot v - f^*(v, y, \omega)\},$$

where, for each $(y, \omega) \in \mathbb{R}^m \times \Omega$, the function $(\mathrm{cl}_x\, l)(\cdot, y, \omega)$ denotes the *closure* of the function $l(\cdot, y, \omega)$; see Sect. A.8.

Given $x \in \mathcal{X}$, the function

$$(y, \omega) \mapsto -l(x(\omega), y, \omega) = \sup_{u \in \mathbb{R}^m} \{u \cdot y - f(x(\omega), u, \omega)\}$$

is a normal integrand, by Theorem 1.20 and Theorem 1.46. Similarly, the function

$$(x, \omega) \mapsto (\mathrm{cl}_x\, l)(x, y(\omega), \omega) = \sup_{v \in \mathbb{R}^n} \{x \cdot v - f^*(v, y(\omega), \omega)\}$$

is normal integrand for any $y \in \mathcal{Y}$. Thus, by Corollary 1.18, the functions

$$\omega \mapsto l(x(\omega), y(\omega), \omega) \quad \text{and} \quad \omega \mapsto (\mathrm{cl}_x\, l)(x(\omega), y(\omega), \omega)$$

are measurable for any $(x, y) \in \mathcal{X} \times \mathcal{Y}$. It follows that the integral functionals

$$El(x, y) := \int_\Omega l(x(\omega), y(\omega), \omega) dP(\omega)$$

and

$$E(\mathrm{cl}_x\, l)(x, y) := \int_\Omega (\mathrm{cl}_x\, l)(x(\omega), y(\omega), \omega) dP(\omega)$$

are well-defined extended real-valued functions on $\mathcal{X} \times \mathcal{Y}$.

We will denote the projection of dom Ef to the x-component by

$$\mathrm{dom}_x\, Ef := \{x \in \mathcal{X} \mid \exists u \in \mathcal{U} : \ Ef(x, u) < \infty\}$$

and the projection of dom Ef^* to the y-component by

$$\mathrm{dom}_y\, Ef^* := \{y \in \mathcal{Y} \mid \exists v \in \mathcal{U} : \ Ef^*(v, y) < \infty\}.$$

Theorem 3.24 *We have*

$$L(x, p, y) = \begin{cases} +\infty & \text{if } x \notin \mathrm{dom}_x\, Ef, \\ El(x, y) - \langle x, p \rangle & \text{if } x \in \mathrm{dom}_x\, Ef \text{ and } p \in \mathcal{X}_a^\perp, \\ -\infty & \text{otherwise.} \end{cases}$$

If $\operatorname{dom} Ef \cap (\mathcal{X} \times \mathcal{U}) \neq \emptyset$, *then*

$$(\operatorname{cl}_x L)(x, p, y) = \begin{cases} E(\operatorname{cl}_x l)(x, y) - \langle x, p \rangle & \text{if } y \in \operatorname{dom}_y Ef^* \text{ and } p \in \mathcal{X}_a^\perp, \\ -\infty & \text{otherwise.} \end{cases}$$

If $\operatorname{dom} Ef \cap (\mathcal{X} \times \mathcal{U}) \neq \emptyset$ *and* $\operatorname{dom} Ef^* \cap (\mathcal{V} \times \mathcal{Y}) \neq \emptyset$, *then all functions between* L *and* $\operatorname{cl}_x L$ *have the same saddle value and saddle points. In this case, the KKTR-conditions*

$$0 \in \partial_x L(x, p, y), \quad (0, \bar{u}) \in \partial_{p,y}[-L](x, p, y)$$

in Theorem 3.19 hold if and only if $x \in \mathcal{X}_a$, $p \in \mathcal{X}_a^\perp$ *and*

$$p \in \partial_x l(x, y), \quad \bar{u} \in \partial_y[-l](x, y) \quad a.s.$$

Proof By Lemma B.3,

$$\begin{aligned} L(x, p, y) &= \inf_{(z,u) \in \mathcal{X} \times \mathcal{U}} \{F(x, z, u) - \langle z, p \rangle - \langle u, y \rangle\} \\ &= \inf_{(z,u) \in \mathcal{X} \times \mathcal{U}} \{E[f(x, u) - z \cdot p - u \cdot y] \mid x - z \in \mathcal{X}_a\} \\ &= \inf_{(z',u) \in \mathcal{X} \times \mathcal{U}} \{E[f(x, u) - (x - z') \cdot p - u \cdot y] \mid z' \in \mathcal{X}_a\}, \end{aligned}$$

so the expression for L follows from Theorem 1.64. By Theorem 3.20 and Lemma B.3,

$$\begin{aligned} (\operatorname{cl}_x L)(x, p, y) &= \sup_{v \in \mathcal{V}} \{\langle x, v \rangle - F^*(v, p, y)\} \\ &= \begin{cases} \sup_{v \in \mathcal{V}} \{\langle x, v \rangle - Ef^*(v + p, y)\} & \text{if } p \in \mathcal{X}_a^\perp, \\ -\infty & \text{otherwise} \end{cases} \\ &= \begin{cases} \sup_{v \in \mathcal{V}} E[x \cdot v - f^*(v + p, y)] & \text{if } p \in \mathcal{X}_a^\perp, \\ -\infty & \text{otherwise} \end{cases} \end{aligned}$$

so the expression for $\operatorname{cl}_x L$ follows from Theorem 1.64 again. When $\operatorname{dom} Ef \neq \emptyset$ and $\operatorname{dom} Ef^* \neq \emptyset$, the function F is closed and proper, by Theorem 3.20, so the claims about saddle value and saddle points follow from Theorem A.78.

By Theorem 3.19, the KKTR-conditions hold if and only if $(0, p, y) \in \partial F(x, 0, \bar{u})$ or, equivalently, if

$$F(x, 0, \bar{u}) + F^*(0, p, y) = \langle \bar{u}, y \rangle.$$

By Theorem 3.20, this means that $x \in \mathcal{X}_a$, $p \in \mathcal{X}_a^\perp$ and

$$Ef(x, \bar{u}) + Ef^*(p, y) = E[\bar{u} \cdot y]$$

or, equivalently,

$$Ef(x, \bar{u}) + Ef^*(p, y) = E[x \cdot p] + E[\bar{u} \cdot y].$$

Since, by Fenchel's inequality,

$$f(x, \bar{u}, \omega) + f^*(v, y, \omega) \geq x \cdot v + \bar{u} \cdot y,$$

this means that $(p, y) \in \partial f(x, \bar{u})$ almost surely. By Theorem A.73, this is equivalent to $v \in \partial_x l(x, y)$ and $\bar{u} \in \partial_y[-l](x, y)$. □

The functions L and $\mathrm{cl}_x L$ are not quite integral functionals because of the constraints on the variables. However, one of the saddle functions between L and $\mathrm{cl}_x L$ is the function

$$\tilde{L}(x, p, y) = \begin{cases} E[l(x, y) - x \cdot p] & \text{if } p \in \mathcal{X}_a^\perp, \\ -\infty & \text{otherwise.} \end{cases}$$

Theorems 3.19, 3.24, and A.78 thus yield the following extension of Theorem 3.22.

Corollary 3.25 *If $(P_{\mathcal{X}})$ and (D) are feasible, the following are equivalent:*

1. *x solves $(P_{\mathcal{X}})$, (p, y) solves (D) and $\inf(P_{\mathcal{X}}) = \sup(D)$;*
2. *$x \in \mathcal{X}_a$, $(p, y) \in \mathcal{X}_a^\perp \times \mathcal{Y}$ and*

$$(p, y) \in \partial f(x, \bar{u}) \quad a.s.;$$

3. *(x, p, y) is a saddle point of the integral functional*

$$(x, p, y) \mapsto E[l(x, y) - x \cdot p + \bar{u} \cdot y],$$

when minimizing over $x \in \mathcal{X}$ and maximizing over $(p, y) \in \mathcal{X}_a^\perp \times \mathcal{Y}$;

4. *$x \in \mathcal{X}_a$, $(p, y) \in \mathcal{X}_a^\perp \times \mathcal{Y}$ and*

$$p \in \partial_x l(x, y), \quad \bar{u} \in \partial_y[-l](x, y) \quad a.s.$$

Similarly, we can augment Corollary 3.23 as follows.

Corollary 3.26 *If $\partial\varphi(0, \bar{u}) \neq \emptyset$, the following are equivalent for an $x \in \mathcal{X}_a$:*

1. *x solves $(P_{\mathcal{X}})$;*
2. *there exists $(p, y) \in \mathcal{X}_a^\perp \times \mathcal{Y}$ with*

$$(p, y) \in \partial f(x, \bar{u}) \quad a.s.;$$

3. *there exists* $(p, y) \in \mathcal{X}_a^\perp \times \mathcal{Y}$ *with*

$$p \in \partial_x l(x, y), \quad \bar{u} \in \partial_y[-l](x, y) \quad a.s.$$

In the deterministic setting, $\mathcal{X}_a^\perp = \{0\}$ so condition 3 in Corollary 3.26 becomes the KKTR-condition in finite-dimensional convex optimization. In the stochastic setting, the shadow price of information $p \in \mathcal{X}_a^\perp$ allows us to write the KKTR-conditions scenariowise.

3.2.2 *Reduced Dual Problems*

In many applications, one can restrict the dual variables (p, y) to a subset of $\mathcal{X}_a^\perp \times \mathcal{Y}$ without lowering the optimum value of the dual problem. This happens, in particular, under the following.

Assumption 3.27 *There is a mapping* $\Pi : \mathcal{V} \times \mathcal{Y} \to \mathcal{V} \times \mathcal{Y}$ *such that*

$$\varphi^* \circ \Pi \leq \varphi^*$$

and $\langle (0, \bar{u}), \Pi(p, y) \rangle = \langle \bar{u}, y \rangle$ *for all* $(p, y) \in \operatorname{dom} \varphi^*$.

Indeed, under Assumption 3.27 the optimum value of (D) equals that of the problem

$$\text{maximize} \quad \langle \bar{u}, y \rangle - \varphi^*(p, y) \quad \text{over } (p, y) \in \operatorname{rge} \Pi \tag{3.8}$$

while

$$\Pi(\operatorname{argmax}(D)) \subseteq \operatorname{argmax}(3.8) = \operatorname{argmax}(D) \cap \operatorname{rge} \Pi.$$

Assumption 3.27 is clearly satisfied if Π is the identity mapping but in many situations, more interesting choices are available.

By Theorem A.1, the function

$$g(y) := \inf_{p \in \mathcal{V}} \varphi^*(p, y)$$

is convex on $\mathcal{Y}$. It is clear that the optimum value of (D) equals that of the problem

$$\text{maximize} \quad \langle \bar{u}, y \rangle - g(y) \qquad \text{over} \quad y \in \mathcal{Y} \tag{3.9}$$

and that a pair (p, y) solves (D) if and only if y solves (3.9) and p attains the infimum in the definition of g. Recall that a mapping $\gamma : \mathcal{Y} \to \mathcal{Y}$ is *idempotent* if $\gamma \circ \gamma = \gamma$.

Theorem 3.28 (Reduced Dual) *Assume that there exist mappings* $\pi : \mathcal{Y} \to \mathcal{V}$ *and* $\gamma : \mathcal{Y} \to \mathcal{Y}$ *such that* γ *is idempotent and the mapping* $\Pi(p, y) = (\pi(y), \gamma(y))$ *satisfies Assumption 3.27. Then*

$$g(y) = \varphi^*(\pi(y), y) \quad \forall y \in \operatorname{rge}\gamma,$$

optimum value of (D) *coincides with that of the problem*

$$\text{maximize} \quad \langle \bar{u}, y \rangle - g(y) \qquad \text{over} \quad y \in \operatorname{rge}\gamma, \tag{3.10}$$

and if $y \in \operatorname{rge}\gamma$ *solves* (3.10) *then* $(\pi(y), y)$ *solves* (D)*. If* (p, y) *solves* (D)*, then* $\gamma(y)$ *solves* (3.10)*. If* $(P_{\mathcal{X}})$ *and* (3.10) *are feasible, then the following are equivalent:*

1. x *solves* $(P_{\mathcal{X}})$*,* y *solves* (3.10) *and* $\inf(P_{\mathcal{X}}) = \sup(3.10)$*;*
2. $x \in \mathcal{X}_a$*,* $y \in \operatorname{rge}\gamma$ *and*

$$(\pi(y), y) \in \partial f(x, \bar{u}) \quad a.s.;$$

3. $x \in \mathcal{X}_a$*,* $y \in \operatorname{rge}\gamma$ *and*

$$\pi(y) \in \partial_x l(x, y), \quad \bar{u} \in \partial_y[-l](x, y) \quad a.s.$$

Proof Given $y \in \mathcal{Y}$, we have

$$\begin{aligned} g(\gamma(y)) &= \inf_{p \in \mathcal{V}} \varphi^*(p, \gamma(y)) \\ &\leq \varphi^*(\pi(y), \gamma(y)) \\ &\leq \inf_{p \in \mathcal{V}} \varphi^*(p, y) \\ &= g(y), \end{aligned}$$

where the second inequality holds by the assumption on Π. Combining the above with $\langle (0, \bar{u}), \Pi(p, y) \rangle = \langle \bar{u}, y \rangle$, we get

$$\sup(3.10) \geq \sup(3.8) \geq \sup(3.9),$$

where equalities must hold since, trivially, $\sup(3.9) \geq \sup(3.10)$. Thus, the optimum value of (3.10) equals that of (3.9) which in turn equals $\sup(D)$. When γ is idempotent and $y \in \operatorname{rge}\gamma$, we have $\gamma(y) = y$ so the above hold with equalities and

$$g(y) = \varphi^*(\pi(y), y).$$

This gives the relations between the optimal solutions. The rest now follows from Corollary 3.25. □

The conditions of Theorem 3.28 may seem rather special, but they are satisfied in many applications; see Sect. 3.4 for examples. In the applications, the mappings π and γ are typically defined in terms of conditional expectations.

Theorem 3.29 *If g is closed at $y \in \mathcal{Y}$ and φ is closed at $(0, u)$ for all $u \in \mathcal{U}$, then $g(y) = \varphi(0, \cdot)^*(y)$. In particular, if g and φ are closed, then $g = \varphi(0, \cdot)^*$.*

Proof By Theorem A.54,

$$\begin{aligned} g^*(u) &= \sup_y \{\langle u, y\rangle - g(y)\} \\ &= \sup_{p,y} \{\langle u, y\rangle - \varphi^*(p, y)\} \\ &= (\operatorname{cl} \varphi)(0, u). \end{aligned}$$

Under the closedness assumptions, another application of Theorem A.54 proves the claim. □

Remark 3.30 Under the assumptions of Theorem 3.29, the reduced dual (3.9) is the dual problem obtained from the general conjugate duality but without the parameter $z \in \mathcal{X}$. Unlike φ^*, however, the function g is not closed, in general.

The function g is closed under the assumptions of Theorem 3.28 if π is continuous and γ is the identity mapping. Indeed, Theorem 3.28 then says that $g(y) = \varphi^*(\pi(y), y)$ which is closed as a composition of a continuous linear mapping and a closed convex function.

The following lemma gives sufficient conditions for Assumption 3.27.

Lemma 3.31 *Let $\Pi : \mathcal{V} \times \mathcal{Y} \to \mathcal{V} \times \mathcal{Y}$ and $\xi : \mathcal{X} \times \mathcal{X} \times \mathcal{U} \to \mathcal{X}$ be such that Π is continuous and linear and*

$$F(\xi(x, z, u), \Pi^*(z, u)) \le F(x, z, u) \quad \forall (x, z, u) \in \mathcal{X} \times \mathcal{X} \times \mathcal{U}.$$

Then

$$\varphi \circ \Pi^* \le \varphi \quad \text{and} \quad \varphi^* \circ \Pi \le \varphi^*.$$

If, in addition, $\Pi^*(0, \bar{u}) \in \mathcal{X}_a \times \{\bar{u}\}$, then Assumption 3.27 holds.

Proof Minimizing both sides of the inequality over $x \in \mathcal{X}$ gives

$$\inf_x F(x, z, u) \le \inf_x F(\xi(x, z, u), \Pi^*(z, u)) \le \inf_x F(x, z, u) \quad \forall (z, u) \in \mathcal{X} \times \mathcal{U}$$

or, in other words, $\varphi \circ \Pi^* \le \varphi$. By Lemma A.47, this implies $\varphi^* \circ \Pi \le \varphi^*$.

Assume now that $\Pi^*(0, \bar{u}) \in \mathcal{X}_a \times \{\bar{u}\}$. Since $\operatorname{dom} \varphi^* \subseteq \mathcal{X}_a^{\perp} \times \mathcal{Y}$, we get

$$\langle (0, \bar{u}), \Pi(p, y) \rangle = \langle \Pi^*(0, \bar{u}), (p, y) \rangle = \langle \bar{u}, y \rangle$$

for all $(p, y) \in \operatorname{dom} \varphi^*$. □

3.3 Duality for General Strategies

We now return to problem (P) where one optimizes over the space $\mathcal{N}$ of all adapted strategies, not just those belonging to $\mathcal{X}$ as in $(P_{\mathcal{X}})$. While problem $(P_{\mathcal{X}})$ in Sect. 3.2 allows for a convenient dualization within the purely functional analytic conjugate duality framework, there are interesting applications where $\inf(P_{\mathcal{X}}) > \inf(P)$ or where the infimum in (P) is attained in L^0 but not in $\mathcal{X}$; see Counterexample 3.37 for a simple illustration. It may even happen that $(P_{\mathcal{X}})$ is infeasible while (P) is not. This section shows that many of the duality relations between $(P_{\mathcal{X}})$ and (D) derived in Sect. 3.2 also hold between (P) and (D).

Recall that the dual of $(P_{\mathcal{X}})$ can be written as

$$\text{maximize} \quad \langle \bar{u}, y \rangle - \varphi^*(p, y) \quad \text{over } (p, y) \in \mathcal{V} \times \mathcal{Y}. \tag{D}$$

One could define a dual problem for (P) simply by replacing φ^* by the conjugate of the function $\bar{\varphi} : \mathcal{X} \times \mathcal{U} \to \overline{\mathbb{R}}$ defined by

$$\bar{\varphi}(z, u) := \inf_{x \in L^0} \{Ef(x, u) \mid x - z \in \mathcal{N}\}.$$

While in the definition of φ, the strategies are sought from the locally convex space $\mathcal{X} \subset L^0$, in the definition of $\bar{\varphi}$, we minimize over all of L^0. Clearly, $\bar{\varphi}(0, \bar{u}) = \inf(P)$ and $\bar{\varphi} \leq \varphi$. Under a mild condition, the conjugates of φ and $\bar{\varphi}$ coincide, so we may regard (D) as the dual problem of both $(P_{\mathcal{X}})$ and (P).

Lemma 3.32 *If* $\operatorname{dom} Ef \cap (\mathcal{X} \times \mathcal{U}) \neq \emptyset$, *then* $\varphi^* = \bar{\varphi}^*$ *and*

$$\partial \varphi(z, u) \subseteq \partial \bar{\varphi}(z, u)$$

for every $(z, u) \in \mathcal{X} \times \mathcal{U}$ *with an equality whenever the left side is nonempty.*

Proof Since $\varphi \geq \bar{\varphi}$, we have $\varphi^* \leq \bar{\varphi}^*$. To prove the converse, let $(p, y) \in \operatorname{dom} \varphi^*$. By Theorem 3.20,

$$\varphi^*(p, y) = Ef^*(p, y) + \delta_{\mathcal{X}_a^{\perp}}(p),$$

so $p \in \mathcal{X}_a^\perp$. Given any $(x, z, u) \in L^0 \times \mathcal{X} \times \mathcal{U}$, Fenchel's inequality gives

$$Ef(x, u) + \delta_{\mathcal{N}}(x - z) + Ef^*(p, y) \geq E[(x - z) \cdot p] + E[z \cdot p] + E[u \cdot y]$$

so, by Lemma 2.67,

$$Ef(x, u) + \delta_{\mathcal{N}}(x - z) + Ef^*(p, y) \geq E[z \cdot p] + E[u \cdot y].$$

Thus, $\bar{\varphi}(z, u) + \varphi^*(p, y) \geq \langle z, p\rangle + \langle u, y\rangle$ for all $(z, u) \in \mathcal{X} \times \mathcal{U}$, which means that $\bar{\varphi}^*(p, y) \leq \varphi^*(p, y)$. This proves the first claim.

Trivially, $\partial\varphi(z, u) \subseteq \partial\bar{\varphi}(z, u)$ if $\partial\varphi(z, u) = \emptyset$, so assume that $\partial\varphi(z, u) \neq \emptyset$. We have, in particular, $(z, u) \in \operatorname{dom}\varphi$ and thus, $\operatorname{dom} Ef \cap (\mathcal{X} \times \mathcal{U}) \neq \emptyset$. By the first claim, $\varphi^* = \bar{\varphi}^*$. Recall that, by Fenchel's inequality,

$$\varphi(z, u) + \varphi^*(p, y) \geq \langle z, p\rangle + \langle u, y\rangle$$

and that the equality holds if and only if $(p, y) \in \partial\varphi(z, u)$. Similarly for $\bar{\varphi}$. The subdifferential inclusion thus follows from the fact that $\varphi \geq \bar{\varphi}$.

By the first claim and Theorem A.54, $\operatorname{cl}\varphi = \operatorname{cl}\bar{\varphi}$. In particular, $\varphi \geq \bar{\varphi} \geq \operatorname{cl}\varphi$. When $\partial\varphi(z, u) \neq \emptyset$, we have $\varphi(z, u) = \operatorname{cl}\varphi(z, u)$ so $\bar{\varphi}(z, u) = \varphi(z, u)$ and thus, $\bar{\varphi}(z, u) + \bar{\varphi}^*(p, y) = \langle x, p\rangle + \langle u, y\rangle$ if and only if $\varphi(z, u) + \varphi^*(p, y) = \langle x, p\rangle + \langle u, y\rangle$. In other words, $(p, y) \in \partial\bar{\varphi}(z, u)$ if and only if $(p, y) \in \partial\varphi(z, u)$. □

The following summarizes the relationships between problems $(P_{\mathcal{X}})$, (P) and (D).

Theorem 3.33 *Assume that* $\operatorname{dom} Ef \cap (\mathcal{X} \times \mathcal{U}) \neq \emptyset$. *We have*

$$\inf(P_{\mathcal{X}}) \geq \inf(P) \geq \sup(D)$$

and

1. $\inf(P) = \sup(D)$ *if and only if* $\bar{\varphi}$ *is closed at* $(0, \bar{u})$,
2. $\inf(P_{\mathcal{X}}) = \inf(P) = \sup(D)$ *if and only if* φ *is closed at* $(0, \bar{u})$,
3. *if* $\bar{\varphi}(0, \bar{u}) < \infty$, *then the following are equivalent:*
   (a) (p, y) *solves* (D) *and* $\inf(P) = \sup(D)$;
   (b) *either* $(p, y) \in \partial\bar{\varphi}(0, \bar{u})$ *or* $\bar{\varphi}(0, \bar{u}) = -\infty$,
4. *if* $\varphi(0, \bar{u}) < \infty$, *then the following are equivalent:*
   (a) (p, y) *solves* (D) *and* $\inf(P_{\mathcal{X}}) = \inf(P) = \sup(D)$;
   (b) *either* $(p, y) \in \partial\varphi(0, \bar{u})$ *or* $\varphi(0, \bar{u}) = -\infty$.

Proof The first inequality is trivial. Fenchel's inequality gives

$$\begin{aligned}\inf(P) &= \bar{\varphi}(0, \bar{u}) \\ &\geq \sup_{(p,y)\in\mathcal{V}\times\mathcal{Y}} \{\langle\bar{u}, y\rangle - \bar{\varphi}^*(p, y)\} \\ &= \sup_{(p,y)\in\mathcal{V}\times\mathcal{Y}} \{\langle\bar{u}, y\rangle - \varphi^*(p, y)\} = \sup(D),\end{aligned}$$

where the second equality holds by Lemma 3.32.

By definition, $\inf(P) = \bar{\varphi}(0, \bar{u})$ and $\sup(D) = \varphi^{**}(0, \bar{u})$. By Lemma 3.32, $\varphi^{**} = \bar{\varphi}^{**}$ so part 1 follows from Theorem A.54. Part 2 follows from Theorem 3.17 while part 4 follows from Theorem 3.18. It remains to prove 3. By Fenchel's inequality,

$$\bar{\varphi}(0, z) \geq \langle\bar{u}, y\rangle - \bar{\varphi}^*(p, y).$$

Condition 3a means that either $\bar{\varphi}(0, \bar{u}) = -\infty$ or $\bar{\varphi}(0, z) = \langle\bar{u}, y\rangle - \varphi^*(p, y)$ while, by the definition of a subgradient, 3b means that either $\bar{\varphi}(0, \bar{u}) = -\infty$ or $\bar{\varphi}(0, z) = \langle\bar{u}, y\rangle - \bar{\varphi}^*(p, y)$. The claim thus follows from Lemma 3.32. □

The condition $\operatorname{dom} Ef \cap (\mathcal{X} \times \mathcal{U}) \neq \emptyset$ in Lemma 3.32 and Theorem 3.33 holds, in particular, if $(P_{\mathcal{X}})$ is feasible. Sufficient conditions for the closedness of $\bar{\varphi}$ will be given in Chap. 4 while Chap. 5 gives sufficient conditions for the subdifferentiability of φ at $(0, \bar{u})$.

The following gives an analogue of Corollary 3.25 for general strategies $x \in L^0$.

Theorem 3.34 *If* $\operatorname{dom} Ef \cap (\mathcal{X} \times \mathcal{U}) \neq \emptyset$ *and* (P) *and* (D) *are feasible, then the following are equivalent:*

1. x *solves* (P)*,* (p, y) *solves* (D) *and* $\inf(P) = \sup(D)$*;*
2. x *is feasible in* (P)*,* (p, y) *is feasible in* (D) *and*

$$(p, y) \in \partial f(x, \bar{u}) \quad a.s.;$$

3. x *is feasible in* (P)*,* (p, y) *is feasible in* (D) *and*

$$p \in \partial_x l(x, y), \quad \bar{u} \in \partial_y[-l](x, y) \quad a.s.$$

Proof The equivalence of 2 and 3 follows by scenariowise application of the equivalence of 3 and 5 in Theorem A.73. Let $x \in \mathcal{N}$ and $(p, y) \in \mathcal{V} \times \mathcal{Y}$ be feasible. By Fenchel's inequality,

$$f(x, \bar{u}) + f^*(p, y) - \bar{u} \cdot y \geq x \cdot p \quad a.s. \tag{3.11}$$

so, by Lemma B.3,

$$Ef(x, \bar{u}) + E[f^*(p, y) - \bar{u} \cdot y] \geq E[x \cdot p]. \tag{3.12}$$

By the feasibility of x and (p, y), the expectations on the left are finite so (3.12) holds as an equality if and only if (3.11) holds as an equality almost surely. Equality in (3.11) means that 2 holds. By Lemma 2.67, $E[x \cdot p] = 0$, so equality in (3.12) means that 1 holds. □

If $\partial\bar{\varphi}(0, \bar{u}) \neq \emptyset$, then, by Theorem 3.33, $\inf(P) = \sup(D)$ and the dual has a solution. Theorem 3.34 thus implies the following optimality conditions for (P).

Corollary 3.35 *If* $\operatorname{dom} Ef \cap (\mathcal{X} \times \mathcal{U}) \neq \emptyset$ *and* $\partial\bar{\varphi}(0, \bar{u}) \neq \emptyset$*, then* $\inf(P) = \sup(D)$*, the dual optimum is attained, and the following are equivalent:*

1. *x solves (P);*
2. *x is feasible in (P) and there exists (p, y) feasible in (D) with*

$$(p, y) \in \partial f(x, \bar{u}) \quad a.s.;$$

3. *x is feasible in (P) and there exists (p, y) feasible in (D) with*

$$p \in \partial_x l(x, y), \quad \bar{u} \in \partial_y[-l](x, y) \quad a.s.$$

Recall that, by Lemma 3.32, the condition $\partial\bar{\varphi}(0, \bar{u}) \neq \emptyset$ is implied by $\partial\varphi(0, \bar{u}) \neq \emptyset$. Sufficient conditions for this will be given in Chap. 5.

Corollary 3.36 *Assume that* $\operatorname{dom} Ef \cap (\mathcal{X} \times \mathcal{U}) \neq \emptyset$ *and* $(p, y) \in \partial\bar{\varphi}(0, \bar{u})$*. Then optimal solutions x of (P) are scenariowise minimizers of the function*

$$x \mapsto l(x, y(\omega), \omega) - x \cdot p(\omega).$$

Conversely, if this function has a unique scenariowise minimizer x and if (P) admits solutions, then x solves (P) and, in particular, x is feasible and adapted.

Proof The first claim follows directly from Corollary 3.35 after observing that the first inclusion in part 3 means that x minimizes the function

$$x \mapsto l(x, y(\omega), \omega) - x \cdot p(\omega).$$

If the primal admits solutions, then it satisfies the inclusions in part 3 of Corollary 3.35. If the minimizer x of the above function is unique, it thus has to be a unique primal solution. □

Counterexample 3.37 It may happen that

$$\inf(P_{\mathcal{X}}) > \inf(P) = \sup(D).$$

Indeed, let

$$f(x, u, \omega) = \delta_{\{0\}}(x_T - u\xi(\omega)).$$

If $\xi \notin \mathcal{X}$ and $\bar{u} = 1$, then (P) is feasible while $(P_{\mathcal{X}})$ is not. Clearly Ef is proper on $\mathcal{X} \times \mathcal{U}$ and $f^*(0, 0) = 0$, so $\inf(P) = \sup(D) = 0$. Another example with finite $\inf(P_{\mathcal{X}}) < \infty$ is obtained by letting

$$f(x, u, \omega) = (x_0 - 1)^2 + \delta_{\{0\}}(x_0\xi(\omega) - x_1),$$

$\mathcal{F}_0 = \{\Omega, \emptyset\}$ and $\xi \in L^0(\mathcal{F}_1)$ with $\xi \notin \mathcal{X}$. Since f is nonnegative, $(1, \xi)$ is optimal for (P) and the optimum value is zero. Here Ef is proper on $\mathcal{X} \times \mathcal{U}$, and, by a direct verification, $f^*(0, 0) = 0$, so the origin is a dual solution and $\inf(P) = \sup(D) = 0$. On the other hand, the only feasible solution of $(P_{\mathcal{X}})$ is the origin, so $\inf(P_{\mathcal{X}}) = 1$.

Recall the reduced dual problems and the function

$$g(y) := \inf_{p \in \mathcal{V}} \varphi^*(p, y)$$

from Sect. 3.2.2. The following extends Theorem 3.28 to general strategies $x \in \mathcal{N}$.

Theorem 3.38 (Reduced Dual) *Assume that there exist mappings $\pi : \mathcal{Y} \to \mathcal{V}$ and $\gamma : \mathcal{Y} \to \mathcal{Y}$ such that γ is idempotent and the mapping $\Pi(p, y) = (\pi(y), \gamma(y))$ satisfies Assumption 3.27. Then*

$$g(y) = \varphi^*(\pi(y), y) \quad \forall y \in \operatorname{rge} \gamma,$$

optimum value of (D) coincides with that of the problem

$$\text{maximize} \quad \langle \bar{u}, y \rangle - g(y) \qquad \text{over} \quad y \in \operatorname{rge} \gamma, \tag{3.13}$$

and if $y \in \operatorname{rge} \gamma$ solves (3.13) then $(\pi(y), y)$ solves (D). If (p, y) solves (D), then $\gamma(y)$ solves (3.13). If (P) and (3.13) are feasible, then the following are equivalent:

1. *x solves (P), y solves (3.13) and $\inf(P) = \sup(3.13)$;*
2. *x is feasible in (P), y is feasible in (3.13) and*

$$(\pi(y), y) \in \partial f(x, \bar{u}) \quad a.s.;$$

3. *x is feasible in (P), y is feasible in (3.13) and*

$$\pi(y) \in \partial_x l(x, y), \quad \bar{u} \in \partial_y[-l](x, y) \quad a.s.$$

Proof The claims up to the equivalences are a repetition of Theorem 3.28. The equivalences follow now from Theorem 3.34. □

The following is analogous to Theorem 3.29.

Theorem 3.39 *Assume that* dom $Ef \cap (\mathcal{X} \times \mathcal{U}) \neq \emptyset$. *If g is closed at $y \in \mathcal{Y}$ and $\bar{\varphi}$ is closed at $(0, u)$ for all $u \in \mathcal{U}$, then $g(y) = \bar{\varphi}(0, \cdot)^*(y)$. In particular, if g and $\bar{\varphi}$ are closed, then $g = \bar{\varphi}(0, \cdot)^*$.*

Proof By Theorem A.54 and Lemma 3.32,

$$\begin{aligned} g^*(u) &= \sup_y \{\langle u, y\rangle - g(y)\} \\ &= \sup_{p,y} \{\langle u, y\rangle - \varphi^*(p, y)\} \\ &= (\operatorname{cl} \varphi)(0, u) = (\operatorname{cl} \bar{\varphi})(0, u). \end{aligned}$$

Under the closedness assumptions, another application of Theorem A.54 proves the claim. □

The closedness of $\bar{\varphi}$ will be the topic of the Chap. 4. The following is the analogue of Lemma 3.31 for general strategies x. The proof is almost identical so it is omitted.

Lemma 3.40 *Let $\Pi : \mathcal{V} \times \mathcal{Y} \to \mathcal{V} \times \mathcal{Y}$ and $\xi : L^0 \times \mathcal{X} \times \mathcal{U} \to L^0$ be such that Π is continuous and linear and*

$$F(\xi(x, z, u), \Pi^*(z, u)) \leq F(x, z, u) \quad \forall (x, z, u) \in L^0 \times \mathcal{X} \times \mathcal{U}.$$

Then

$$\bar{\varphi} \circ \Pi^* \leq \bar{\varphi} \quad \textit{and} \quad \bar{\varphi}^* \circ \Pi \leq \bar{\varphi}^*.$$

If, in addition, $\Pi^*(0, \bar{u}) \in \mathcal{X}_a \times \{\bar{u}\}$, then Assumption 3.27 holds.

3.4 Applications

This section applies the general duality results of this chapter to the five examples considered in the earlier chapters. We find explicit expressions for the involved functions and conditions but only give selected statements as examples of how the general results can be applied.

3.4.1 Mathematical Programming

Consider again problem (1.8)

$$\begin{aligned} &\text{minimize} \quad && Ef_0(x) \quad \text{over} \quad x \in \mathcal{N}, \\ &\text{subject to} && f_j(x) \leq 0 \quad j = 1, \dots, l \; a.s., \\ & && f_j(x) = 0 \quad j = l+1, \dots, m \; a.s. \end{aligned}$$

from Sect. 1.3.1. This fits the general duality framework with $\bar{u} = 0$ and

$$f(x, u, \omega) = \begin{cases} f_0(x, \omega) & \text{if } x \in \operatorname{dom} H(\cdot, \omega), \; H(x, \omega) + u \in K, \\ +\infty & \text{otherwise,} \end{cases}$$

where $K = \mathbb{R}^l_- \times \{0\}$ and H is the random K-convex function defined by

$$\operatorname{dom} H(\cdot, \omega) = \bigcap_{j=1}^{m} \operatorname{dom} f_j(\cdot, \omega) \quad \text{and} \quad H(x, \omega) = (f_i(x, \omega))_{j=1}^{m}.$$

That f is a convex normal integrand follows from arguments similar to those used in Sect. 1.3.1 to show that h is a convex normal integrand.

The Lagrangian integrand becomes

$$\begin{aligned} l(x, y, \omega) &= \inf\{f(x, u, \omega) - u \cdot y\} \\ &= \inf\{f_0(x, \omega) - u \cdot y \mid x \in \operatorname{dom} H(\cdot, \omega), \; H(x, \omega) + u \in K\} \\ &= \begin{cases} +\infty & \text{if } x \notin \operatorname{dom} H(\cdot, \omega), \\ f_0(x, \omega) + y \cdot H(x, \omega) & \text{if } x \in \operatorname{dom} H(\cdot, \omega) \text{ and } y \in K^\circ, \\ -\infty & \text{otherwise.} \end{cases} \end{aligned}$$

The conjugate of f is given by

$$\begin{aligned} f^*(p, y, \omega) &= \sup_{x \in \mathbb{R}^n} \{x \cdot p - l(x, y, \omega)\} \\ &= \sup_{x \in \mathbb{R}^n} \{x \cdot p - f_0(x, \omega) - y \cdot H(x, \omega) \mid x \in \operatorname{dom} H(\cdot, \omega)\} \end{aligned}$$

for $y \in K^\circ$ and $f^*(p, y, \omega) = +\infty$ for $y \notin K^\circ$. If $\operatorname{dom} Ef \cap (\mathcal{X} \times \mathcal{U}) \neq \emptyset$, Corollary 3.21 says that the dual problem can be written as

$$\begin{aligned} &\text{maximize} \;\; E[\inf_{x \in \mathbb{R}^n} \{f_0(x) + y \cdot H(x) - x \cdot p\}] \;\; \text{over} \;\; (p, y) \in \mathcal{X}_a^\perp \times \mathcal{Y} \\ &\text{subject to} \qquad\qquad y \in K^\circ \quad a.s. \end{aligned} \tag{3.14}$$

To get more explicit expressions for f^* and the dual problem, additional structure is needed; see e.g Example 3.42 below. Theorem 3.34 gives the following.

Theorem 3.41 *If* $\operatorname{dom} Ef \cap (\mathcal{X} \times \mathcal{U}) \neq \emptyset$ *and* (1.8) *and* (3.14) *are feasible, then the following are equivalent:*

1. x *solves* (1.8), (p, y) *solves* (3.14) *and* $\inf(1.8) = \sup(3.14)$*;*
2. x *is feasible in* (1.8), (p, y) *is feasible in* (3.14) *and*

$$p \in \partial_x[f_0 + y \cdot H](x),$$
$$H(x) \in K, \quad y \in K^\circ, \quad y \cdot H(x) = 0$$

almost surely.

Proof By Theorem 3.34, it suffices to note that, when $(x, y) \in \operatorname{dom} l$, we have

$$0 \in \partial_y[-l](x, y) = -H(x) + N_{K^\circ}(y),$$

if and only if $H(x) \in \partial\delta_{K^\circ}(y)$. By Example A.43, this is equivalent to the given complementarity condition. □

The more general composite format in Example 1.82 can be treated in an analogous way. In case of linear stochastic programming, the dual can be written down explicitly in terms of the problem data.

Example 3.42 (Linear Stochastic Programming) Consider the problem

$$\begin{aligned} &\text{minimize} \quad && E[x \cdot c] \quad \text{over } x \in \mathcal{N} \\ &\text{subject to} \quad && Ax + b \in K \quad a.s. \end{aligned} \tag{3.15}$$

from Example 1.83 and assume that there exists $(x, u) \in \mathcal{X} \times \mathcal{U}$ such that $E[x \cdot c] < \infty$ and $Ax + u + b \in K$ almost surely. The dual problem becomes

$$\begin{aligned} &\text{maximize} \quad && E[b \cdot y] \quad \text{over } p \in \mathcal{X}_a^\perp,\ y \in \mathcal{Y}, \\ &\text{subject to} \quad && A^*y + c = p,\ y \in K^\circ \quad a.s. \end{aligned}$$

and the scenariowise KKTR-conditions

$$A^*y + c = p,$$
$$Ax + b \in K, \quad y \in K^\circ, \quad (Ax + b) \cdot y = 0,$$

where A^* is the scenariowise transpose of A.

Proof This is a special case of (1.8) with $f_0(x,\omega) = c(\omega)\cdot x$ and $f_j(x,\omega) = a_j(\omega)\cdot x + b_j(\omega)$ for $j = 1,\dots,m$. We get

$$l(x,y,\omega) = x\cdot c(\omega) + y\cdot A(\omega)x + y\cdot b(\omega) - \delta_{K^\circ}(y)$$

and

$$\begin{aligned} f^*(p,y,\omega) &= \sup_{x\in\mathbb{R}^n}\{x\cdot p - l(x,y,\omega)\} \\ &= \begin{cases} -y\cdot b(\omega) & \text{if } y\in K^\circ \text{ and } A^*(\omega)y + c(\omega) = p, \\ +\infty & \text{otherwise.} \end{cases} \end{aligned}$$

This gives the dual problem while the KKTR-conditions follow directly from Theorem 3.41. □

The stochastic linear programming problem satisfies the assumptions of Theorem 3.38 under natural conditions. We will denote the *adapted projection* of an integrable process $u = (u_t)_{t=0}^T$ by

$$^a u := (E_t u_t)_{t=0}^T.$$

Example 3.43 (Linear Stochastic Programming, Reduced Dual) In the setting of Example 3.42 assume that $c \in \mathcal{V}$ and $A^* y \in \mathcal{V}$ for all $y \in \mathcal{Y}$. Then, the optimum value of the dual problem equals that of the *reduced dual problem*

$$\begin{aligned} &\text{maximize} && E[b\cdot y] \quad \text{over } y\in\mathcal{Y}, \\ &\text{subject to} && {}^a(A^*y + c) = 0,\ y\in K^\circ \quad a.s. \end{aligned} \tag{3.16}$$

and a pair (p, y) solves the dual if and only if y solves (3.16) and

$$p = A^*y + c - {}^a(A^*y + c).$$

If (3.15) and (3.16) are feasible, then the following are equivalent:

1. x solves (3.15), y solves (3.16) and $\inf(P) = \sup(3.13)$;
2. x is feasible in (3.15), y is feasible in (3.16) and

$$\begin{gathered} {}^a(A^*y + c) = 0, \\ Ax + b \in K, \quad y\in K^\circ, \quad (Ax + b)\cdot y = 0. \end{gathered}$$

Proof This fits the format of Theorem 3.38 with $\pi(y) := A^*y + c - {}^a(A^*y + c)$ and $\gamma(y) = y$. Indeed, using the expression for f^* in the proof of Example 3.42, gives

$$\varphi^*(p, y) = \begin{cases} E[-b \cdot y] & \text{if } p \in \mathcal{X}_a^\perp \text{ and } y \in K^\circ \text{ and } A^*y + c = p, \\ +\infty & \text{otherwise.} \end{cases}$$

Since any $p \in \mathcal{X}_a^\perp$ has ${}^a p = 0$, it is clear that Assumption 3.27 is satisfied so the claims follow from Theorem 3.38. □

Remark 3.44 If, in Example 3.43, the elements of c_t and the columns A_t of A corresponding to x_t are $\mathcal{F}_t$-measurable, then by Lemma B.13, the reduced dual can be written as

$$\begin{aligned} &\text{maximize} && E[b \cdot y] \quad \text{over } \ y \in \mathcal{Y}, \\ &\text{subject to} && c_t + A_t^* \cdot E_t y = 0 \ \forall t, \ y \in K^\circ \quad a.s. \end{aligned}$$

If, in addition, $b \in \mathcal{U}$ and b is $\mathcal{F}_T$-measurable and the space $\mathcal{Y}$ is such that $E_t y \in \mathcal{Y}$ for every $y \in \mathcal{Y}$, then by Lemma B.13, we can write this as

$$\begin{aligned} &\text{maximize} && E[b \cdot y_T] \quad \text{over } \ (y_t)_{t=0}^T \in \mathcal{M}^{\mathcal{Y}}, \\ &\text{subject to} && c_t + A_t^* \cdot y_t = 0, \ y_t \in K^\circ \ \forall t \quad a.s., \end{aligned}$$

where $\mathcal{M}^{\mathcal{Y}}$ is the linear space of martingales $(y_t)_{t=0}^T$ with $y_t \in \mathcal{Y}$ for all t.

3.4.2 Optimal Stopping

Consider again the relaxed optimal stopping problem

$$\begin{aligned} &\text{maximize} && E \sum_{t=0}^T R_t x_t \quad \text{over } x \in \mathcal{N}, \\ &\text{subject to} && x \geq 0, \ \sum_{t=0}^T x_t \leq 1 \quad a.s. \end{aligned}$$

from Sects. 1.3.2 and 2.3.2. This fits the general duality framework with $n_t = 1$, $m = 1$,

$$f(x, u, \omega) = \begin{cases} -\sum_{t=0}^T x_t R_t(\omega) & \text{if } x \geq 0 \text{ and } \sum_{t=0}^T x_t + u \leq 0, \\ +\infty & \text{otherwise} \end{cases}$$

and $\bar{u} = -1$. Recall that we have flipped signs here to make the maximization problem (1.10) fit the general format (P). We get

$$
\begin{aligned}
l(x, y, \omega) &= \inf_{u \in \mathbb{R}^n} \{f(x, u, \omega) - uy\} \\
&= \inf_{u \in \mathbb{R}^n} \{-\sum_{t=0}^{T} x_t R_t(\omega) - uy \mid x \geq 0, \sum_{t=0}^{T} x_t + u \leq 0\} \\
&= \begin{cases} -\sum_{t=0}^{T} x_t R_t(\omega) + y \sum_{t=0}^{T} x_t + \delta_{\mathbb{R}^n_+}(x) & \text{if } y \geq 0, \\ -\infty & \text{otherwise} \end{cases} \\
&= \begin{cases} \sum_{t=0}^{T} x_t [y - R_t(\omega)] + \delta_{\mathbb{R}^n_+}(x) & \text{if } y \geq 0, \\ -\infty & \text{otherwise} \end{cases}
\end{aligned}
$$

and

$$
\begin{aligned}
f^*(p, y, \omega) &= \sup_{x \in \mathbb{R}^n} \{x \cdot p - l(x, y, \omega)\} \\
&= \sup_{x \in \mathbb{R}^n_+} \sum_{t=0}^{T} x_t [p_t - y + R_t(\omega)] \\
&= \begin{cases} 0 & \text{if } y \geq 0 \text{ and } p_t + R_t(\omega) \leq y,\ t = 0, \ldots, T, \\ +\infty & \text{otherwise.} \end{cases}
\end{aligned}
$$

Since $\operatorname{dom} Ef \cap (\mathcal{X} \times \mathcal{U}) \neq \emptyset$, Corollary 3.21 says that the dual of (1.10) can be written as

$$
\begin{aligned}
&\text{minimize} && Ey \quad \text{over } (p, y) \in \mathcal{X}_a^{\perp} \times \mathcal{Y}_+ \\
&\text{subject to} && p_t + R_t \leq y \quad t = 0, \ldots, T \ a.s.
\end{aligned}
\tag{3.17}
$$

Again, we have changed the sign to conform to the tradition of writing the optimal stopping problem as a maximization problem. It is clear that (1.10) is feasible, and (3.17) is feasible as soon as the pathwise maximum $\max_t R_t$ of R belongs $\mathcal{Y}$. Theorem 3.34 thus gives the following.

Theorem 3.45 *Assume that* $\max_t R_t \in \mathcal{Y}$. *The following are equivalent:*

1. x *solves* (1.10), (p, y) *solves* (3.17) *and there is no duality gap;*
2. $x \in \mathcal{N}$ *and* $(p, y) \in \mathcal{X}_a^{\perp} \times \mathcal{Y}$ *and*

$$
x_t \geq 0,\ p_t + R_t \leq y,\ x_t(p_t + R_t - y) = 0 \quad t = 0, \ldots, T,
$$

$$
y \geq 0,\ \sum_{t=0}^{T} x_t \leq 1,\ y(\sum_{t=0}^{T} x_t - 1) = 0
$$

almost surely.

In particular, a stopping time $\tau \in \mathcal{T}$ solves (1.9) *and* $(p, y) \in \mathcal{X}_a^\perp \times \mathcal{Y}_+$ *solves the* (3.17) *if and only if* $p_t + R_t \leq y$ *for all* t *and* $p_\tau + R_\tau = y$ *almost surely. Here,* $p_{T+1} := 0$.

Proof The scenariowise KKTR-condition in Theorem 3.34 can be written as

$$p_t + R_t - y \in N_{\mathbb{R}_+}(x_t) \quad t = 0, \dots, T,$$

$$\sum_{t=0}^{T} x_t - 1 \in N_{\mathbb{R}_+}(y).$$

This is equivalent to the conditions given in the statement; see Example A.43. The second claim thus follows from Theorem 3.22 and Corollary 3.25. The last claim follows from the fact that a $\tau \in \mathcal{T}$ solves the optimal stopping problem (1.9) if and only if the process $x \in \mathcal{N}$ given by

$$x_t = \begin{cases} 1 & \text{if } t = \tau, \\ 0 & \text{if } t \neq \tau \end{cases}$$

is optimal in (1.10); see the beginning of Sect. 1.3.2. □

Under mild conditions, the assumptions of Theorem 3.38 on the reduced dual problem hold with the mappings $\pi : \mathcal{Y} \to \mathcal{V}$ and $\gamma : \mathcal{Y} \to \mathcal{Y}$ given by $\pi(y) = (y - E_t y)_{t=0}^T$ and $\gamma(y) = E_T y$, respectively. Combining Theorem 3.45 with Theorem 3.38 thus gives the following.

Corollary 3.46 (Reduced Dual) *Assume that* $R_t \in \mathcal{Y}$ *and* $E_t\mathcal{Y} \subseteq \mathcal{Y} \subseteq \mathcal{V}_t$ *for all* t. *The optimum value of* (3.17) *equals that of*

$$\begin{aligned} &\text{minimize} \quad Ey \quad \text{over } y \in \mathcal{Y}_+(\mathcal{F}_T) \\ &\text{subject to} \quad R_t \leq E_t y \quad t = 0, \dots, T \text{ a.s.} \end{aligned} \tag{3.18}$$

If (p, y) *solves* (3.17), *then* $E_T y$ *solves* (3.18). *If* y *solves* (3.18), *then* $((y - E_t y)_{t=0}^T, y)$ *solves* (3.17). *An* $x \in \mathcal{N}$ *solves* (1.10), $y \in \mathcal{Y}_+(\mathcal{F}_T)$ *solves* (3.18) *and there is no duality gap if and only if*

$$x_t \geq 0,\ R_t \leq E_t y,\ x_t(R_t - E_t y) = 0 \quad t = 0, \dots, T,$$

$$y \geq 0,\ \sum_{t=0}^{T} x_t \leq 1,\ y(\sum_{t=0}^{T} x_t - 1) = 0.$$

Remark 3.47 The reduced dual in Corollary 3.46 can be written as

$$\begin{aligned} &\text{minimize} \quad Ey_0 \quad \text{over } y \in \mathcal{M}_+^{\mathcal{Y}} \\ &\text{subject to} \quad R_t \leq y_t \quad t = 0, \dots, T \text{ a.s.}, \end{aligned} \tag{3.19}$$

where $\mathcal{M}_+^{\mathcal{Y}}$ is the cone of nonnegative martingales y with $y_t \in \mathcal{Y}$ for all $t = 0, \dots, T$. Thus, $x \in \mathcal{N}$ solves the primal, $y \in \mathcal{M}_+^{\mathcal{Y}}$ solves (3.19) and there is no duality gap if and only if

$$x_t \geq 0,\ R_t \leq y_t,\ x_t(R_t - y_t) = 0 \quad t = 0, \dots, T,$$

$$y_T \geq 0,\ \sum_{t=0}^{T} x_t \leq 1,\ y_T(\sum_{t=0}^{T} x_t - 1) = 0.$$

In particular, a stopping time $\tau \in \mathcal{T}$ is optimal in (1.9) and $y \in \mathcal{M}_+^{\mathcal{Y}}$ solves (3.19) if and only if $R_t \leq y_t$ for all t and $R_\tau = y_\tau$, where $y_{T+1} := y_T$.

Chapters 4 and 5 below give sufficient conditions for the absence of a duality gap and the existence of dual solutions, respectively, in the general formulation of (P). In optimal stopping, absence of a duality gap and the existence of dual solutions can be proved directly using Theorem 2.88. The argument is based on the Doob decomposition of the Snell envelope of the reward process R. A stochastic process A is said to be *predictable* if A_t is $\mathcal{F}_{t-1}$-measurable for all t.

Lemma 3.48 (Doob Decomposition) *Assume that $E_t\mathcal{Y} \subset \mathcal{Y}$ for all t. Given an adapted process $y = (y_t)_{t=0}^T$ with $y_t \in \mathcal{Y}$ for all t, there exist unique processes M and A such that*

$$y = M + A,$$

M is a martingale, A is predictable $A_0 = 0$ and $M_t, A_t \in \mathcal{Y}$ for all t. If y is a supermartingale, A is nonincreasing.

Proof It suffices to define M and A recursively by $A_0 = 0$, $\Delta A_t = E_{t-1}\Delta y_t$ and $M_0 = y_0$, $\Delta M_t = \Delta y_t - \Delta A_t$. The uniqueness follows from the fact that a process that is both predictable and a martingale is necessarily a constant process. □

Recall from Sect. 2.3.2 that the Snell envelope S of R is the smallest supermartingale that dominates R.

Corollary 3.49 (Snell Envelope) *Assume that $R_t \in \mathcal{Y}$ and $E_t\mathcal{Y} \subseteq \mathcal{Y} \subseteq \mathcal{V}_t$ for all t. Then there is no duality gap and the martingale part M of the Snell envelope of R solves* (3.19). *The optimal stopping times τ are characterised by the condition $R_\tau = M_\tau$, where $M_{T+1} := M_T$.*

Proof Let S be the Snell envelope of the reward process R. By decomposability of $\mathcal{Y}$, $S_t \in \mathcal{Y}$ and, since S is a supermartingale, it admits the decomposition $S = M + A$ from Lemma 3.48. Since A is nonincreasing and $A_0 = 0$, the martingale M dominates R and $ES_0 = EM_0$ while, by Theorem 2.88, ES_0 equals the optimum value of (1.10). Thus, M solves (3.19) and there is no duality gap. The last claim now follows from that of Remark 3.47. □

Remark 3.50 (Davis–Karatzas Duality) Trivially, the normal integrand

$$\tilde{f}(x,u,\omega)=\begin{cases}-\sum_{t=0}^T x_tR_t(\omega) & \text{if } x\geq 0, \sum_{t=0}^T x_t+u\leq 0 \text{ and } \sum_{t=0}^T x_t\leq 1,\\ +\infty & \text{otherwise}\end{cases}$$

gives rise to the same primal problem as f introduced above. The corresponding Lagrangian integrand becomes

$$\tilde{l}(x,y,\omega)=\begin{cases}\sum_{t=0}^T x_t(y-R_t(\omega))+\delta_{\mathbb{R}^n_+}(x)+\delta_{\mathbb{R}_-}(\sum_{t=0}^T x_t-1) & \text{if } y\geq 0,\\ -\infty & \text{otherwise}\end{cases}$$

and the conjugate integrand

$$\begin{aligned}\tilde{f}^*(p,y,\omega)&=\sup_x\{x\cdot p-\tilde{l}(x,y,\omega)\}\\ &=\sup_{x\in\mathbb{R}^n_+}\left\{\sum_{t=0}^T x_t(p_t-y+R_t(\omega))\,\middle|\,\sum_{t=0}^T x_t\leq 1\right\}\\ &=\sup_{t=0,\dots,T}\{p_t-y+R_t(\omega)\}.\end{aligned}$$

Much like in Remark 3.47, we find the reduced dual problem

$$\text{minimize}\quad E\left[\sup_{t=0,\dots,T}\{R_t+y_T-y_t\}\right]\quad\text{over } y\in\mathcal{M}^y_+. \tag{3.20}$$

Note that, unlike f^*, the normal integrand $\tilde{f}^*$ is everywhere finite and it is dominated by f^*. It follows that the optimum value of the dual problem associated with $\tilde{f}$ lies between the original primal and dual optimum values. In particular, if $\inf(1.9)=\sup(3.17)$, then the optimum value of (3.20) equals that of (3.17). The finiteness of $\tilde{f}^*$ may make (3.20) easier to solve numerically. Being sandwiched between the primal and the original dual, its values provide tighter upper bounds for the optimum value of the primal problem.

3.4.3 *Optimal Control*

Consider the optimal control problem (1.11)

$$\begin{aligned}&\text{minimize}\quad E\left[\sum_{t=0}^T L_t(X_t,U_t)\right]\quad\text{over } (X,U)\in\mathcal{N},\\ &\text{subject to}\quad \Delta X_t=A_tX_{t-1}+B_tU_{t-1}+W_t\quad t=1,\dots,T \text{ a.s.}\end{aligned}$$

from Sect. 1.3.3 and recall that X_t takes values in $\mathbb{R}^N$ and U_t in $\mathbb{R}^M$. Problem (1.11) fits the general duality framework with $x = (X, U)$, $\bar{u} = (W_t)_{t=1}^T$, $n_t = N + M$, $m = T$ and

$$f(x, u, \omega) = \sum_{t=0}^{T} L_t(X_t, U_t, \omega) + \sum_{t=1}^{T} \delta_{\{0\}}(\Delta X_t - A_t(\omega)X_{t-1} - B_t(\omega)U_{t-1} - u_t).$$

We thus assume that $\mathcal{X}$ and $\mathcal{U}$ are spaces of $\mathbb{R}^{(T+1)(N+M)}$- and $\mathbb{R}^{TM}$-valued random variables, respectively, and that $(W_1, \dots, W_T) \in \mathcal{U}$. By solidity,

$$\mathcal{U} = \mathcal{U}_1 \times \cdots \times \mathcal{U}_T, \quad \mathcal{Y} = \mathcal{Y}_1 \times \cdots \times \mathcal{Y}_T,$$

where $\mathcal{U}_t$ and $\mathcal{Y}_t$ are solid decomposable spaces of $\mathbb{R}^M$-valued random variables in separating duality under the bilinear form $(u_t, y_t) \mapsto E[u_t \cdot y_t]$. It follows that

$$\langle u, y \rangle = \sum_{t=1}^{T} E[u_t \cdot y_t].$$

For simplicity, we assume further that, for all t,

$$\begin{aligned} \mathcal{X}_t &= \mathcal{S} \times \mathcal{C}, & \mathcal{U}_t &= \mathcal{S}, \\ \mathcal{V}_t &= \mathcal{S}' \times \mathcal{C}', & \mathcal{Y}_t &= \mathcal{S}', \end{aligned}$$

where $\mathcal{S}$ and $\mathcal{C}$ are solid decomposable spaces in separating duality with solid decomposable spaces $\mathcal{S}'$ and $\mathcal{C}'$, respectively.

The Lagrangian integrand becomes

$$\begin{aligned} l(x, y, \omega) &= \inf_{u \in \mathbb{R}^m} \{f(x, u, \omega) - u \cdot y\} \\ &= \sum_{t=0}^{T} L_t(X_t, U_t, \omega) - \sum_{t=1}^{T} (\Delta X_t - A_t(\omega)X_{t-1} - B_t(\omega)U_{t-1}) \cdot y_t. \end{aligned}$$

Using the *integration by parts* formula,

$$\sum_{t=1}^{T} \Delta X_t \cdot y_t = -\sum_{t=0}^{T} X_t \cdot \Delta y_{t+1},$$

where $y_0 := y_{T+1} := 0$, we get

$$-\sum_{t=1}^{T}(\Delta X_t - A_t(\omega)X_{t-1} - B_t(\omega)U_{t-1}) \cdot y_t$$
$$= \sum_{t=0}^{T} X_t \cdot (\Delta y_{t+1} + A_{t+1}^*(\omega)y_{t+1}) + U_t \cdot B_{t+1}^*(\omega)y_{t+1}), \qquad (3.21)$$

where $A_{T+1} := 0$ and $B_{T+1} := 0$. Thus, the Lagrangian integrand can be written as

$$l(x, y, \omega) = \sum_{t=0}^{T}[L_t(X_t, U_t, \omega)$$
$$+ (X_t, U_t) \cdot (\Delta y_{t+1} + A_{t+1}^*(\omega)y_{t+1}, B_{t+1}^*(\omega)y_{t+1})]$$

and the conjugate of f becomes

$$f^*(v, y, \omega) = \sup_{x\in\mathbb{R}^n}\{x \cdot v - l(x, y, \omega)\}$$
$$= \sum_{t=0}^{T} L_t^*(v_t - (\Delta y_{t+1} + A_{t+1}^*(\omega)y_{t+1}, B_{t+1}^*(\omega)y_{t+1}), \omega).$$

As soon as $\operatorname{dom} Ef \cap (\mathcal{X} \times \mathcal{U}) \neq \emptyset$, Theorem 3.20 says that the conjugate of the optimum value function can be written as

$$\varphi^*(p, y) = E\left[\sum_{t=0}^{T} L_t^*(p_t - (\Delta y_{t+1} + A_{t+1}^* y_{t+1}, B_{t+1}^* y_{t+1}))\right] + \delta_{\mathcal{X}_a^\perp}(p) \qquad (3.22)$$

and, by Corollary 3.21, the dual problem becomes

$$\begin{array}{ll} \text{maximize} & E\left[\sum_{t=1}^{T} W_t \cdot y_t - \sum_{t=0}^{T} L_t^*(p_t - (\Delta y_{t+1} + A_{t+1}^* y_{t+1}, B_{t+1}^* y_{t+1}))\right] \\ \text{over} & (p, y) \in \mathcal{X}_a^\perp \times \mathcal{Y}. \end{array} \qquad (3.23)$$

Theorem 3.34 gives the following.

Theorem 3.51 *If* $\operatorname{dom} Ef \cap (\mathcal{X} \times \mathcal{U}) \neq \emptyset$ *and* (1.11) *and* (3.23) *are feasible, then the following are equivalent:*

1. (X, U) *solves* (1.11), (p, y) *solves* (3.23) *and there is no duality gap;*

2. *(X, U) is feasible in* (1.11), *(p, y) is feasible in* (3.23) *and*

$$p_t - (\Delta y_{t+1} + A^*_{t+1} y_{t+1}, B^*_{t+1} y_{t+1}) \in \partial L_t(X_t, U_t),$$

for all t almost surely.

The optimality conditions in Theorem 3.51 yield a characterization of the optimal control U as a pointwise minimizer of a "Hamiltonian" function associated with problem (1.11).

Remark 3.52 (Maximum Principle) The scenariowise KKTR-conditions in Theorem 3.51 mean that (X, U) satisfies the system equations and that

$$-(\Delta y_{t+1}, 0) \in \partial_{(X_t,U_t)} H_t(X_t, U_t, y_{t+1}) - p_t,$$

where

$$H_t(X_t, U_t, y_{t+1}) := L_t(X_t, U_t) + y_{t+1} \cdot (A_{t+1} X_t + B_{t+1} U_t).$$

This can be written equivalently as

$$U_t \in \operatorname*{argmin}_{U_t \in \mathbb{R}^M} \{H_t(X_t, U_t, y_{t+1}) - (X_t, U_t) \cdot p_t\},$$

$$-\Delta y_{t+1} \in \partial_{X_t} \bar{H}_t(X_t, p_t, y_{t+1}),$$

where

$$\bar{H}_t(X_t, p_t, y_{t+1}) := \inf_{U_t \in \mathbb{R}^M} \{H_t(X_t, U_t, y_{t+1}) - (X_t, U_t) \cdot p_t\}.$$

If, for all $(X_t, U_t, y_{t+1}) \in \mathbb{R}^N \times \mathbb{R}^M \times \mathbb{R}^N$,

$$\partial_{(X_t,U_t)} H_t(X_t, U_t, y_{t+1}) = \partial_{X_t} H_t(X_t, U_t, y_{t+1}) \times \partial_{U_t} H_t(X_t, U_t, y_{t+1}), \tag{3.24}$$

this can be written as

$$U_t \in \operatorname*{argmin}_{U_t \in \mathbb{R}^M} \{H_t(X_t, U_t, y_{t+1}) - (X_t, U_t) \cdot p_t\},$$

$$-\Delta y_{t+1} \in \partial_{X_t} \{H_t(X_t, U_t, y_{t+1}) - (X_t, U_t) \cdot p_t\}$$

almost surely. Condition (3.24) holds, by Corollary A.80, in particular, if L_t is of the form

$$L_t(X, U) = L^0_t(X, U) + L^1_t(X) + L^2_t(U),$$

where all the functions are convex and L^0_t is finite and differentiable.

Proof The optimality conditions in Theorem 3.51 mean that

$$-(\Delta y_{t+1}, 0) \in \partial f_t(X_t, U_t), \tag{3.25}$$

where $f_t(X_t, U_t) := H_t(X_t, U_t, y_{t+1}) - (X_t, U_t) \cdot p_t$. The first claim thus follows from Theorem A.73 with $v = 0$, $x = U_t$, $u = X_t$ and $F = f_t$. Under (3.24), condition (3.25) can be written as

$$\begin{aligned} -\Delta y_{t+1} &\in \partial_{X_t} f_t(X_t, U_t), \\ 0 &\in \partial_{U_t} f_t(X_t, U_t), \end{aligned}$$

which is the second condition. □

Remark 3.59 below gives a version of the maximum principle which does not involve the shadow price of information p. This will require some extra assumptions on the problem data. Recall that the Köthe dual of a space $\mathcal{U}$ of random variables is the linear space

$$\{y \in L^0 \mid u \cdot y \in L^1 \ \forall u \in \mathcal{U}\}.$$

Assumption 3.53 *The spaces $\mathcal{S}'$ and $\mathcal{C}'$ are the Köthe duals of $\mathcal{S}$ and $\mathcal{C}$, respectively, and, for all t,*

1. $E_t\mathcal{S} \subseteq \mathcal{S}$ *and* $E_t\mathcal{C} \subseteq \mathcal{C}$,
2. $A_t\mathcal{S} \subseteq \mathcal{S}$ *and* $B_t\mathcal{C} \subseteq \mathcal{S}$.

Except for condition 2, Assumption 3.53 holds automatically e.g. in Lebesgue and Orlicz spaces; see the examples in Sect. 3.1.1. Condition 2 imposes natural integrability conditions on the matrices. It holds e.g. in spaces of finite moments if the elements of A_t and B_t have finite moments; see Example 5.12. By Lemmas 3.6 and 3.5, Assumption 3.53 implies that, for all t,

1. $E_t\mathcal{S}' \subseteq \mathcal{S}'$ and $E_t\mathcal{C}' \subseteq \mathcal{C}'$,
2. $A_t^*\mathcal{S}' \subseteq \mathcal{S}'$ and $B_t^*\mathcal{S}' \subseteq \mathcal{C}'$.

Note that, under Assumption 3.53, $\operatorname{dom} Ef \cap (\mathcal{X} \times \mathcal{U}) \neq \emptyset$ means that $\operatorname{dom} E[\sum_{t=0}^T L_t] \cap \mathcal{X} \neq \emptyset$ for every t. This holds in particular, if each EL_t is proper on $\mathcal{S} \times \mathcal{C}$.

Under Assumption 3.53, the dual problem (3.23) can be simplified using the general techniques of Sect. 3.2.2. We will find that problem (1.11) satisfies Assumption 3.27 with different choices for the mapping Π resulting in different restrictions in the dual problem.

Remark 3.54 Let $(p, y) \in \mathcal{X}_a^\perp \times \mathcal{Y}$. Since $x = (X, U)$, the shadow price of information $p \in \mathcal{X}_a^\perp$ can be decomposed as $p = (V, Y)$, where $V \in (\mathcal{S}')^{T+1}$ and $Y \in (\mathcal{C}')^{T+1}$. Under Assumption 3.53, there exists $(\tilde{p}, \tilde{y}) \in \mathcal{X}_a^\perp \times \mathcal{Y}$ that achieves the same dual objective value as (p, y) but the component of $\tilde{p}_t$ corresponding to the state X_t is zero for $t \geq 1$. This is quite natural given that the information constraint on the state X_t for $t \geq 1$ is redundant as observed in Remark 1.85.

Proof Define $\Pi^* : \mathcal{X} \times \mathcal{U} \to \mathcal{X} \times \mathcal{U}$ by $\Pi^*(Z, R, u) = (\tilde{Z}, R, u)$, where $\tilde{Z}$ is given by

$$\tilde{Z}_0 = Z_0,$$
$$\Delta\tilde{Z}_t = A_t\tilde{Z}_{t-1} + B_tR_{t-1} + u_t.$$

Given $(x, z, u) \in \operatorname{dom} F$ and $(\tilde{x}, \tilde{z}, \tilde{u}) = \Pi^*(x, z, u)$, we have

$$\Delta X_t = A_tX_{t-1} + B_tU_{t-1} + u_t$$

so

$$\Delta(X_t - \tilde{Z}_t) = A_t(X_{t-1} - \tilde{Z}_{t-1}) + B_t(U_{t-1} - R_{t-1}).$$

It follows that $x - \tilde{z} \in \mathcal{N}$ so $F(\tilde{x}, \tilde{z}, \tilde{u}) = F(x, z, u)$. We thus have

$$F(x, \Pi^*(z, u)) \leq F(x, z, u) \quad (x, z, u) \in L^0 \times \mathcal{X} \times \mathcal{U}.$$

Recursive application of Lemma 3.5 shows that, under Assumption 3.53, Π^* is continuous so it has a continuous adjoint Π. Clearly, $\Pi^*(0, \bar{u}) \in \mathcal{X}_a \times \{\bar{u}\}$, so Assumption 3.27 holds, by Lemma 3.40. Thus, as observed at the beginning of Sect. 3.2.2, we may restrict the dual problem to $\operatorname{rge} \Pi$. Since

$$\ker \Pi^* = \{(Z, R, u) \mid Z_0 = 0,\ R = 0,\ u = 0\},$$

we have

$$\operatorname{cl}\operatorname{rge} \Pi = (\ker \Pi^*)^{\perp} = \{(V, Y, y) \mid V_t = 0 \quad t = 1, \dots, T\}$$

which completes the proof. □

Recall that the adapted projection ${}^a y$ of a process $y \in \mathcal{Y}$ is defined by ${}^a y_t := E_t y_t$. We will denote the set of adapted processes in $\mathcal{Y}$ by $\mathcal{Y}_a$.

Remark 3.55 Let Assumption 3.53 hold and $(p, y) \in \mathcal{X}_a^{\perp} \times \mathcal{Y}$. There exists $(\tilde{p}, \tilde{y}) \in \mathcal{X}_a^{\perp} \times \mathcal{Y}_a$ that achieves the same dual objective value as (p, y). In particular, if dual solutions exist, then there exists one where y is adapted. If, in addition, each L_t is $\mathcal{F}_{t+1}$-measurable and each EL_t is closed and proper on $\mathcal{S} \times \mathcal{C}$, then there exists $(\tilde{p}, \tilde{y}) \in \mathcal{X}_a^{\perp} \times \mathcal{Y}_a$ such that $\tilde{p}_t$ is $\mathcal{F}_{t+1}$-measurable and $(\tilde{p}, \tilde{y})$ achieves dual objective value at least as good as (p, y).

Proof Let $\tilde{y} = {}^a y$ and

$$\tilde{p}_t = p_t + (\Delta\tilde{y}_{t+1} + A^*_{t+1}\tilde{y}_{t+1}, B^*_{t+1}\tilde{y}_{t+1}) - (\Delta y_{t+1} + A^*_{t+1}y_{t+1}, B^*_{t+1}y_{t+1}).$$

It is clear that the dual objective values are the same at (p, y) and $(\tilde{p}, \tilde{y})$, while Lemma 3.7 implies $\tilde{p} \in \mathcal{X}_a^\perp$. Under the additional assumptions, for any $(p, y) \in \mathcal{X}_a^\perp \times \mathcal{Y}_a$, Theorem 3.13 implies

$$E\left[\sum_{t=0}^T L_t^*(p_t - (\Delta y_{t+1} + A^*_{t+1}y_{t+1}, B^*_{t+1}y_{t+1}))\right]$$

$$\geq E\left[\sum_{t=0}^T L_t^*(E_{t+1}[p_t] - (\Delta y_{t+1} + A^*_{t+1}y_{t+1}, B^*_{t+1}y_{t+1}))\right],$$

which completes the proof. □

Theorem 3.38 yields the following.

Corollary 3.56 (Reduced Dual) *Assume that each L_t is $\mathcal{F}_t$-measurable, each EL_t is closed and proper on $\mathcal{S} \times \mathcal{C}$ and that Assumption 3.53 holds. Then the optimum value of the dual problem* (3.23) *equals that of the* reduced dual problem

$$\underset{y\in\mathcal{Y}_a}{\text{maximize}}\; E\left[\sum_{t=1}^T W_t \cdot y_t - \sum_{t=0}^T L_t^*(-E_t(\Delta y_{t+1} + A^*_{t+1}y_{t+1}, B^*_{t+1}y_{t+1}))\right] \tag{3.26}$$

and the dual has a solution if and only if the reduced dual has a solution. If dom $Ef \cap (\mathcal{X} \times \mathcal{U}) \neq \emptyset$ *and* (1.11) *and* (3.26) *are feasible, then the following are equivalent:*

1. *(X, U) solves* (1.11), *y solves* (3.26) *and* inf*(1.11)* = sup*(3.26);*
2. *(X, U) is feasible in* (1.11), *y is feasible in* (3.26) *and*

$$-E_t(\Delta y_{t+1} + A^*_{t+1}y_{t+1}, B^*_{t+1}y_{t+1}) \in \partial L_t(X_t, U_t)$$

   for all t almost surely.

Proof By Assumption 3.53, the mappings $\pi : \mathcal{Y} \to \mathcal{V}$ and $\gamma : \mathcal{Y} \to \mathcal{Y}$ are well-defined by $\gamma(y) = {}^a y$ and

$$\pi(y)_t = (\Delta\, {}^a y_{t+1} + A^*_{t+1}\, {}^a y_{t+1}, B^*_{t+1}\, {}^a y_{t+1}) - E_t(\Delta y_{t+1} + A^*_{t+1}y_{t+1}, B^*_{t+1}y_{t+1}).$$

It suffices to show that they satisfy the assumptions of Theorem 3.38. Applying the Jensen's inequality in Theorem 3.13 to the expression of φ^* in (3.22) gives, for every $(p, y) \in \mathcal{X}_a^\perp \times \mathcal{Y}$,

$$\begin{aligned}\varphi^*(p, y) &= E\left[\sum_{t=0}^{T} L_t^*(p_t - (\Delta y_{t+1} + A_{t+1}^* y_{t+1}, B_{t+1}^* y_{t+1}))\right] \\ &\geq E\left[\sum_{t=0}^{T} L_t^*(-E_t[\Delta y_{t+1} + A_{t+1}^* y_{t+1}, B_{t+1}^* y_{t+1}])\right] \\ &= \varphi^*(\pi(y), \gamma(y)),\end{aligned}$$

which is the first condition in Assumption 3.27. Since $\bar{u} = W$ is adapted, the last condition in Assumption 3.27 holds by Lemma B.13. □

The following gives a more functional analytic proof of Corollary 3.56. The argument is based on establishing the properties of the primal problem in Lemma 3.31.

Remark 3.57 We saw in the derivation of the Lagrangian integrand that the random linear mapping

$$K(x, \omega) := (\Delta X_t - A_t(\omega) X_{t-1} - B_t(\omega) U_{t-1})_{t=1}^T$$

has scenariowise adjoint given by

$$K^*(y, \omega) = -(\Delta y_{t+1} + A_{t+1}^* y_{t+1}, B_{t+1}^* y_{t+1})_{t=0}^T.$$

By Lemma 3.5, Assumption 3.53 implies that the pointwise application of K induces a continuous linear mapping $\mathcal{K} : \mathcal{X} \to \mathcal{U}$ and that its adjoint $\mathcal{K}^* : \mathcal{Y} \to \mathcal{V}$ is given by pointwise application of K^*.

Given $(x, z, u) \in \operatorname{dom} F$, we have $x - z \in \mathcal{N}$ and $\mathcal{K}(x) = u$. Jensen's inequality gives

$$EL_t(E_t X_t, E_t U_t) \leq EL_t(X_t, U_t)$$

while

$$\mathcal{K}({}^a x) = \mathcal{K}(x) + \mathcal{K}({}^a x - x) = u + \mathcal{K}({}^a z - z) = u + \mathcal{K}({}^a z) - \mathcal{K}(z).$$

Since $\mathcal{K}({}^a x)$ and $\mathcal{K}({}^a z)$ are adapted, we have

$$\mathcal{K}({}^a x) = {}^a u + \mathcal{K}({}^a z) - {}^a \mathcal{K}(z).$$

Thus, the conditions of Lemma 3.31 are satisfied with $\xi(x, z, u) = {}^a x$ and

$$\Pi^*(z, u) = (0, {}^a u + \mathcal{K}({}^a z) - {}^a\mathcal{K}(z)).$$

We have

$$\begin{aligned}\langle \Pi^*(z, u), (p, y)\rangle &= \langle {}^a u + \mathcal{K}({}^a z) - {}^a\mathcal{K}(z), y\rangle \\ &= \langle z, {}^a\mathcal{K}^*(y) - \mathcal{K}^*({}^a y)\rangle + \langle u, {}^a y\rangle\end{aligned}$$

so the adjoint of Π^* is given by $\Pi(p, y) = ({}^a\mathcal{K}^*(y) - \mathcal{K}^*({}^a y), {}^a y)$. This satisfies the assumptions of Theorem 3.38.

Remark 3.58 Even if the normal integrands L_t are not $\mathcal{F}_t$-measurable, one can apply Corollary 3.56 to the optimal control problem where each L_t has been replaced by its conditional expectation $E_t L_t$. The corresponding optimality conditions would then become

$$-E_t(\Delta y_{t+1} + A^*_{t+1} y_{t+1}, B^*_{t+1} y_{t+1}) \in \partial(E_t L_t)(X_t, U_t)$$

for all t almost surely. By Fenchel's inequality, dual feasibility implies that $L_t(X_t, U_t)$ are quasi-integrable for every $(X, U) \in \mathcal{X}$ so, by Lemmas B.13 and B.14,

$$E\left[\sum_{t=0}^{T} L_t(X_t, U_t)\right] = E\left[\sum_{t=0}^{T} (E_t L_t)(X_t, U_t)\right] \quad \forall (X, U) \in \mathcal{X}_a.$$

The two control problems thus coincide over the space $\mathcal{X}_a$.

Remark 3.59 (Maximum Principle in Reduced Form) The scenariowise optimality condition in Corollary 3.56 can be written as

$$-(E_t \Delta y_{t+1}, 0) \in \partial_{(X,U)} H_t(X_t, U_t, y_{t+1}),$$

where

$$H_t(X_t, U_t, y_{t+1}) := L_t(X_t, U_t) + E_t[A^*_{t+1} y_{t+1}] \cdot X_t + E_t[B^*_{t+1} y_{t+1}] \cdot U_t.$$

As in Remark 3.52, the optimality conditions can thus be written as

$$\begin{aligned} U_t &\in \operatorname*{argmin}_{U_t \in \mathbb{R}^M} H_t(X_t, U_t, y_{t+1}), \\ -E_t \Delta y_{t+1} &\in \partial_X \bar{H}_t(X_t, y_{t+1}),\end{aligned}$$

where

$$\bar{H}_t(X_t, y_{t+1}) := \inf_{U_t \in \mathbb{R}^M} H_t(X_t, U_t, y_{t+1}).$$

Under mild conditions, the optimal dual solutions provide subgradients of the cost-to-go functions in the dynamic programming recursion (2.24)

$$\begin{aligned} I_T &= 0, \\ J_t(X_t) &= \inf_{U_t \in \mathbb{R}^M} E_t(L_t + I_t)(X_t, U_t), \\ I_{t-1}(X_{t-1}, U_{t-1}) &= J_t(X_{t-1} + A_t X_{t-1} + B_t U_{t-1} + W_t) \end{aligned}$$

studied in Sect. 2.3.3.

Corollary 3.60 (Cost-to-Go Functions) *Assume that each L_t is $\mathcal{F}_{t+1}$-measurable, each EL_t is closed and proper on $\mathcal{S} \times \mathcal{C}$, Assumption 3.53 holds and that $(J_t, I_t)_{t=0}^T$ is an L-bounded solution of* (2.24). *Let $(X, U) \in \mathcal{N}$ and $(p, y) \in \mathcal{X}_a^{\perp} \times \mathcal{Y}$ be such that $(L_t + I_t)(X_t, U_t)$ is integrable and p_t is $\mathcal{F}_{t+1}$-measurable for all t. If (X, U) solves* (1.11), *(p, y) solves* (3.23) *and there is no duality gap, then*

$$y_t \in \partial J_t(X_t), \tag{3.27}$$

$$(y_t, 0) \in \partial(E_t(L_t + I_t))(X_t, U_t), \tag{3.28}$$

$$p_t + (y_t, 0) \in \partial(L_t + I_t)(X_t, U_t) \tag{3.29}$$

for all t almost surely. Conversely, if each J_t is almost surely real-valued and differentiable and (X, U) and (p, y) are feasible in (1.11) *and* (3.23), *respectively, then the inclusions* (3.29) *imply that (X, U) solves* (1.11), *(p, y) solves* (3.23) *and there is no duality gap.*

Proof By Theorem 2.26, (3.29) implies (3.28) which in turn, by the equivalence of 2 and 3 in Theorem A.73, implies (3.27). If (X, U) and (p, y) satisfy the optimality conditions in Theorem 3.51, then $p_T + (y_T, 0) \in \partial L_T(X_T, U_T)$, so (3.29) holds for $t = T$. Assume now that $y_{t+1} \in \partial J_{t+1}(X_{t+1})$. By the second claim in Remark A.44,

$$(y_{t+1} + A_{t+1}^* y_{t+1}, B_{t+1}^* y_{t+1}) \in \partial I_t(X_t, U_t).$$

Combined with the optimality conditions in Theorem 3.51 and the first claim in Remark A.44, this gives

$$\begin{aligned} p_t - (\Delta y_{t+1} &+ A_{t+1}^* y_{t+1}, B_{t+1}^* y_{t+1}) \\ &+ (y_{t+1} + A_{t+1}^* y_{t+1}, B_{t+1}^* y_{t+1}) \in \partial(L_t + I_t)(X_t, U_t). \end{aligned}$$

Cancelling terms gives (3.29). By induction, (3.29) and thus (3.28) and (3.27) hold for all t.

As to the converse, assume that (3.27)–(3.29) hold for all t. Since J_t is real-valued almost surely, Corollaries A.80 and A.81 give

$$\begin{aligned}
&p_t + (y_t, 0) \in \partial(L_t + I_t)(X_t, U_t)\\
&= \partial L_t(X_t, U_t) + (1 + A^*_{t+1}, B^*_{t+1})\partial J_{t+1}(X_t + A_{t+1}X_t + B_{t+1}U_t + W_{t+1})\\
&= \partial L_t(X_t, U_t) + (1 + A^*_{t+1}, B^*_{t+1})\partial J_{t+1}(X_{t+1})\\
&= \partial L_t(X_t, U_t) + (1 + A^*_{t+1}, B^*_{t+1})y_{t+1},
\end{aligned}$$

where the last equality holds by differentiability of J_t. This is the optimality condition in Theorem 3.51. □

The following counterexample shows that, even in a deterministic setting, the converse in Corollary 3.60 does not necessarily hold if J_t are not differentiable.

Counterexample 3.61 Consider a one-dimensional deterministic setting with $T = 1$, $L_0 = L_1 = \delta_{\mathbb{R}_+\times\mathbb{R}}$, $A_1 = 0$, $B_1 = 0$ and $W_1 = 0$. Then $J_2 = 0$, $I_2 = 0$, $J_1 = J_0 = \delta_{\mathbb{R}_+}$ and $I_1 = \delta_{\mathbb{R}_+\times\mathbb{R}}$. Clearly $(X, U) := (0, 0)$ is primal optimal. Here (3.29) becomes

$$(y_t, 0) \in \partial\delta_{\mathbb{R}_+\times\mathbb{R}}(0, 0) \quad \forall t = 0, 1$$

while (the first) KKTR-condition in Theorem 3.51 becomes

$$(-\Delta y_t, 0) \in \partial\delta_{\mathbb{R}_+\times\mathbb{R}}(0, 0) \quad \forall t = 0, 1.$$

Since $\partial\delta_{\mathbb{R}_+\times\mathbb{R}}(0, 0) = \mathbb{R}_- \times \{0\}$, any y with $y_0, y_1 < 0$ satisfies the former but not necessarily the latter.

3.4.4 Problems of Lagrange

Consider again problem (1.13)

$$\text{minimize} \quad E\left[\sum_{t=0}^{T} K_t(x_t, \Delta x_t)\right] \quad \text{over } x \in \mathcal{N}$$

from Sect. 1.3.4. This fits the general duality framework with $\bar{u} = 0$ and

$$f(x, u, \omega) = \sum_{t=0}^{T} K_t(x_t, \Delta x_t + u_t, \omega).$$

We thus assume that both $\mathcal{X}$ and $\mathcal{U}$ are spaces of $\mathbb{R}^{(T+1)d}$-valued random variables. For simplicity, we assume that

$$\mathcal{X}_t = \mathcal{S}, \quad \mathcal{V}_t = \mathcal{S}', \quad \mathcal{U} = \mathcal{X}, \quad \mathcal{Y} = \mathcal{V},$$

where $\mathcal{S}$ and $\mathcal{S}'$ are solid decomposable spaces in separating duality under the bilinear form $\langle x_t, v_t \rangle := E[x_t \cdot v_t]$.

The Lagrangian integrand becomes

$$\begin{aligned} l(x, y, \omega) &= \inf_{u \in \mathbb{R}^m} \{f(x, u, \omega) - u \cdot y\} \\ &= \sum_{t=0}^{T} (\Delta x_t \cdot y_t + H_t(x_t, y_t, \omega)) \\ &= \sum_{t=0}^{T} (-x_t \cdot \Delta y_{t+1} + H_t(x_t, y_t, \omega)), \end{aligned}$$

where we define $y_{T+1} := 0$ and

$$H_t(x_t, y_t, \omega) := \inf_{u_t \in \mathbb{R}^d} \{K_t(x_t, u_t, \omega) - u_t \cdot y_t\}.$$

The function H_t is called the *Hamiltonian* associated with K_t. It has similar properties as the Lagrangian integrand in Sect. 3.2.1. In particular, $H_t(x_t, y_t, \omega)$ is convex in x_t and concave in y_t. The conjugate integrand can be written as

$$\begin{aligned} f^*(v, y, \omega) &= \sup_{x \in \mathbb{R}^n} \{x \cdot v - l(x, y, \omega)\} \\ &= \sup_{x \in \mathbb{R}^n} \left\{ \sum_{t=0}^{T} x_t \cdot v_t - \sum_{t=0}^{T} (-x_t \cdot \Delta y_{t+1} + H_t(x_t, y_t, \omega)) \right\} \\ &= \sum_{t=0}^{T} K_t^*(v_t + \Delta y_{t+1}, y_t, \omega). \end{aligned}$$

If $\operatorname{dom} Ef \cap (\mathcal{X} \times \mathcal{U}) \neq \emptyset$, Theorem 3.20 says that the conjugate of the optimum value function can be written as

$$\varphi^*(p, y) = E\left[\sum_{t=0}^{T} K_t^*(p_t + \Delta y_{t+1}, y_t)\right] + \delta_{\mathcal{X}_a^\perp}(p) \tag{3.30}$$

and, by Corollary 3.21, the dual problem becomes

$$\text{maximize} \qquad E\left[-\sum_{t=0}^{T} K_t^*(p_t + \Delta y_{t+1}, y_t)\right] \quad \text{over } (p, y) \in \mathcal{X}_a^\perp \times \mathcal{Y}. \tag{3.31}$$

Theorem 3.22 and Corollary 3.25 now give the following.

Theorem 3.62 *If* dom $Ef \cap (\mathcal{X} \times \mathcal{U}) \neq \emptyset$ *and* (1.13) *and* (3.31) *are feasible, then the following are equivalent:*

1. x *solves* (1.13), (p, y) *solves* (3.31) *and there is no duality gap;*
2. x *is feasible in* (1.13), (p, y) *is feasible in* (3.31) *and, for all* t,
$$(p_t + \Delta y_{t+1}, y_t) \in \partial K_t(x_t, \Delta x_t)$$
almost surely;
3. x *is feasible in* (1.13), (p, y) *is feasible in* (3.31) *and, for all* t,
$$\begin{aligned} p_t + \Delta y_{t+1} &\in \partial_x H_t(x_t, y_t), \\ \Delta x_t &\in \partial_y[-H_t](x_t, y_t), \end{aligned}$$
almost surely.

The subgradient condition in part 2 of Theorem 3.62 is known as *Euler-Lagrange condition* for (1.13) while the conditions in 3 are known as *Hamiltonian conditions*. To clarify the connection with classical formulations, consider the deterministic case where $p = 0$ and assume that both K_t and H_t are differentiable. Condition 3 can then be written as

$$\Delta y_{t+1} = \nabla_x H_t(x_t, y_t), \quad \Delta x_t = -\nabla_y H_t(x_t, y_t)$$

while the conditions in 2 become

$$\Delta y_{t+1} = \nabla_x K_t(x_t, \Delta x_t), \quad y_t = \nabla_u K_t(x_t, \Delta x_t),$$

which is a discrete-time version of the classical Euler-Lagrange equation

$$\nabla_x K_t(x_t, \dot{x}_t) = \frac{d}{dt} \nabla_u K_t(x_t, \dot{x}_t).$$

Much as in Remark 3.52, one could formulate the optimality conditions in part 3 of Theorem 3.62 in the form of a maximum principle, where the optimal solutions x are pointwise minimizers of a linearly tilted Hamiltonian.

Under the following assumption, the optimality conditions can be written without the shadow price of information p, much as in Corollary 3.56.

Assumption 3.63 *The space $\mathcal{S}'$ is the Köthe dual of $\mathcal{S}$ and $E_t\mathcal{S} \subseteq \mathcal{S}$ for all t.*

By Lemma 3.6, Assumption 3.63 implies that $E_t\mathcal{S}' \subseteq \mathcal{S}'$ for all t. As before, we will denote the set of adapted processes in $\mathcal{Y}$ by $\mathcal{Y}_a$. The adapted projection ${}^a y$ of a process $y \in \mathcal{Y}$ is defined by ${}^a y_t = E_t y_t$.

Theorem 3.38 yields the following.

Corollary 3.64 (Reduced Dual) *Assume that* $\operatorname{dom} Ef \cap (\mathcal{X} \times \mathcal{U}) \neq \emptyset$, *Assumption 3.63 holds, that* (1.13) *and* (3.31) *are feasible and that, for all t, K_t is $\mathcal{F}_t$-measurable and EK_t is proper on $\mathcal{S} \times \mathcal{S}$. Then the optimum value of the dual problem* (3.31) *equals that of the reduced dual problem*

$$\text{maximize} \qquad E[-\sum_{t=0}^{T} K_t^*(E_t \Delta y_{t+1}, y_t)] \quad \text{over } y \in \mathcal{Y}_a$$

and the dual has a solution if and only if the reduced dual has a solution. If the reduced dual has a solution, then an x is optimal if and only if it is feasible and there is a y feasible in the reduced dual such that

$$\begin{aligned} E_t \Delta y_{t+1} &\in \partial_x H_t(x_t, y_t), \\ \Delta x_t &\in \partial_y[-H_t](x_t, y_t) \end{aligned}$$

almost surely.

Proof Under Assumption 3.63, the mappings $\pi : \mathcal{Y} \to \mathcal{V}$ and $\gamma : \mathcal{Y} \to \mathcal{Y}$ are well-defined by $\gamma(y) = {}^a y$ and

$$\pi(y)_t = E_t[\Delta y_{t+1}] - \Delta\, {}^a y_{t+1}.$$

It suffices to show that they satisfy the assumptions of Theorem 3.38. Applying the Jensen's inequality in Theorem 3.13 to the expression of φ^* in (3.30) gives, for every $(p, y) \in \mathcal{X}_a^\perp \times \mathcal{Y}$,

$$\begin{aligned} \varphi^*(p, y) &= E\left[\sum_{t=0}^{T} K_t^*(p_t + \Delta y_{t+1}, y_t)\right] \\ &\geq E\left[\sum_{t=0}^{T} K_t^*(E_t[\Delta y_{t+1}], E_t[y_t])\right] \\ &= \varphi^*(\pi(y), \gamma(y)), \end{aligned}$$

which is the first condition in Assumption 3.27. The last condition in Assumption 3.27 holds trivially since $\bar{u} = 0$. □

Remark 3.65 By Lemma 3.48, any $y \in \mathcal{Y}_a$ has a unique decomposition $y = M+A$ where M is a martingale and A is predictable with $A_0 = 0$. Defining $A_{T+1} := -M_T$ the reduced dual problem in Corollary 3.64 can thus be written without conditional expectations as

$$\text{maximize} \quad E[-\sum_{t=0}^{T} K_t^*(\Delta A_{t+1}, M_t + A_t)] \quad \text{over } (M, A) \in \mathcal{M}^{\mathcal{Y}} \times \mathcal{Y}_p,$$

where $\mathcal{M}^{\mathcal{Y}} \subset \mathcal{Y}$ and $\mathcal{Y}_p \subset \mathcal{Y}$ are the linear spaces of martingales and predictable processes, respectively. Accordingly, the optimality conditions in Corollary 3.64 can be written as

$$\Delta A_{t+1} \in \partial_x H_t(x_t, M_t + A_t),$$
$$\Delta x_t \in \partial_y[-H_t](x_t, M_t + A_t).$$

Example 3.66 (Optimal Stopping) Let Assumption 3.63 hold and let R be an adapted process with $R_t \in \mathcal{S}'$ for all t. The relaxed optimal stopping problem from Sect. 1.3.2 can be written as

$$\underset{x \in \mathcal{N}_+}{\text{maximize}} \quad E\sum_{t=0}^{T} R_t \Delta x_t \quad \text{subject to} \quad \Delta x \geq 0,\ x_T \leq 1\ a.s.$$

This is a problem of Lagrange with

$$K_t(x_t, u_t) = -R_t u_t + \delta_{\mathbb{R}_-}(x_t - 1) + \delta_{\mathbb{R}_+}(u_t).$$

We get

$$\begin{aligned} K_t^*(v_t, y_t) &= \sup_{x_t, u_t \in \mathbb{R}} \{x_t \cdot v_t + u_t \cdot y_t - K_t(x_t, u_t)\} \\ &= \sup_{x_t, u_t \in \mathbb{R}} \{x_t \cdot v_t + u_t \cdot y_t + R_t u_t \mid x_t \leq 1,\ u_t \geq 0\} \\ &= \begin{cases} v_t & \text{if } v_t \geq 0 \text{ and } R_t + y_t \leq 0, \\ +\infty & \text{otherwise,} \end{cases} \end{aligned}$$

so the reduced dual in Corollary 3.64 becomes

$$\begin{aligned} &\text{maximize} \quad Ey_0 \text{ over } y \in \mathcal{Y}_a \\ &\text{subject to} \quad E_t[\Delta y_{t+1}] \geq 0, \\ &\qquad\qquad\qquad R_t + y_t \leq 0, \end{aligned}$$

where $y_{T+1} := 0$. With the change of variables $S = -y$, this can be written as

$$\begin{aligned} &\text{minimize} \quad ES_0 \text{ over } S \in \mathcal{Y}_a \\ &\text{subject to} \quad E_t[\Delta S_{t+1}] \leq 0, \\ &\qquad\qquad\qquad R_t \leq S_t. \end{aligned}$$

Here one minimizes the expectation of the initial value of a supermartingale that dominates the reward process. In the dual problem in Corollary 3.46, one minimizes expectation of the initial value of a martingale that dominates the reward process. By Corollary 3.49, the martingale part of the Snell envelope S of R solves the dual in Corollary 3.46 while the above dual is solved by the Snell envelope itself. Indeed, if the dual had a solution S better than the Snell envelope, then S would be a supermartingale that dominates R (recall that $S_{T+1} = y_{T+1} = 0$) and it would be strictly smaller than the Snell envelope at $t = 0$, contradicting the definition of the Snell envelope.

The following is an analogue of the first part of Corollary 3.60.

Corollary 3.67 (Cost-to-Go Functions) *Recall the Bellman equations*

$$\begin{aligned} \tilde{V}_T &= 0, \\ V_t &= E_t \tilde{V}_t, \\ \tilde{V}_{t-1}(x_{t-1}, \omega) &= \inf_{x_t \in \mathbb{R}^d} \{(E_t K_t)(x_t, \Delta x_t, \omega) + V_t(x_t, \omega)\} \end{aligned}$$

from Sect. 2.3.4. If x and (p, y) satisfy the optimality conditions in Theorem 3.62 and if $K_t(x_t, \Delta x_t)$ and $\tilde{V}_t(x_t)$ are integrable, then

$$-E_t y_{t+1} \in \partial V_t(x_t)$$

for all t almost surely.

Proof Since $y_{T+1} = 0$, by definition, the claim holds trivially for $t = T$. Assume that $-E_t y_{t+1} \in \partial V_t(x_t)$ and let

$$F_t(x_t, x_{t-1}, \omega) := (E_t K_t)(x_t, \Delta x_t, \omega) + V_t(x_t, \omega).$$

If x and (p, y) satisfy the optimality conditions in Theorem 3.62 and if $K_t(x_t, \Delta x_t)$ is integrable, Theorem 2.26 gives

$$(E_t \Delta y_{t+1}, E_t y_t) \in \partial(E_t K_t)(x_t, \Delta x_t).$$

By Remark A.44,

$$(0, -E_t y_t) \in \partial F_t(x_t, x_{t-1}).$$

By Theorem A.73, this implies $-E_t y_t \in \partial \tilde{V}_{t-1}(x_{t-1})$. If $\tilde{V}_{t-1}(x_{t-1})$ is integrable, Theorem 2.26 gives $-E_{t-1} y_t \in \partial V_{t-1}(x_{t-1})$. The claim now follows by induction on t. □

Much as in Corollary 3.60, converse of Corollary 3.67 holds under additional conditions.

3.4.5 Financial Mathematics

Consider again problem (1.15)

$$\begin{aligned} &\text{minimize} \quad EV\left(c - \sum_{t=0}^{T-1} x_t \cdot \Delta s_{t+1}\right) \quad \text{over} \quad x \in \mathcal{N}, \\ &\text{subject to} \quad x_t \in D_t \quad t = 0, \dots, T \ a.s. \end{aligned}$$

from Sect. 1.3.5. This fits the general duality framework with $\bar{u} = c$ and

$$f(x, u, \omega) = V\left(u - \sum_{t=0}^{T-1} x_t \cdot \Delta s_{t+1}(\omega), \omega\right) + \sum_{t=0}^{T} \delta_{D_t(\omega)}(x_t).$$

We thus assume that $\mathcal{X}_t$ and $\mathcal{U}$ are spaces of $\mathbb{R}^J$- and $\mathbb{R}$-valued random variables, respectively, and that $c \in \mathcal{U}$. For simplicity, we also assume that, for all t,

$$\mathcal{X}_t = \mathcal{S} \quad \text{and} \quad \mathcal{V}_t = \mathcal{S}'$$

where $\mathcal{S}$ and $\mathcal{S}'$ are solid decomposable spaces in separating duality. We continue with the assumptions of Sect. 1.3.5 on the model. In particular, the loss function V is convex, nondecreasing and nonconstant, $0 \in D_t$ almost surely for all t, and $D_T = \{0\}$. For convenience, we will also assume that $V(0) = 0$.

The Lagrangian integrand can be written as

$$l(x, y, \omega) = \inf_{u \in \mathbb{R}}\{f(x, u, \omega) - uy\}$$
$$= -y \sum_{t=0}^{T-1} x_t \cdot \Delta s_{t+1}(\omega) - V^*(y, \omega) + \sum_{t=0}^{T} \delta_{D_t(\omega)}(x_t).$$

and, since $D_T = \{0\}$, the conjugate of f becomes

$$f^*(v, y, \omega) = \sup_{x \in \mathbb{R}^J} \{x \cdot v - l(x, y, \omega)\}$$
$$= V^*(y, \omega) + \sum_{t=0}^{T-1} \sigma_{D_t(\omega)}(v_t + y\Delta s_{t+1}(\omega)),$$

where

$$\sigma_{D_t(\omega)}(v) := \sup_{x \in D_t(\omega)} x \cdot v$$

is the support function of $D_t(\omega)$. Since Ef is finite at the origin, Theorem 3.20 says that the conjugate of the optimum value function can be written as

$$\varphi^*(p, y) = E\left[V^*(y) + \sum_{t=0}^{T-1} \sigma_{D_t}(p_t + y\Delta s_{t+1})\right] + \delta_{\mathcal{X}_a^\perp}(p) \tag{3.32}$$

and, by Corollary 3.21, the dual problem becomes

$$\text{maximize } E\left[cy - V^*(y) - \sum_{t=0}^{T-1} \sigma_{D_t}(p_t + y\Delta s_{t+1})\right] \text{ over } (p, y) \in \mathcal{X}_a^\perp \times \mathcal{Y}. \tag{3.33}$$

Theorem 3.22 and Corollary 3.25 now give the following.

Theorem 3.68 *If* (1.15) *and* (3.33) *are feasible, then the following are equivalent:*

1. x *solves* (1.15), (p, y) *solves* (3.33) *and there is no duality gap;*
2. x *is feasible in* (1.15), (p, y) *is feasible in* (3.33) *and*

$$y \in \partial V(u - \sum_{t=0}^{T-1} x_t \cdot \Delta s_{t+1}),$$

$$p_t + y\Delta s_{t+1} \in N_{D_t}(x_t) \quad t = 0, \dots, T-1$$

almost surely.

For the rest of the section, we focus on the reduced dual problem. The following assumption implies the existence of the mappings π and γ in Theorem 3.38.

Assumption 3.69 *$\mathcal{Y}$ is the Köthe dual of $\mathcal{U}$ and, for all t, $\mathcal{X}_t = L^\infty$, $\mathcal{V}_t = L^1$ and, for every $t = 0, \dots, T$*

1. *$E_t\mathcal{U} \subseteq \mathcal{U}$,*
2. *$\Delta s_{t+1} \in \mathcal{U}$,*

where $s_{T+1} := 0$.

Part 1 of Assumption 3.69 holds automatically, e.g., in Lebesgue and Orlicz spaces; see Remark 5.9 in Sect. 5.1.1. By Lemmas 3.6 and 3.5, Assumption 3.69 implies that, for all t,

1. $E_t\mathcal{Y} \subseteq \mathcal{Y}$,
2. $y\Delta s_{t+1} \in L^1$ for all $y \in \mathcal{Y}$.

Lemma 3.70 *Under Assumption 3.69, the mappings $\pi : \mathcal{Y} \to \mathcal{V}$ and $\gamma : \mathcal{Y} \to \mathcal{Y}$ defined by*

$$\pi(y) := (E_t[y\Delta s_{t+1}] - y\Delta s_{t+1})_{t=0}^T,$$
$$\gamma(y) := y$$

are continuous and satisfy the assumptions of Theorem 3.38.

Proof By Lemmas 3.5 and 3.6, Assumption 3.69 implies that the mapping π is continuous. Applying the Jensen's inequality in Theorem 3.13 to the expression of φ^* in (3.32) gives, for every $(p, y) \in \mathcal{X}_a^\perp \times \mathcal{Y}$,

$$\begin{aligned}
\varphi^*(p, y) &= E[V^*(y) + \sum_{t=0}^{T-1} \sigma_{D_t}(p_t + y\Delta s_{t+1})] \\
&\geq E[V^*(y) + \sum_{t=0}^{T-1} \sigma_{D_t}(E_t[y\Delta s_{t+1}])] \\
&= \varphi^*(\pi(y), \gamma(y))
\end{aligned}$$

which is the first condition in Assumption 3.27. The last condition in Assumption 3.27 holds trivially since γ is the identity mapping. □

The following is a direct consequence of Lemma 3.70 and Theorem 3.38.

Theorem 3.71 (Reduced Dual) *Let Assumption 3.69 hold and π be as in Lemma 3.70. Then the optimum value of* (3.33) *coincides with that of the problem*

$$\text{maximize} \quad E\left[cy - V^*(y) - \sum_{t=0}^{T-1} \sigma_{D_t}(E_t[y\Delta s_{t+1}])\right] \quad \text{over } y \in \mathcal{Y} \qquad (3.34)$$

and if $y \in \mathcal{Y}$ *solves* (3.34)*, then* $(\pi(y), y)$ *solves* (3.33)*. If* $(p, y) \in \mathcal{V} \times \mathcal{Y}$ *solves* (3.33) *then* y *solves* (3.33)*. If* (1.15) *and* (3.34) *are feasible, then the following are equivalent:*

1. x *solves* (1.15)*,* y *solves* (3.34) *and* $\inf(1.15) = \sup(3.34)$*;*
2. x *is feasible in* (1.15)*,* y *is feasible in* (3.34) *and*

$$y \in \partial V(c - \sum_{t=0}^{T-1} x_t \cdot \Delta s_{t+1}),$$

$$E_t[y\Delta s_{t+1}] \in N_{D_t}(x_t) \quad t = 0, \dots, T$$

almost surely.

Since V is nondecreasing, the first subgradient condition in Theorem 3.71 implies that y is almost surely nonnegative. In the absence of portfolio constraints, the last condition reads $E_t[y\Delta s_{t+1}] = 0$. When $y \neq 0$, this is means, by Theorem B.15, that s is a martingale under the probability measure Q defined by $dQ/dP := y/Ey$.

Combining Theorems 3.38, 3.39, and Lemma 3.70 gives the following.

Theorem 3.72 *Assume that* $\bar{\varphi}$ *is closed at* $(0, c)$ *for all* $c \in \mathcal{U}$ *and that Assumption 3.69 holds. Then*

$$\bar{\varphi}(0, \cdot)^*(y) = E\left[V^*(y) + \sum_{t=0}^{T-1} \sigma_{D_t}(E_t[y\Delta s_{t+1}])\right].$$

Proof By Lemma 3.70 and Theorem 3.38,

$$g(y) := \inf_{p \in \mathcal{V}} \varphi^*(p, y) = \bar{\varphi}^*(\pi(y), y) = E\left[V^*(y) + \sum_{t=0}^{T-1} \sigma_{D_t}(E_t[y\Delta s_{t+1}])\right],$$

where π is continuous. Thus, g is closed as a composition of a continuous linear mapping with a closed convex function. Thus, the claim follows from Theorem 3.39. □

When $V = \delta_{\mathbb{R}_-}$, the optimum value function of (1.15) becomes $\delta_{\tilde{C}}$, where

$$\tilde{C} := \left\{(z, u) \in \mathcal{X} \times \mathcal{U} \,\middle|\, \exists x \in L^0 : x - z \in \mathcal{N}, \right.$$
$$\left. x_t \in D_t, \sum_{t=0}^{T-1} x_t \cdot \Delta s_{t+1} \geq u \quad a.s.\right\}.$$

The set

$$\begin{aligned}\mathcal{C} &:= \{c \in \mathcal{U} \mid (0, c) \in \tilde{\mathcal{C}}\}\\ &= \left\{u \in \mathcal{U} \,\middle|\, \exists x \in \mathcal{N},\ x_t \in D_t,\ \sum_{t=0}^{T-1} x_t \cdot \Delta s_{t+1} \geq u \quad a.s.\right\}\end{aligned} \tag{3.35}$$

consists of claims that can be superhedged without a cost by dynamic trading in the market.

Example 3.73 Assume that $V = \delta_{\mathbb{R}_-}$ and let Assumption 3.69 hold. We then have $\bar{\varphi} = \delta_{\tilde{\mathcal{C}}}$ so, by Lemma 3.32, $\bar{\varphi}^* = \varphi^* = \sigma_{\tilde{\mathcal{C}}}$ and Eq. (3.32) becomes

$$\sigma_{\tilde{\mathcal{C}}}(p, y) = E\left[\delta_{\mathbb{R}_+}(y) + \sum_{t=0}^{T-1} \sigma_{D_t}(p_t + y\Delta s_{t+1})\right] + \delta_{\mathcal{X}_a^\perp}(p).$$

If the portfolio constraints D_t are conical, then σ_{D_t} is the indicator of the polar D_t° of D_t and $\varphi^* = \delta_{\tilde{\mathcal{C}}^\circ}$, where

$$\tilde{\mathcal{C}}^\circ = \{(p, y) \in \mathcal{X}_a^\perp \times \mathcal{Y}_+ \mid p_t + y\Delta s_{t+1} \in D_t^\circ\}$$

is the polar of $\tilde{\mathcal{C}}$; see Example A.43. If $(p, y) \in \tilde{\mathcal{C}}^\circ$, then $E_t[y\Delta s_{t+1}] \in D_t^\circ$. If there are no portfolio constraints, we have $D_t = \mathbb{R}^J$ so that $D_t^\circ = \{0\}$ and the condition becomes $E_t[y\Delta s_{t+1}] = 0$ which, for nonzero $y \in \mathcal{Y}_+$ means that s is a martingale under the probability measure Q defined by $dQ/dP := y/Ey$; see Theorem B.15. Similarly, if $D_t = \mathbb{R}_+^J$ (prohibition of short selling), the condition means that s is a supermartingale under Q.

If $\tilde{\mathcal{C}}$ is closed, then by Theorem 3.72, the support function of $\mathcal{C}$ has the expression

$$\sigma_{\mathcal{C}}(y) = \delta_{\mathcal{C}}^*(y) = \bar{\varphi}(0, \cdot)^*(y) = E\left[\delta_{\mathbb{R}_+}(y) + \sum_{t=0}^{T-1} \sigma_{D_t}(E_t[y\Delta s_{t+1}])\right].$$

If, in addition, D_t are conical, we get

$$\mathcal{C}^\circ = \{y \in \mathcal{Y}_+ \mid E_t[y\Delta s_{t+1}] \in D_t^\circ\ t = 0, \dots, T-1\ a.s\}.$$

Again, if there are no portfolio constraints, the elements of $\mathcal{C}^\circ$ are nonnegative multiples of martingale densities. Sufficient conditions for the closedness of $\tilde{\mathcal{C}}$ will be given in Sect. 4.3.5.

The following example is concerned with a variant of problem (1.15) where the objective is replaced by the "optimized certainty equivalent" discussed briefly in Remark 1.77. Its proof is a direct application of Theorem 3.38 and Theorem 3.33.

Example 3.74 (Optimized Certainty Equivalent) Consider the problem

$$\begin{aligned} &\text{minimize} \quad \mathcal{V}(c - \sum_{t=0}^{T-1} x_t \cdot \Delta s_{t+1}) \ \text{ over } \ x \in \mathcal{N} \\ &\text{subject to} \quad x_t \in D_t \ t = 0, \dots, T \quad a.s, \end{aligned} \tag{3.36}$$

where $\mathcal{V} : L^0 \to \overline{\mathbb{R}}$ is the *optimized certainty equivalent* associated with the normal integrand V, i.e.

$$\mathcal{V}(c) := \inf_{\alpha \in \mathbb{R}} \{\alpha + EV(c - \alpha)\}$$

see Remark 1.77. This fits the general duality framework with time t running from -1 to T, $x_{-1} = \alpha$, $\mathcal{F}_{-1} = \{\emptyset, \Omega\}$, $\bar{u} = c$ and

$$f(\hat{x}, u, \omega) = \alpha + \sum_{t=0}^{T} \delta_{D_t(\omega)}(x_t) + V(u - \alpha - \sum_{t=0}^{T-1} x_t \cdot \Delta s_{t+1}(\omega)),$$

where $\hat{x} = (\alpha, x)$. Indeed, problem (1.16) can then be written as

$$\text{minimize} \quad Ef(\hat{x}, u) \ \text{ over } \ \hat{x} \in \hat{\mathcal{N}}, \tag{3.37}$$

where $\hat{\mathcal{N}} := L^0(\mathcal{F}_{-1}) \times \mathcal{N}$. In addition to Assumption 3.69, we set $\mathcal{X}_{-1} = L^\infty$ and $\mathcal{V}_{-1} = L^1$.

The dual problem can be written as

$$\begin{aligned} &\text{maximize} \ E[V^*(y) + \sum_{t=0}^{T-1} \sigma_{D_t}(p_t + y\Delta s_{t+1})] \ \text{ over } \ (p, y) \in \mathcal{X}_a^\perp \in \mathcal{Y} \\ &\text{subject to} \ p_{-1} + y = 1 \quad a.s. \end{aligned} \tag{3.38}$$

If Assumption 3.69 holds, the reduced dual becomes

$$\begin{aligned} &\text{maximize} \quad E\left[yc - V^*(y) - \sum_{t=0}^{T-1} \sigma_{D_t}(E_t[y\Delta s_{t+1}])\right] \ \text{ over } \ y \in \mathcal{Y} \\ &\text{subject to} \quad Ey = 1. \end{aligned} \tag{3.39}$$

Note that the dual problems are the same as (3.33) and (3.34) except that they have the additional constraint requiring $Ey = 1$. The optimum value of (3.36) equals that of (3.38) and (3.39) if and only if the function

$$\bar{\varphi}(z, u) := \inf_{x \in L^0} \{Ef(\hat{x}, u) \mid \hat{x} - z \in \hat{\mathcal{N}}\}$$

is closed at $(0, c)$ with respect to $\sigma(\hat{\mathcal{X}} \times \mathcal{U}, \hat{\mathcal{V}} \times \mathcal{Y})$, where $\hat{\mathcal{X}} := \mathcal{X}_{-1} \times \mathcal{X}$ and $\hat{\mathcal{V}} := \mathcal{V}_{-1} \times \mathcal{V}$. Sufficient conditions for the closedness of $\bar{\varphi}$ will be given in Sect. 4.3.5. If (3.36) and (3.39) are feasible, then the following are equivalent:

1. (α, x) solves (3.37), y solves (3.39) and $\inf(3.37) = \sup(3.39)$;
2. (α, x) is feasible in (3.37), y is feasible in (3.39) and

$$y \in \partial V(c - \sum_{t=0}^{T-1} x_t \cdot \Delta s_{t+1} - \alpha),$$

$$E_t[y \Delta s_{t+1}] \in N_{D_t}(x_t) \quad t = 0, \ldots, T.$$

Given any $y \in \mathcal{Y}_+$ with $Ey = 1$, Theorem B.15 gives

$$E_t[y \Delta s_{t+1}] = E_t[y] E_t^Q[\Delta s_{t+1}],$$

where the measure Q is defined by $dQ/dP = y$. Thus, by positive homogeneity of the support function and by Lemma B.13, (3.39) can be written as

$$\underset{Q \in \mathcal{Q}^{\mathcal{Y}}}{\text{maximize}} \; E^Q[c] - EV^*(dQ/dP) - E^Q\left[\sum_{t=0}^{T-1} \sigma_{D_t}(E_t^Q[\Delta s_{t+1}])\right],$$

where $\mathcal{Q}^{\mathcal{Y}}$ is the set of probability measures with $dQ/dP \in \mathcal{Y}$. If there are no portfolio constraints, this can be written as

$$\text{maximize} \quad E^Q[c] - E[V^*(dQ/dP)] \quad \text{over} \quad Q \in \mathcal{Q}_s^{\mathcal{Y}},$$

where $\mathcal{Q}_s^{\mathcal{Y}}$ is the set of martingale measures in $\mathcal{Q}^{\mathcal{Y}}$. When $V(u) = \delta_{\mathbb{R}_-}$, problem (3.36) becomes the superhedging problem (1.16) and $V^* = \delta_{\mathbb{R}_+}$. When $V(u) = \exp(u) - 1$, the optimized certainty equivalent becomes the *entropic risk measure*

$$\mathcal{V}(c) = \ln Ee^c.$$

In this case, $V^*(y) = y \log y - y + 1 + \delta_{\mathbb{R}_+}(y)$, so that

$$EV^*(dQ/dP) = E^Q \ln(dQ/dP),$$

the *entropy* of Q relative to P. When $V(u) = \beta u^+$ for $\beta > 1$, the optimized certainty equivalent becomes the Conditional Value at Risk and $V^* = \delta_{[0,\beta]}$.

Proof The Lagrangian integrand becomes

$$\begin{aligned} l(\hat{x}, y, \omega) &= \inf_{u \in \mathbb{R}} \{f(\hat{x}, u, \omega) - uy\} \\ &= \alpha - y(\alpha + \sum_{t=0}^{T-1} x_t \cdot \Delta s_{t+1}(\omega)) + \sum_{t=0}^{T} \delta_{D_t(\omega)}(x_t) - V^*(y) \end{aligned}$$

and the conjugate of f,

$$\begin{aligned} f^*(v, y, \omega) &= \sup_{\hat{x} \in \mathbb{R}^n} \{\hat{x} \cdot v - l(\hat{x}, y, \omega)\} \\ &= V^*(y) + \delta_{\{0\}}(v_{-1} + y - 1) + \sum_{t=0}^{T-1} \sigma_{D_t(\omega)}(v_t + y\Delta s_{t+1}(\omega)). \end{aligned}$$

Since Ef is finite at the origin, Theorem 3.20 says that

$$\varphi^*(p, y) = \begin{cases} E[V^*(y) + \sum_{t=0}^{T-1} \sigma_{D_t}(p_t + y\Delta s_{t+1})] & \text{if } (p, y) \in \hat{\mathcal{X}}_a^\perp \times \mathcal{Y} \\ & \text{and } p_{-1} + y = 1, \\ +\infty & \text{otherwise,} \end{cases}$$

where $\hat{\mathcal{X}}_a^\perp := \{p \in \hat{\mathcal{V}} \mid E[z \cdot p] = 0 \, \forall z \in \hat{\mathcal{N}} \cap \hat{\mathcal{X}}\}$. Defining a mapping $\pi : \mathcal{Y} \to \hat{\mathcal{V}}$ by $\pi(y)_{-1} := Ey - y$ and $\pi(y)_t := E_t[y\Delta s_{t+1}] - y\Delta s_{t+1}$ for $t \geq 0$, Jensen's inequality in Theorem 3.13 gives, for any $(p, y) \in \hat{\mathcal{X}}_a^\perp \times \mathcal{Y}$,

$$\begin{aligned} \varphi^*(p, y) &= E[V^*(y) + \delta_{\{0\}}(p_{-1} + y - 1) + \sum_{t=0}^{T-1} \sigma_{D_t}(p_{t+1} + y\Delta s_{t+1})] \\ &\geq E[V^*(y) + \delta_{\{0\}}(Ey - 1) + \sum_{t=0}^{T-1} \sigma_{D_t}(E_t[y\Delta s_{t+1}])] \\ &= \varphi^*(\pi(y), y). \end{aligned}$$

Thus, by Theorem 3.38, $\sup (3.39) = \sup_{(p,y) \in \hat{\mathcal{V}} \times \mathcal{Y}} \{\langle \bar{u}, y \rangle - \varphi^*(p, y)\}$, so the claim follows from Theorem 3.33. □

The next example studies the semi-static portfolio optimization problem from Example 1.90.

Example 3.75 (Semi-Static Hedging) Consider again problem (1.17)

$$\begin{aligned}&\text{minimize} \quad EV\left(c-\sum_{t=0}^{T-1} x_t\cdot \Delta s_{t+1} - \bar{c}\cdot\bar{x} + S_0(\bar{x})\right) \quad \text{over} \quad (x,\bar{x})\in\mathcal{N}\times\mathbb{R}^{\bar{J}},\\ &\text{subject to} \qquad x_t\in D_t \; t=0,\dots,T \; a.s.\end{aligned}$$

from Example 1.90 and assume that the claim c as well as the components of $\bar{c}$ belong to $\mathcal{U}$. We start by rewriting the problem in the equivalent form

$$\begin{aligned}&\text{minimize} \quad EV\left(c-\sum_{t=0}^{T-1} x_t\cdot \Delta s_{t+1} - \bar{c}\cdot\bar{x} + \alpha\right) \quad \text{over} \; (x,\bar{x},\alpha)\in\mathcal{N}\times\mathbb{R}^{\bar{J}}\times\mathbb{R},\\ &\text{subject to} \qquad x_t\in D_t \quad t=0,\dots,T \; a.s.\\ &\qquad\qquad\qquad\qquad S_0(\bar{x})\le\alpha.\end{aligned}$$

This fits the general duality framework with time t running from -1 to T, $\mathcal{F}_{-1} = \{\Omega,\emptyset\}$, $x_{-1}=(\bar{x},\alpha)\in\mathbb{R}^{\bar{J}}\times\mathbb{R}$, $\bar{u}=c$ and

$$\begin{aligned}f(\hat{x},u,\omega) = {}& V\left(u-\sum_{t=0}^{T-1} x_t\cdot\Delta s_{t+1}(\omega) - \bar{c}(\omega)\cdot\bar{x}+\alpha,\omega\right)\\ &+\sum_{t=0}^{T}\delta_{D_t(\omega)}(x_t,\omega)+\delta_{\operatorname{epi} S_0}(\bar{x},\alpha),\end{aligned}$$

where $\hat{x}=(x_{-1},x)$. Indeed, since V is nondecreasing, $(\bar{x},x)$ solves (1.17) if and only if $\hat{x}=(\bar{x},S_0(\bar{x}),x)$ solves

$$\text{minimize} \quad Ef(\hat{x},u) \quad \hat{x}\in\hat{\mathcal{N}},$$

where $\hat{\mathcal{N}} := L^0(\Omega,\mathcal{F}_{-1},P;\mathbb{R}^{\bar{J}}\times\mathbb{R})\times\mathcal{N}$. In addition to Assumption 3.69, we set $\mathcal{X}_{-1}=L^\infty$ and $\mathcal{V}_{-1}=L^1$.

The optimum value of (1.17) equals that of

$$\underset{y\in\mathcal{Y}}{\text{maximize}} \quad E\left[cy - V^*(y) - \sum_{t=0}^{T-1}\sigma_{D_t}(E_t[y\Delta s_{t+1}])\right] - \sigma_{\operatorname{epi} S_0}(E[y\bar{c}], -E[y]) \tag{3.40}$$

if and only if the function

$$\bar{\varphi}(z,u) := \inf_{x\in L^0}\{Ef(\hat{x},u)\mid \hat{x}-z\in\hat{\mathcal{N}}\}$$

is closed at $(0, c)$ with respect to $\sigma(\hat{\mathcal{X}} \times \mathcal{U}, \hat{\mathcal{V}} \times \mathcal{Y})$, where $\hat{\mathcal{X}} := \mathcal{X}_{-1} \times \mathcal{X}$ and $\hat{\mathcal{V}} := \mathcal{V}_{-1} \times \mathcal{V}$. If (1.17) and (3.40) are feasible, then the following are equivalent:

1. $(\bar{x}, x)$ solves (1.17), y solves (3.40) and $\inf(P) = \sup(3.40)$;
2. $(\bar{x}, x)$ is feasible in (1.17), y is feasible in (3.40) and

$$y \in \partial V\left(c - \sum_{t=0}^{T-1} x_t \cdot \Delta s_{t+1} - \bar{c} \cdot \bar{x} + S_0(\bar{x})\right),$$

$$E_t[y\Delta s_{t+1}] \in N_{D_t}(x_t) \quad t = 0, \dots, T,$$

$$E[y\bar{c}] \in \partial(E[y]S_0)(\bar{x})$$

almost surely.

Proof The corresponding Lagrangian integrand becomes

$$\begin{aligned}
l(\hat{x}, y, \omega) &= \inf_{u \in \mathbb{R}}\{f(\hat{x}, u, \omega) - uy\} \\
&= \inf_{u \in \mathbb{R}}\left\{V\left(u - \sum_{t=0}^{T-1} x_t \cdot \Delta s_{t+1}(\omega) - \bar{c}(\omega) \cdot \bar{x} + \alpha, \omega\right) - uy\right\} \\
&\quad + \sum_{t=0}^{T} \delta_{D_t(\omega)}(x_t, \omega) + \delta_{\mathrm{epi}\, S_0}(\bar{x}, \alpha) \\
&= y\left(\alpha - \bar{c}(\omega) \cdot \bar{x} - \sum_{t=0}^{T-1} x_t \cdot \Delta s_{t+1}(\omega)\right) - V^*(y, \omega) \\
&\quad + \sum_{t=0}^{T} \delta_{D_t(\omega)}(x_t) + \delta_{\mathrm{epi}\, S_0}(\bar{x}, \alpha),
\end{aligned}$$

and the conjugate of f,

$$\begin{aligned}
f^*(v, y, \omega) &= \sup_{\hat{x} \in \mathbb{R}^n}\{\hat{x} \cdot v - l(x, y, \omega)\} \\
&= V^*(y, \omega) + \sum_{t=0}^{T-1} \sup_{x_t \in \mathbb{R}^J}\{x_t \cdot (v_t + y\Delta s_{t+1}(\omega)) - \delta_{D_t(\omega)}(x_t)\} \\
&\quad + \sup_{(\bar{x}, \alpha) \in \mathbb{R}^{\bar{J}} \times \mathbb{R}}\{(\bar{x}, \alpha) \cdot (v_{-1} + (y\bar{c}(\omega), -y) - \delta_{\mathrm{epi}\, S_0}(\bar{x}, \alpha)\} \\
&= V^*(y, \omega) + \sum_{t=0}^{T-1} \sigma_{D_t(\omega)}(v_t + y\Delta s_{t+1}(\omega)) \\
&\quad + \sigma_{\mathrm{epi}\, S_0}(v_{-1} + (y\bar{c}(\omega), -y)).
\end{aligned}$$

By Theorem 3.20,

$$\varphi^*(p, y) = E\left[V^*(y) + \sum_{t=0}^{T-1} \sigma_{D_t}(p_t + y\Delta s_{t+1}) + \sigma_{\text{epi}\, S_0}(p_{-1} + (y\bar{c}, -y))\right]$$

if $(p, y) \in \hat{\mathcal{X}}_a^{\perp} \times \mathcal{Y}$ and $\varphi^*(p, y) = \infty$ otherwise. Here

$$\mathcal{X}_a^{\perp} := \{p \in \bar{\mathcal{V}} \mid E[z \cdot p] = 0\, \forall z \in \hat{\mathcal{N}} \times \hat{\mathcal{X}}\}.$$

Defining $\pi : \mathcal{Y} \to \hat{\mathcal{V}}$ by

$$\pi(y)_{-1} := (E[y\bar{c}] - y\bar{c}, -E[y] + y)$$
$$\pi(y)_t := E_t[y\Delta s_{t+1}] - y\Delta s_{t+1} \quad t = 0, \ldots, T-1,$$

Jensen's inequality in Theorem 3.13 gives, for any $(p, y) \in \hat{\mathcal{X}}_a^{\perp} \times \mathcal{Y}_+$,

$$\begin{aligned}\varphi(p, y) &= E\left[V^*(y) + \sum_{t=0}^{T-1} \sigma_{D_t}(p_t + y\Delta s_{t+1}) + \sigma_{\text{epi}\, S_0}(p_{-1} + (y\bar{c}, -y))\right] \\ &\geq E\left[V^*(y) + \sum_{t=0}^{T-1} \sigma_{D_t}(E_t[y\Delta s_{t+1}]) + \sigma_{\text{epi}\, S_0}(E[y\bar{c}], -E[y])\right] \\ &= \varphi(\pi(y), y). \end{aligned} \tag{3.41}$$

Thus, the claims follow from Theorem 3.38 by choosing γ as the identity mapping. Indeed, the conditions

$$\pi(y) \in \partial_x l(\hat{x}, y), \quad \bar{u} \in \partial_y[-l](\hat{x}, y)$$

can be written as

$$\begin{aligned}&(E[y\bar{c}], -E[y]) \in N_{\text{epi}\, S_0}(\bar{x}, \alpha), \\ &\pi(y)_t \in -y\Delta s_{t+1} + N_{D_t}(x_t) \quad t = 0, \ldots, T, \\ &c \in -\left(\alpha - \bar{c}(\omega) \cdot \bar{x} - \sum_{t=0}^{T-1} x_t \cdot \Delta s_{t+1}(\omega)\right) + \partial V^*(y).\end{aligned}$$

By Corollary A.55, the last condition can be written as

$$y \in \partial V\left(c - \sum_{t=0}^{T-1} x_t \cdot \Delta s_{t+1} - \bar{c} \cdot \bar{x} + \alpha\right). \tag{3.42}$$

If $E[y] > 0$, then the first condition says that $\alpha = S_0(\bar{x})$ and

$$\frac{E[y\bar{c}]}{E[y]} \in \partial S_0(\bar{x})$$

which is equivalent to the last subdifferential condition in the statement. If $E[y] = 0$, then $y = 0$ and the first condition simply means that $S_0(\bar{x}) \leq \alpha$. Since V is nondecreasing, (3.42) implies the first condition in the statement. If optimality conditions in the statement hold, the above conditions holds with $\alpha = S_0(\bar{x})$. □

The last term in the objective of (3.40) simplifies into a constraint in the case of a sublinear cost function S_0.

Remark 3.76 (Calibration of Martingale Measures) The support function of epi S_0 in the dual problem (3.40) in Example 3.75 can be expressed, by Corollary A.59, as

$$\sigma_{\text{epi}\, S_0}(v, -\alpha) = \begin{cases} \alpha S_0^*(v/\alpha) & \text{if } \alpha > 0, \\ (S_0^*)^\infty(v) & \text{if } \alpha = 0, \\ +\infty & \text{otherwise.} \end{cases}$$

If S_0 is sublinear, then $S_0^*(\cdot, \omega)$ is the indicator of a closed convex set. Denoting this set by $C_0(\omega)$, we have

$$\sigma_{\text{epi}\, S_0}(v, -\alpha) = \begin{cases} \delta_{C_0}(v/\alpha) & \text{if } \alpha > 0, \\ \delta_{C_0^\infty}(v) & \text{if } \alpha = 0, \\ +\infty & \text{otherwise.} \end{cases}$$

For $\alpha \geq 0$, this equals $\delta_{\alpha_+ C_0}(v)$ where

$$\alpha_+ C_0 := \begin{cases} \alpha C_0 & \text{if } \alpha > 0, \\ C_0^\infty & \text{if } \alpha = 0. \end{cases}$$

The reduced dual (3.40) can then be written with explicit constraints as

$$\begin{aligned} &\text{maximize} && E\left[cy - V^*(y) - \sum_{t=0}^{T-1} \sigma_{D_t}(E_t[y\Delta s_{t+1}])\right] \quad \text{over } y \in \mathcal{Y} \\ &\text{subject to} && E[y\bar{c}] \in E[y]_+ C_0. \end{aligned}$$

If $E[y] > 0$, the constraint means that

$$E^Q \bar{c} \in C_0,$$

where Q is the probability measure defined by $dQ/dP = y/Ey$. The constraint thus requires that the measure Q be "calibrated" to the observed market prices of the claims $\bar{c}$. For example, if infinite quantities are available to buy and sell at prices $s^a \in \mathbb{R}^{\bar{J}}$ and $s^b \in \mathbb{R}^{\bar{J}}$, respectively (see Example 1.90), then $C_0 = [s^b, s^a]$ and the constraint can be written as the vector inequalities

$$s^b \leq E^Q[\bar{c}] \leq s^a$$

saying that the measure Q "prices" the claims between the bid-ask spread.

We end this section by some functional analytic reformulations of the studied problems.

Remark 3.77 The optimum value of (3.36) can be expressed as

$$\varphi_{OCE}(c) = \inf_{\alpha \in \mathbb{R}} \{\alpha + \varphi_0(c - \alpha)\},$$

where φ_0 is the optimum value function of problem Eq. 1.15 with respect to c. The conjugate of φ_{OCE} can thus be written as

$$\begin{aligned}
\varphi^*_{OCE}(y) &= \sup_{c,\alpha}\{\langle c, y\rangle + \alpha - \varphi_0(c - \alpha)\} \\
&= \sup_{c,\alpha}\{\langle c, y\rangle - \alpha - \varphi_0(c - \alpha)\} \\
&= \sup_{c,\alpha}\{\langle c', y\rangle - \alpha(1 - E[y]) - \varphi_0(c')\} \\
&= \begin{cases} \varphi_0^*(y) & \text{if } E[y] = 1, \\ +\infty & \text{otherwise.} \end{cases}
\end{aligned}$$

If φ_{OCE} is closed and proper then, by Theorem A.54,

$$\varphi_{OCE}(c) = \sup_{y}\{\langle c, y\rangle - \varphi_0^*(c) \mid E[y] = 1\}.$$

If the assumptions of Theorem 3.72 are satisfied, we thus get

$$\varphi_{OCE}(c) = \sup_{Q \in \mathcal{Q}^{\mathcal{Y}}} \{E^Q[c] - EV^*(dQ/dP) - E^Q \sum_{t=0}^{T-1} \sigma_{D_t}(E_t^Q[\Delta s_{t+1}])\}.$$

Note that if $V = \delta_{\mathbb{R}_-}$, then φ_{OCE} coincides with the function

$$\varphi_{SH}(c) := \inf_{\alpha \in \mathbb{R}} \{\alpha \mid c - \alpha \in \mathcal{C}\}$$

where $\mathcal{C}$ is the set of claims that can be superhedged without a cost; see (3.35). Under the above assumptions, we thus have the dual representation

$$\varphi_{SH}(c) = \sup_{Q \in \mathcal{Q}^{\mathcal{Y}}} \{E^Q[c] - E^Q \sum_{t=0}^{T-1} \sigma_{D_t}(E_t^Q[\Delta s_{t+1}])\},$$

where $\mathcal{Q}^{\mathcal{Y}}$ is the set of probability measures with $dQ/dP \in \mathcal{Y}$. If, in addition, there are no portfolio constraints, this becomes

$$\varphi_{SH}(c) = \sup_{Q \in \mathcal{Q}_s^{\mathcal{Y}}} E^Q[c],$$

where $\mathcal{Q}_s^{\mathcal{Y}}$ is the set of elements of $\mathcal{Q}^{\mathcal{Y}}$ under which s is a martingale.

The following reformulates the optimum value function of problem (1.17) in terms of that of (1.15).

Remark 3.78 The optimum value of problem (1.17) can be expressed as

$$\varphi_{SSH}(c) = \inf_{\bar{x} \in \operatorname{dom} S_0} \varphi_0(c + S_0(\bar{x}) - \bar{x} \cdot \bar{c}),$$

where φ_0 is the optimum value function of Eq. 1.15 with respect to c. Thus, the conjugate of φ_{SSH} can be written as

$$\begin{aligned}
\varphi_{SSH}^*(y) &= \sup_{c, \bar{x} \in \operatorname{dom} S_0} \{\langle c, y \rangle - \varphi_0(c + S_0(\bar{x}) - \bar{x} \cdot \bar{c})\} \\
&= \sup_{c', \bar{x} \in \operatorname{dom} S_0} \{\langle c' - S_0(\bar{x}) + \bar{x} \cdot \bar{c}, y \rangle - \varphi_0(c')\} \\
&= \varphi_0^*(y) + \sup_{\bar{x} \in \operatorname{dom} S_0} \{\langle \bar{x} \cdot \bar{c}, y \rangle - \langle S_0(\bar{x}), y \rangle\} \\
&= \varphi_0^*(y) + \sup_{\bar{x} \in \operatorname{dom} S_0} \{\bar{x} \cdot E[y\bar{c}] - E[y] S_0(\bar{x})\} \\
&= \varphi_0^*(y) + \sigma_{\operatorname{epi} S_0}(E[y\bar{c}], -E[y]) \\
&= \varphi_0^*(y) + \begin{cases} E[y] S_0^*(E[y\bar{c}]/E[y]) & \text{if } E[y] > 0, \\ (S_0^*)^\infty(E[y\bar{c}]) & \text{if } E[y] = 0, \\ +\infty & \text{otherwise.} \end{cases}
\end{aligned}$$

Much as in Remark 3.77, one can then use Theorem A.54 and the dual representation of φ_0 in Theorem 3.72 to derive a dual representation for φ_{SSH}.

3.4.6 Subdifferentials and Conditional Expectations

This section applies the results of Sect. 3.2 to the general theory of conditional expectations of normal integrands. More precisely, Theorem 3.79 below gives sufficient conditions for the commutation of conditional expectation and subdifferentiation of a normal integrand. The result is interesting in its own right but it will not be needed for subsequent developments in this book. Sufficient conditions for the assumptions of Theorem 3.79 will be given in Sect. 5.4.6.

Given a convex normal integrand h and an $x \in L^0(\mathbb{R}^n)$, Theorem 1.51 says that the set valued mapping $\partial h(x)(\omega) := \partial h(x(\omega), \omega)$ is closed convex-valued and measurable. We denote the $\mathcal{G}$-measurable selections of a set-valued mapping $S : \Omega \rightrightarrows \mathbb{R}^n$ by $\mathcal{X}(\mathcal{G}; S)$ or, if $S = \mathbb{R}^n$ almost surely, simply $\mathcal{X}(\mathcal{G})$. Recall from Theorem 2.27, that if S is closed convex-valued and measurable, then it has a $\mathcal{G}$-conditional expectation for any σ-algebra $\mathcal{G} \subseteq \mathcal{F}$. The $\mathcal{G}$-conditional expectation is the unique closed convex-valued $\mathcal{G}$-measurable mapping $E^{\mathcal{G}} S$ such that

$$L^1(\mathcal{G}; E^{\mathcal{G}} S) = \text{cl}\{E^{\mathcal{G}} v \mid v \in L^1(\mathcal{F}; S)\},$$

where the closure is taken in the L^1-topology.

Theorem 3.79 *Let h be a convex normal integrand and $\mathcal{G} \subseteq \mathcal{F}$ a σ-algebra such that Eh^* is proper on $\mathcal{V}$ and, for all $v \in \mathcal{V}$, the function $\phi_v : \mathcal{X} \to \overline{\mathbb{R}}$ given by*

$$\phi_v(z) := \inf_{x \in \mathcal{X}} \{E[h(x) - x \cdot v] \mid x - z \in \mathcal{X}(\mathcal{G})\}$$

is either subdifferentiable or equal to $-\infty$ at the origin. We have

$$\mathcal{V}(\mathcal{G}; E^{\mathcal{G}}[\partial h(x)]) = \mathcal{V}(\mathcal{G}; \partial(E^{\mathcal{G}} h)(x))$$

for all $x \in \text{dom}\, Eh \cap \mathcal{X}(\mathcal{G})$. If there exists an $\bar{x} \in \text{dom}\, Eh \cap \mathcal{X}(\mathcal{G})$ such that $\mathcal{V}(\partial h(\bar{x})) \neq \emptyset$, then

$$\partial(E^{\mathcal{G}} h)(x) = E^{\mathcal{G}}[\partial h(x)] \quad a.s.$$

for every $x \in L^0(\mathcal{G})$ such that $h(x) \in L^1$ and $\mathcal{V}(\partial h(x)) \neq \emptyset$.

Proof Let $x \in \text{dom}\, Eh \cap \mathcal{X}(\mathcal{G})$. By Fenchel's inequality, $h(x) + h^*(v) \geq x \cdot v$, so the properness of Eh^* on $\mathcal{V}$ implies that h is L-bounded and $h(x)^- \in L^1$. Thus, by Theorem 2.26,

$$E^{\mathcal{G}} v \in \partial(E^{\mathcal{G}} h)(x) \; a.s.$$

for every $v \in \mathcal{V}(\partial h(x))$. Since L^1-convergence implies almost sure convergence and since $\partial(E^{\mathcal{G}}h)(x)$ is closed-valued, we thus get

$$L^1(\mathcal{G}; E^{\mathcal{G}}[\partial h(x)]) \subseteq L^1(\mathcal{G}; \partial(E^{\mathcal{G}}h)(x))$$

and, in particular,

$$\mathcal{V}(\mathcal{G}; E^{\mathcal{G}}[\partial h(x)]) \subseteq \mathcal{V}(\mathcal{G}; \partial(E^{\mathcal{G}}h)(x)).$$

To prove the converse, let $v \in \mathcal{V}(\mathcal{G}; \partial(E^{\mathcal{G}}h)(x))$. Since h is L-bounded and $h(x)^- \in L^1$, Theorems 1.64 and 2.19 give

$$\begin{aligned}
\phi_v(0) &= \inf_{x' \in \mathcal{X}(\mathcal{G})} E[h(x') - x' \cdot v] \\
&= \inf_{x' \in \mathcal{X}(\mathcal{G})} E[(E^{\mathcal{G}}h)(x') - x' \cdot v] \\
&= E[\inf_{x' \in \mathbb{R}^n} \{(E^{\mathcal{G}}h)(x') - x' \cdot v\}] \\
&= E[(E^{\mathcal{G}}h)(x) - x \cdot v] \\
&= E[h(x) - x \cdot v],
\end{aligned}$$

which is finite since $x \in \operatorname{dom} Eh$ and $h(x)^- \in L^1$.

We apply Theorem 3.18 to the one-period instance of $(P_{\mathcal{X}})$ where $T = 0$, $\mathcal{F}_0 = \mathcal{G}$, $\bar{u} = 0$ and

$$f(x, u, \omega) := h(x, \omega) - x \cdot v.$$

The assumption on ϕ_v implies $\phi_v(0) < \infty$ so $\operatorname{dom} Ef \cap (\mathcal{X} \times \mathcal{U}) \neq \emptyset$ and, by Theorem 3.20,

$$F^*(v, p, y) = Ef^*(v + p, y) + \delta_{\mathcal{X}_a^{\perp}}(p)$$

and, in particular,

$$\varphi^*(p, y) = Ef^*(p, y) + \delta_{\mathcal{X}_a^{\perp}}(p),$$

where $f^*(p, y, \omega) = h^*(v + p, \omega) + \delta_{\{0\}}(y)$ and $\mathcal{X}_a^{\perp} = \{p \in \mathcal{V} \mid E^{\mathcal{G}} p = 0\}$. Thus, by Theorem 3.18,

$$\inf_{x \in \mathcal{X}(\mathcal{G}; \mathbb{R}^n)} E[h(x) - x \cdot v] = \sup_{p \in \mathcal{V}} \{-Eh^*(v + p) \mid E^{\mathcal{G}} p = 0\},$$

where the supremum is attained by some $\bar{p} \in \mathcal{V}$. Since Eh is proper on $\mathcal{X}(\mathcal{G})$, Theorem 2.19 and the interchange rule in Theorem 1.64 give

$$\begin{aligned}\inf_{x \in \mathcal{X}(\mathcal{G};\mathbb{R}^n)} E[h(x) - x \cdot v] &= \inf_{x \in \mathcal{X}(\mathcal{G};\mathbb{R}^n)} \{E(E^{\mathcal{G}}h)(x) - \langle x, v\rangle\} \\ &= -E(E^{\mathcal{G}}h)^*(v)\end{aligned}$$

so $Eh^*(v + \bar{p}) = E(E^{\mathcal{G}}h)^*(v)$. By Theorems 2.31 and 2.35, $E^{\mathcal{G}}h^*(v + \bar{p}) \geq (E^{\mathcal{G}}h)^*(v)$ almost surely, so we must have,

$$E^{\mathcal{G}}h^*(v + \bar{p}) = (E^{\mathcal{G}}h)^*(v) \quad a.s. \tag{3.43}$$

Since $v \in \partial(E^{\mathcal{G}}h)(x)$ almost surely,

$$(E^{\mathcal{G}}h)(x) + (E^{\mathcal{G}}h)^*(v) = x \cdot v \quad a.s.$$

so, by (3.43),

$$(E^{\mathcal{G}}h)(x) + E^{\mathcal{G}}h^*(v + \bar{p}) = x \cdot v \quad a.s.$$

Since $E[x \cdot \bar{p}] = 0$ and $h(x)$ is integrable, Theorem 2.19 gives

$$E[h(x) + h^*(v + \bar{p}) - x \cdot (v + \bar{p})] = 0.$$

This implies that the Fenchel's inequality $h(x) + h^*(v + \bar{p}) - x \cdot (v + \bar{p}) \geq 0$ must hold as an equality almost surely. The equality means that $v + \bar{p} \in \partial h(x)$. Taking conditional expectations, we get $v \in E^{\mathcal{G}}[\partial h(x)]$ almost surely so

$$v \in \mathcal{V}(\mathcal{G}; E^{\mathcal{G}}[\partial h(x)])$$

which completes the proof of the first claim.

Let $x \in L^0(\mathcal{G})$ such that $h(x) \in L^1$ and $\mathcal{V}(\partial h(x)) \neq \emptyset$. Define, for every $\nu \in \mathbb{N}$, $x^\nu := 1_{A^\nu}x + 1_{\Omega\setminus A^\nu}\bar{x}$, where $A^\nu = \{|x| \leq \nu\}$. We have $x^\nu \in \operatorname{dom} Eh \cap \mathcal{X}(\mathcal{G})$ and $\mathcal{V}(\partial h(x^\nu)) \neq \emptyset$. By Theorem 1.72, the first claim implies

$$\partial(E^{\mathcal{G}}h)(x^\nu) = E^{\mathcal{G}}[\partial h(x^\nu)] \quad a.s.$$

By Lemma 2.30 and Corollary 2.29.1,

$$E^{\mathcal{G}}[\partial h(x^\nu)] = \begin{cases} E^{\mathcal{G}}[\partial h(x)] & \text{on } A^\nu, \\ E^{\mathcal{G}}[\partial h(\bar{x})] & \text{on } \Omega \setminus A^\nu. \end{cases}$$

In particular, $\partial(E^{\mathcal{G}}h)(x) = E^{\mathcal{G}}[\partial h(x)]$ on A^ν. It now suffices to note that $\bigcup_\nu A^\nu = \Omega$. □

3.5 Bibliographical Notes

Duality theory of convex stochastic optimization goes back to [59, 83, 156] where a two-stage instance of the linear problem of Example 3.42 was studied. Our approach owes a lot to the works of Rockafellar and Wets in [131–134, 136, 138] where dual problems and optimality conditions were derived from the general conjugate duality framework much like here. Other early references on stochastic optimization duality include [49, 127]. In all the above references, decision strategies were sought from locally convex spaces of random variables. While [49] was concerned with L^p for $p \in (1, \infty)$, the articles of Rockafellar and Wets assumed that the strategies belong to the space L^∞ of essentially bounded random vectors. The strong topology of L^∞ allows for simple conditions that suffice for the subdifferentiability of the optimum value function and thus, for the existence of dual solutions; see Theorem 3.33 and Chap. 5 below.

Soon after the developments in stochastic optimization duality theory in the 1970s, a closely related branch of research started to develop independently in financial mathematics after the influential publications of Harrison and Kreps [60] and Harrison and Pliska [61]. Applications of convex duality to financial economics was discussed in Pliska [114] but that seems to have gone largely unnoticed. One of the most famous results in financial mathematics is the "fundamental theorem of asset pricing" of Dalang et al. [35]; see [144] and Theorem 4.29 for a more functional analytic proof. Generalizations to market models with transaction costs can be found e.g. in [146], [100] and [96]. The proofs of such results require the use of general adapted strategies so the financial models were beyond the scope of the stochastic programming duality theory of the 70s that allowed only for locally convex spaces of decision strategies.

The duality theory of convex stochastic optimization was extended to general adapted processes in [17, 95, 99]. The approach with two parameters (z, u) and the corresponding dual variables (p, y) presented in this chapter is from [104]. The double parameterization greatly simplified the dualization as it allows for the explicit dual problem in terms of the scenariowise conjugate of a normal integrand. In the case of the mathematical programming format of Sect. 3.4.1, the resulting dual problem is essentially the same as that found in [136] for problems over bounded strategies and with inequality constraints only. The main result of [136] gives sufficient conditions for the existence of a dual solution, a topic that will be studied in Chap. 5. The approach of [136] involved two steps where one first establishes the existence of a shadow price of information p using the results of [131] and then the existence of the variable y is proved via measurable selection arguments. The duality framework presented in this chapter as well as the existence arguments of Chap. 5 combine the two steps into one and allow for more general spaces of strategies and dualizing parameterizations.

The concept of a shadow price of information first appeared in [130, 131, 159] in continuous time, single period and multi-period settings, respectively. While the continuous time setting of [159] assumed a special structure of the objective, the objectives studied in [131] were integral functionals of the whole path of the decision strategy like here. The shadow price of information was later studied in continuous time settings in [4, 37]. While [37] was concerned with a stochastic optimal control model, the format of [4] was closer to that of [159] where the associated normal integrand could be expressed scenariowise as an integral over time. This extra assumption makes the dualization considerably easier as one can then use the theory of integral functionals with respect to the measurable space $\Omega \times [0, T]$. Special instances of the shadow price of information can also be found in [38] in the context of optimal stopping.

Most of the results of this chapter are from [104] with few exceptions. Lemma 3.4 is from [101] while the results of Sect. 3.1.2 can be found in [122, 124, 128] except for Theorem 3.11 which seems to be new. The reduced dual problem (3.10) in Sect. 3.2.2 was discussed in [104] but the subsequent formal statements are new.

The dual problem of the mathematical programming model derived in Sect. 3.4.1 was first obtained in [136] under more restrictive assumptions that were geared towards the existence of dual solutions. Most notably, we have relaxed the boundedness assumptions and added equality constraints. The alternative dual problem associated with the normal integrand $\tilde{f}$ in Remark 3.50 was obtained in [38] and later used in a numerical technique for bounding the optimal stopping values in [140]. The optimality conditions for stochastic optimal control in Corollary 3.56 extend, in the convex case, those obtained in [29, Section 5.4.3] by allowing e.g. for nondifferentiable objectives and constraints involving the state variable. A duality framework for continuous time stochastic optimal control was developed already in [20] but without attempts to prove the existence of primal or dual solutions.

The problem of Lagrange studied Sect. 3.4.4 is closely related to the stochastic problem of Bolza studied in [138]. The main difference is that the objective in the problem of Bolza involves a term that is a general convex function of both the starting point x_0 and the end point x_T. The strategy of dualization in [138] was somewhat more involved than the approach taken in Sect. 3.4.4. Optimal investment problems of the form (1.15) have been extensively studied in continuous time models of financial markets; see e.g. [26, 41, 80, 145] and their references. In these references, however, the dual problems are somewhat different from ours as they are not derived by conjugating the primal optimum value function with respect to the random claim c. Embedding the optimal investment problems in general optimization duality framework, yields dual problems and optimality conditions as simple consequences of general theory. Moreover, our approach allows for general random claims/endowments and nondifferentiable loss functions like in the models of [26, 41]. Extensions of the duality theory to illiquid market models can be found e.g. in [34, 102]. The dual problem in Example 3.74 was studied in [58] as a problem of independent interest. Example 3.74 provides a direct link to the optimal investment problem with the optimized certainty equivalent objective. Semi-static hedging problems akin to that in Example 3.75 are popular in the literature of "model-independent" mathematical finance; see e.g. [8]. In that literature, however, rather than looking at expected utilities, it is more common to look for trading strategies that superhedge a given claim with respect to several measures at once. Theorem 3.79 was stated in a different but equivalent form in [153, page 142]. The result extends the main result of [137] which, in turn, extends [22, Theorem 4], on the commutation of subdifferentiation and conditional expectation. More concrete conditions that imply those of Theorem 3.79 will be given in Sect. 5.4.6.

Chapter 4
Absence of a Duality Gap

We saw in the previous chapter that the closedness of the optimum value function $\bar{\varphi}$ of (P) at $(0, \bar{u})$ is equivalent to the absence of a duality gap; see Theorem 3.33. Even when dual solutions do not exist, the absence of a duality gap is relevant in dual expressions of the optimum value and, e.g. in the "fundamental theorem of asset pricing"; see Theorem 4.29 below. Dual expressions form the basis of various numerical techniques for computing tight lower bounds for the optimum value. This chapter gives sufficient conditions for the existence of optimal primal solutions and for the closedness of the optimum value function $\bar{\varphi}$ of (P) and thus, for the absence of a duality gap. In general, the conditions given here do not imply the existence of dual solutions; that will be the topic of Chap. 5. Existence of dual solutions fails e.g. in many popular models of financial mathematics where, nevertheless, one obtains the existence of optimal trading strategies and the absence of a duality gap.

Classical results on the existence of solutions and closedness of the optimum value function are based on topological arguments involving some form of compactness in the primal problem. For example, if the objective Ef is inf-compact with respect to x uniformly in (z, u) in a neighborhood of $(0, \bar{u})$, then primal solutions do exist and $\bar{\varphi}$ is lsc at $(0, \bar{u})$. In most practical applications, such conditions are perfectly reasonable but they fail e.g. in classical models of financial mathematics where the level-sets of the objective are typically unbounded. This chapter gives more general conditions that hold under classical topological conditions as well as under various no-arbitrage conditions often employed in financial mathematics. We replace the usual compactness assumption by an algebraic condition involving the scenariowise recession cone of the normal integrand f. In classical models of financial mathematics, the condition becomes the "no-arbitrage condition" on the market model. In market models with transaction costs, it becomes the so called "robust no-arbitrage condition". Like the no-arbitrage conditions, our condition is purely algebraic. It is essentially the same as the linearity condition used for the existence of solutions for the Bellman equations in Sect. 2.2.

T. Pennanen, A.-P. Perkkiö, *Convex Stochastic Optimization*, Probability Theory and Stochastic Modelling 107, https://doi.org/10.1007/978-3-031-76432-5_4

The closedness of the optimum value function of (P) is established in two steps. We first assume that the normal integrand in (P) is lower bounded (see Theorem 4.8 below) which allows us to use versions of Komlós' theorem for nonnegative random variables; see Theorem B.25. The proof of Theorem 4.8 uses an inductive argument employing versions of Komlós' theorem and the recession properties of the solution of the Bellman equations obtained in Theorem 2.62. The extension to more general normal integrands in Sect. 4.2.2 is based on a reduction argument using Lemma 2.67. In the context of portfolio optimization, the conditions that allow for the reduction are implied by the well-known asymptotic elasticity conditions on the utility function; see Sect. 4.3.5.

4.1 A Substitute for Compactness in $\mathcal{N}$

A classical argument for establishing the existence of solutions to an optimization problem is to prove that the lower level-sets of the objective are compact; see Lemma A.22. We saw in Theorem 1.81 that, in convex stochastic optimization, the compactness assumption can be relaxed to almost sure boundedness of the level-sets of the associated normal integrand. Indeed, Komlós' theorem in Theorem B.25 then gives, for every minimizing sequence, a sequence of convex combinations that converge almost surely. By convexity, such a sequence achieves objective values at least as good as the original sequence while Fatou's lemma guarantees that the limit of the sequence achieves the optimum value.

The present chapter gives a far reaching extension of Theorem 1.81 that covers both classical topological conditions as well as various no-arbitrage conditions used in financial mathematics. The current section derives one of the key ingredients used in the proof. Theorem 4.3 below, on almost sure boundedness of adapted sequences of random variables can be seen as a stochastic version of the classical boundedness criterion for convex sets in $\mathbb{R}^n$; see Theorem A.27. Combined with Komlós' theorem and the existence results on dynamic programming from Chap. 2 it will yield the existence of solutions as well as the closedness of the value function $\bar{\varphi}$ of problem (P).

A sequence $(x^\nu)_{\nu\in\mathbb{N}}$ of random vectors is said to be *almost surely bounded* if

$$\sup_\nu |x^\nu| < \infty \quad a.s.,$$

where the supremum is taken scenariowise. Recall that, by the Bolzano-Weierstrass theorem, a bounded sequence in $\mathbb{R}^n$ contains a converging subsequence. The following extends this to sequences of random vectors.

Lemma 4.1 *Given an almost surely bounded sequence $(\eta^\nu)_{\nu\in\mathbb{N}}$ in $L^0(\mathbb{R}^d)$, there exists $\eta \in L^0(\mathbb{R}^d)$ and $\mathcal{F}$-measurable $\mathbb{N}$-valued functions μ^ν such that $\mu^\nu \to \infty$ and $\eta^{\mu^\nu} \to \eta$ almost surely as $\nu \to \infty$.*

Proof We will denote the ith component of the random vector η^ν by η_i^ν. Let $\bar{\eta}_1 := \limsup_\nu \eta_1^\nu$ and define a random $\mathbb{N}$-valued sequence $(\mu_1^\nu)_{\nu=1}^\infty$ recursively by $\mu_1^0 := 0$ and

$$\mu_1^{\nu+1} := \inf\{\nu' > \nu \mid |\eta_1^{\nu'} - \bar{\eta}_1| \le 1/\nu\}.$$

We have $\mu_1^\nu \to \infty$ and $\eta_1^{\mu_1^\nu} \to \bar{\eta}_1$. Repeating the argument recursively for components $i = 2, \dots, d$ of the sequence $(\eta^{\mu_1^\nu})_{\nu=1}^\infty$, we find an $\bar{\eta} \in L^0(\mathbb{R}^d)$ and a sequence $(\mu_d^\nu)_{\nu=1}^\infty$ of $\mathbb{N}$-valued measurable functions such that $\mu_d^\nu \to \infty$ and $\eta^{\mu_d^\nu} \to \bar{\eta}$ almost surely. □

Lemma 4.1 can be extended to adapted processes as follows.

Lemma 4.2 *Given an almost surely bounded sequence $(x^\nu)_{\nu\in\mathbb{N}}$ in $\mathcal{N}$, there exists an $x \in \mathcal{N}$ and $\mathcal{F}_T$-measurable $\mathbb{N}$-valued functions μ^ν such that $\mu^\nu \to \infty$ and $x^{\mu^\nu} \to x$ almost surely as $\nu \to \infty$.*

Proof Applying Lemma 4.1 to the sequence $(x_0^\nu)_{\nu\in\mathbb{N}}$ gives a sequence of $\mathcal{F}_0$-measurable $\mathbb{N}$-valued functions μ_0^ν such that $\mu_0^\nu \to \infty$ and $x_0^{\mu_0^\nu} \to x_0$ for an $x_0 \in L^0(\Omega, \mathcal{F}_0, P; \mathbb{R}^{n_0})$. Applying Lemma 4.1 next to $(x_1^{\mu_0^\nu})_{\nu\in\mathbb{N}}$ we get an $\mathcal{F}_1$-measurable subsequence μ_1^ν of μ_0^ν such that $\mu_1^\nu \to \infty$ and $x_1^{\mu_1^\nu} \to x_1$ for an $x_1 \in L^0(\Omega, \mathcal{F}_1, P; \mathbb{R}^{n_1})$. Since $x_0^{\mu_0^\nu} \to x_0$ we also have $x_0^{\mu_1^\nu} \to x_0$. Extracting further subsequence similarly for $t = 2, \dots, T$ we arrive at the conclusion. □

Recall that a closed convex set in $\mathbb{R}^n$ is bounded if and only if its recession cone only contains the origin; see Theorem A.27. The following theorem extends this to adapted selections of a closed convex random set C. We will denote the adapted selections of a set-valued mapping $C : \Omega \rightrightarrows \mathbb{R}^n$ by

$$\mathcal{N}(C) := \{x \in \mathcal{N} \mid x \in C \text{ a.s.}\}.$$

Recall that, by Theorem 1.38, the recession cone mapping C^∞ of C, defined scenariowise as $C^\infty(\omega) := C(\omega)^\infty$, is measurable.

Theorem 4.3 *Let $C : \Omega \rightrightarrows \mathbb{R}^n$ be an $\mathcal{F}$-measurable closed convex-valued mapping. Every sequence in $\mathcal{N}(C)$ is almost surely bounded if and only if $\mathcal{N}(C^\infty) = \{0\}$.*

Proof If $\mathcal{N}(C^\infty) \neq \{0\}$, then $\mathcal{N}(C)$ contains a half-line and thus an unbounded sequence. Assume now that $\mathcal{N}(C^\infty) = \{0\}$ and let (x^ν) be a sequence in $\mathcal{N}(C)$.

Translating with an adapted process, if necessary, we may assume that $0 \in C$ almost surely. Indeed, the translation does not affect any of the conditions of the statement.

We will proceed by induction. Assume that the claim holds for any $(T-1)$-period model. Assume next that $\rho := \sup |x_0^\nu| < \infty$ almost surely and let

$$\mathcal{N}_1 := \{(x_1, \dots, x_T) \mid x_t \in L^0(\mathcal{F}_t)\}$$
$$C_1(\omega) : = \{(x_1, \dots, x_T) \mid \exists x_0 \in \rho(\omega)\mathbb{B} : \; (x_0, \dots, x_T) \in C(\omega)\}.$$

The set $C_1(\omega)$ is the projection of $(\rho(\omega)\mathbb{B} \times \mathbb{R}^{n_1} \times \cdots \times \mathbb{R}^{n_T}) \cap C(\omega)$, where the latter is closed convex-valued and measurable, by Theorem 1.7. Thus, by Corollary 1.41, C_1 is a measurable closed convex-valued mapping with

$$C_1^\infty(\omega) = \{(x_1, \dots x_T) \mid (0, x_1, \dots, x_T) \in C^\infty(\omega)\}.$$

Hence, by the induction hypothesis, $(x_1^\nu, \dots, x_T^\nu)$ is bounded since the condition $\mathcal{N}(C^\infty) = \{0\}$ implies

$$\{(x_1, \dots, x_T) \in \mathcal{N}_1 \mid (x_1, \dots, x_T) \in C_1^\infty \text{ a.s.}\} = \{0\}.$$

Assume now that the set $A(\omega) := \{\omega \mid \sup_\nu |x_0^\nu(\omega)| = \infty\}$ has positive probability. Let $\alpha^\nu := 1_A/(|x_0^\nu| \vee 1)$ and $\bar{x}^\nu := \alpha^\nu x^\nu$. The random variables

$$\nu^\mu(\omega) := \inf\{\nu \mid \alpha^\nu(\omega) \leq 1/\mu\}.$$

are $\mathcal{F}_0$-measurable, $\mathbb{N}$-valued and such that $\alpha^{\nu^\mu} \searrow 0$ almost surely as $\mu \to \infty$. We have $\bar{x}^{\nu^\mu} \in \mathcal{N}$, $\bar{x}^{\nu^\mu} \in \alpha^{\nu^\mu} C$ and $|\bar{x}_0^{\nu^\mu}| \leq 1$. Since $\alpha^{\nu^\mu} \leq 1$, $\alpha^{\nu^\mu} C \subset C$ by convexity. The argument in the previous paragraph shows that $(\bar{x}^{\nu^\mu})_{\mu \in \mathbb{N}}$ is almost surely bounded and thus, by Lemma 4.2, there is a further subsequence, still denoted by (ν^μ), such that $\bar{x}^{\nu^\mu} \to \bar{x} \in \mathcal{N}$ almost surely. By Theorem A.25, $\bar{x} \in C^\infty$, so $\bar{x} = 0$ by assumption. This is a contradiction, since $|\bar{x}_0| = 1$ on A by construction.

To complete the induction argument, it suffices to prove the claim for $T = 0$. The argument is the same as in the previous paragraph except that we do not need to refer to the earlier paragraph for almost sure boundedness of $(\bar{x}^{\nu^\mu})_{\mu \in \mathbb{N}}$. □

Corollary 4.4 *If $C : \Omega \rightrightarrows \mathbb{R}^n$ is an $\mathcal{F}$-measurable closed convex-valued mapping such that $\mathcal{N}(C^\infty) = \{0\}$, then $\mathcal{N}(C)$ is bounded in probability.*

Proof If $\mathcal{N}(C)$ is not bounded in probability, there exists an $\epsilon > 0$ and, for every $\nu \in \mathbb{N}$, an $x^\nu \in \mathcal{N}(C)$ such that $P(\{|x^\nu| \geq \nu\}) \geq \epsilon$. This is impossible since $(x^\nu)_{\nu \in \mathbb{N}}$ is almost surely bounded, by Theorem 4.3. □

4.2 Closedness of the Value Function

This section gives sufficient conditions for the closedness of the optimum value function $\bar{\varphi}$ from Sect. 3.3 and for the existence of solutions to (P). We follow the strategy of Sect. 2.2 and first assume in Sect. 4.2.1 that the normal integrand f is lower bounded in the sense that there exists an $m \in L^1$ such that $f \geq -m$ almost surely everywhere. Section 4.2.2 gives more general conditions without lower boundedness but under which problem (P) problem is equivalent to another problem that satisfies the assumptions of Sect. 4.2.1. The sufficient conditions for the closedness of $\bar{\varphi}$ and for the existence of solutions to (P) given below are analogous to the sufficient conditions for the existence of solutions to the Bellman equations obtained in Sect. 2.2.

4.2.1 Lower Bounded Objectives

This section establishes the closedness of the optimum value function $\bar{\varphi}$ of (P) under the following assumption.

Assumption 4.5 $\operatorname{dom}\bar{\varphi} \neq \emptyset$, *$f$ is lower bounded and the set*

$$\mathcal{L} := \{x \in \mathcal{N} \mid f^\infty(x, 0) \leq 0 \; a.s.\}$$

is linear.

The condition $\operatorname{dom}\bar{\varphi} \neq \emptyset$ in Assumption 4.5 holds, in particular, if (P) is feasible. It also holds under the condition $\operatorname{dom} Ef \cap (\mathcal{X} \times \mathcal{U}) \neq \emptyset$ that was used in the main results of Chap. 3. Since the set $\mathcal{L}$ is a convex cone, the linearity condition in Assumption 4.5 simply means that $\mathcal{L} = -\mathcal{L}$. This clearly holds if the sets $\{x \in \mathbb{R}^n \mid f^\infty(x, 0, \omega) \leq 0\}$ are linear for almost every ω. This happens, in particular, if the functions $f(\cdot, 0, \omega)$ are inf-compact since then, $\{x \in \mathbb{R}^n \mid f^\infty(x, 0, \omega) \leq 0\} = \{0\}$, by Theorems A.17 and A.27. The linearity condition may even hold when the sets $\{x \in \mathbb{R}^n \mid f^\infty(x, 0, \omega) \leq 0\}$ are nonlinear. A notable example is the "no-arbitrage" condition in classical models of financial mathematics; see Sect. 4.3.5 below. Assumption 4.5 makes good sense also in more realistic financial models with nonlinear illiquidity effects and constraints.

The following lemma shows that Assumption 4.5 is essentially the same as Assumption 2.64 that was used in the context of dynamic programming in Chap. 2.

Lemma 4.6 *Let $\bar{u} \in \mathcal{U}$ and $h(x, \omega) := f(x, \bar{u}(\omega), \omega)$. Then*

$$h^\infty(x, \omega) = f^\infty(x, 0, \omega).$$

for every ω such that $h(\cdot, \omega)$ is proper.

Proof Assume that $h(\cdot,\omega)$ is proper and let $\bar{x} \in \operatorname{dom} h(\cdot,\omega)$. A lsc convex function is proper as soon as it takes a finite value somewhere. Thus, $f(\cdot,\cdot,\omega)$ is proper and, by Theorem A.16,

$$\begin{aligned} h^\infty(x,\omega) &= \sup_{\lambda>0} \frac{h(\bar{x}+\lambda x,\omega) - h(\bar{x},\omega)}{\lambda} \\ &= \sup_{\lambda>0} \frac{f(\bar{x}+\lambda x,\bar{u}(\omega),\omega) - f(\bar{x},\bar{u}(\omega),\omega)}{\lambda} \\ &= f^\infty(x,0,\omega), \end{aligned}$$

which completes the proof. □

Recall that the absence of a duality gap is equivalent to the closedness of the optimum value function $\bar{\varphi}$ of problem (P) with respect to the pairing of $\mathcal{U} \times \mathcal{X}$ with $\mathcal{Y} \times \mathcal{V}$; see Theorem 3.17.

Lemma 4.7 *If $\bar{\varphi}$ is proper and L^0-lsc, then it is closed.*

Proof By Lemma 3.4, the L^0-topology is weaker than the Mackey-topology $\tau(\mathcal{X} \times \mathcal{U}, \mathcal{V} \times \mathcal{Y})$ on $\mathcal{X} \times \mathcal{U}$. Thus, L^0-lower semicontinuity of $\bar{\varphi}$ implies its lower semicontinuity with respect to the Mackey-topology. The claim now follows from the fact that a convex function on a locally convex vector space is weakly closed if and only if it is Mackey closed; see Theorem A.54. □

The following is the main result of this section.

Theorem 4.8 *Under Assumption 4.5, the optimum value function*

$$\bar{\varphi}(z,u) = \inf_{x\in L^0}\{Ef(x,u) \mid x - z \in \mathcal{N}\}$$

is closed and proper, its recession function is given by

$$\bar{\varphi}^\infty(z,u) = \inf_{x\in L^0}\{Ef^\infty(x,u) \mid x - z \in \mathcal{N}\}$$

and both infimums are attained for every $(z,u) \in \mathcal{X} \times \mathcal{U}$.

Proof Adding an integrable random variable to f, if necessary, we may assume that $f \geq 0$. By Lemma 4.7, it suffices to prove L^0-lower semicontinuity. By Theorem B.1, L^0 is a metric space, so L^0-lower semicontinuity is equivalent to sequential L^0-lower semicontinuity which means that for any $\gamma \in \mathbb{R}$ and for any sequence $(z^\nu, u^\nu)_{\nu\in\mathbb{N}}$ such that

$$\bar{\varphi}(z^\nu, u^\nu) \leq \gamma$$

and $(z^\nu, u^\nu) \to (z, u)$ in L^0, we have $\bar{\varphi}(z, u) \le \gamma$. Note that

$$\begin{aligned}\bar{\varphi}(z, u) &= \inf_{x \in L^0} \{Ef(x, u) \mid x - z \in \mathcal{N}\} \\ &= \inf_{x \in L^0} \{Ef(x + z, u) \mid x \in \mathcal{N}\}.\end{aligned}$$

To prove the closedness of $\bar{\varphi}$ and the attainment of the infimum in its definition, it suffices to establish the existence of an $x \in \mathcal{N}$ such that $Ef(x + z, u) \le \gamma$.

For any $(z, u) \in \mathcal{X} \times \mathcal{U}$,

$$\begin{aligned}\bar{\varphi}(z, u) &= \inf_{x \in \mathcal{N}} Ef(x + z, u) \\ &= \inf_{x \in \mathcal{N}} Eh_{z,u}(x),\end{aligned}$$

where $h_{z,u}(x, \omega) := f(x + z(\omega), u(\omega), \omega)$. Given $(z, u) \in \operatorname{dom} \bar{\varphi}$, Lemma 4.6 and Theorem A.17 give $h^\infty_{z,u}(x, \omega) = f^\infty(x, 0, \omega)$ for almost every ω. Thus, by Theorem 2.65, Assumption 4.5 implies that the infimum in $\bar{\varphi}(z, u)$ is attained by an $x \in \mathcal{N}$ with

$$x_t(\omega) \perp N_t(\omega) := \{x_t \in \mathbb{R}^{n_t} \mid h^\infty_t(x^t, \omega) \le 0,\ x^{t-1} = 0\},$$

where h_t is the solution of the Bellman equations for $h_{z,u}$. By Corollary 2.63, h^∞_t is the solution of the Bellman equations for $h^\infty_{z,u}(x, \omega) = f^\infty(x, 0, \omega)$. Thus, N_t is independent of z and u.

Let $\gamma \in \mathbb{R}$ and $(z^\nu, u^\nu)_{\nu \in \mathbb{N}}$ a sequence such that

$$\bar{\varphi}(z^\nu, u^\nu) \le \gamma$$

and $(z^\nu, u^\nu) \to (z, u)$ in L^0. Passing to a subsequence if necessary, we may assume that $(z^\nu, u^\nu) \to (z, u)$ almost surely. As observed above, there exists, for every ν, an $x^\nu \in \mathcal{N}$ such that $x^\nu_t \perp N_t$ and

$$Ef(x^\nu + z^\nu, u^\nu) \le \gamma.$$

Since f is lower bounded, the sequence $f(x^\nu + z^\nu, u^\nu)$ must be bounded in L^1 so Lemma B.25 gives a sequence of convex combinations

$$\phi^\nu := \sum_{\mu=\nu}^{\infty} \alpha^{\nu,\mu} f(x^\mu + z^\mu, u^\mu)$$

that converges almost surely to a real-valued random variable. In particular, the function $\phi(\omega) := \sup_\nu \phi^\nu(\omega)$ is almost surely finite. Defining

$$(\bar{x}^\nu, \bar{z}^\nu, \bar{u}^\nu) = \sum_{\mu=\nu}^{\infty} \alpha^{\nu,\mu}(x^\mu, z^\mu, u^\mu),$$

convexity of f gives

$$f(\bar{x}^\nu + \bar{z}^\nu, \bar{u}^\nu) \leq \phi^\nu \leq \phi \quad a.s.$$

and $Ef(\bar{x}^\nu + \bar{z}^\nu, \bar{u}^\nu) \leq \gamma$. Moreover, we still have $\bar{x}_t^\nu \in N_t^\perp$ and $(\bar{z}^\nu, \bar{u}^\nu) \to (z, u)$ almost surely.

We show next that there is an almost surely bounded subsequence of $(\bar{x}^\nu, \bar{z}^\nu, \bar{u}^\nu)_{\nu \in \mathbb{N}}$. Since $(\bar{z}^\nu, \bar{u}^\nu) \to (z, u)$ almost surely, the measurable function $\rho(\omega) := \sup_\nu |(\bar{z}^\nu(\omega), \bar{u}^\nu(\omega))|$ is almost surely finite. Each $(\bar{x}^\nu, \bar{z}^\nu, \bar{u}^\nu)$ then belongs to the set

$$\mathcal{C} = \{(x, z, u) \in \mathcal{N} \times L^0 \times L^0 \mid (x, z, u) \in C \ a.s.\},$$

where

$$C(\omega) := \{(x, z, u) \in \mathbb{R}^n \times \mathbb{R}^n \times \mathbb{R}^m \mid x_t \in N_t^\perp(\omega), \\ |(z, u)| \leq \rho(\omega),\ f(x + z, u, \omega) \leq \phi(\omega)\}.$$

By Theorem 4.3, the sequence $(\bar{x}^\nu, \bar{z}^\nu, \bar{u}^\nu)_{\nu \in \mathbb{N}}$ is almost surely bounded if

$$\{(x, z, u) \in \mathcal{N} \times L^0 \times L^0 \mid (x, z, u) \in C^\infty \ a.s.\} = \{(0, 0, 0)\}. \tag{4.1}$$

By Theorem A.17,

$$C^\infty(\omega) = \{(x, 0, 0) \in \mathbb{R}^n \times \mathbb{R}^n \times \mathbb{R}^m \mid x_t \in N_t^\perp(\omega),\ f^\infty(x, 0, \omega) \leq 0\}.$$

If $x \in \mathcal{N}$ is such that $f^\infty(x(\omega), 0, \omega) \leq 0$ then, by the last part of Theorem 2.62, we have $x_0 \in N_0$. Thus, if $(x, z, u) \in \mathcal{N} \times L^0 \times L^0$ is such that $(x, z, u) \in C^\infty$ almost surely, then $x_0 \in N_0 \cap N_0^\perp = \{0\}$. By the last part of Theorem 2.62 again, $x_1 \in N_1$ almost surely so $x_1 = 0$. Repeating the argument for $t = 2, \ldots, T$, we find that $x_t = 0$ almost surely for all $t = 0, \ldots, T$ and thus, (4.1) holds so $(\bar{x}^\nu, \bar{z}^\nu, \bar{u}^\nu)_{\nu \in \mathbb{N}}$ is almost surely bounded by Theorem 4.3.

By Theorem B.25, there is a sequence $(\hat{x}^\nu, \hat{z}^\nu, \hat{u}^\nu)_{\nu\in\mathbb{N}}$ of convex combinations of $(\bar{x}^\nu, \bar{z}^\nu, \bar{u}^\nu)_{\nu\in\mathbb{N}}$ that converges almost surely to a point $(x, \hat{z}, \hat{u})$, where necessarily, $(\hat{z}, \hat{u}) = (z, u)$ since $(\bar{z}^\nu, \bar{u}^\nu) \to (z, u)$ almost surely. By convexity, $Ef(\hat{x}^\nu + \hat{z}^\nu, \hat{u}^\nu) \le \gamma$ while, by Fatou's lemma,

$$Ef(x+z,u) \le \liminf_{\nu\to\infty} Ef(\hat{x}^\nu + \hat{z}^\nu, \hat{u}^\nu)$$

and thus, $\bar{\varphi}(z, u) \le \gamma$. This proves the lower semicontinuity of $\bar{\varphi}$ and the attainment of the infimum in its definition.

As to the recession function, let $(\bar{z}, \bar{u}) \in \operatorname{dom}\bar{\varphi}$ and $\bar{x} \in \mathcal{N}$ be such that $\bar{\varphi}(\bar{z}, \bar{u}) = Ef(\bar{x} + \bar{z}, \bar{u})$. We have

$$\bar{\varphi}^\infty(z,u) = \sup_{\lambda>0} \frac{\bar{\varphi}(\bar{z}+\lambda z, \bar{u}+\lambda u) - \bar{\varphi}(\bar{z},\bar{u})}{\lambda} = \sup_{\lambda>0}\inf_{x\in\mathcal{N}} Ef_\lambda(x+z,u),$$

where

$$f_\lambda(x,u,\omega) := \frac{f(\bar{x}(\omega)+\bar{z}(\omega)+\lambda x, \bar{u}(\omega)+\lambda u, \omega) - f(\bar{x}(\omega)+\bar{z}(\omega), \bar{u}(\omega), \omega)}{\lambda}.$$

By definition of the recession function, $f_\lambda \le f^\infty$ for all $\lambda > 0$, so

$$\bar{\varphi}^\infty(z,u) \le \inf_{x\in\mathcal{N}} Ef^\infty(x+z,u).$$

To prove the converse, let $\gamma \ge \sup_{\lambda>0}\inf_{x\in\mathcal{N}} Ef_\lambda(x,u)$. Similarly to the beginning of the proof, defining

$$h_\lambda(x,\omega) := \frac{f(\bar{x}(\omega)+\lambda x+\bar{z}(\omega)+\lambda z, \bar{u}(\omega)+\lambda u) - f(\bar{x}(\omega)+\bar{z}(\omega), \bar{u}(\omega))}{\lambda},$$

we have that $h_\lambda^\infty(x,\omega) = f^\infty(x, 0, \omega)$. As in the first part of the proof, we can find, for every positive integer λ, an $x^\lambda \in \mathcal{N}$ with $Ef_\lambda(x^\lambda + z, u) \le \gamma$ and $x_t^\lambda \in N_t^\perp$ for every t. The functions f_λ are non-decreasing in λ, so $Ef_1(x^\lambda + z, u) \le \gamma$ and we may proceed as in the first part of the proof to obtain a sequence of convex combinations $\tilde{x}^\lambda = \operatorname{co}\{x^{\lambda'} \mid \lambda' \ge \lambda\}$ and $\tilde{x}$ such that $\tilde{x}^\lambda \to \tilde{x}$ almost surely. By convexity and the fact that f_λ is pointwise nondecreasing in λ, we still have $Ef_\lambda(\tilde{x}^\lambda + z, u) \le \gamma$. By Corollary A.26,

$$f^\infty(\tilde{x}+z,u) \le \liminf_{\lambda\nearrow\infty} f_\lambda(\tilde{x}^\lambda + z, u)$$

almost surely, so Fatou's lemma gives

$$\begin{aligned} Ef^\infty(\tilde{x}+z,u) &\le E[\liminf_{\lambda\nearrow\infty} f_\lambda(\tilde{x}^\lambda+z,u)] \\ &\le \liminf_{\lambda\nearrow\infty} Ef_\lambda(\tilde{x}^\lambda+z,u) \\ &\le \gamma. \end{aligned}$$

This proves the claimed expression for $\bar{\varphi}^\infty$ as well as the attainment of the infimum.

□

4.2.2 L-Bounded Objectives

This section extends Theorem 4.8 to integrands that need not be lower bounded. More precisely, we relax Assumption 4.5 into Assumption 4.9 below which is analogous to Assumption 2.72. The assumption may seem somewhat abstract, but more concrete sufficient conditions will be given at the end of this section.

Assumption 4.9 $\operatorname{dom}\bar{\varphi} \neq \emptyset$ *and there exists* $(p, y) \in \mathcal{X}_a^\perp \times \mathcal{Y}$ *such that*

1. $k(x, u, \omega) := f(x, u, \omega) - x \cdot p(\omega) - u \cdot y(\omega)$ *is lower bounded,*
2. $\{x \in \mathcal{N} \mid k^\infty(x, 0) \le 0\ a.s.\}$ *is linear,*
3. $Ek(x, u) = E[f(x, u) - z \cdot p - u \cdot y]$ *for all* $(x, z, u) \in L^0 \times \mathcal{X} \times \mathcal{U}$ *with* $x - z \in \mathcal{N}$.

The main result of this section, Theorem 4.11 below, is based on applying Theorem 4.8 to the normal integrand k in Assumption 4.9. Note that Assumption 4.5 implies Assumption 4.9 with $(p, y) = (0, 0)$. The first condition in Assumption 4.9 means that $f^*(p(\omega), y(\omega), \omega) \le m(\omega)$, where f^* is the conjugate integrand of f. Thus, if $\operatorname{dom} Ef \cap (\mathcal{X} \times \mathcal{U}) \neq \emptyset$, then, by Corollary 3.21, $(p, y) \in \mathcal{X}_a^\perp \times \mathcal{Y}$ satisfies Assumption 4.9.1 if and only if (p, y) is dual feasible. When Ω is finite, we have $\mathcal{X}_a = \mathcal{N}$ and Assumption 4.9.3 holds automatically. More general sufficient conditions for Assumption 4.9 will be given after Theorem 4.11 below. Note also that Assumption 4.9 implies the assumptions of Theorems 2.71 and 2.70 when $h(x, \omega) = f(x, \bar{u}(\omega), \omega)$. Condition 3 in Assumption 4.9 can be written as

3'. $Ef(x + z, u) = E[f(x + z, u) - x \cdot p]$ for all $(x, z, u) \in \mathcal{N} \times \mathcal{X} \times \mathcal{U}$.

Choosing $z = 0$ and $u = \bar{u}$, condition 3 gives Assumption 2.72.3.

It is clear that if f satisfies Assumption 4.9, then its recession function f^∞ satisfies parts 1 and 2 of the same assumption. Lemma 4.10 below says that f^∞ satisfies condition 3 as well.

Lemma 4.10 *Under Assumption 4.9,*

$$Ek^{\infty}(x,u) = E[f^{\infty}(x,u) - z \cdot p - u \cdot y]$$

for all $(x, z, u) \in L^0 \times \mathcal{X} \times \mathcal{U}$ *with* $x - z \in \mathcal{N}$.

Proof By Theorem 1.55, Ek is lsc on $L^0 \times \mathcal{X} \times \mathcal{U}$ so, by Theorem 1.58, $(Ek)^{\infty} = Ek^{\infty}$. By Assumption 4.9.3, Ef is then lsc on the linear subspace

$$\mathcal{L} := \{(x, z, u) \in L^0 \times \mathcal{X} \times \mathcal{U} \mid x - z \in \mathcal{N}\}$$

so, by Theorem 1.58, $(Ef)^{\infty} = Ef^{\infty}$ on $\mathcal{L}$. Thus, by Theorem A.17, Lemma B.3 and Assumption 4.9.3,

$$\begin{aligned} Ek^{\infty}(x,u) &= (Ek)^{\infty}(x,u) \\ &= (Ef)^{\infty}(x,u) - \langle z, p\rangle - \langle u, y\rangle \\ &= Ef^{\infty}(x,u) - \langle z, p\rangle - \langle u, y\rangle \\ &= E[f^{\infty}(x,u) - x \cdot p - u \cdot y] \end{aligned}$$

on $\mathcal{L}$, which completes the proof. □

The following is the main theorem of this chapter. It extends Theorem 4.8 beyond lower bounded integrands.

Theorem 4.11 *Under Assumption 4.9, the function*

$$\bar{\varphi}(z,u) = \inf_{x \in L^0} \{Ef(x,u) \mid x - z \in \mathcal{N}\}$$

is closed and proper, its recession function can be expressed as

$$\bar{\varphi}^{\infty}(z,u) = \inf_{x \in L^0} E\{f^{\infty}(x,u) \mid x - z \in \mathcal{N}\}$$

and both infimums are attained for every $(z, u) \in \mathcal{X} \times \mathcal{U}$.

Proof We apply Theorem 4.8 to the normal integrand k and denote

$$\hat{\varphi}(z,u) := \inf_{x \in L^0} \{Ek(x,u) \mid x - z \in \mathcal{N}\}.$$

By Assumption 4.9.3, $\operatorname{dom} \hat{\varphi} = \operatorname{dom} \bar{\varphi}$. By Theorem 4.8, Assumption 4.9 implies that $\hat{\varphi}$ is closed and proper,

$$\hat{\varphi}^{\infty}(z,u) = \inf_{x \in L^0} \{Ek^{\infty}(x,u) \mid x - z \in \mathcal{N}\}$$

and that the infimums above are attained. By Lemma B.3 and Assumption 4.9.3,

$$
\begin{aligned}
\bar{\varphi}(z,u) &= \inf_{x\in L^0}\{Ef(x,u) - \langle z,p\rangle - \langle u,y\rangle \mid x-z\in\mathcal{N}\} + \langle z,p\rangle + \langle u,y\rangle\\
&= \inf_{x\in L^0}\{Ek(x,u) \mid x-z\in\mathcal{N}\} + \langle z,p\rangle + \langle u,y\rangle\\
&= \hat{\varphi}(z,u) + \langle z,p\rangle + \langle u,y\rangle,
\end{aligned}
$$

so the closedness of $\bar{\varphi}$ follows from that of $\hat{\varphi}$ and the attainment of the infimum in $\bar{\varphi}$ follows from that in $\hat{\varphi}$ and from Assumption 4.9.3. Combining the above with Theorem A.17,

$$
\begin{aligned}
\bar{\varphi}^\infty(z,u) &= \hat{\varphi}^\infty(z,u) + \langle z,p\rangle + \langle u,y\rangle\\
&= \inf_{x\in L^0}\{Ek^\infty(x,u) \mid x-z\in\mathcal{N}\} + \langle z,p\rangle + \langle u,y\rangle\\
&= \inf_{x\in L^0}\{Ef^\infty(x,u) \mid x-z\in\mathcal{N}\},
\end{aligned}
$$

where the last equality holds by that of Lemma 4.10. □

The rest of this section gives various sufficient conditions for Assumption 4.9. The following is an analogue of Lemma 2.74.

Lemma 4.12 *Assumption 4.9 holds if* $\operatorname{dom}\bar{\varphi} \neq \emptyset$,

1. $\{x \in \mathcal{N} \mid f^\infty(x,0) \le 0 \text{ a.s.}\}$ *is linear,*
2. *there exist* $p \in \mathcal{X}_a^\perp$, $m \in L^1$ *and* $\epsilon > 0$ *such that, for every* $\lambda \in [1-\epsilon, 1+\epsilon]$, *there exists* $y \in \mathcal{Y}$ *such that*

$$
f(x,u,\omega) \ge \lambda x\cdot p(\omega) + u\cdot y(\omega) - m(\omega) \quad (x,u)\in\mathbb{R}^n\times\mathbb{R}^m
$$

for almost every $\omega \in \Omega$.

Condition 2 can be written equivalently as

2'. *there exist* $p \in \mathcal{X}_a^\perp$ *and* $\epsilon > 0$ *such that*

$$
\inf_{y\in\mathcal{Y}} Ef^*(\lambda p, y) < \infty \quad \forall \lambda \in [1-\epsilon, 1+\epsilon].
$$

If $\operatorname{dom} Ef \cap (\mathcal{X}\times\mathcal{U}) \neq \emptyset$, *this means that*

2". *there exist* $p \in \mathcal{X}_a^\perp$ *and* $\epsilon > 0$ *such that for all* $\lambda \in [1-\epsilon, 1+\epsilon]$ *there exists* $y \in \mathcal{Y}$ *such that* $(\lambda p, y)$ *is feasible in the dual problem* (*D*).

Proof Condition 2 gives the existence of $y, y' \in \mathcal{Y}$ such that

$$
\begin{aligned}
f(x,u,\omega) &\geq x \cdot p(\omega) + u \cdot y(\omega) - m(\omega),\\
f(x,u,\omega) &\geq (1+\epsilon)x \cdot p(\omega) + u \cdot y'(\omega) - m(\omega).
\end{aligned} \tag{4.2}
$$

We will show that Assumption 4.9 holds with the above (p, y). Assumption 4.9.1 is clear. The inequalities (4.2) can be written as

$$
\begin{aligned}
f(x,u,\omega) &\geq x \cdot p(\omega) + u \cdot y(\omega) - m(\omega),\\
k(x,u,\omega) &\geq \epsilon x \cdot p(\omega) + u \cdot (y'(\omega) - y(\omega)) - m(\omega).
\end{aligned} \tag{4.3}
$$

By Remark A.20, these imply

$$
\begin{aligned}
f^\infty(x,0,\omega) &\geq x \cdot p(\omega),\\
k^\infty(x,0,\omega) &\geq \epsilon x \cdot p(\omega).
\end{aligned}
$$

Given $x \in \mathcal{N}$, if either $f^\infty(x,0) \leq 0$ or $k^\infty(x,0) \leq 0$ almost surely, then $x \cdot p \leq 0$ almost surely, so Lemma 2.67 implies $x \cdot p = 0$ almost surely. Thus

$$
\{x \in \mathcal{N} \mid k^\infty(x,0) \leq 0\, a.s.\} = \{x \in \mathcal{N} \mid f^\infty(x,0) \leq 0\, a.s.\},
$$

so condition 1 implies Assumption 4.9.2.

As to Assumption 4.9.3, let $(x, z, u) \in L^0 \times \mathcal{X} \times \mathcal{U}$ with $x - z \in \mathcal{N}$. We may assume that either $Ek(x,u) < \infty$ or $Ef(x,u) < \infty$, since otherwise the equality in Assumption 4.9.3 holds trivially. The inequalities in (4.3) then imply $E[x \cdot p] < \infty$ so, by Lemma 2.67,

$$
Ek(x,u) = E[f(x,u) - (x-z) \cdot p + z \cdot p - u \cdot y] = E[f(x,u) - z \cdot p - u \cdot y],
$$

so Assumption 4.9.3 holds.

The lower bound in condition 2 means that $f^*(\lambda p(\omega), y(\omega), \omega)) \leq m(\omega)$ so 2 implies 2'. Assuming 2', there exist, for every $\lambda \in [1-\epsilon, 1+\epsilon]$, a $y_\lambda \in \mathcal{Y}$ and an $m_\lambda \in L^1$ such that $f^*(\lambda p(\omega), y_\lambda(\omega), \omega)) \leq m_\lambda(\omega)$. By convexity,

$$
f^*(\lambda p(\omega), y_\lambda(\omega), \omega)) \leq \max\{m_{1-\epsilon}(\omega), m_{1+\epsilon}(\omega)\}
$$

so 2 holds with $m := \max\{m_{1-\epsilon}(\omega), m_{1+\epsilon}(\omega)\}$. The last claim now follows from Corollary 3.21. □

By Lemma 4.6, assumptions of Lemma 4.12 imply that the function $h(x,\omega) := f(x, \bar{u}(\omega), \omega)$ satisfies the assumptions of Lemma 2.74.

Remark 4.13 Condition 2 in Lemma 4.12 holds if $\operatorname{dom} Ef \cap (\mathcal{X} \times \mathcal{U}) \neq \emptyset$ and there exist $(p, y) \in \mathcal{X}_a^\perp \times \mathcal{Y}$ and $\epsilon > 0$ such that $\lambda(p, y)$ is dual feasible for every $\lambda \in [1-\epsilon, 1+\epsilon]$. This holds, in particular, if $\operatorname{dom} Ef^* \cap (\mathcal{X}_a^\perp \times \mathcal{Y})$ is a nonempty cone. By Lemma 1.59, this holds if Ef^* is proper on $\mathcal{X}_a^\perp \times \mathcal{Y}$ and f^* is p-homogeneous.

The following gives a sufficient condition for condition 2 in Lemma 4.12 in terms of the reduced dual problem from Theorem 3.38.

Lemma 4.14 (Reduced Dual) *Let π and γ be as in Theorem 3.38 and assume that π is linear,* $\operatorname{dom} Ef \cap (\mathcal{X} \times \mathcal{U}) \neq \emptyset$ *and there exist $y \in \mathcal{Y}$ and $\epsilon > 0$ such that λy is feasible in the reduced dual* (3.10) *for every $\lambda \in [1-\epsilon, 1+\epsilon]$. Then condition 2 in Lemma 4.12 holds. If $\bar{\varphi}$ is closed and proper, then*

$$\bar{\varphi}^\infty(z,u) \geq \langle z, \pi(y)\rangle + \langle u, y\rangle + \epsilon|\langle z, \pi(y)\rangle + \langle u, y\rangle|.$$

Proof By Theorem 3.38, the assumptions here imply $\varphi^*(\pi(y), y) < \infty$ for every $\lambda \in [1-\epsilon, 1+\epsilon]$. Since $\operatorname{dom} Ef \cap (\mathcal{X} \times \mathcal{U}) \neq \emptyset$, Theorem 3.20 implies $\varphi^*(\lambda\pi(y), y) = Ef^*(\lambda\pi(y), \lambda y)$ and $\pi(y) \in \mathcal{X}_a^\perp$, which proves the first claim. Assume now that $\bar{\varphi}$ is closed and proper. By Lemma A.45, $\bar{\varphi}^\infty = \sigma_{\operatorname{dom}\varphi^*}$, so

$$\bar{\varphi}^\infty(u) \geq \langle \lambda\pi(y), z\rangle + \langle u, \lambda y\rangle \quad \forall \lambda \in [1-\epsilon, 1+\epsilon],$$

which gives the lower bound in the statement. □

The following is an analogue of Lemma 2.75.

Lemma 4.15 *Assumption 4.9 holds if* $\operatorname{dom}\bar{\varphi} \neq \emptyset$ *and there exists $(p, y) \in \mathcal{X}_a^\perp \times \mathcal{Y}$ such that*

1. $k(x,u,\omega) := f(x,u,\omega) - x \cdot p(\omega) - u \cdot y(\omega)$ *is lower bounded,*
2. $\{x \in \mathcal{N} \mid k^\infty(x,0) \leq 0\ a.s.\}$ *is linear,*
3. $(x,u) \in \operatorname{dom} Ef$ *for all $(x,z,u) \in L^0 \times \mathcal{X} \times \mathcal{U}$ with $x - z \in \mathcal{N}$ and $(x,u) \in$* $\operatorname{dom} f$ *almost surely.*

Proof It suffices to show that Assumption 4.9.3 holds, so let $(x,z,u) \in L^0 \times \mathcal{X} \times \mathcal{U}$ with $x - z \in \mathcal{N}$. We may assume that either $E[k(x,u)] < \infty$ or $E[f(x,u)] < \infty$ since otherwise the equality in Assumption 4.9.3 holds trivially. Then $(x,u) \in \operatorname{dom} f$ almost surely, so $(x,u) \in \operatorname{dom} Ef$, by condition 3. Condition 1 then gives $E[x \cdot p] < \infty$. By Lemma 2.67,

$$Ek(x,u) = E[f(x,u) - (x-z)\cdot p + z \cdot p - u \cdot y] = E[f(x,u) - z \cdot p - u \cdot y],$$

so Assumption 4.9.3 holds. □

By Lemma 4.6, assumptions of Lemma 4.15 imply that the function $h(x,\omega) := f(x, \bar{u}(\omega), \omega)$ satisfies the assumptions of Lemma 2.75.

4.3 Applications

This section applies the main results of this chapter to prove the existence of primal solutions and the absence of a duality gap in the examples studied in the earlier chapters. In most cases, we apply the Lemma 4.12 to verify Assumption 4.9. In the example on optimized certainty equivalents in Sect. 4.3.5, we establish the sufficient conditions in Lemma 4.15 instead.

4.3.1 Mathematical Programming

Consider again problem (1.8)

$$\begin{aligned} &\text{minimize} && Ef_0(x) \quad \text{over } x \in \mathcal{N}, \\ &\text{subject to} && f_j(x) \le 0 \quad j = 1, \dots, l \ a.s., \\ & && f_j(x) = 0 \quad j = l+1, \dots, m \ a.s. \end{aligned}$$

from Sect. 1.3.1 and recall from Sect. 3.4.1 that this fits the general duality framework with $\bar{u} = 0$ and

$$f(x, u, \omega) = \begin{cases} f_0(x, \omega) & \text{if } x \in \operatorname{dom} H(\cdot, \omega),\ H(x, \omega) + u \in K, \\ +\infty & \text{otherwise,} \end{cases}$$

where $K = \mathbb{R}^l_- \times \{0\}$ and H is the random K-convex function defined by

$$\operatorname{dom} H(\cdot, \omega) = \bigcap_{j=1}^{m} \operatorname{dom} f_j(\cdot, \omega) \quad \text{and} \quad H(x, \omega) = (f_i(x, \omega))_{j=1}^m.$$

It was shown in Sect. 3.4.1 that if $\operatorname{dom} Ef \cap (\mathcal{X} \times \mathcal{U}) \neq \emptyset$, then the dual problem can be written as

$$\begin{aligned} &\text{maximize } E \inf_{x \in \mathbb{R}^n} \{f_0(x) + y \cdot H(x) - x \cdot p\} \text{ over } (p, y) \in \mathcal{X}_a^{\perp} \times \mathcal{Y} \\ &\text{subject to} \qquad y \in K^{\circ} \quad a.s. \end{aligned}$$

We will prove the existence of primal solutions and the absence of a duality gap under the following.

Assumption 4.16 *Problem* (1.8) *is feasible,* $\operatorname{dom} Ef \cap (\mathcal{X} \times \mathcal{U}) \neq \emptyset$,

1. *the set*

$$\{x \in \mathcal{N} \mid f_j^\infty(x) \leq 0 \; j = 0, \ldots l, \; f_j^\infty(x) = 0 \; j = l+1, \ldots, m \; a.s.\}$$

is linear,

2. *there exist* $p \in \mathcal{X}_a^\perp$, $m \in L^1$ *and* $\epsilon > 0$ *such that, for all* $\lambda \in [1-\epsilon, 1+\epsilon]$, *there exists* $y \in \mathcal{Y}(K^\circ)$ *such that*

$$f_0(x,\omega) + y(\omega) \cdot H(x,\omega) \geq \lambda x \cdot p(\omega) - m(\omega) \quad \forall x \in \mathbb{R}^n$$

for almost every $\omega \in \Omega$.

The conditions in Assumption 4.16 are sufficient for Lemma 4.12. One could alternatively use Lemma 4.15 or the general conditions in Assumption 4.9 to establish the existence of primal solutions and the absence of a duality gap. The last condition in Assumption 4.16 holds, in particular, if f_0 is lower bounded as one can then take $p = 0$ and $y = 0$.

Note that condition 1 of Assumption 4.16 is the same as that of Assumption 2.85. Note also that, by the definition of the polar cone K°, condition 2 of Assumption 4.16 implies

$$f_0(x,\omega) \geq f_0(x,\omega) + y(\omega) \cdot H(x,\omega)$$

for every $x \in \mathbb{R}^n$ and for almost every $\omega \in \Omega$ such that $H(x,\omega) \in K$, so condition 2 of Assumption 2.85 holds under that of Assumption 4.16.

Theorem 4.17 *Under Assumption 4.16,* $\inf(1.8) = \sup(3.14)$ *and* (1.8) *has a solution. In this case, a dual feasible* (p, y) *solves* (3.14) *if and only if there exists a primal feasible* x *with*

$$p \in \partial_x[f_0 + y \cdot H](x),$$
$$H(x) \in K, \quad y \in K^\circ, \quad y \cdot H(x) = 0$$

almost surely.

Proof The claims are direct applications of Lemma 4.12, Theorems 4.11 and 3.41. □

In the case of linear programming, the dual formulation of condition 2 in Lemma 4.12 can be written explicitly in terms of problems data.

Example 4.18 (Linear Stochastic Programming) Consider the problem

$$\begin{aligned} &\text{minimize} \quad E[x \cdot c] \quad \text{over } x \in \mathcal{N} \\ &\text{subject to} \quad Ax + b \in K \quad a.s. \end{aligned}$$

and its dual

$$\begin{aligned}&\text{minimize} \quad E[b \cdot y] \quad \text{over} \quad p \in \mathcal{X}_a^\perp,\ y \in \mathcal{Y},\\&\text{subject to} \quad c + A^* y = p,\ y \in K^\circ \quad a.s.\end{aligned}$$

from Example 3.42. If the primal problem is feasible, the set

$$\{x \in \mathcal{N} \mid x \cdot c \leq 0,\ Ax \in K\ a.s.\}$$

is linear and there exist $p \in \mathcal{N}^\perp$ and $\epsilon > 0$ such that the problem

$$\begin{aligned}&\text{minimize} \quad E[y \cdot b] \quad \text{over} \quad y \in \mathcal{Y},\\&\text{subject to} \quad c + A^* y = \lambda p,\ y \in K^\circ \quad a.s.\end{aligned}$$

is feasible for all $\lambda \in [1 - \epsilon, 1 + \epsilon]$, then there is no duality gap and the primal has a solution.

4.3.2 Optimal Stopping

Consider again the relaxed optimal stopping problem (1.10)

$$\begin{aligned}&\text{maximize} \quad E\left[\sum_{t=0}^{T} R_t x_t\right] \quad \text{over } x \in \mathcal{N},\\&\text{subject to} \quad x \geq 0,\ \sum_{t=0}^{T} x_t \leq 1 \quad a.s.\end{aligned}$$

from Sect. 1.3.2. Recall from Sect. 3.4.2 that this fits the general duality framework with $n_t = 1$, $m = 1$,

$$f(x, u, \omega) = \begin{cases} -\sum_{t=0}^T x_t R_t(\omega) & \text{if } x \geq 0 \text{ and } \sum_{t=0}^T x_t + u \leq 0,\\ +\infty & \text{otherwise}\end{cases}$$

and $\bar{u} = -1$, and that the dual problem can be written as

$$\begin{aligned}&\text{minimize} \quad Ey \quad \text{over } (p, y) \in \mathcal{X}_a^\perp \times \mathcal{Y}_+\\&\text{subject to} \quad p_t + R_t \leq y \quad t = 0, \dots, T\ a.s.;\end{aligned}$$

see (3.17).

Theorem 4.19 *If* $\max_t R_t \in \mathcal{Y}$, *then* $\sup(1.9) = \sup(1.10) = \inf(3.17)$, *and both* (1.9) *and* (1.10) *have a solution. In this case, a dual feasible* (p, y) *solves* (3.17) *if and only if there exists a stopping time* $\tau \in \mathcal{T}$ *with* $p_\tau + R_\tau = y$ *almost surely.*

Proof Assumption 4.9 holds with $p = 0$, $y = \max_t R_t$ and $m = 0$. Indeed, condition 3 is clear, condition 2 holds, since $f^\infty = f$ so $f^\infty(x, 0, \omega) = \delta_{\{0\}}(x)$ while, for every $(x, u) \in \operatorname{dom} f(\cdot, \cdot, \omega)$,

$$f(x, u, \omega) = -\sum_{t=0}^{T} x_t R_t(\omega) \geq -\sum_{t=0}^{T} x_t y(\omega) \geq u y(\omega),$$

which is condition 1. Thus, the claims follow from Theorems 4.11 and 3.45. □

4.3.3 Optimal Control

Consider again the optimal control problem (1.11)

$$\begin{aligned} &\text{minimize} && E\left[\sum_{t=0}^{T} L_t(X_t, U_t)\right] \quad \text{over } (X, U) \in \mathcal{N}, \\ &\text{subject to} && \Delta X_t = A_t X_{t-1} + B_t U_{t-1} + u_t \quad t = 1, \dots, T \ a.s. \end{aligned}$$

from Sect. 1.3.3 and recall from Sect. 3.4.3 that, as soon as $W \in \mathcal{U}$, this fits the general duality framework with $x = (X, U)$, $\bar{u} = (W_t)_{t=1}^T$ and

$$f(x, u, \omega) = \sum_{t=0}^{T} L_t(X_t, U_t, \omega) + \sum_{t=1}^{T} \delta_{\{0\}}(\Delta X_t - A_t(\omega)X_{t-1} - B_t(\omega)U_{t-1} - u_t).$$

As in Sect. 3.4.3, we assume that, for all t,

$$\begin{aligned} \mathcal{X}_t &= \mathcal{S} \times \mathcal{C}, & \mathcal{U}_t &= \mathcal{S}, \\ \mathcal{V}_t &= \mathcal{S}' \times \mathcal{C}', & \mathcal{Y}_t &= \mathcal{S}', \end{aligned}$$

where $\mathcal{S}$ and $\mathcal{C}$ are solid decomposable spaces in separating duality with solid decomposable spaces $\mathcal{S}'$ and $\mathcal{C}'$, respectively. If $\operatorname{dom} Ef \cap (\mathcal{X} \times \mathcal{U}) \neq \emptyset$, then the dual problem can be written as

$$\begin{aligned} &\text{maximize} && E\left[\sum_{t=1}^{T} W_t \cdot y_t - \sum_{t=0}^{T} L_t^*(p_t - (\Delta y_{t+1} + A_{t+1}^* y_{t+1}, B_{t+1}^* y_{t+1}))\right] \\ &\text{over} && (p, y) \in \mathcal{X}_a^\perp \times \mathcal{Y}; \end{aligned}$$

see Sect. 3.4.3.

We will prove the existence of primal solutions and the absence of a duality gap under the following.

Assumption 4.20 *Problem* (1.11) *is feasible,* $\operatorname{dom} Ef \cap (\mathcal{X} \times \mathcal{U}) \neq \emptyset$ *and*

1. *the set*

$$\{(X, U) \in \mathcal{N} \mid \sum_{t=0}^{T} L_t^\infty(X_t, U_t) \leq 0,\ \Delta X_t = A_t X_{t-1} + B_t U_{t-1}\ \forall t\ a.s.\}$$

is linear,

2. *there exist* $p \in \mathcal{X}_a^\perp$ *and* $\epsilon > 0$ *such that for all* $\lambda \in [1-\epsilon, 1+\epsilon]$ *there exists* $y \in \mathcal{Y}$ *such that* $(\lambda p, y)$ *is feasible in* (3.23).

The following gives sufficient conditions for the assumptions of Sect. 2.3.3.

Remark 4.21 If the integral functionals EL_t are proper and closed on $\mathcal{S} \times \mathcal{C}$ and Assumption 3.53 holds, then Assumption 4.20 implies the conditions of Example 2.95 which, in turn, imply Assumptions 2.90 and 2.93.

Proof Condition Assumption 4.20.1 is the same as condition 1 of Example 2.95. Since $W \in \mathcal{U}$, by assumption, $W_t \cdot y_t$ is integrable. Assumption 3.53 implies that $A_t^* y_t \in \mathcal{S}'$, $B_t^* y_t \in \mathcal{C}'$ and, by Lemma 3.7, that

$$E_t[A_t^* y_t] = A_t^* E_t y_t, \quad E_t[B_t^* E_t y_t] = B_t^* E_t y_t, \quad E_t[W_t \cdot y] = W_t \cdot E_t y_t$$

for all t. Let $\lambda \in [1-\epsilon, 1+\epsilon]$ and $y \in \mathcal{Y}$ be such that $(\lambda p, y)$ is feasible in (3.23). Since, by assumption, EL_t are proper and closed on $\mathcal{S} \times \mathcal{C}$, Corollary 3.9 says that EL_t^* are proper on $\mathcal{S}' \times \mathcal{C}'$. Thus, the dual feasibility implies that each $m_t := L_t^*(\lambda p_t - (\Delta y_{t+1} + A_{t+1}^* y_{t+1}, B_{t+1}^* y_{t+1}))$ is integrable. By Fenchel's inequality (we omit ω from the notation),

$$\begin{aligned} L_t(X_t, U_t) &\geq (X_t, U_t) \cdot (\lambda p_t - (\Delta y_{t+1} + A_{t+1}^* y_{t+1}, B_{t+1}^* y_{t+1}) - m_t \\ &= \lambda (X_t, U_t) \cdot p_t - X_t \cdot \Delta y_{t+1} - (A_{t+1} X_t + B_{t+1} U_t) \cdot y_{t+1} - m_t \\ &= \lambda (X_t, U_t) \cdot p_t - X_t \cdot \Delta y_{t+1} - \Delta X_{t+1} \cdot y_{t+1} + W_{t+1} \cdot y_{t+1} - m_t \end{aligned}$$

for every t and for every $(X, U) \in \mathbb{R}^n$ satisfying the system equations. Thus condition 2 of Assumption 4.20 implies that of Example 2.95. □

The following is the main result of this section.

Theorem 4.22 *Under Assumption 4.20,* $\inf(1.11) = \sup(3.23)$ *and* (1.11) *has a solution. In this case, a dual feasible* (p, y) *solves* (3.23) *if and only if there exists a primal feasible* x *such that*

$$p_t - (\Delta y_{t+1} + A^*_{t+1} y_{t+1}, B^*_{t+1} y_{t+1}) \in \partial L_t(X_t, U_t),$$
$$\Delta X_t = A_t X_{t-1} + B_t U_{t-1} + W_t$$

for all t *almost surely.*

Proof By Theorem A.17,

$$f^\infty(x, u, \omega) = \sum_{t=0}^{T} L_t^\infty(X_t, U_t, \omega)$$
$$+ \sum_{t=1}^{T} \delta_{\{0\}}(\Delta X_t - A_t(\omega) X_{t-1} - B_t(\omega) U_{t-1} - u_t).$$

Thus, Assumption 4.20 implies the assumptions of Lemma 4.12 so the claims follow from Theorem 4.11 and Theorem 3.51. □

Assumption 4.20.2 can also be written in terms of the reduced dual problem from Corollary 3.56.

Remark 4.23 Assume that $\operatorname{dom} Ef \cap (\mathcal{X} \times \mathcal{U}) \neq \emptyset$, each L_t is $\mathcal{F}_t$-measurable, each EL_t is closed and proper on $\mathcal{S} \times \mathcal{C}$ and that Assumption 3.53 holds. By Remark 3.56, the reduced dual problem can then be written as

$$\underset{y \in \mathcal{Y}_a}{\text{maximize}}\ E\left[\sum_{t=1}^{T} W_t \cdot y_t - \sum_{t=0}^{T} L_t^*(-E_t(\Delta y_{t+1} + A^*_{t+1} y_{t+1}, B^*_{t+1} y_{t+1}))\right].$$

If there exist $y \in \mathcal{Y}$ and $\epsilon > 0$ such that λy is feasible in the reduced dual for every $\lambda \in [1 - \epsilon, 1 + \epsilon]$, then Assumption 4.20.2 holds.

Proof As shown in the proof of Remark 3.56, the mappings $\pi : \mathcal{Y} \to \mathcal{V}$ and $\gamma : \mathcal{Y} \to \mathcal{Y}$ defined by $\gamma(y) = {}^a y$ and

$$\pi(y)_t = (\Delta\, {}^a y_{t+1} + A^*_{t+1}\, {}^a y_{t+1}, B^*_{t+1}\, {}^a y_{t+1}) - E_t(\Delta y_{t+1} + A^*_{t+1} y_{t+1}, B^*_{t+1} y_{t+1})$$

satisfy the assumptions of Theorem 3.38. Thus, the claim follows from Lemma 4.14. □

4.3.4 Problems of Lagrange

Consider again problem (1.13)

$$\text{minimize} \quad E\left[\sum_{t=0}^{T} K_t(x_t, \Delta x_t)\right] \quad \text{over } x \in \mathcal{N}$$

from Sect. 1.3.4 and recall from Sect. 3.4.4 that this fits the general duality framework with $\bar{u} = 0$ and

$$f(x, u, \omega) = \sum_{t=0}^{T} K_t(x_t, \Delta x_t + u_t, \omega).$$

As in Sect. 3.4.4, we assume that $\mathcal{X}_t = \mathcal{S}$, $\mathcal{V}_t = \mathcal{S}'$, $\mathcal{U} = \mathcal{X}$ and $\mathcal{Y} = \mathcal{V}$, where $\mathcal{S}$ and $\mathcal{S}'$ are solid decomposable spaces in separating duality under the bilinear form $\langle x_t, v_t\rangle := E[x_t \cdot v_t]$.

It was shown in Sect. 3.4.4, that if $\operatorname{dom} Ef \cap (\mathcal{X} \times \mathcal{U}) \neq \emptyset$, then the dual problem becomes (3.31):

$$\text{maximize} \qquad E\left[-\sum_{t=0}^{T} K_t^*(p_t + \Delta y_{t+1}, y_t)\right] \quad \text{over } (p, y) \in \mathcal{X}_a^{\perp} \times \mathcal{Y}.$$

We will prove the existence of primal solutions and the absence of a duality gap under the following.

Assumption 4.24 *Problem* (1.13) *is feasible,* $\operatorname{dom} Ef \cap (\mathcal{X} \times \mathcal{U}) \neq \emptyset$ *and*

1. *the set* $\{x \in \mathcal{N} \mid \sum_{t=0}^{T} K_t^{\infty}(x_t, \Delta x_t) \leq 0 \; a.s.\}$ *is linear,*
2. *there exist* $p \in \mathcal{X}_a^{\perp}$ *and* $\epsilon > 0$ *such that for all* $\lambda \in [1-\epsilon, 1+\epsilon]$ *there exists* $y \in \mathcal{Y}$ *such that* $(\lambda p, y)$ *is feasible in* (3.31).

The last condition in Assumption 4.24 holds, in particular, if the dual is feasible and the functions K_t^* are p-homogeneous; see the discussion after Lemma 4.12.

Remark 4.25 If the integral functionals EK_t are proper and closed on $\mathcal{S} \times \mathcal{S}$, then Assumption 4.24 implies the assumptions of Example 2.109 which, in turn, imply Assumptions 2.105 and 2.107.

Proof It suffices to verify condition 2 in Example 2.109. Let $y \in \mathcal{Y}$ be such that $(\lambda p, y)$ is feasible in (3.31). By Corollary 3.9, EK_t^* are proper on $\mathcal{S}' \times \mathcal{S}'$, so each $m_t := K_t^*(\lambda p_t + \Delta y_{t+1}, y_t)$ is integrable. By Fenchel's inequality,

$$K_t(x_t, \Delta x_{t+1}, \omega) \geq x_t \cdot (\lambda p_t(\omega) + \Delta y_{t+1}(\omega)) + \Delta x_t \cdot y_t(\omega) - m_t(\omega),$$

which is condition 2 in Example 2.109. □

Recall from Sect. 3.4.4 that the *Hamiltonian* associated with K_t is the extended real-valued function

$$H_t(x_t, y_t, \omega) := \inf_{u_t \in \mathbb{R}^d} \{K_t(x_t, u_t, \omega) - u_t \cdot y_t\},$$

which is convex in x_t and concave in y_t.

Theorem 4.26 *Under Assumption 4.24,* $\inf(1.13) = \sup(3.31)$ *and* (1.11) *has a solution. In this case, a dual feasible* (p, y) *solves* (3.31) *if and only if there exists a primal feasible* x *such that*

$$\begin{aligned} p_t + \Delta y_{t+1} &\in \partial_x H_t(x_t, y_t), \\ \Delta x_t &\in \partial_y[-H_t](x_t, y_t) \end{aligned}$$

for all t almost surely.

Proof By Theorem A.17,

$$f^\infty(x, u, \omega) = \sum_{t=0}^{T} K_t^\infty(x_t, \Delta x_t + u_t, \omega).$$

Thus, Assumption 4.24 implies the assumptions of Lemma 4.12 so the claims follow from Theorems 4.11 and 3.62. □

Much like in Remark 4.23, one can give sufficient conditions for Assumption 4.24.2 in terms of the reduced dual problem from (3.64).

4.3.5 Financial Mathematics

Consider again the problem (1.15)

$$\begin{aligned} &\text{minimize} \quad EV\left(c - \sum_{t=0}^{T-1} x_t \cdot \Delta s_{t+1}\right) \quad \text{over} \quad x \in \mathcal{N}, \\ &\text{subject to} \quad x_t \in D_t \quad t = 0, \dots, T \ a.s. \end{aligned}$$

from Sect. 1.3.5 and assume, as in Sect. 3.4.5, that $c \in \mathcal{U}$, $V(0) = 0$ almost surely, $\mathcal{X}_t = \mathcal{S}$ and that $\mathcal{V}_t = \mathcal{S}'$, where $\mathcal{S}$ and $\mathcal{S}'$ are solid decomposable spaces in separating duality. Recall from Sect. 3.4.5 that the problem fits the general duality framework with $\bar{u} = c$ and

$$f(x, u, \omega) = V\left(u - \sum_{t=0}^{T-1} x_t \cdot \Delta s_{t+1}(\omega), \omega\right) + \sum_{t=0}^{T} \delta_{D_t(\omega)}(x_t), \tag{4.4}$$

and that the dual problem can be written as

$$\text{maximize} \quad E\left[cy - V^*(y) - \sum_{t=0}^{T-1} \sigma_{D_t}(p_t + y\Delta s_{t+1})\right] \text{ over } (p, y) \in \mathcal{X}_a^\perp \times \mathcal{Y}.$$

Since V is a nondecreasing, nonconstant convex normal integrand, its recession function V^∞ is nondecreasing and strictly positive on strictly positive reals. It turns out that, in the absence of portfolio constraints, the linearity condition in Lemma 4.12 becomes the classical *no-arbitrage* condition

$$x \in \mathcal{N}, \sum_{t=0}^{T-1} x_t \cdot \Delta s_{t+1} \geq 0 \; a.s. \implies \sum_{t=0}^{T-1} x_t \cdot \Delta s_{t+1} = 0 \quad a.s.$$

encountered already in Remark 2.120 in Sect. 2.3.5. The lower bound in Lemma 4.12 holds, in particular, if there exists a martingale measure $Q \ll P$ such that

$$dQ/dP \in \mathcal{Y} \cap \operatorname{dom} EV^*,$$

V is deterministic and satisfies one of the asymptotic elasticity conditions

$$AE_-(V) := \limsup_{u \to -\infty} \frac{uV'(u)}{V(u)} < 1 \quad \text{or} \quad AE_+(V) := \liminf_{u \to +\infty} \frac{uV'(u)}{V(u)} > 1;$$

see Remark 2.121 and Theorem A.87. More generally, Lemma 4.12 is implied by the following.

Assumption 4.27 *Problem* (1.15) *is feasible and*

1. *the set*

$$\{x \in \mathcal{N} \mid \sum_{t=0}^{T-1} x_t \cdot \Delta s_{t+1} \geq 0, \; x_t \in D_t^\infty \; \forall t \; a.s.\}$$

   is linear,
2. *there exist* $p \in \mathcal{X}_a^\perp$ *and* $\epsilon > 0$ *such that, for all* $\lambda \in [1-\epsilon, 1+\epsilon]$, *there exists* $y \in \mathcal{Y}$ *such that* $(\lambda p, y)$ *is feasible in* (3.33).

Recall from Sect. 2.3.5 that Assumption 4.27.1 is a generalization of the classical no-arbitrage condition. If the function V in (1.15) is bounded from below, as happens e.g. with the exponential loss function, we have $EV^*(0) < \infty$ so Assumption 4.27.2 holds trivially with $(p, y) = (0, 0)$. More general sufficient conditions for Assumption 4.27.2 will be given in Lemma 4.30 below.

Theorem 4.28 *Under Assumption 4.27, $\bar{\varphi}$ is closed and the infimum in its definition is attained for every $(z,u) \in \mathcal{X} \times \mathcal{U}$. In particular,* $\inf(1.15) = \sup(3.33)$ *and* (1.15) *has a solution. In this case, a dual feasible (p, y) solves* (3.33) *if and only if there is a primal feasible x such that*

$$y \in \partial V\Big(u - \sum_{t=0}^{T-1} x_t \cdot \Delta s_{t+1}\Big),$$

$$p_t + y\Delta s_{t+1} \in N_{D_t}(x_t) \quad t = 0, \dots, T$$

almost surely.

Proof By Theorem A.17,

$$f^\infty(x,u,\omega) = V^\infty\left(u - \sum_{t=0}^{T-1} x_t \cdot \Delta s_{t+1}(\omega), \omega\right) + \sum_{t=0}^{T} \delta_{D_t^\infty(\omega)}(x_t).$$

As observed at the beginning of the section, V^∞ is nondecreasing and strictly positive on strictly positive reals. Thus, Assumption 4.27 implies the assumptions of Lemma 4.12 so the claims follow from Theorems 4.11 and 3.68. □

Note that Assumption 4.27.1 is the same as that in Assumption 2.117. It extends the classical no-arbitrage condition

$$x \in \mathcal{N},\ \sum_{t=0}^{T-1} x_t \cdot \Delta s_{t+1} \geq 0\ a.s. \implies \sum_{t=0}^{T-1} x_t \cdot \Delta s_{t+1} = 0\ a.s.$$

by allowing for portfolio constraints; see Remark 2.120. The no-arbitrage condition can also be written as

$$\mathcal{C} \cap L^0_+ = \{0\}, \tag{4.5}$$

where

$$\mathcal{C} := \{u \in \mathcal{U} \mid \exists x \in \mathcal{N} : \sum_{t=0}^{T-1} x_t \cdot \Delta s_{t+1} \geq u\ a.s.\}$$

is the set of claims that can be superhedged at zero cost; see Example 3.73.

The following result is sometimes referred to as the "fundamental theorem of asset pricing". Its proof is based on the closedness result in Theorem 4.28 and on Theorem A.67 which, for convex cones, provides a form of strict separation.

Theorem 4.29 (Dalang–Morton–Willinger) *If there are no portfolio constraints, then the no-arbitrage condition* (4.5) *(or, equivalently, Assumption 4.27.1) holds if* *and only if there exists a probability measure $Q \sim P$ under which s martingale. Moreover, when it exists, the measure Q can be chosen so that $dQ/dP \in L^\infty$.*

Proof Note first that the statement is invariant with respect to an equivalent changes of measure P. We may thus assume that $s \in L^1$. Indeed, s is integrable with respect to e.g. the measure $\tilde{P} \sim P$ defined by

$$\frac{d\tilde{P}}{dP} = \frac{e^{-|s|}}{Ee^{-|s|}}.$$

Assume that s is a martingale under a $Q \sim P$ and let $u \in \mathcal{C} \cap L^0_+$. Then there exists $x \in \mathcal{N}$ such that $\sum_{t=0}^{T-1} x_t \cdot \Delta s_{t+1} \geq u$. Since $E_t^Q[\Delta s_{t+1}] = 0$ for all t, Lemma 2.67 gives $E_Q[\sum_{t=0}^{T-1} x_t \cdot \Delta s_{t+1}] = 0$, so $u = 0$ and thus, (4.5) holds.

To prove the converse, we will apply Theorem 4.28 with $V = \delta_{\mathbb{R}_-}$, $\mathcal{U} = L^1$, $\mathcal{Y} = L^\infty$, $\mathcal{X} = L^\infty(\mathbb{R}^n)$ and $\mathcal{V} = L^1(\mathbb{R}^n)$. Observe first that (4.5) is equivalent to

$$\tilde{\mathcal{C}} \cap (\{0\} \times L^0_+) = \{0\}, \tag{4.6}$$

where

$$\tilde{\mathcal{C}} := \{(z, u) \in \mathcal{X} \times \mathcal{U} \mid \exists x \in L^0 : x - z \in \mathcal{N}, \sum_{t=0}^{T-1} x_t \cdot \Delta s_{t+1} \geq u \quad a.s.\}.$$

Since $V = \delta_{\mathbb{R}_-}$, we have $\bar{\varphi} = \delta_{\tilde{\mathcal{C}}}$ and $V^* = \delta_{\mathbb{R}^+}$. In particular, the point $(p, y) = (0, 0)$ is feasible in (3.33), so condition 2 of Assumption 4.27 holds. As noted after Theorem 4.28, the no-arbitrage condition (4.5) and thus, (4.6) is equivalent to the first condition of Assumption 4.27. Theorem 4.28 now says that $\delta_{\tilde{\mathcal{C}}}$ is lsc or, equivalently, $\tilde{\mathcal{C}}$ is closed. Applying Theorem A.67 and Example A.69 with $U = L^\infty \times L^1$, $Y = L^1 \times L^\infty$, $C = \tilde{\mathcal{C}}$ and $K = \{0\} \times L^1_+$, we get the existence of $(p, y) \in \tilde{\mathcal{C}}^\circ$ such that y is strictly positive. The random variable y/Ey is the density of an equivalent martingale measure; see Example 3.73. □

Recall the reduced dual problem

$$\text{maximize} \quad E\left[cy - V^*(y) - \sum_{t=0}^{T-1} \sigma_{D_t}(E_t[y\Delta s_{t+1}])\right] \quad \text{over } y \in \mathcal{Y}$$

from Theorem 3.71. The following gives sufficient conditions for Assumption 4.27.2.

Lemma 4.30 (Reduced Dual) *Let Assumption 3.69 hold and assume that there exist y feasible in the reduced dual and $\epsilon > 0$ such that*

$$EV^*(\lambda y) < \infty \quad \forall \lambda \in [1 - \epsilon, 1 + \epsilon]. \tag{4.7}$$

Then Assumption 4.27.2 holds. If, in addition, Assumption 4.27.1 holds, then

$$\bar{\varphi}^{\infty}(z,u) \geq \langle z, \pi(y)\rangle + \langle u, y\rangle + \epsilon|\langle z, \pi(y)\rangle + \langle u, y\rangle|,$$

where $\pi(y) := (E_t[y\Delta s_{t+1}] - y\Delta s_{t+1})_{t=0}^T$.

Proof By Lemma 3.70, π in the statement and $\gamma(y) := y$ satisfy the assumptions of Theorem 3.38. The first claim thus follows from the first claim of Lemma 4.14. If, in addition, Assumption 4.27.1 holds, then $\bar{\varphi}$ is closed and proper, by Theorem 4.28, so the second claim follows from the second claim of Lemma 4.14. □

Note that the assumptions of Lemma 4.30 imply Assumption 2.117. Condition (4.7) holds, in particular, if one of the elasticity conditions in Remark 2.121 is satisfied.

Unlike problem (1.15), the problem of minimizing the optimized certainty equivalent in Example 3.74 fails condition 2 of Lemma 4.12. The following gives sufficient conditions that allow us to verify Assumption 4.9 directly.

Example 4.31 (Optimized Certainty Equivalent) Consider again the problem

$$\begin{aligned} &\text{minimize} \quad \mathcal{V}(c - \sum_{t=0}^{T-1} x_t \cdot \Delta s_{t+1}) \quad \text{over} \quad x \in \mathcal{N} \\ &\text{subject to} \quad x_t \in D_t \; t = 0, \ldots, T \quad a.s, \end{aligned}$$

from Example 3.74 and recall that the function $\mathcal{V} : L^0 \to \overline{\mathbb{R}}$ is defined by

$$\mathcal{V}(c) := \inf_{\alpha \in \mathbb{R}} \{\alpha + EV(c - \alpha)\}.$$

Let Assumption 3.69 hold and assume that

1. the set

$$\{x \in \mathcal{N} \mid \sum_{t=0}^{T-1} x_t \cdot \Delta s_{t+1} \geq 0, \; x_t \in D_t^{\infty} \; \forall t \; a.s.\}$$

   is linear,
2. there exist y feasible in (3.34) and an $\epsilon > 0$ such that $E[y] = 1$ and

$$EV^*(\lambda y) < \infty \quad \forall \lambda \in [1 - \epsilon, 1 + \epsilon].$$

Then $\inf(3.36) = \inf(3.37) = \sup(3.38) = \sup(3.39)$ and (3.37) has a solution. In this case, a dual feasible y solves (3.39) if and only if there exists (α, x) feasible in (3.37) such that

$$y \in \partial V(c - \sum_{t=0}^{T-1} x_t \cdot \Delta s_{t+1} - \alpha),$$

$$E_t[y\Delta s_{t+1}] \in N_{D_t}(x_t) \quad t = 0, \ldots, T$$

almost surely.

Proof As observed in Example 3.74, this fits the general framework with time t running from -1 to T, $x_{-1} = \alpha$, $\mathcal{F}_{-1} = \{\emptyset, \Omega\}$, $\bar{u} = c$ and

$$\hat{f}(\hat{x}, u, \omega) := \alpha + V(u - \alpha - \sum_{t=0}^{T-1} x_t \cdot \Delta s_{t+1}(\omega), \omega) + \sum_{t=0}^{T} \delta_{D_t(\omega)}(x_t),$$

where $\hat{x} = (\alpha, x)$. As in Example 3.74, we denote $\hat{\mathcal{N}} := L^0(\Omega, \mathcal{F}_{-1}, P; \mathbb{R}) \times \mathcal{N}$, $\hat{\mathcal{X}} := L^\infty \times \mathcal{X}$ and $\hat{\mathcal{X}}_a := \hat{\mathcal{X}} \cap \hat{\mathcal{N}}$. We will show that Assumption 4.9 is satisfied with $(\hat{p}, y)$, where $\hat{p} := (p_{-1}, p)$, $p_{-1} := E[y] - y$ and $p_t := E_t[y\Delta s_{t+1}] - y\Delta s_{t+1}$. The claims then follow from Theorem 4.11 and Example 3.74.

We have

$$\hat{f}(\hat{x}, u, \omega) = \alpha + f(x, u - \alpha, \omega),$$

where f is the normal integrand corresponding to (1.15); see (4.4). As observed in the proof of Theorem 4.28, f satisfies the assumptions of Lemma 4.12, so Lemma 4.10 implies, for every $(x, u) \in \mathcal{N} \times \mathcal{U}$,

$$Ek^\infty(x, u) = E[f^\infty(x, u) - u \cdot y], \tag{4.8}$$

where $k(x, u, \omega) := f(x, u, \omega) - x \cdot p(\omega) - uy(\omega)$.

Since y is feasible in (3.34), $(\hat{p}, y)$ is feasible in (3.38). Thus, by Fenchel's inequality, the function

$$\hat{k}(\hat{x}, u, \omega) := \hat{f}(\hat{x}, u, \omega) - \hat{x} \cdot \hat{p}(\omega) - uy(\omega)$$

is lower bounded so Assumption 4.9.1 holds. Theorem A.17 gives

$$\hat{k}^\infty(\hat{x}, u, \omega) = \hat{f}^\infty(\hat{x}, u, \omega) - \hat{x} \cdot \hat{p}(\omega) - uy(\omega),$$

$$\hat{f}^\infty(\hat{x}, u) = \alpha + f^\infty(x, u - \alpha)$$

so

$$\begin{aligned}\hat{\mathcal{L}} :&= \{\hat{x} \in \hat{\mathcal{N}} \mid \hat{k}^{\infty}(\hat{x}, 0) \leq 0 \, a.s.\} \\ &= \{\hat{x} \in \hat{\mathcal{N}} \mid \alpha + f^{\infty}(x, -\alpha) - \alpha p_{-1} - x \cdot p \leq 0 \, a.s.\} \\ &= \{\hat{x} \in \hat{\mathcal{N}} \mid \alpha(1 - y - p_{-1}) + k^{\infty}(x, -\alpha) \leq 0 \, a.s.\} \\ &= \{\hat{x} \in \hat{\mathcal{N}} \mid k^{\infty}(x, -\alpha) \leq 0 \, a.s.\},\end{aligned}$$

where $k(x, u, \omega) := f(x, u, \omega) - x \cdot p(\omega) - uy(\omega)$. Let $\hat{x} \in \hat{\mathcal{L}}$. Taking expectations in $k^{\infty}(x, -\alpha) \leq 0$ and recalling that $E[y] = 1$, we get

$$0 \geq Ek^{\infty}(x, -\alpha) = E[\alpha y + f^{\infty}(x, -\alpha)] \geq \alpha + \bar{\varphi}^{\infty}(0, -\alpha) \geq \epsilon|\alpha|,$$

where the equality follows from (4.8), the second inequality from Theorem 4.11 and the third from Lemma 4.30. Thus $\alpha = 0$. By Lemma 4.30, Assumption 4.27 holds, so $\{x \in \hat{\mathcal{N}} \mid k^{\infty}(x, 0) \leq 0 \, a.s.\}$ is linear, by Lemma 4.12. Thus $\hat{\mathcal{L}}$ is linear, i.e. Assumption 4.9.2 holds.

It remains to verify Assumption 4.9.3. Let $(\hat{x}, \hat{z}, u) \in L^0 \times \hat{\mathcal{X}} \times \mathcal{U}$ with $\hat{x} - \hat{z} \in \hat{\mathcal{N}}$. Since $\mathcal{X} = L^{\infty}$ and $\alpha - z_{-1}$ is $\mathcal{F}_0$ and thus a constant, we have $\alpha \in \mathcal{U}$. Since f satisfies Assumption 4.9.3., we have

$$Ek(x, u - \alpha) = E[f(x, u - \alpha) - z \cdot p - (u - \alpha)y].$$

Thus

$$\begin{aligned}E\hat{k}(\hat{x}, u) &= E[\hat{f}(\hat{x}, u) - \hat{x} \cdot \hat{p} - uy] \\ &= E[\alpha + f(x, u - \alpha) - \alpha p_{-1} - x \cdot p - uy] \\ &= E[\alpha + k(x, u - \alpha) - \alpha p_{-1} - \alpha y] \\ &= E[\alpha + f(x, u - \alpha) - z \cdot p - \alpha p_{-1} - uy] \\ &= E[\hat{f}(\hat{x}, u) - \hat{z} \cdot \hat{p} - uy],\end{aligned}$$

where the last equality follows from Lemma B.3, since $E[(\alpha - z_{-1})p_{-1}] = 0$. Thus Assumption 4.9.3. holds. □

Note that, if there are no portfolio constraints, the first condition in Example 4.31 becomes the no-arbitrage condition in Remark 2.120 and feasible dual variables become martingale densities of the price process; see Sect. 3.4.5. When $V = \delta_{\mathbb{R}_-}$, the problem in Example 4.31 becomes the superhedging problem in Example 1.88; see also the end of Example 3.74. Condition 2 in Example 4.31 is then trivially satisfied so Examples 4.31 and 3.74 give the following.

Example 4.32 (Superhedging) If there are no portfolio constraints and if the price process s satisfies the no-arbitrage condition (see Remark 2.120), then the superhedging cost of a claim $c \in \mathcal{U}$ can be expressed as

$$\sup_{Q \in \mathcal{Q}^{\mathcal{Y}}} E^Q[c],$$

where $\mathcal{Q}^{\mathcal{Y}}$ is the set of martingale measures with densities in $\mathcal{Y}$.

Remark 4.33 (Inada Conditions) A differentiable loss function V is said to satisfy the *Inada conditions* if

$$\lim_{c \nearrow \infty} V'(c) = \infty \quad \text{and} \quad \lim_{c \searrow -\infty} V'(c) = 0.$$

By Remark A.19, this can be written equivalently as $V^\infty = \delta_{\mathbb{R}_-}$. A benefit of the latter formulation is that it makes good sense also for nondifferentiable loss functions.

The following characterizes the recession cone $\mathcal{C}^\infty$ of the set $\mathcal{C}$ of claims that can be superhedged without a cost. It also shows that, under the Inada condition, the recession function of the optimum value function φ of (1.15) becomes the indicator function of $\mathcal{C}^\infty$.

Remark 4.34 If Assumption 4.27.1 holds, then the recession cone of the set

$$\mathcal{C} = \left\{ u \in \mathcal{U} \;\middle|\; \exists x \in \mathcal{N} : u \leq \sum_{t=0}^{T-1} x_t \cdot \Delta s_{t+1},\ x_t \in D_t\ \forall t\ a.s. \right\}$$

can be expressed as

$$\mathcal{C}^\infty = \left\{ u \in \mathcal{U} \;\middle|\; \exists x \in \mathcal{N} : u \leq \sum_{t=0}^{T-1} x_t \cdot \Delta s_{t+1},\ x_t \in D_t^\infty\ \forall t\ a.s. \right\}.$$

If, in addition, Assumption 4.27.2 holds and $V^\infty = \delta_{\mathbb{R}_-}$, then the recession function of $\bar{\varphi}(0, \cdot)$ can be expressed as

$$\bar{\varphi}(0, \cdot)^\infty = \delta_{\mathcal{C}^\infty}.$$

Proof When $V^\infty = \delta_{\mathbb{R}_-}$, Theorem A.17 gives

$$\begin{aligned} f^\infty(x, u, \omega) &= V^\infty\left(u - \sum_{t=0}^{T-1} x_t \cdot \Delta s_{t+1}(\omega), \omega\right) + \sum_{t=0}^{T} \delta_{D_t^\infty(\omega)}(x_t) \\ &= \delta_{\mathbb{R}_-}\left(u - \sum_{t=0}^{T-1} x_t \cdot \Delta s_{t+1}(\omega), \omega\right) + \sum_{t=0}^{T} \delta_{D_t^\infty(\omega)}(x_t). \end{aligned}$$

By Theorem 4.11, Assumption 4.27 then implies that $\bar{\varphi}$ is closed and

$$\begin{aligned}\bar{\varphi}^\infty(0,u) &= \inf_{x\in\mathcal{N}} Ef^\infty(x,u)\\ &= \inf_{x\in\mathcal{N}} E\left[\delta_{\mathbb{R}_-}\left(u-\sum_{t=0}^{T-1} x_t\cdot\Delta s_{t+1}\right)+\sum_{t=0}^{T}\delta_{D_t^\infty(\omega)}(x_t)\right]\\ &= \delta_{\mathcal{K}}(u),\end{aligned}$$

where

$$\mathcal{K} := \left\{u\in\mathcal{U}\;\middle|\;\exists x\in\mathcal{N}:\ u\le\sum_{t=0}^{T-1} x_t\cdot\Delta s_{t+1},\ x_t\in D_t^\infty\ \forall t\ a.s.\right\}.$$

When $V=\delta_{\mathbb{R}_-}$, we have $\bar{\varphi}(0,\cdot)=\delta_{\mathcal{C}}$ and, by the definition of the recession function, $\bar{\varphi}(0,\cdot)^\infty=\delta_{\mathcal{C}^\infty}$. Since $\bar{\varphi}$ is closed and $\operatorname{dom}\bar{\varphi}(0,\cdot)\neq\emptyset$, Theorem A.15 gives $\bar{\varphi}(0,\cdot)^\infty=\bar{\varphi}^\infty(0,\cdot)$ which proves both claims. □

Example 4.35 (Semi-Static Hedging) Consider again problem (1.17)

$$\begin{aligned}&\text{minimize}\quad EV\left(c-\sum_{t=0}^{T-1} x_t\cdot\Delta s_{t+1}-\bar{c}\cdot\bar{x}+S_0(\bar{x})\right)\quad\text{over}\quad x\in\mathcal{N},\ \bar{x}\in\mathbb{R}^{\bar{J}},\\ &\text{subject to}\quad x_t\in D_t\quad t=0,\dots,T-1\ a.s.\end{aligned}$$

and its reduced dual (3.40),

$$\underset{y\in\mathcal{Y}}{\text{maximize}}\quad E\left[cy-V^*(y)-\sum_{t=0}^{T-1}\sigma_{D_t}(E_t[y\Delta s_{t+1}])\right]-\sigma_{\operatorname{epi} S_0}(E[y\bar{c}],-E[y])$$

from Example 3.75. Assume that

1. The set

$$\left\{(\bar{x},x)\in\mathbb{R}^{\bar{J}}\times\mathcal{N}\;\middle|\;S_0^\infty(\bar{x})-\sum_{t=0}^{T-1} x_t\cdot\Delta s_{t+1}-\bar{c}\cdot\bar{x}\le 0,\ x_t\in D_t^\infty\ \forall t\ a.s.\right\}$$

is linear.

2. there exist $y\in\mathcal{Y}$ and $\epsilon>0$ such that λy is feasible in (3.40) for all $\lambda\in[1-\epsilon,1+\epsilon]$.

Then $\inf(1.17) = \sup(3.40)$, and (1.17) has a solution. If (1.17) is feasible, then a dual feasible y solves (3.40) if and only if there is a primal feasible x such that

$$y \in \partial V\left(c - \sum_{t=0}^{T-1} x_t \cdot \Delta s_{t+1} - \bar{c} \cdot \bar{x} + S_0(\bar{x})\right),$$

$$E_t[y \Delta s_{t+1}] \in N_{D_t}(x_t) \quad t = 0, \ldots, T,$$

$$E[y\bar{c}] \in \partial(E[y]S_0)(\bar{x})$$

almost surely.

Proof As in Example 3.75, problem (1.17) fits the general duality framework with time t running from -1 to T, $\mathcal{F}_{-1} = \{\Omega, \emptyset\}$, $x_{-1} = (\bar{x}, \alpha) \in \mathbb{R}^{\bar{J}} \times \mathbb{R}$, $\bar{u} = c$ and

$$f(\hat{x}, u, \omega) = V\left(u - \sum_{t=0}^{T-1} x_t \cdot \Delta s_{t+1}(\omega) - \bar{c}(\omega) \cdot \bar{x} + \alpha, \omega\right) + \sum_{t=0}^{T-1} \delta_{D_t(\omega)}(x_t, \omega) + \delta_{\operatorname{epi} S_0}(\bar{x}, \alpha)$$

where $\hat{x} := (\bar{x}, \alpha, x)$. As in Example 3.75, we denote $\hat{\mathcal{N}} := L^0(\Omega, \mathcal{F}_{-1}, P; \mathbb{R}^{\bar{J}} \times \mathbb{R}) \times \mathcal{N}$.

It suffices to show that the assumptions of Lemma 4.12 are satisfied since then, the claims follow from Theorem 4.11 and Example 3.75. As shown in the proof of Example 3.75, the assumptions of Theorem 3.38 are satisfied by linear mappings π and γ. Thus, by Lemma 4.14, condition 2 of Lemma 4.12 holds. It remains to show that the set

$$\mathcal{L} := \{\hat{x} \in \hat{\mathcal{N}} \mid f^\infty(\hat{x}, 0) \leq 0 \; a.s.\}$$

is linear. By Theorem A.17,

$$f^\infty(\hat{x}, u, \omega) = V^\infty\left(u - \sum_{t=0}^{T-1} x_t \cdot \Delta s_{t+1}(\omega) - \bar{c}(\omega) \cdot \bar{x} + \alpha, \omega\right) + \sum_{t=0}^{T-1} \delta_{D_t^\infty(\omega)}(x_t, \omega) + \delta_{\operatorname{epi} S_0^\infty}(\bar{x}, \alpha).$$

Since V^∞ is strictly positive on strictly positive reals, we thus get

$$\mathcal{L} = \left\{ \hat{x} \in \hat{\mathcal{N}} \;\middle|\; \alpha - \sum_{t=0}^{T-1} x_t \cdot \Delta s_{t+1} - \bar{c} \cdot \bar{x} \leq 0, \right.$$

$$\left. x_t \in D_t^\infty \;\forall t,\; S_0^\infty(\bar{x}) \leq \alpha \quad a.s. \right\}.$$

Given $x \in \mathcal{L}$, we have

$$S_0^\infty(\bar{x}) - \sum_{t=0}^{T-1} x_t \cdot \Delta s_{t+1} - \bar{c} \cdot \bar{x} \leq 0,\; x_t \in D_t^\infty \quad \forall t \; a.s.,$$

where strict inequality holds unless $S_0^\infty(\bar{x}) = \alpha$. Condition 1 gives

$$S_0^\infty(-\bar{x}) + \sum_{t=0}^{T-1} x_t \cdot \Delta s_{t+1} + \bar{c} \cdot \bar{x} \leq 0,\; -x_t \in D_t^\infty \quad \forall t \; a.s.$$

Adding the inequalities together gives

$$S_0^\infty(\bar{x}) + S_0^\infty(-\bar{x}) \leq 0 \quad a.s.$$

By convexity, this cannot hold as a strict inequality so we must have $S_0^\infty(\bar{x}) = \alpha$ and

$$-\alpha + \sum_{t=0}^{T-1} x_t \cdot \Delta s_{t+1} + \bar{c} \cdot \bar{x} \leq 0,\; -x_t \in D_t^\infty \;\forall t,\; S_0^\infty(-\bar{x}) \leq -\alpha \quad a.s.$$

Thus, the set $\mathcal{L}$ is linear. □

The following example gives a sufficient condition for the linearity condition in Example 4.35. Given a convex function g on $\mathbb{R}^n$, the set

$$\operatorname{lin} g = \{x \in \mathbb{R}^n \mid g^\infty(x) = -g^\infty(-x)\}$$

is called the *lineality space* of g.

Example 4.36 (Robust No-Arbitrage) Consider Example 3.75. Assume that there are no portfolio constraints and that there exists a sublinear cost function $\hat{S}_0$ such that

$$\hat{S}_0(\bar{x}, \omega) \leq S_0^\infty(\bar{x}, \omega) \quad \forall \bar{x} \in \mathbb{R}^{\bar{J}},$$

$$\hat{S}_0(\bar{x}, \omega) < S_0^\infty(\bar{x}, \omega) \quad \forall \bar{x} \notin \operatorname{lin} S_0(\cdot, \omega)$$

and that the market model described by $\hat{S}_0$ and s satisfies the no-arbitrage condition, i.e.

$$\hat{C} \cap L^0_+ = \{0\},$$

where

$$\hat{C} := \{u \in \mathcal{U} \mid \exists(\bar{x}, x) \in \mathbb{R}^{\bar{J}} \times \mathcal{N} : \sum_{t=0}^{T-1} x_t \cdot \Delta s_{t+1} + \bar{x} \cdot c - \hat{S}_0(\bar{x}) \geq u \quad a.s.\}.$$

Then the linearity condition in Example 4.35 holds, i.e. the set

$$\mathcal{L} := \{(\bar{x}, x) \in \mathbb{R}^{\bar{J}} \times \mathcal{N} \mid S_0^\infty(\bar{x}) - \sum_{t=0}^{T-1} x_t \cdot \Delta s_{t+1} - \bar{x} \cdot \bar{c} \leq 0 \quad a.s.\}$$

is linear.

Proof If the linearity condition fails, there is a $(\bar{x}, x) \in \mathcal{L}$ such that

$$S_0^\infty(-\bar{x}) + \sum_{t=0}^{T-1} x_t \Delta s_{t+1} + \bar{x} \cdot \bar{c} > 0$$

on a set $A \in \mathcal{F}$ with $P(A) > 0$. It suffices to show that $(\bar{x}, x)$ is an arbitrage strategy for $\hat{S}_0$. Since $(\bar{x}, x) \in \mathcal{L}$, and $\hat{S}_0 \leq S_0^\infty$, we have

$$\hat{S}_0(\bar{x}) - \sum_{t=0}^{T-1} x_t \Delta s_{t+1} - \bar{x} \cdot \bar{c} \leq 0.$$

If $\bar{x} \notin \operatorname{lin} S_0$, then $\hat{S}_0(x) < S_0^\infty(x)$ and the inequality is strict so $(\bar{x}, x)$ is an arbitrage strategy for $\hat{S}_0$. If $\bar{x} \in \operatorname{lin} S_0$, then $\hat{S}_0(\bar{x}) \leq S_0^\infty(\bar{x}) = -S_0^\infty(-\bar{x})$ so

$$\hat{S}_0(\bar{x}) - \sum_{t=0}^{T-1} x_t \Delta s_{t+1} - \bar{x} \cdot \bar{c} \leq -S_0^\infty(-\bar{x}) - \sum_{t=0}^{T-1} x_t \Delta s_{t+1} - \bar{x} \cdot \bar{c} < 0$$

on A so $(\bar{x}, x)$ is an arbitrage strategy in this case too. □

4.4 Bibliographical Notes

The main results of this chapter are essentially from [95, 101, 110] with slight generalizations and simplifications. Lemma 4.1 can be found, e.g., in [57, Lemma 1.64] and [69, Lemma 2]. Theorem 4.3 is from [95] where it was used to give a sufficient condition for the L^0-closedness of adapted selections of a closed measurable set (which is a special case of Theorem 4.8 applied to the indicator of a closed random set). In the case $T = 0$, Theorem 4.3 gives [31, Theorem 3.13] for measurable selections of random sets. The closedness result of [95] extended [100, Theorem 3.3] which, in turn, extended earlier arguments from [70, 146] on the currency market model of [67]. The first part of Theorem 4.8 is from [101] while the second part concerning the recession function is from [110]. The closedness of the optimum value function for integrands without lower bounds was first given in [110] under the conditions of Lemma 4.12. The more abstract sufficient conditions in Sect. 4.2.2 are new.

The applications in 4.3 are essentially from [104] where the absence of duality gap was obtained under the assumption of Lemma 4.12. Example 4.31 is new. More references on the treated applications can be found in the previous chapters. Section 4.3.1 on mathematical programming extends earlier existence results by relaxing the compactness and boundedness assumptions; see e.g. [136] and the references there. The existence result in Sect. 4.3.2 can be viewed as a discrete-time version of continuous-time optimal stopping problems; see e.g. [23, 105] and their references. The existence and closedness results in Sects. 4.3.3 and 4.3.4 are from [104]. The "fundamental theorem of asset pricing" in the form of Theorem 4.29 was first obtained in [35]. The Inada conditions in Remark 4.33 are classical in financial mathematics (see e.g. [57, Section 3.3]) but as we saw in Sect. 4.3.5, much weaker assumptions suffice for most of the duality results in finite discrete time. If there are no portfolio constraints, the dual representation of the superhedging cost in Example 4.32 is classical as well; see e.g. [57, Section 5.3]. The "robust no-arbitrage" condition in Example 4.36 was inspired by the main result of [146] on the currency market model of [67]; see also [96, 100, 102].

Chapter 5
Existence of Dual Solutions

The aim of this chapter is to give sufficient conditions for subdifferentiability of the optimum value function $\bar{\varphi}$ of (P) introduced in Sect. 3.3. As we saw there, the subdifferentiability implies the absence of a duality gap and the existence of dual solutions; see Theorem 3.18. By Corollary 3.35, the scenariowise KKTR-condition $(p, y) \in \partial f(x, \bar{u})$ is then necessary as well as sufficient for optimality of feasible solutions of (P). To establish the subdifferentiability of $\bar{\varphi}$, it suffices by Lemma 3.32, to show that the optimum value function φ of $(P_{\mathcal{X}})$ is subdifferentiable.

A simple sufficient condition for subdifferentiability of φ at $(0, \bar{u})$ is that φ be Mackey-continuous at $(0, \bar{u})$; see Corollary A.62. In some applications this is easy to establish by showing that φ has a Mackey continuous upper bound; see Theorem A.29 as well as the discussion after Remark 5.58 for an example. In others, however, Mackey continuity fails. This is typical in problems with pointwise constraints such as those studied in the applications of this book.

This chapter gives sufficient conditions for subdifferentiability of the value function φ without requiring Mackey-continuity. Instead, we first establish, in Theorem 5.1, relative continuity of φ with respect to a topology that may be strictly stronger than the Mackey topology. Strengthening the topology makes it possible to establish the continuity in many applications. However, continuity with respect to such a topology does not, in general, yield subgradients in the Mackey-dual $\mathcal{V} \times \mathcal{Y}$ of $\mathcal{X} \times \mathcal{U}$ but in a strictly larger dual space $\mathcal{X}^* \times \mathcal{U}^*$. Under an additional condition that extends classical 'relatively complete recourse' conditions, we will show, in Sect. 5.3, that the subdifferential of φ with respect to the strong topology at $(0, \bar{u})$ intersects the space $\mathcal{V} \times \mathcal{Y}$ thus giving a subgradient with respect to the original dual pairing. As noted earlier, this implies the absence of a duality gap and the existence of solutions to the dual problem (D).

The strategy outlined above is based on endowing the spaces $\mathcal{X}$ and $\mathcal{U}$ with topologies under which they become Fréchet spaces and under which their topological duals $\mathcal{X}^*$ and $\mathcal{U}^*$ can be expressed as direct sums of $\mathcal{V}$ and $\mathcal{Y}$, respectively,

T. Pennanen, A.-P. Perkkiö, *Convex Stochastic Optimization*, Probability Theory and Stochastic Modelling 107, https://doi.org/10.1007/978-3-031-76432-5_5

with linear spaces of certain singular functionals; see Sect. 5.1.1 below. Much as in Sect. 3.2, we apply the general conjugate duality framework to $(P_{\mathcal{X}})$ with the Rockafellian

$$F(x, z, u) = Ef(x, u) + \delta_{\mathcal{N}}(x - z),$$

but this time, with respect to the pairings of $\mathcal{X}$ and $\mathcal{U}$ with $\mathcal{X}^*$ and $\mathcal{U}^*$, respectively. This gives rise to the *strong dual problem*

$$\text{maximize} \quad \langle \bar{u}, y \rangle - F^*(0, p, y) \quad \text{over } (p, y) \in \mathcal{X}^* \times \mathcal{U}^*, \tag{D_s}$$

where, as opposed to Sect. 3.2, we define the conjugate F^* with respect to the pairings of $\mathcal{X}$ and $\mathcal{U}$ with their strong duals $\mathcal{X}^*$ and $\mathcal{U}^*$, respectively. More explicit expressions for the conjugate $F^* : \mathcal{X}^* \times \mathcal{U}^* \to \overline{\mathbb{R}}$ will be given in Sect. 5.2 below.

Theorem 5.1 below gives sufficient conditions for the relative continuity of φ with respect to the Fréchet space topologies of $\mathcal{X}$ and $\mathcal{U}$. This implies the absence of a duality gap between $(P_{\mathcal{X}})$ and the strong dual problem (D_s) and the existence of solutions to (D_s). The proof is based on the general existence results on conjugate duality in Fréchet spaces; see Sect. A.9.

Recall that the *relative core* rcore C of a set C in a linear space is the set of points $x \in C$ such that the *positive hull*

$$\operatorname{pos}(C - x) := \bigcup_{\alpha > 0} \alpha(C - x)$$

of the set $(C - x)$ is a linear space; see Sect. A.2. Recall also that the *affine hull* aff C of a set C is the smallest affine set containing C. We will say that a strategy $\bar{x} \in \mathcal{X}_a$ is *strictly feasible* if $(\bar{x}, \bar{u}) \in \operatorname{rcore} \operatorname{dom} Ef$. The problem $(P_{\mathcal{X}})$ is said to be *strictly feasible* if it admits a strictly feasible strategy.

We say that a function g is *relatively continuous* at $u \in \operatorname{dom} g$ if it is continuous at u relative to aff dom g; see Sect. A.5. We will refer to the Fréchet space topologies of $\mathcal{X}$ and $\mathcal{U}$ as the *strong topologies*. A set is said to be *strongly closed* if it closed in the strong topology and a function is said to be *strongly lsc* is it is lower semicontinuous in the strong topology. As in Sect. 3.2, we denote the optimum value function of problem $(P_{\mathcal{X}})$ by

$$\varphi(z, u) := \inf_{x \in \mathcal{X}} \{Ef(x, u) \mid x - z \in \mathcal{N}\}.$$

Theorem 5.1 *Assume that Ef is proper and strongly lsc on $\mathcal{X} \times \mathcal{U}$, $(P_{\mathcal{X}})$ is strictly feasible and the set*

$$\operatorname{aff} \operatorname{dom} Ef - \mathcal{X}_a \times \{0\}$$

is strongly closed in $\mathcal{X} \times \mathcal{U}$. Then φ is either relatively strongly continuous at $(0, \bar{u})$ or $\varphi(0, \bar{u}) = -\infty$. In particular, the strong dual problem (D_s) has a solution and

$$\inf(P_{\mathcal{X}}) = \sup(D_s).$$

Proof By Lemma 3.16, the set $\mathcal{X}_a$ is $\sigma(\mathcal{X}, \mathcal{V})$-closed. Since the strong topology of $\mathcal{X}$ is stronger than the Mackey topology, $\mathcal{X}_a$ is strongly closed as well. Since Ef is proper and strongly lsc, it thus follows that the function F is proper and strongly lsc. By Theorem A.75, it suffices to show that $(0, \bar{u}) \in \operatorname{rcore} \operatorname{dom} \varphi$ and that $\operatorname{aff} \operatorname{dom} \varphi$ is strongly closed.

By definition,

$$\operatorname{dom} \varphi = \operatorname{dom} Ef - \mathcal{X}_a \times \{0\}. \tag{5.1}$$

By strict feasibility, there exists an $\bar{x} \in \mathcal{X}_a$ with $(\bar{x}, \bar{u}) \in \operatorname{rcore} \operatorname{dom} Ef$ so, by Lemma A.5,

$$\operatorname{rcore} \operatorname{dom} \varphi = \operatorname{rcore} \operatorname{dom} Ef - \mathcal{X}_a \times \{0\}.$$

Thus, $(0, \bar{u}) \in \operatorname{rcore} \operatorname{dom} \varphi$. By Lemma A.9,

$$\operatorname{aff} \operatorname{dom} \varphi = \operatorname{aff} \operatorname{dom} Ef - \mathcal{X}_a \times \{0\}$$

which is closed by assumption. □

Section 5.1 below reviews the theory of Fréchet spaces of random variables and gives useful expressions for conjugates of integral functionals with respect to the associated dual pairings. Section 5.2 then gives an explicit expression for the objective of the strong dual problem (D_s). Section 5.3 gives sufficient conditions for the assumptions of Theorem 5.1 as well as for the existence of solutions to the original dual problem (D_s) from Chap. 3. Section 5.4 applies the general results to the examples studied in the previous chapters.

5.1 Integral Functionals in Strong Duality

This section studies convex integral functionals on Fréchet spaces and, in particular, Banach spaces of random variables. The Fréchet space structure is convenient in that it allows for simple conditions for the relative continuity of convex functions; see Sect. A.5. By Corollary A.63, relative continuity, in turn, implies subdifferentiability which is intimately related to the existence of dual solutions; see Sect. A.9, Chap. 3 and Sect. 5.3 below.

5.1.1 *Fréchet Spaces of Random Variables*

As in Chaps. 3 and 4, we assume that $\mathcal{U}$ and $\mathcal{Y}$ are solid decomposable spaces of $\mathbb{R}^m$-valued random variables in separating duality under the bilinear form

$$(u, y) \mapsto E[u \cdot y].$$

From now on, we will also assume that $\mathcal{U}$ is endowed with a topology under which it is a *Fréchet space* and under which the linear functionals $u \mapsto \langle u, y\rangle$ are continuous for all $y \in \mathcal{Y}$. Recall that a *Fréchet space* is a complete locally convex topological vector space whose topology is generated by a countable collection of seminorms. In particular, Banach spaces are Fréchet spaces with the topology generated by a single norm.

If the Fréchet topology of $\mathcal{U}$ is strictly stronger than the Mackey topology $\tau(\mathcal{U}, \mathcal{Y})$ that $\mathcal{U}$ has under the pairing with the space $\mathcal{Y}$ in Chap. 3, then by the Mackey-Arens theorem, the corresponding topological dual of $\mathcal{U}$ is strictly larger than $\mathcal{Y}$; see Sect. A.7. In general, the extra elements in the dual space cannot be represented by measurable functions but, often, they have properties that can be employed in the duality theory of integral functionals and, as we will see, in establishing the existence of solutions to the dual problem (D).

We will refer to the Fréchet space topology of $\mathcal{U}$ as the *strong topology* and the corresponding topological dual $\mathcal{U}^*$ as the *strong dual* space. Recall that $\mathcal{U}^*$ is the linear space of all strongly continuous linear functionals on $\mathcal{U}$. We will assume that the dual can be expressed as

$$\mathcal{U}^* = \mathcal{Y} \oplus \mathcal{Y}^s$$

in the sense that for every $y \in \mathcal{U}^*$ there exist unique $y^c \in \mathcal{Y}$ and $y^s \in \mathcal{Y}^s$ such that

$$\langle u, y\rangle = E[u \cdot y^c] + \langle u, y^s\rangle \quad \forall u \in \mathcal{U}.$$

Here $\mathcal{Y}^s \subseteq \mathcal{U}^*$ is the set of strongly continuous linear functionals that are *singular* in the sense that, for every $u \in \mathcal{U}$, there exists a decreasing sequence $(A^\nu)_{\nu\in\mathbb{N}}$ in $\mathcal{F}$ such that $P(A^\nu)\searrow 0$ and

$$\langle 1_{\Omega\setminus A^\nu} u, y^s\rangle = 0 \quad \forall \nu \in \mathbb{N}.$$

Most familiar topological spaces of random variables are Fréchet and their duals are of the above form. Examples include the classical L^p and Orlicz spaces; see Examples 5.7 and 5.11 below.

The following lemma says that the assumptions above imply that $\mathcal{Y}$ equals the Köthe dual

$$\mathcal{U}' := \{y \in L^0 \mid u \cdot y \in L^1\ \forall u \in \mathcal{U}\}$$

of $\mathcal{U}$; see Remark 3.2.

Lemma 5.2 $\mathcal{Y} = \mathcal{U}'$.

Proof It is clear that $\mathcal{Y}$ cannot be larger than $\mathcal{U}'$. Since $\mathcal{U}^* = \mathcal{Y} \oplus \mathcal{Y}^s$, it thus suffices to show that $\mathcal{U}' \subseteq \mathcal{U}^*$ in the sense that, for every $y \in \mathcal{U}'$, the linear functional

$$u \mapsto E[u \cdot y]$$

is strongly continuous on $\mathcal{U}$. Let $y \in \mathcal{U}'$. The function $u \mapsto E[u \cdot y]$ is majorized by

$$p_y(u) := E[\sum_{i=1}^{m} |u_i||y_i|]$$

so it suffices, by Theorem A.29, to show that p_y is continuous on $\mathcal{U}$. The solidity of $\mathcal{U}$ implies that the random vector $u' \in L^0$ defined componentwise by $u'_i := |u_i| \operatorname{sign}(y_i)$ belongs to $\mathcal{U}$ so $u' \cdot y \in L^1$. Since $u' \cdot y = \sum_{i=1}^{m} |u_i||y_i|$, the function p_y is finite on $\mathcal{U}$. By Theorem 1.55, p_y is L^0-lsc. Since $\mathcal{U}^* = \mathcal{Y} \oplus \mathcal{Y}^s$, the strong topology on $\mathcal{U}$ is stronger than $\tau(\mathcal{U}, \mathcal{Y})$ which, by Lemma 3.4, is stronger than the L^0-topology. The function p_y is thus strongly lsc. Since $\mathcal{U}$ is Fréchet, the continuity of p_y now follows from Theorem A.33. □

The following gives conditions for the strong dual to coincide with the Köthe dual.

Lemma 5.3 *The following are equivalent:*

1. $\mathcal{Y}^s = \{0\}$;
2. *the strong topology of* $\mathcal{U}$ *coincides with the Mackey topology* $\tau(\mathcal{U}, \mathcal{Y})$;
3. $u1_{A^\nu} \to 0$ *strongly for all* $u \in \mathcal{U}$ *and all decreasing sequences* $(A^\nu)_{\nu \in \mathbb{N}}$ *in* $\mathcal{F}$ *with* $P(A^\nu) \searrow 0$.

Proof The equivalence of 1 and 2 follows from Remark A.48. Assume 3, let $u \in \mathcal{U}$, $y \in \mathcal{U}^*$ and let A^ν be the sets in the definition of the singular components y^s of y. We get

$$\langle u, y^s \rangle = \langle 1_{A^\nu} u, y^s \rangle + \langle 1_{\Omega \setminus A^\nu} u, y^s \rangle \to 0,$$

by 3. Since $u \in \mathcal{U}$ was arbitrary, this implies $y^s = 0$ so 1 holds. If 2 holds, then the strong topology is generated by seminorms of the form σ_C, where $C \subset \mathcal{Y}$ is $\sigma(\mathcal{Y}, \mathcal{U})$-compact. We have

$$\sigma_C(u1_{A^\nu}) = \sup_{y \in C} E[u \cdot y1_{A^\nu}]$$

so it suffices to show, by Theorem B.2, that the set $D := \{u \cdot y \mid y \in C\} \subset L^1$ is $\sigma(L^1, L^\infty)$-compact. Given any $u \in \mathcal{U}$, $y \in \mathcal{Y}$ and $\eta \in L^\infty$, we have

$$E[(u \cdot y)\eta] = E[(\eta u) \cdot y],$$

where $\eta u \in \mathcal{U}$, by solidity of $\mathcal{U}$. It follows that the mapping $y \mapsto u \cdot y$ is continuous from $(\mathcal{Y}, \sigma(\mathcal{Y}, \mathcal{U}))$ to $(L^1, \sigma(L^1, L^\infty))$. Thus, D is the image of a compact set under a continuous mapping and thus, compact. □

The conditions in Lemma 5.3 hold e.g. in L^p with $p \in [1, \infty)$ but not in L^∞. The following is an analogue of Lemma 3.6 which characterized the adjoint of the conditional expectation operator with respect to the pairing of $\mathcal{U}$ with $\mathcal{Y}$.

Lemma 5.4 *Let $\mathcal{G} \subset \mathcal{F}$ be a σ-algebra such that $E^{\mathcal{G}} u \in \mathcal{U}$ for all $u \in \mathcal{U}$. Then the linear mapping $E^{\mathcal{G}} : \mathcal{U} \to \mathcal{U}$ is strongly continuous and its adjoint is given by*

$$(E^{\mathcal{G}})^* y = E^{\mathcal{G}} y^c + (E^{\mathcal{G}})^* y^s$$

for every $y \in \mathcal{U}^*$.

Proof By Lemma 5.2, $\mathcal{Y}$ is the Köthe dual of $\mathcal{U}$. Thus, by Lemma 3.6, $E^{\mathcal{G}}$ is continuous with respect to $\sigma(\mathcal{U}, \mathcal{Y})$-topology. In particular, the graph of $E^{\mathcal{G}}$ is $\sigma(\mathcal{U}, \mathcal{Y}) \times \sigma(\mathcal{U}, \mathcal{Y})$-closed. The weak closedness implies strong closedness. Since $\mathcal{U}$ is Fréchet, the strong continuity follows now from the closed graph theorem, Example A.40. Given $y \in \mathcal{Y}$, Lemma 3.6 gives $E^{\mathcal{G}} y \in \mathcal{Y}$ and

$$\langle u, (E^{\mathcal{G}})^* y \rangle = \langle E^{\mathcal{G}} u, y \rangle = \langle u, E^{\mathcal{G}} y \rangle$$

for every $u \in \mathcal{U}$. Since $\mathcal{U}$ separates points in $\mathcal{U}^*$, we have $(E^{\mathcal{G}})^* y = E^{\mathcal{G}} y$. □

By Lemma 5.4, $(E^{\mathcal{G}})^* y \in \mathcal{Y}$ for every $y \in \mathcal{Y}$ but the image under $(E^{\mathcal{G}})^*$ of a singular functional need not be singular. For instance, if $(E^{\mathcal{G}})^* y = 0$, then $(E^{\mathcal{G}})^* y^s = -E^{\mathcal{G}} y^c$.

As in Chaps. 3 and 4, we assume that $(\mathcal{X}, \mathcal{V})$ is another pair of solid decomposable spaces of random variables in separating duality under the bilinear form $E[x \cdot v]$. As with $\mathcal{U}$, we will also assume that $\mathcal{X}$ is a Fréchet space whose topological dual $\mathcal{X}^*$ can be identified with $\mathcal{V} \oplus \mathcal{V}^s$, where $\mathcal{V}^s$ is the set of singular functionals on $\mathcal{X}$ defined analogously to the singular elements of $\mathcal{U}^*$. The following gives an analogue of the first part of Lemma 3.5.

Lemma 5.5 *Let $A \in L^0(\Omega, \mathcal{F}, P; \mathbb{R}^{m \times n})$ be a random matrix such that $Ax \in \mathcal{U}$ for all $x \in \mathcal{X}$. Then the linear mapping $\mathcal{A} : \mathcal{X} \to \mathcal{U}$ defined pointwise by*

$$\mathcal{A}x = Ax \quad a.s.$$

is strongly continuous and its adjoint is given by

$$\mathcal{A}^* y = A^* y^c + \mathcal{A}^* y^s.$$

Proof By Lemma 5.2, $\mathcal{V}$ is the Köthe dual of $\mathcal{X}$ so the $(\sigma(\mathcal{X}, \mathcal{V}), \sigma(\mathcal{U}, \mathcal{Y}))$-continuity follows from Lemma 3.5. In particular, $\operatorname{gph} \mathcal{A}$ is $\sigma(\mathcal{X}, \mathcal{V}) \times \sigma(\mathcal{U} \times \mathcal{Y})$-closed. The weak closedness implies strong closedness. Since $\mathcal{U}$ and $\mathcal{X}$ are Fréchet, the strong continuity follows now from the closed graph theorem, Example A.40. Given $y \in \mathcal{Y}$, Lemma 3.5 gives $A^* y \in \mathcal{V}$ and

$$\langle x, \mathcal{A}^* y\rangle = \langle \mathcal{A}x, y\rangle = \langle Ax, y\rangle = \langle x, A^* y\rangle$$

for every $x \in \mathcal{X}$. Since $\mathcal{X}$ separates points in $\mathcal{X}^*$, we have $\mathcal{A}^* y = A^* y$. □

The rest of the section gives examples of spaces $\mathcal{U}$ and $\mathcal{Y}$ satisfying the assumptions of this section. Let

$$L_{\mathcal{P}} := \bigcap_{p \in \mathcal{P}} \operatorname{dom} p,$$

where $\mathcal{P}$ is a countable collection of symmetric sublinear functionals on L^0 such that, for every $p \in \mathcal{P}$,

(A1) there exists $c > 0$ such that $\frac{1}{c}\|u\|_{L^1} \le p(u) \le c\|u\|_{L^\infty}$ for all $u \in L^1$,
(A2) if $u', u \in L^1$ with $|u'_j| \le |u_j|$ almost surely for all j, then $p(u') \le p(u)$,
(A3) p is L^1-lsc.

Consider also the following extra property: for every $p \in \mathcal{P}$,

(A4) $p(u 1_{A^\nu}) \searrow 0$ for all $u \in \mathcal{U}$ and decreasing sequence $(A^\nu)_{\nu \in \mathbb{N}}$ in $\mathcal{F}$ with $P(A^\nu) \searrow 0$.

Without loss of generality we may assume that $\mathcal{P}$ is a nondecreasing sequence. The assumption that p are symmetric and sublinear implies that $L_{\mathcal{P}}$ is a linear space of random variables. Let

$$L^{\mathcal{P}'} = \bigcup_{p \in \mathcal{P}} \operatorname{dom} p',$$

where p' is defined on L^1 by

$$p'(y) := \sup_{u \in L^\infty} \{E[u \cdot y] \mid p(u) \le 1\}.$$

Since $\mathcal{P}$ is nondecreasing, the sets $\operatorname{dom} p'$ are increasing, so $L^{\mathcal{P}'}$ is a linear space. The following is proved in Sect. B.1.

Theorem 5.6 *Let $\mathcal{U} = L_{\mathcal{P}}$ and $\mathcal{Y} = L^{\mathcal{P}'}$.*

1. *$\mathcal{U}$ and $\mathcal{Y}$ are solid decomposable spaces of random variables in separating duality under the bilinear form $(u, y) \mapsto E[u \cdot y]$.*
2. *$\mathcal{U}$ is a Fréchet space under the topology generated by $\mathcal{P}$, the corresponding strong dual can be expressed as $\mathcal{U}^* = \mathcal{Y} \oplus \mathcal{Y}^s$ and the Köthe dual of $\mathcal{U}$ is $\mathcal{Y}$.*

3. *For every $p \in \mathcal{P}$,*

$$E[u \cdot y] \leq p(u)p'(y) \quad \forall u \in \mathcal{U}, y \in \mathcal{Y}.$$

In particular, p' coincides with the polar seminorm

$$p^\circ(y) := \sup_{u \in \mathcal{U}}\{E[u \cdot y] \mid p(u) \leq 1\},$$

4. $\mathcal{Y}^s = \{0\}$ *if and only if (A4) holds.*

We say that a convex normal integrand $\Phi : \mathbb{R}^m \times \Omega \to \overline{\mathbb{R}}$ is a *random Young function* if, for almost every $\omega \in \Omega$,

1. $\Phi(0, \omega) = 0$,
2. $\Phi(\cdot, \omega)$ is inf-compact,
3. $0 \in \operatorname{int} \operatorname{dom} \Phi(\cdot, \omega)$,
4. $\Phi(u', \omega) \leq \Phi(u, \omega)$ for every $u', u \in \mathbb{R}^m$ with $|u'_j| \leq |u_j|$ for all j.

The *Musielak-Orlicz space* associated with Φ is the normed linear space $L^\Phi := \{u \in L^0 \mid \|u\|_{L^\Phi} < \infty\}$ where

$$\|u\|_{L^\Phi} := \inf\{\alpha > 0 \mid E\Phi(u/\alpha) \leq 1\}.$$

It is easily checked that, if Φ satisfies 1 and 4, then Φ^* satisfies 1 and 4 as well. If Φ satisfies 2 and 3, then by Theorem A.60, Φ^* satisfies 2 and 3 as well. Thus, by Theorem 1.46, a convex normal integrand is a random Young function if and only if Φ^* is a random Young function. Note that in the scalar case where $m = 1$, condition 2 can be replaced by the assumption that $\Phi(\cdot, \omega)$ be nonconstant and 4 by the condition that $\Phi(\cdot, \omega)$ be symmetric. If Φ is a random Young function on $\mathbb{R}$, then the function $\Phi(|\cdot|)$ satisfies the axioms above. The above multivariate definition also covers functions of the form $\Phi(u) = \Phi_1(u_1) + \Phi_2(u_2)$, where $u = (u_1, u_2)$ and Φ_1 and Φ_2 are random Young functions of u_1 and u_2.

The Musielak-Orlicz space associated with Φ^* is defined analogously to L^Φ. As before, the Euclidean ball with radius $\epsilon > 0$ will be denoted by $\mathbb{B}_\epsilon$.

Example 5.7 (Musielak-Orlicz Spaces) *Let Φ be a random Young function such that*

$$\Phi(u),\ \Phi^*(y) \leq m \quad \forall u, y \in \mathbb{B}_\epsilon\ a.s.$$

for some $\epsilon > 0$ and $m \in L^1$. Let $\mathcal{U} = L^\Phi$ and $\mathcal{Y} = L^{\Phi^*}$.

1. *$\mathcal{U}$ and $\mathcal{Y}$ are solid decomposable spaces of random variables in separating duality under the bilinear form $(u, y) \mapsto E[u \cdot y]$.*
2. *$\mathcal{U}$ is a Banach space, the corresponding strong dual can be expressed as $\mathcal{U}^* = \mathcal{Y} \oplus \mathcal{Y}^s$ and the Köthe dual of $\mathcal{U}$ is $\mathcal{Y}$.*

3. *We have*

$$E[u \cdot y] \le \|u\|_{L^\Phi} \|y\|^\circ_{L^\Phi} \quad \forall u \in \mathcal{U}, y \in \mathcal{Y},$$

where

$$\|y\|^\circ_{L^\Phi} := \sup_{u \in \mathcal{U}} \{E[u \cdot y] \mid \|u\|_{L^\Phi} \le 1\} = \inf_{\beta > 0} E[\beta \Phi^*(y/\beta) + \beta].$$

4. *Denoting* $A := \{\omega \in \Omega \mid \operatorname{dom} \Phi(\cdot, \omega) = \mathbb{R}^m\}$, *we have* $\mathcal{Y}^s = \{0\}$ *if and only if, for every* $u \in \mathcal{U}$, $u 1_A \in \operatorname{dom} E\Phi$ *and* $\{\Phi(u) = \infty\}$ *is a finite union of atoms.*

Proof We apply Theorem 5.6 with $\mathcal{P} = \{\| \cdot \|_{L^\Phi}\}$. We will first verify that $\| \cdot \|_{L^\Phi}$ satisfies (A1)–(A3). For any $\alpha > 0$, Fenchel's inequality gives

$$u \cdot y/\alpha \le \Phi(u/\alpha) + \Phi^*(y).$$

Maximizing both sides over $y \in \mathbb{B}_\epsilon$ gives $\epsilon |u|/\alpha \le \Phi(u/\alpha) + m$ so

$$\|u\|_{L^\Phi} \ge \inf\{\alpha > 0 \mid \epsilon \|u\|_{L^1}/\alpha - Em \le 1\} = \frac{\epsilon}{1 + Em} \|u\|_{L^1}.$$

On the other hand, if $u \in L^\infty$ with $\|u\|_{L^\infty} \le \epsilon$, we have $E\Phi(u) \le Em$, by assumption. Since $E\Phi(0) = 0$, convexity gives $E\Phi(u/\alpha) \le Em/\alpha$ for every $\alpha > 1$, so

$$\|u\|_{L^\Phi} \le \inf\{\alpha > 1 \mid Em/\alpha \le 1\} = \max\{Em, 1\}.$$

It follows that $\|u\|_{L^\Phi} \le (\max\{Em, 1\}/\epsilon) \|u\|_{L^\infty}$ for every $u \in L^\infty$. Thus, (A1) holds. Assumption (A2) holds since Φ has the corresponding properties on $\mathbb{R}^m$. By Theorem 1.55, $E\Phi$ is L^1-lsc so the set $\{u \in L^1 \mid E\Phi(u) \le 1\}$ is L^1-closed. Example A.21 thus gives (A3).

Let $y \in L^1$. By Example A.21, the infimum in the definition of the norm is attained. This, together with Example A.76 and the interchange rule in Theorem 1.64 give

$$\begin{aligned}
\|y\|'_{L^\Phi} &:= \sup_{u \in L^\infty} \{E[u \cdot y] \mid \|u\|_{L^\Phi} \le 1\} \\
&= \sup_{u \in L^\infty} \{E[u \cdot y] \mid E\Phi(u) \le 1\} \\
&= \inf_{\beta > 0} \sup_{u \in L^\infty} \{E[u \cdot y] - \beta E\Phi(u) + \beta\} \\
&= \inf_{\beta > 0} E[\sup_{u \in \mathbb{R}^m} \{u \cdot y - \beta \Phi(u)\} + \beta] \\
&= \inf_{\beta > 0} E[\beta \Phi^*(y/\beta) + \beta],
\end{aligned}$$

so

$$\operatorname{dom}\|\cdot\|'_{L^\Phi} = \{y \in L^0 \mid \exists \beta > 0 : E\Phi^*(y/\beta) < \infty\} = L^{\Phi^*}.$$

The first three claims now follow from Theorem 5.6.

Assume now that $\mathcal{Y}^s = \{0\}$, and, for a contradiction, that there exists $u \in L^\Phi$ such that either $E\Phi(u1_A) = \infty$ or $\{\Phi(u) = \infty\}$ is not a finite union of atoms. In both cases, there exists a sequence $(A^\nu)_{\nu\in\mathbb{N}}$ in $\mathcal{F}$ with $E\Phi(u1_{A^\nu}) = \infty$ and $P(A^\nu)\searrow 0$. Indeed, in the first case, we can take $A^\nu := \{\omega \in A \mid \Phi(u(\omega),\omega) \geq \nu\}$ while in the second case, there exist $A^\nu \subset \{\Phi(u) = \infty\}$ with $P(A^\nu) > 0$ and $P(A^\nu)\searrow 0$. Since $E\Phi(u/\beta)$ is nonincreasing in $\beta > 0$, we have $\|u1_{A^\nu}\|_{L^\Phi} \geq 1$ for every ν so $\mathcal{Y}^s \neq \{0\}$, by Lemma 5.3.

To prove the converse, let $u \in L^\Phi$ and $(A^\nu)_{\nu\in\mathbb{N}}$ a decreasing sequence in $\mathcal{F}$ with $P(A^\nu)\searrow 0$. Since $\Omega \setminus A$ is a finite union of atoms, there exists $\bar{\nu}$ such that $P((\Omega \setminus A) \cap A^\nu) = 0$ for every $\nu \geq \bar{\nu}$. Given any $\beta > 0$, we thus have $u1_{A^\nu}/\beta \in \operatorname{dom} E\Phi$ for $\nu \geq \bar{\nu}$ so, by the dominated convergence theorem, $E\Phi(u1_{A^\nu}/\beta)\searrow 0$. Since $\beta > 0$ was arbitrary, $\|u1_{A^\nu}\|_{L^\Phi}\searrow 0$ as well. By Lemma 5.3, this is equivalent to $\mathcal{Y}^s = \{0\}$. □

We will denote the closure of L^∞ in L^Φ by M^Φ.

Example 5.8 (Orlicz Heart) *Let Φ be as in Example 5.7, $\mathcal{U} = M^\Phi$ and $\mathcal{Y} = L^{\Phi^*}$.*

1. *$\mathcal{U}$ and $\mathcal{Y}$ are solid decomposable spaces of random variables in separating duality under the bilinear form $(u, y) \mapsto E[u \cdot y]$.*
2. *$\mathcal{U}$ is a Banach space, the corresponding strong dual can be expressed as $\mathcal{U}^* = \mathcal{Y} \oplus \mathcal{Y}^s$ and the Köthe dual of $\mathcal{U}$ is $\mathcal{Y}$.*
3. *We have*

$$E[u \cdot y] \leq \|u\|_{L^\Phi}\|y\|^\circ_{L^\Phi} \quad \forall u \in \mathcal{U}, y \in \mathcal{Y},$$

where

$$\|y\|^\circ_{L^\Phi} := \sup_{u\in\mathcal{U}}\{E[u \cdot y] \mid \|u\|_{L^\Phi} \leq 1\} = \inf_{\beta>0} E[\beta\Phi^*(y/\beta) + \beta].$$

4. *$\mathcal{Y}^s = \{0\}$ if and only if, for every $u \in \mathcal{U}$, $u1_A \in \operatorname{dom} E\Phi$ and $\{\Phi(u) = \infty\}$ is a finite union of atoms. Here, A is defined as in Example 5.7,*
5. *if $\Phi(u) \in L^1$ for all $u \in \mathbb{R}^m$, then*

$$\mathcal{U} = \{u \in L^0 \mid \forall\alpha > 0 : E\Phi(u/\alpha) < \infty\}$$

and $\mathcal{Y}^s = \{0\}$. This happens, in particular, when Φ is nonrandom and real-valued. A space $\mathcal{U}$ of the above form is known as the Orlicz heart *associated with Φ.*

Proof To show that M^Φ is solid, let $u \in M^\Phi$ and $u' \in L^0$ such that $|u'_j| \le |u_j|$ for $j = 1, \dots, n$. Let $(u^\nu)_{\nu\in\mathbb{N}}$ in L^∞ such that $\|u - u^\nu\|_{L^\Phi} \searrow 0$ and let D be the random diagonal matrix with elements u'_j/u_j, where $0/0 := 0$. As observed in the proof of Example 5.7, the norm $\|\cdot\|_{L^\Phi}$ satisfies (A2), so

$$\|u' - Du^\nu\|_{L^\Phi} = \|D(u - u^\nu)\|_{L^\Phi} \le \|u - u^\nu\|_{L^\Phi}.$$

Since $Du^\nu \in L^\infty$, we thus have $u' \in M^\Phi$ so M^Φ is solid. Since M^Φ contains L^∞, it is also decomposable. Since $L^\infty \subset M^\Phi$ and L^∞ separates points in L^{Φ^*}, the spaces M^Φ and L^{Φ^*} are in separating duality by Example 5.7.1.

Since M^Φ is a closed subspace of the Banach space L^Φ (see Example 5.7), it is Banach. By Corollary A.14, every continuous linear functional on M^Φ can be represented by an element of $(L^\Phi)^*$. By Example 5.7, $(M^\Phi)^*$ is thus contained in $L^{\Phi^*} \oplus \mathcal{Y}^s$. On the other hand, every element of $(L^\Phi)^*$ is continuous on M^Φ. Since $L^\infty \subset M^\Phi$ and L^∞ separates points in L^{Φ^*}, we have $(M^\Phi)^* = L^{\Phi^*} \oplus \mathcal{Y}^s$. Part 2 follows now from Lemma 5.2. Part 3 is a direct consequence of part 3 in Example 5.7. Part 4 is proved like that of Example 5.7.

As to 5, assume that $E\Phi(u) \in L^1$ for every $u \in \mathbb{R}^m$. To prove that $\mathcal{U}$ coincides with the space

$$M := \{u \in L^0 \mid \forall \alpha > 0 : E\Phi(u/\alpha) < \infty\},$$

let $u \in M$. Dominated convergence theorem gives

$$E\Phi((u 1_{|u|\le\nu} - u)/\beta) \searrow 0$$

for any $\beta > 0$, so $u 1_{|u|\le\nu} \to u$ in L^Φ. Thus, $M \subseteq \mathcal{U}$. To prove the converse, note first that $L^\infty \subseteq M$. It thus suffices to show that M is closed in L^Φ. If $(u^\nu)_{\nu\in\mathbb{N}} \subset M$ converges to u in L^Φ, then, for any $\beta > 0$,

$$E\Phi(u/(2\beta)) \le \frac{1}{2}E\Phi(u^\nu/\beta) + \frac{1}{2}E\Phi(u - u^\nu/\beta) \le \frac{1}{2}E\Phi(u^\nu/\beta) + \frac{1}{2}$$

for ν large enough, so $E\Phi(u/(2\beta)) < \infty$ and thus $u \in M$. □

Remark 5.9 *Some reduced formulations of the dual problems in Sect. 3.4 assumed that the conditional expectations map the involved spaces into themselves. In the setting of Theorem 5.6, we have $E^{\mathcal{G}} L_{\mathcal{P}} \subseteq L_{\mathcal{P}}$ if $p(E^{\mathcal{G}} u) \le p(u)$ for every $p \in \mathcal{P}$ and $u \in L^1$. This holds in L^Φ if Φ is $\mathcal{G}$-measurable. Indeed, by Jensen's inequality in Theorem 3.13, $E\Phi(E^{\mathcal{G}} u) \le E\Phi(u)$, so $\|E^{\mathcal{G}} u\|_{L^\Phi} \le \|u\|_{L^\Phi}$. If Φ is $\mathcal{G}$-measurable, we also have $E^{\mathcal{G}} M^\Phi \subseteq M^\Phi$ since, by Lemma 5.4, $E^{\mathcal{G}} : L^\Phi \to L^\Phi$ is strongly continuous, so if $u^\nu \to u$, then $E^{\mathcal{G}} u^\nu \to E^{\mathcal{G}} u$.*

Remark 5.10 *Let Φ be as in Example 5.7. We have*

$$L^{\Phi} = \bigcup_{\alpha>0} \alpha \operatorname{dom} E\Phi$$

and, if $\Phi(u) \in L^1$ for all $u \in \mathbb{R}^m$, then, by Example 5.8.5,

$$M^{\Phi} = \bigcap_{\alpha>0} \alpha \operatorname{dom} E\Phi.$$

If, in addition, $\operatorname{dom} E\Phi$ *is a cone, then by Examples 5.7.4 and 5.8.5, $L^{\Phi} = M^{\Phi} =$* $\operatorname{dom} E\Phi$ *and the topological duals of both L^{Φ} and M^{Φ} coincide with L^{Φ^*}.*

The domain of $E\Phi$ is a cone if Φ satisfies the Δ_2-condition: there exist $\bar{u} \in$ $\operatorname{dom} E\Phi$ *and $K > 0$ such that*

$$\Phi(2u, \omega) \leq K\Phi(u, \omega) \quad \forall u \notin S_{\bar{u}}(\omega),$$

where $S_{\bar{u}}(\omega) := \{u \in \mathbb{R}^m \mid |u^j| \leq |\bar{u}^j(\omega)| \, \forall j\}$.

The domain of $E\Phi$ fails to be a cone e.g. for $\Phi(u) = \exp(|u|) - 1$. Indeed, if $u \in L^0$ has the standard normal distribution, we have

$$E[\Phi(\alpha u^2)] = \frac{1}{\sqrt{2\pi}} \int_{\mathbb{R}} \exp(\alpha u^2 - \frac{1}{2}u^2) du - 1$$

which is finite if and only if $\alpha < 1/2$. This also shows that $\operatorname{dom} E\Phi$ *need not be closed in L^{Φ}.*

Proof Assume the Δ_2-condition and let $u \in \operatorname{dom} E\Phi$ and $A := \{\omega \in \Omega \mid u(\omega) \in S_{\bar{u}}(\omega)\}$. We get

$$\begin{aligned}
\Phi(2u) &= \Phi(2u)1_A + \Phi(2u)1_{\Omega\setminus A} \\
&\leq \Phi(2u_0)1_A + \Phi(2u)1_{\Omega\setminus A} \\
&\leq K\Phi(u_0)1_A + K\Phi(u)1_{\Omega\setminus A},
\end{aligned}$$

where the first inequality follows from property 4 of the definition of a random Young function and the second inequality from the Δ_2-condition. Thus, $2u \in \operatorname{dom} E\Phi$. □

The following is a special case of Example 5.7.

Example 5.11 (Lebesgue Spaces) *Let $p \in [1, \infty]$ and $q \in [1, \infty]$ be such that $\frac{1}{p} + \frac{1}{q} = 1$. Equipped with the norm*

$$\|u\|_{L^p} := \begin{cases} (E|x|^p)^{1/p} & \text{if } p < \infty, \\ \operatorname{esssup} |x| & \text{if } p = \infty, \end{cases}$$

the Lebesgue space $L^p = \{u \in L^0 \mid \|u\|_{L^p} < \infty\}$ *is a Banach space, its topological dual can be expressed as*

$$(L^p)^* = \begin{cases} L^q & \text{if } p < \infty, \\ L^1 \oplus (L^1)^s & \text{if } p = \infty, \end{cases}$$

where $(L^1)^s$ *is the set of singular elements of* $(L^\infty)^*$, *the Köthe dual of* L^p *is* L^q,

$$E[u \cdot y] \leq \|u\|_{L^q} \|y\|_{L^q} \quad \forall u \in L^p, y \in L^q$$

and

$$\|y\|'_{L^p} = \|y\|_{L^q}.$$

We have $(L^1)^s = \{0\}$ *if and only if* Ω *is a union of finitely many atoms of* P.

Proof The L^p-norm is the Orlicz norm associated with

$$\Phi(u) := \begin{cases} |u|^p & \text{if } p \in [1, \infty), \\ \delta_{\mathbb{B}}(u) & \text{if } p = \infty. \end{cases}$$

We have

$$\Phi^*(u) := \begin{cases} \delta_{\mathbb{B}}(u) & \text{if } p = 1, \\ \frac{p^{1-q}}{q}|y|^q & \text{if } p \in (1, \infty), \\ |u| & \text{if } p = \infty. \end{cases}$$

An elementary calculation gives

$$\inf_{\beta > 0} E[\beta \Phi^*(y/\beta) + \beta] = \|y\|_{L^q},$$

so the claims follow from Example 5.7. □

Given an increasing sequence $S \subset [1, \infty)$, let

$$L^S := \{u \in L^1 \mid \forall p \in S : \|u\|_{L^p} < \infty\}.$$

If S contains its supremum $\bar{p} := \sup S$, then L^S coincides with the Lebesgue space $L^{\bar{p}}$. If $\bar{p} \notin S$, L^S is the space of random variables with finite pth moments $E[|u|^p]$ for p strictly less than $\bar{p}$. When S is unbounded, L^S is the *space of random variables with finite moments*. Let S' be the set of conjugate exponents of S, i.e.

$$S' := \{q \in (1, \infty] \mid \exists p \in S : \frac{1}{p} + \frac{1}{q} = 1\}$$

and define

$$L_{S'} := \{y \in L^1 \mid \exists q \in S' : \|y\|_{L^q} < \infty\}.$$

The L^p-norms with $p < \infty$ satisfy 5.1.1–5.1.1. The following example is thus a consequence of Theorem 5.6 and the identity $\|y\|'_{L^p} = \|y\|_{L^q}$ from Example 5.11.

Example 5.12 (Spaces with Finite Moments) *Equipped with the collection of seminorms $\{\|\cdot\|_{L^p} \mid p \in S\}$, the space L^S is Fréchet and both its topological dual and Köthe dual may be identified with $L_{S'}$ under the bilinear form $\langle u, y\rangle := E[u\cdot y]$. If S is unbounded, then by Hölder's inequality in Example 5.11, L^S contains pointwise products of its elements.*

Besides the above examples, Theorem 5.6 covers many other spaces of random variables. Examples include, e.g., Lorentz and Marcinkiewicz spaces that have found applications in the theory of spectral risk measures.

5.1.2 Conjugates of Integral Functionals

The goal of this section is to calculate conjugates and subdifferentials of integral functionals with respect to the strong dual $\mathcal{U}^*$ of $\mathcal{U}$. We continue with the assumptions of Sect. 5.1.1 so that the strong dual of $\mathcal{U}$ can be identified with $\mathcal{U}^* = \mathcal{Y} \oplus \mathcal{Y}^s$ in the sense that every $y \in \mathcal{U}^*$ can be expressed uniquely as the sum $y = y^c + y^s$ of a random variable $y^c \in \mathcal{Y}$ and a singular functional $y^c \in \mathcal{Y}^s$.

Let h be a convex normal integrand on $\mathbb{R}^m \times \Omega$. We saw in Sect. 3.1.2 that, as soon as $\operatorname{dom} Eh \neq \emptyset$, the conjugate of the integral functional $Eh : \mathcal{U} \mapsto \overline{\mathbb{R}}$, in the pairing of $\mathcal{U}$ with $\mathcal{Y}$, is the integral functional $Eh^* : \mathcal{Y} \mapsto \overline{\mathbb{R}}$, where h^* is the conjugate integrand of h. With respect to the strong dual $\mathcal{U}^*$, the conjugate is not just an integral functional, in general, but the structure $\mathcal{U}^* = \mathcal{Y} \oplus \mathcal{Y}^s$ allows us to still come up with a fairly explicit expression for $(Eh)^*$.

Lemma 5.13 *Let Eh be proper. If $\bar{u} \in \mathcal{U}$, $y \in \mathcal{U}^*$ and $\epsilon \geq 0$ are such that*

$$Eh(u) \geq Eh(\bar{u}) + \langle u - \bar{u}, y\rangle - \epsilon \quad \forall u \in \mathcal{U}, \tag{5.2}$$

then

$$Eh(u) \geq Eh(\bar{u}) + \langle u - \bar{u}, y^c\rangle - \epsilon \quad \forall u \in \mathcal{U}, \tag{5.3}$$

and

$$0 \geq \langle u - \bar{u}, y^s\rangle - \epsilon \quad \forall u \in \operatorname{dom} Eh. \tag{5.4}$$

Proof Let $u \in \operatorname{dom} Eh$ and let $(A^\nu)_{\nu\in\mathbb{N}}$ be the decreasing sequence of sets in the characterization of the singular component y^s of y. Let $u^\nu := 1_{A^\nu}\bar{u} + 1_{\Omega\setminus A^\nu}u$. Inequality (5.2) implies $\bar{u} \in \operatorname{dom} Eh$ and

$$Eh(u^\nu) \geq Eh(\bar{u}) + \langle u^\nu - \bar{u}, y\rangle - \epsilon.$$

Since

$$h(u^\nu) \leq \max\{h(u), h(\bar{u})\},$$

where the right side is integrable, we have

$$\limsup_{\nu\to\infty} Eh(u^\nu) \leq Eh(u),$$

by Fatou's lemma. Combining the above gives

$$Eh(u) \geq Eh(\bar{u}) + \limsup\langle u^\nu - \bar{u}, y\rangle - \epsilon.$$

Since $u^\nu - \bar{u} = 1_{\Omega\setminus A^\nu}(u - \bar{u})$, we have

$$\langle u^\nu - \bar{u}, y\rangle = \langle u^\nu - \bar{u}, y^c\rangle \to \langle u - \bar{u}, y^c\rangle$$

so (5.3) holds.

Now let $u^\nu := 1_{A^\nu}u + 1_{\Omega\setminus A^\nu}\bar{u}$. By Fatou's lemma and (5.2) again,

$$Eh(\bar{u}) \geq \limsup Eh(u^\nu) \geq Eh(\bar{u}) + \limsup\langle u^\nu - \bar{u}, y\rangle - \epsilon.$$

Since $u^\nu - \bar{u} = 1_{A^\nu}(u - \bar{u})$, we have

$$\langle u^\nu - \bar{u}, y\rangle = \langle u^\nu - \bar{u}, y^c\rangle + \langle u^\nu - \bar{u}, y^s\rangle \to \langle u - \bar{u}, y^s\rangle$$

so $\langle u - \bar{u}, y^s\rangle - \epsilon \leq 0$. Since $u \in \operatorname{dom} Eh$ was arbitrary, (5.4) holds. □

Using Theorem 3.8, Lemma 5.13 can be restated as follows.

Corollary 5.14 *Let Eh be proper. If $\bar{u} \in \mathcal{U}$, $y \in \mathcal{U}^*$ and $\epsilon \geq 0$ are such that*

$$Eh(\bar{u}) + (Eh)^*(y) \leq \langle \bar{u}, y\rangle + \epsilon \tag{5.5}$$

then

$$Eh(\bar{u}) + Eh^*(y^c) \leq \langle \bar{u}, y^c\rangle + \epsilon \tag{5.6}$$

and

$$\sigma_{\operatorname{dom} Eh}(y^s) \leq \langle \bar{u}, y^s\rangle + \epsilon. \tag{5.7}$$

We now come to the main result of this section which states that, at each $y = y^c + y^s \in \mathcal{U}^*$, the conjugate of Eh equals the sum of the integral functional Eh^* evaluated at y^c and the recession function of $(Eh)^*$ evaluated at y^s; see Lemma A.45.

Theorem 5.15 *If Eh is proper, then*

1. $(Eh)^*(y) = Eh^*(y^c) + \sigma_{\mathrm{dom}\, Eh}(y^s) \quad \forall y \in \mathcal{U}^*$,
2. *Eh is strongly lsc if and only if it is $\sigma(\mathcal{U}, \mathcal{Y})$-lsc,*
3. *$y \in \partial Eh(\bar{u})$ if and only if $y^c \in \partial Eh(\bar{u})$ and $y^s \in N_{\mathrm{dom}\, Eh}(\bar{u})$.*

Proof Given $y \in \mathcal{U}^*$,

$$
\begin{aligned}
(Eh)^*(y) &= \sup_u \{\langle u, y\rangle - Eh(u)\} \\
&= \sup_u \{\langle u, y^c\rangle - Eh(u) + \langle u, y^s\rangle - \delta_{\mathrm{dom}\, Eh}(u)\} \\
&\le Eh^*(y^c) + \sigma_{\mathrm{dom}\, Eh}(y^s).
\end{aligned}
$$

If $y \in \mathrm{dom}(Eh)^*$ and $\epsilon > 0$, there exists $\bar{u} \in \mathrm{dom}\, Eh$ such that (5.5) holds. Adding (5.6) and (5.7) together,

$$
\begin{aligned}
(Eh)^*(y^c) + \sigma_{\mathrm{dom}\, Eh}(y^s) &\le \langle \bar{u}, y^c\rangle - Eh(\bar{u}) + \epsilon + \langle \bar{u}, y^s\rangle + \epsilon \\
&\le (Eh)^*(y) + 2\epsilon.
\end{aligned}
$$

This proves the first claim since $\epsilon > 0$ was arbitrary.

It is clear that if Eh is $\sigma(\mathcal{U}, \mathcal{Y})$-lsc it is strongly lsc. When Eh is proper and strongly lsc, the biconjugate theorem says that it has a proper conjugate. The first claim then implies that Eh^* is proper on $\mathcal{Y}$. The second claim thus follows from Corollary 3.9.

When $\epsilon = 0$, Lemma 5.13 says that if $y \in \partial Eh(\bar{u})$ then $y^c \in \partial Eh(\bar{u})$ and $y^s \in N_{\mathrm{dom}\, Eh}(\bar{u})$. The converse implication follows simply by adding (5.3) and (5.4) together. □

Theorem 5.15 says, in particular, that subdifferentiability of Eh with respect to the strong dual $\mathcal{U}^*$ implies subdifferentiability of Eh with respect to the pairing of $\mathcal{U}$ with $\mathcal{Y}$. Combined with Theorem A.33 and Corollary A.62, Theorem 5.15 yields the following

Corollary 5.16 *If Eh is proper and $u \in \mathrm{core}\,\mathrm{dom}\, Eh$, then Eh is strongly continuous at u,*

$$
\partial Eh(u) = \{y \in \mathcal{Y} \mid y \in \partial h(u) \ a.s.\} \neq \emptyset
$$

and Eh is strongly lsc throughout $\mathcal{U}$.

Proof By Theorem 3.11, Eh is $\sigma(\mathcal{U}, \mathcal{Y})$-lsc so is also strongly lsc. The continuity thus follows from Theorem A.33 and subdifferentiability from Corollary A.62. By Theorem 5.15, it now suffices to note that $N_{\mathrm{dom}\, Eh}(u) = \{0\}$ for all $u \in \mathrm{core}\,\mathrm{dom}\, Eh$. □

Combining Corollary 5.16 with Theorem A.60 (an extension of the Banach-Alaoglu theorem), gives the following.

Corollary 5.17 *If Eh is finite on $\mathcal{U}$, then it is $\tau(\mathcal{U}, \mathcal{Y})$-continuous throughout $\mathcal{U}$ and $y \mapsto E[h^*(y) - u \cdot y]$ is $\sigma(\mathcal{Y}, \mathcal{U})$-inf-compact on $\mathcal{Y}$ for every $u \in \mathcal{U}$.*

Proof By Corollary 5.16, Eh is strongly continuous throughout $\mathcal{U}$. By Theorem A.60, level-sets of $y \mapsto (Eh)^*(y) - \langle u, y\rangle$ are $\sigma(\mathcal{U}^*, \mathcal{U})$-compact. By Theorem 5.15, the finiteness of Eh implies that $(Eh)^* = Eh^*$ and that all the level-sets are contained in $\mathcal{Y}$, so they are $\sigma(\mathcal{Y}, \mathcal{U})$-compact. By Theorem A.60 again, this implies that Eh is $\tau(\mathcal{U}, \mathcal{Y})$-continuous. □

The following theorem says that, if $\mathcal{Y}^s = \{0\}$, then $\mathrm{core}\,\mathrm{dom}\, Eh$ tends to be empty unless it equals the whole space $\mathcal{U}$. By Lemma 5.3, $\mathcal{Y}^s = \{0\}$ if and only if strong topology coincides with the Mackey topology $\tau(\mathcal{U}, \mathcal{Y})$.

Theorem 5.18 *Assume that $\mathcal{Y}^s = \{0\}$, Eh is proper and that the set $\{\omega \in \Omega \mid \mathrm{dom}\, h(\cdot, \omega) \neq \mathbb{R}^m\}$ is atomless. Then* $\mathrm{core}\,\mathrm{dom}\, Eh \neq \emptyset$ *if and only if* $\mathrm{dom}\, Eh = \mathcal{U}$.

Proof Assume, for contradiction, that $\mathrm{core}\,\mathrm{dom}\, Eh \neq \emptyset$ but $Eh(u) = +\infty$ for some $u \in \mathcal{U}$. Translating if necessary, we may assume that $0 \in \mathrm{core}\,\mathrm{dom}\, Eh$. By Corollary 5.16, Eh is strongly continuous at the origin. Let $A := \{h(u) < \infty\}$. We have either $Eh(u1_A) = \infty$ or $P(\{h(u) = \infty\}) > 0$. In both cases, there exists a sequence $(A^\nu)_{\nu \in \mathbb{N}}$ in $\mathcal{F}$ with $Eh(u1_{A^\nu}) = \infty$ and $P(A^\nu) \searrow 0$. Indeed, in the first case, we can take $A^\nu := \{\omega \in A \mid h(u(\omega), \omega) \geq \nu\}$ while in the second case, there exist $A^\nu \subset \{h(u) = \infty\}$ with $P(A^\nu) > 0$ and $P(A^\nu) \searrow 0$. We then have $Eh(u1_{A^\nu}) = +\infty$ for every ν while, by Lemma 5.3, $u1_{A^\nu} \to 0$ strongly. This contradicts the continuity of Eh at the origin. □

Under the assumptions of Theorem 5.18, $\mathrm{core}\,\mathrm{dom}\, Eh = \emptyset$ unless $\mathrm{dom}\, Eh = \mathcal{U}$. In the context of stochastic optimization, this would mean that, in the presence of pointwise constraints, the domain of the objective of (P) has empty core. In such applications, the core can be nonempty only in Fréchet spaces $\mathcal{U}$ whose duals contain nonzero singular functionals.

If $\mathcal{U}$ is the Musielak-Orlicz space associated with a random Young function Φ and if $\mathcal{Y}^s \neq \{0\}$, then there always exist integral functionals on $\mathcal{U}$ with $\mathrm{core}\,\mathrm{dom}\, Eh \neq \emptyset$ but $\mathrm{dom}\, Eh \neq \mathcal{U}$. Indeed, $\mathrm{dom}\, E\Phi$ contains the unit ball of L^Φ but, by Remark 5.10, $\mathrm{dom}\, E\Phi \neq L^\Phi$ when $\mathcal{Y}^s \neq \{0\}$.

When $\mathcal{U} = L^\infty$, $\mathrm{core}\,\mathrm{dom}\, Eh$ can be given a pointwise characterization in terms of $\mathrm{dom}\, h$.

Remark 5.19 *If* $\mathrm{int}\,\mathcal{U}(\mathrm{dom}\, h) \subseteq \mathrm{dom}\, Eh$*, then*

$$\mathrm{int}\,\mathrm{dom}\, Eh = \mathrm{int}\,\mathcal{U}(\mathrm{dom}\, h).$$

If $\mathcal{U} = L^\infty$, *then*

$$\operatorname{int}\mathcal{U}(\operatorname{dom} h) = \{u \in L^\infty \mid \exists \epsilon > 0 : \mathbb{B}_\epsilon(u) \in \operatorname{dom} h \ a.s.\}.$$

Proof The second claim is obvious while the first claim follows from the fact that $\operatorname{dom} Eh \subseteq \mathcal{U}(\operatorname{dom} h)$. □

If Eh is strongly lsc and proper with $\operatorname{aff}\operatorname{dom} Eh$ strongly closed, then by Corollaries A.36 and A.64, it is subdifferentiable throughout the relative core of its domain. A simple sufficient condition for the closedness of $\operatorname{aff}\operatorname{dom} Eh$ is that it equals $\mathcal{U}(\operatorname{aff}\operatorname{dom} h)$. Indeed, the closedness then follows from Corollary 3.10.

Example 5.20 *Let* $h(u, \omega) = h_0(u, \omega) + \delta_{H(\omega)}(u)$, *where* H *is a random affine set and* h_0 *is a convex normal integrand with* $\operatorname{core}\operatorname{dom} Eh_0 \cap \mathcal{U}(H) \neq \emptyset$. *Then* $\operatorname{rcore}\operatorname{dom} Eh = \operatorname{core}\operatorname{dom} Eh_0 \cap \mathcal{U}(H)$ *and* $\operatorname{aff}\operatorname{dom} Eh = \mathcal{U}(H)$ *which is strongly closed. If* Eh_0 *is proper, then* Eh *is relatively strongly continuous throughout* $\operatorname{rcore}\operatorname{dom} Eh$.

Proof By Theorem 1.53, $Eh = Eh_0 + \delta_{\mathcal{U}(H)}$ so $\operatorname{dom} Eh = \operatorname{dom} Eh_0 \cap \mathcal{U}(H)$. The first claim thus follows from Lemma A.5. Given $\bar{u} \in \operatorname{core}\operatorname{dom} Eh_0 \cap \mathcal{U}(H)$, Lemmas A.4 and A.3 give

$$\begin{aligned}\operatorname{aff}(\operatorname{dom} Eh - \bar{u}) &= \operatorname{pos}(\operatorname{dom} Eh - \bar{u})\\ &= \operatorname{pos}(\operatorname{dom} Eh_0 - \bar{u}) \cap \operatorname{pos}(\mathcal{U}(H) - \bar{u}))\\ &= \mathcal{U}(H) - \bar{u}.\end{aligned}$$

Thus, $\operatorname{aff}\operatorname{dom} Eh = \mathcal{U}(H)$ which is $\sigma(\mathcal{U}, \mathcal{Y})$-lsc, by Corollary 3.10. By Corollary 5.16, Eh_0 is strongly lsc. It follows that Eh is strongly lsc as well. Since $\bar{u} \in \operatorname{dom} Eh_0$, the properness of Eh_0 implies that of Eh so the last claim follows from Corollary A.36. □

The following gives a simple example where $\operatorname{pos}\operatorname{dom} Eh$ is linear but not closed. In other words, we get $0 \in \operatorname{rcore}\operatorname{dom} Eh$ but $\operatorname{aff}\operatorname{dom} Eh$ is not closed. Remark A.37 thus implies that the function Eh is not relatively continuous on $\operatorname{rcore}\operatorname{dom} Eh$.

Counterexample 5.21 *Let* $\mathcal{U} = L^\infty$, $h(u, \omega) = \delta_S(u, \omega)$, *where* $S(\omega) := \{u \mid |u| \le \eta(\omega)\}$ *for strictly positive* $\eta \in L^\infty$ *such that* $1/\eta \notin L^\infty$. *Then*

$$\operatorname{pos}\operatorname{dom} Eh = \{u \in L^\infty \mid \exists \lambda > 0 : |u| \le \lambda\eta\}$$

is linear but not closed in L^∞. *Indeed, defining* $u^\nu := \min\{\sqrt{\eta}, \nu\eta\}$, *we have* $u^\nu \in \nu \operatorname{dom} Eh \subseteq \operatorname{pos}\operatorname{dom} Eh$ *for all* ν *and* $u^\nu \to \sqrt{\eta}$ *in* L^∞ *but, since* $1/\sqrt{\eta} \notin L^\infty$, *we have* $\sqrt{\eta} \notin \operatorname{pos}\operatorname{dom} Eh$.

5.2 The Strong Dual Problem

For the remainder of this chapter, we will assume that both $\mathcal{X}$ and $\mathcal{U}$ satisfy the assumptions of Sect. 5.1.2. That is, $\mathcal{X}$ and $\mathcal{U}$ are endowed with *strong topologies* under which they are Fréchet and such that the dual $\mathcal{U}^*$ of $\mathcal{U}$ can be identified with $\mathcal{Y} \oplus \mathcal{Y}^s$ and the dual $\mathcal{X}^*$ of $\mathcal{X}$ can be identified with $\mathcal{V} \oplus \mathcal{V}^s$. More precisely, for each $v \in \mathcal{X}^*$, there exist unique $v^c \in \mathcal{V}$ and $v^s \in \mathcal{V}^s$ such that

$$\langle x, v\rangle = E[x \cdot v^c] + \langle x, v^s\rangle$$

for all $x \in \mathcal{X}$. Here, $\mathcal{V}^s$ denotes the space of singular elements of $\mathcal{X}^*$, defined analogously to $\mathcal{Y}^s$ in Sect. 5.1.2. Note again that the above assumptions imply, by Lemma 5.2, that $\mathcal{V}$ is the Köthe dual of $\mathcal{X}$.

Much like in Sect. 3.2, we apply the general conjugate duality framework to $(P_{\mathcal{X}})$ with the Rockafellian

$$F(x, z, u) = Ef(x, u) + \delta_{\mathcal{N}}(x - z),$$

the optimization variable $x \in \mathcal{X}$, the parameter space $\mathcal{X} \times \mathcal{U}$ and $\bar{v} = 0$. This time, however, the spaces $\mathcal{X}$ and $\mathcal{U}$ are paired with $\mathcal{X}^*$ and $\mathcal{U}^*$, respectively. According to the general conjugate duality theory from Sect. A.9, this results in the dual problem

$$\text{maximize} \quad \langle \bar{u}, y\rangle - F^*(0, p, y) \quad \text{over } (p, y) \in \mathcal{X}^* \times \mathcal{U}^*, \tag{D_s}$$

where F^* is the conjugate of F with respect to the pairings of $\mathcal{X}$ and $\mathcal{U}$ with their strong duals $\mathcal{X}^*$ and $\mathcal{U}^*$, respectively. We refer to (D_s) as the *strong dual problem*. Note that the original dual problem (D) introduced in Sect. 3.2 is obtained from (D_s) by restricting the variables (p, y) to $\mathcal{V} \times \mathcal{Y}$ which is a subspace of the strong dual space $\mathcal{X}^* \times \mathcal{U}^*$. If the strong dual spaces don't have singular elements, then $\mathcal{X}^* \times \mathcal{U}^* = \mathcal{X} \times \mathcal{Y}$ and the strong dual problem (D_s) becomes (D).

As long as the optimum value of $(P_{\mathcal{X}})$ is finite, the subgradients of the optimum value function

$$\varphi(z, u) := \inf_{x\in\mathcal{X}}\{Ef(x, u) \mid x - z \in \mathcal{N}\}$$

in the space $\mathcal{X}^* \times \mathcal{U}^*$ are solutions of (D_s); see Theorem A.72. Establishing the existence of a subgradient in the strong dual may be significantly easier than establishing the existence in the smaller space $\mathcal{V} \times \mathcal{Y} \subseteq \mathcal{X}^* \times \mathcal{U}^*$. Indeed, since $\mathcal{X}$ and $\mathcal{U}$ are Fréchet, we can use Theorem A.75 to derive sufficient conditions for the subdifferentiability. The special structure of conjugates of integral functionals given in Sect. 5.1.2 will then allow us to show that the existence of solutions to the strong dual problem (D_s) implies the existence of solutions of the original dual problem (D) and, moreover, that the optimum values are equal. This will be done in Sect. 5.3 below.

Clearly,

$$F(x, z, u) = Ef(x, u) + \delta_{\mathcal{X}_a}(x - z) \quad \forall (x, z, u) \in \mathcal{X} \times \mathcal{X} \times \mathcal{U},$$

where, as in Sect. 3.2, $\mathcal{X}_a := \mathcal{X} \cap \mathcal{N}$. By Lemma 3.16, $\mathcal{X}_a$ is weakly closed and thus, strongly closed as well. We will denote the orthogonal complement of $\mathcal{X}_a$ in the strong dual of $\mathcal{X}$ by

$$\mathcal{X}_a^\circ := \{p \in \mathcal{X}^* \mid \langle x, p \rangle = 0 \,\forall x \in \mathcal{X}_a\}.$$

Note that the orthogonal complement

$$\mathcal{X}_a^\perp = \{p \in \mathcal{V} \mid \langle x, p \rangle = 0 \,\forall x \in \mathcal{X}_a\}$$

of $\mathcal{X}_a$ in $\mathcal{V}$ (see Sect. 3.2) is contained in $\mathcal{X}_a^\circ$. The following is an analogue of Lemma 3.20.

Theorem 5.22 *If* dom $Ef \cap (\mathcal{X} \times \mathcal{U}) \neq \emptyset$, *then*

$$F^*(v, p, y) = Ef^*((v + p)^c, y^c) + \sigma_{\mathrm{dom}\, Ef}((v + p)^s, y^s) + \delta_{\mathcal{X}_a^\circ}(p)$$

and, in particular,

$$\varphi^*(p, y) = Ef^*(p^c, y^c) + \sigma_{\mathrm{dom}\, Ef}(p^s, y^s) + \delta_{\mathcal{X}_a^\circ}(p).$$

If, in addition, dom $Ef^* \cap (\mathcal{V} \times \mathcal{Y}) \neq \emptyset$, *then F is proper and strongly lsc on $\mathcal{X} \times \mathcal{U}$.*

Proof By Theorem 5.15,

$$\begin{aligned} F^*(v, p, y) &= \sup_{x \in \mathcal{X}, z \in \mathcal{X}, u \in \mathcal{U}} \{\langle x, v \rangle + \langle z, p \rangle + \langle u, y \rangle - Ef(x, u) \,|\, x - z \in \mathcal{X}_a\} \\ &= \sup_{x \in \mathcal{X}, z' \in \mathcal{X}, u \in \mathcal{U}} \{\langle x, v + p \rangle + \langle u, y \rangle - Ef(x, u) - \langle z', p \rangle] \,|\, z' \in \mathcal{X}_a\} \\ &= Ef^*((v + p)^c, y^c) + \sigma_{\mathrm{dom}\, Ef}((v + p)^s, y^s) + \delta_{\mathcal{X}_a^\circ}(p). \end{aligned}$$

When dom $Ef^* \neq \emptyset$, Lemma 3.20 says that F is proper and lsc in the weak topology $\sigma(\mathcal{X}, \mathcal{V}) \times \sigma(\mathcal{U}, \mathcal{Y})$. This clearly implies strong lower semicontinuity. □

As an immediate corollary, we get the following.

Corollary 5.23 *If* dom $Ef \cap (\mathcal{X} \times \mathcal{U}) \neq \emptyset$, *the strong dual problem* (D_s) *can be written as*

$$\text{maximize } \langle \bar{u}, y \rangle - Ef^*(p^c, y^c) - \sigma_{\mathrm{dom}\, Ef}(p^s, y^s) \text{ over } (p, y) \in \mathcal{X}_a^\circ \times \mathcal{U}^*.$$

Restricting the dual variables (p, y) to $\mathcal{V} \times \mathcal{Y}$, problem (D_s) reduces to (D), so

$$\sup (D_s) \geq \sup (D).$$

In particular, if there is no duality gap between $(P_{\mathcal{X}})$ and (D), the same holds between $(P_{\mathcal{X}})$ and (D_s). As noted earlier, the two dual problems coincide if $\mathcal{Y}^s = \{0\}$ and $\mathcal{X}^s = \{0\}$. The problems coincide also in the special case where $\operatorname{dom} Ef = \mathcal{X} \times \mathcal{U}$ since then, $\sigma_{\operatorname{dom} Ef} = \delta_{\{(0,0)\}}$. The next section gives more general conditions under which the optimum values of (D) and (D_s) coincide and, moreover, one has a solution if and only if the other one does.

5.3 Existence of Dual Solutions

The set of solutions of the dual problem

$$\text{maximize} \quad \langle \bar{u}, y \rangle - F^*(0, p, y) \quad \text{over } (p, y) \in \mathcal{V} \times \mathcal{Y} \tag{D}$$

from Sect. 3.2 is the intersection of the solution set of the strong dual problem (D_s) with $\mathcal{V} \times \mathcal{Y}$. In general, it may happen that (D) does not admit solutions even if (D_s) does and there is no duality gap; see Example 5.60. This section gives conditions under which, for each point $(p, y) \in \mathcal{X}^* \times \mathcal{U}^*$, there is a $(\tilde{p}, \tilde{y}) \in \mathcal{V} \times \mathcal{Y}$ that achieves a dual objective value at least as good as (p, y). Combined with Theorem 5.1, on the existence of strong dual solutions, we then obtain sufficient conditions for the existence of solutions in (D).

Recall that the solidity of $\mathcal{X}$ implies that

$$\mathcal{X} = \mathcal{X}_0 \times \cdots \times \mathcal{X}_T,$$

where $\mathcal{X}_t$ are solid decomposable spaces. We assume that $\mathcal{X}_t$ and $\mathcal{U}$ are as in Sect. 5.1.1 and we equip $\mathcal{X}$ with the corresponding product topology so that

$$\mathcal{X}^* = \mathcal{X}_0^* \times \cdots \times \mathcal{X}_T^*.$$

We assume, in addition, that $E_t \mathcal{X} \subseteq \mathcal{X}$ for all t. By Lemma 5.4, this implies that the conditional expectations E_t are both strongly and weakly continuous and that, when restricted to $\mathcal{V}$, the adjoints $E_t^* : \mathcal{X}^* \to \mathcal{X}^*$ are given by the conditional expectations E_t. By Lemma 3.6, $E_t \mathcal{X} \subseteq \mathcal{X}$ implies $E_t \mathcal{V} \subseteq \mathcal{V}$. Most familiar spaces of random variables satisfy the above assumptions; see e.g. the examples in Sect. 5.1.1.

The following is an analogue of Lemma 3.16.

Lemma 5.24 *The set $\mathcal{X}_a$ is strongly closed and*

$$\mathcal{X}_a^\circ = \{p \in \mathcal{X}^* \mid E_t^* p_t = 0 \quad t = 0, \ldots, T\}.$$

Proof By Lemma 3.16. $\mathcal{X}_a$ weakly and thus, strongly closed. Since $\mathcal{X}_t(\mathcal{F}_t) = E_t\mathcal{X}_t$ and

$$\mathcal{X}_a = \mathcal{X}_0(\mathcal{F}_0) \times \cdots \times \mathcal{X}_T(\mathcal{F}_T),$$

we have $p \in \mathcal{X}_a^\circ$ if and only if $0 = \langle E_t x_t, p_t\rangle = \langle x_t, E_t^* p_t\rangle$ for every t and $x_t \in \mathcal{X}_t$. □

We will denote the projection of the set $\operatorname{dom} Ef \cap (\mathcal{X} \times \mathcal{U})$ on $\mathcal{U}$ by

$$\operatorname{dom}_u Ef := \{u \in \mathcal{U} \mid \exists x \in \mathcal{X} : \ (x, u) \in \operatorname{dom} Ef\}.$$

Given $x \in \mathcal{X}$, we denote $x^t := (x_s)_{s=0}^t$ and

$$\mathcal{D}^t := \{z^t \in \mathcal{X}^t \mid z \in \operatorname{dom} Ef(\cdot, \bar{u})\}.$$

Assumption 5.25

1. $\bar{u} \in \operatorname{rcore} \operatorname{dom}_u Ef$ *and* $\operatorname{aff} \operatorname{dom}_u Ef$ *is strongly closed.*
2. $E_t z^t \in \mathcal{D}^t$ *for every* t *and* $z^t \in \mathcal{D}^t$*.*

Part 2 of Assumption 5.25 means that for every t and $z \in \mathcal{X}$ with $(z, \bar{u}) \in \operatorname{dom} Ef$, there exists $\bar{z} \in \mathcal{X}$ with $(\bar{z}, \bar{u}) \in \operatorname{dom} Ef$ and $\bar{z}^t = E_t z^t$. Assumption 5.25 clearly holds if $\operatorname{dom} Ef = \mathcal{X} \times \mathcal{U}$. In this case, the support function of $\operatorname{dom} Ef$ is the indicator of the origin so feasible solutions in (D_s) have $(p^s, u^s) = (0, 0)$. In general, the condition $\bar{u} \in \operatorname{rcore} \operatorname{dom}_u Ef$ implies $\operatorname{dom} Ef \cap (\mathcal{X} \times \mathcal{U}) \neq \emptyset$, and thus, by Corollary 5.23, that the strong dual problem (D_s) can be written as

$$\text{maximize } \langle \bar{u}, y\rangle - Ef^*(p^c, y^c) - \sigma_{\operatorname{dom} Ef}(p^s, y^s) \ \text{ over } \ (p, y) \in \mathcal{X}_a^\circ \times \mathcal{U}^*.$$

The condition $\bar{u} \in \operatorname{rcore} \operatorname{dom}_u Ef$ also implies $\bar{u} \in \operatorname{dom}_u Ef$ which means that there is a $z \in \mathcal{X}$ with $(z, \bar{u}) \in \operatorname{dom} Ef$. Applying condition 2 recursively for $t = 0, \dots, T$, then gives the existence of an $x \in \mathcal{X}_a$ with $(x, \bar{u}) \in \operatorname{dom} Ef$. Thus, Assumption 5.25 implies, in particular, that problem (P) is feasible.

We will show that, under Assumption 5.25, one can restrict dual variables to $\mathcal{V} \times \mathcal{Y}$ without worsening the objective value in the strong dual problem (D_s). The argument is based on a recursive application of the following somewhat technical lemma, the proof of which employs conjugate duality reviewed in Sect. A.9. We will denote the closure operation in the strong topology of the Fréchet space $\mathcal{X}^t := \mathcal{X}_0 \times \cdots \times \mathcal{X}_t$ by cl_s.

Lemma 5.26 *Under Assumption 5.25.2,*

$$E_t z^t \in \operatorname{cl}_s \mathcal{D}^t \quad \forall z^t \in \operatorname{cl}_s \mathcal{D}^t \tag{5.8}$$

which is equivalent to

$$\sigma_{\mathcal{D}^t}(E_t^* p^t) \le \sigma_{\mathcal{D}^t}(p^t) \quad \forall p \in \mathcal{X}^*. \tag{5.9}$$

Under Assumption 5.25.1,

$$\sigma_{\mathcal{D}^t}(p^t) = \inf_{y\in\mathcal{U}^*}\{\sigma_{\operatorname{dom} Ef}((p^t,0),y) - \langle \bar{u}, y\rangle\} \quad \forall p \in \mathcal{X}^*,$$

where the infimum is attained.

Proof By Lemma 5.4, E_t is strongly continuous, so Assumption 5.25.2 implies (5.8). The equivalence of (5.8) and (5.9) follows from Lemma A.47 and Theorem A.54.

We have $\sigma_{\mathcal{D}^t}(p^t) = \sigma_{\operatorname{dom} Ef(\cdot,\bar{u})}(p^t,0)$ while $\operatorname{dom} Ef = \operatorname{dom} Ef^+$. Since $Ef^+(\cdot,\bar{u})$ is closed and proper, its conjugate $Ef^+(\cdot,\bar{u})^*$ is closed and proper as well, by Theorem A.54. Thus, by Lemma A.45, $\sigma_{\operatorname{dom} Ef(\cdot,\bar{u})}$ is the recession function of $Ef^+(\cdot,\bar{u})^*$, which in turn fits the format of Theorem A.75. Indeed, under Assumption 5.25.1, the assumptions of Theorem A.75 are satisfied with $X = \mathcal{X}$, $U = \mathcal{U}$ and $F = Ef^+$ and thus,

$$\sigma_{\operatorname{dom} Ef(\cdot,\bar{u})}(p) = \inf_y\{(F^*)^\infty(p,y) - \langle \bar{u}, y\rangle\},$$

where the infimum is attained. By Lemma A.45 again, $(F^*)^\infty = \sigma_{\operatorname{dom} Ef}$, which completes the proof. □

The following is the main result of this section. Its proof is based on Theorem 5.22 and Lemma 5.26.

Theorem 5.27 *Under Assumption 5.25, the optimum values of* (D_s) *and* (D) *coincide and one has a solution if the other one does.*

Proof It suffices to show that, for every $(p,y) \in \mathcal{X}_a^\circ \times \mathcal{U}^*$, there exists a $(\tilde{p},\tilde{y}) \in \mathcal{X}_a^\perp \times \mathcal{Y}$ with

$$\varphi^*(\tilde{p},\tilde{y}) - \langle \bar{u}, \tilde{y}\rangle \le \varphi^*(p,y) - \langle \bar{u}, y\rangle. \tag{5.10}$$

We will prove this by first showing that if $p^s_{t'} = 0$ for $t' \in \{t+1,\dots,T\}$, then there exists $(\tilde{p},\tilde{y}) \in \mathcal{X}_a^\circ \times \mathcal{U}^*$ satisfying (5.10) and $\tilde{p}^s_{t'} = 0$ for $t' \ge t$. We then get, by induction, a $(\tilde{p},\tilde{y}) \in \mathcal{X}_a^\perp \times \mathcal{U}^*$ satisfying (5.10). As we will see, the singular part of $\tilde{y}$ can be then dropped as well.

By Theorem 5.22,

$$\varphi^*(p,y) - \langle \bar{u}, y\rangle = Ef^*(p^c,y^c) - \langle \bar{u}, y^c\rangle + \sigma_{\operatorname{dom} Ef}(p^s,y^s) - \langle \bar{u}, y^s\rangle.$$

Assume that $p^s_{t'} = 0$ for $t' \in \{t+1,\dots,T\}$. By Lemma 5.26, there is a $\hat{y} \in \mathcal{U}^*$ such that

$$\sigma_{\operatorname{dom} Ef}(E_t^* p^s, \hat{y}) - \langle \bar{u}, \hat{y}\rangle \le \sigma_{\operatorname{dom} Ef}(p^s,y^s) - \langle \bar{u}, y^s\rangle.$$

Thus,

$$\varphi^*(p, y) - \langle \bar{u}, y\rangle \geq Ef^*(p^c, y^c) - \langle \bar{u}, y^c\rangle + \sigma_{\text{dom}\, Ef}(E_t^* p^s, \hat{y}) - \langle \bar{u}, \hat{y}\rangle.$$

By Fenchel's inequality,

$$Ef^*(p^c, y^c) + Ef(x, u) \geq \langle x, p^c\rangle + \langle u, y^c\rangle,$$
$$\delta_{\text{dom}\, Ef}(x, u) + \sigma_{\text{dom}\, Ef}(E_t^* p^s, \hat{y}) \geq \langle x, E_t^* p^s\rangle + \langle u, \hat{y}\rangle$$

for all $(x, u) \in \mathcal{X} \times \mathcal{U}$. Combining the above inequalities, gives

$$\varphi^*(p, y) - \langle \bar{u}, y\rangle \geq \langle x, p^c + E_t^* p^s\rangle + \langle u, y^c + \hat{y}\rangle - Ef(x, u) - \langle \bar{u}, y^c + \hat{y}\rangle.$$

Denoting $\tilde{p} = p^c + E_t^* p^s$ and $\tilde{y} = y^c + \hat{y}$ and maximizing over $(x, u) \in \mathcal{X} \times \mathcal{U}$ gives

$$\begin{aligned}\varphi^*(p, y) - \langle \bar{u}, y\rangle &\geq \sup_{x\in\mathcal{X}, u\in\mathcal{U}} \{\langle x, \tilde{p}\rangle + \langle u, \tilde{y}\rangle - Ef(x, u)\} - \langle \bar{u}, \tilde{y}\rangle \\ &= Ef^*(\tilde{p}^c, \tilde{y}^c) + \sigma_{\text{dom}\, Ef}(\tilde{p}^s, \tilde{y}^s) - \langle \bar{u}, \tilde{y}\rangle,\end{aligned}$$

where the equality holds by Theorem 5.15. Since, $p \in \mathcal{X}_a^\circ$, we have $E_t^* p_t = 0$ so, by Lemma 3.6, $E_t^* p_t^s = -E_t^* p_t^c = -E_t p_t^c$ and thus, $\tilde{p}_{t'}^s = 0$ for every $t' \geq t$ as desired. Since $p_{t'}^s = 0$ for $t' > t$, we have for every $x \in \mathcal{X}_a$,

$$\begin{aligned}\langle x, \tilde{p}\rangle &= \langle x, p^c\rangle + \langle x, E_t^* p^s\rangle \\ &= \langle x, p^c\rangle + \langle E_t x, p^s\rangle \\ &= \langle x, p^c\rangle + \langle x, p^s\rangle = \langle x, p\rangle\end{aligned}$$

so $p \in \mathcal{X}_a^\circ$ implies $\tilde{p} \in \mathcal{X}_a^\circ$. By Theorem 5.22 again,

$$\varphi^*(\tilde{p}, \tilde{y}) - \langle \bar{u}, \tilde{y}\rangle \leq \varphi^*(p, y) - \langle \bar{u}, y\rangle.$$

By induction on t, we will thus get a $(\tilde{p}, \tilde{y}) \in \mathcal{X}_a^\perp \times \mathcal{U}^*$ satisfying (5.10) as claimed at the beginning of the proof.

Since the singular part of $\tilde{p}$ is zero, Theorem 5.22 gives

$$\varphi^*(\tilde{p}, \tilde{y}) - \langle \bar{u}, \tilde{y}\rangle = Ef^*(\tilde{p}, \tilde{y}^c) - \langle \bar{u}, \tilde{y}^c\rangle + \sigma_{\text{dom}\, Ef}(0, \tilde{y}^s) - \langle \bar{u}, \tilde{y}^s\rangle.$$

Under Assumption 5.25.1 there exists an $\bar{x} \in \mathcal{X}$ with $(\bar{x}, \bar{u}) \in \text{dom}\, Ef$ so

$$\langle \bar{u}, y\rangle \leq \sigma_{\text{dom}\, Ef}(0, y) \quad \forall y \in \mathcal{U}^*.$$

We can thus replace $\tilde{y}^s$ by zero without increasing the value of $\varphi^*(\tilde{p}, \tilde{y}) - \langle \bar{u}, \tilde{y}\rangle$. □

It is clear from the above proof that, instead of Assumption 5.25, it would suffice to assume Assumption 5.25.1 and (5.9) in Theorem 5.27. Condition (5.9) is a priori a weaker condition than Assumption 5.25.2

Remark 5.28 *By Theorems A.71 and A.72, Theorem 5.27 implies that, under Assumption 5.25, the weak and strong closures of φ coincide at $(0, \bar{u})$ and that φ is weakly subdifferentiable at $(0, \bar{u})$ if and only if it is strongly subdifferentiable at $(0, \bar{u})$.*

Combining Theorems 5.1 and 5.27 gives a two-step strategy for establishing the existence of solutions to (D). While Theorem 5.1 gives sufficient conditions for the existence of solutions to the strong dual problem (D_s), Theorem 5.27 says that, under Assumption 5.25, existence of solutions to (D_s) implies the existence of solutions to (D). Combining this with the optimality conditions from Chap. 3 gives the following theorem. Recall the notation

$$\mathcal{D}^t := \{z^t \in \mathcal{X}^t \mid z \in \operatorname{dom} Ef(\cdot, \bar{u})\}$$

from Assumption 5.25.

Theorem 5.29 *Assume that*

1. *Ef is proper and strongly lsc,*
2. *$(P_{\mathcal{X}})$ is strictly feasible,*
3. $\operatorname{aff} \operatorname{dom} Ef - \mathcal{X}_a \times \{0\}$ *and* $\operatorname{aff} \operatorname{dom}_u Ef$ *are strongly closed,*
4. *$E_t z^t \in \mathcal{D}^t$ for every t and $z^t \in \mathcal{D}^t$.*

Then

$$\inf(P) = \inf(P_{\mathcal{X}}) = \sup(D)$$

and the dual optimum is attained. If the optimum value is finite, the following are equivalent:

1. *x solves (P);*
2. *x is feasible in (P) and there exists (p, y) feasible in (D) with*

$$(p, y) \in \partial f(x, \bar{u}) \quad a.s.;$$

3. *x is feasible in (P) and there exists (p, y) feasible in (D) with*

$$p \in \partial_x l(x, y), \quad \bar{u} \in \partial_y[-l](x, y) \quad a.s.$$

and imply that (p, y) is dual optimal.

Proof By Lemma A.5, strict feasibility implies that $\bar{u} \in \operatorname{rcore} \operatorname{dom}_u Ef$. Combining Theorems 3.33, 5.1 with 5.27 gives the two equalities and the dual attainment. The remaining claims follow from Theorem 3.34. □

Remark 5.30 *Combining Theorems 5.29, 3.38 and 3.28, gives sufficient conditions for the existence of solutions to reduced dual problems studied in Sects. 3.2.2 and 3.3.*

5.3.1 Existence of Strong Dual Solutions

This section gives sufficient conditions for the assumptions of Theorem 5.1. Recall that the problem $(P_{\mathcal{X}})$ is *strictly feasible* if there exists $\bar{x} \in \mathcal{X}_a$ such that $(\bar{x}, \bar{u}) \in \operatorname{rcore}\operatorname{dom} Ef$. This happens, in particular, if there exists an $\bar{x} \in \mathcal{X}_a$ such that $(\bar{x}, \bar{u}) \in \operatorname{int}\operatorname{dom} Ef$. This corresponds to the classical *Slater condition* in optimization theory. Example 5.20 yields the following sufficient condition for strict feasibility.

Example 5.31 *Let* $f(x, u, \omega) = f_0(u, \omega) + \delta_{H(\omega)}(x, u)$*, where* H *is a random affine set and* f_0 *is a convex normal integrand with* $\operatorname{core}\operatorname{dom} Ef_0 \cap (\mathcal{X} \times \mathcal{U})(H) \neq \emptyset$. *Then* $\operatorname{rcore}\operatorname{dom} Ef = \operatorname{core}\operatorname{dom} Ef_0 \cap (\mathcal{X} \times \mathcal{U})(H)$ *and* $\operatorname{aff}\operatorname{dom} Ef = (\mathcal{X} \times \mathcal{U})(H)$ *which is strongly closed.*

Theorem 5.1 also requires that the set

$$\operatorname{aff}\operatorname{dom} Ef - \mathcal{X}_a \times \{0\} \tag{5.11}$$

be strongly closed in $\mathcal{X} \times \mathcal{U}$. This clearly holds if there exists $\bar{x} \in \mathcal{X}_a$ with $(\bar{x}, \bar{u}) \in \operatorname{int}\operatorname{dom} Ef$. As noted above, this implies strict feasibility as well and thus, the existence of strong dual solutions, by Theorem 5.1.

More generally, if there exists a strictly feasible $\bar{x}$ such that $\mathcal{X}_a \times \{0\} \subseteq \operatorname{aff}\operatorname{dom} Ef$, then closedness of $\operatorname{aff}\operatorname{dom} Ef$ implies the closedness of (5.11). In general, the sum of closed affine sets need not be closed; see Example 5.34 below. Closedness of sums of closed linear subspaces is a well-studied question in functional analysis and various sufficient as well as necessary conditions for it are available. Many practical instances of stochastic optimization problems have additional structure that can be used to establish the closedness of the set in (5.11). The following lemma gives a sufficient condition that may look awkward at first glance but turns out to be convenient in many applications.

Lemma 5.32 *Let* $\bar{x} \in \mathcal{X}_a$ *be strictly feasible and assume that* $\operatorname{aff}\operatorname{dom} Ef$ *is closed. If there exists a strongly continuous linear idempotent mapping* π *on* $\mathcal{X} \times \mathcal{U}$ *such that* $\pi \operatorname{aff}\operatorname{dom} Ef \subseteq \operatorname{aff}\operatorname{dom} Ef$ *and*

$$\operatorname{aff}\operatorname{dom} Ef - \mathcal{X}_a \times \{0\} = \operatorname{aff}\operatorname{dom} Ef - \operatorname{rge} \pi,$$

then $\operatorname{aff}\operatorname{dom} Ef - \mathcal{X}_a \times \{0\}$ *is strongly closed. The above equality holds, in particular, if*

$$\mathcal{X}_a \times \{0\} \subseteq \operatorname{rge} \pi \subseteq \mathcal{X}_a \times \{0\} + (\bar{x}, \bar{u}) - \operatorname{aff}\operatorname{dom} Ef.$$

Proof Denoting $\mathcal{L} := \operatorname{aff}\operatorname{dom} Ef - (\bar{x}, \bar{u})$, we have $\pi\mathcal{L} \subseteq \mathcal{L}$ and

$$\operatorname{aff}\operatorname{dom} Ef - \operatorname{rge} \pi = \mathcal{L} - \operatorname{rge} \pi + (\bar{x}, \bar{u}),$$

so the first claim follows from Lemma A.41. The additional inclusions give

$$\mathcal{L} - \mathcal{X}_a \times \{0\} \subseteq \mathcal{L} - \operatorname{rge} \pi \subseteq \mathcal{L} - \mathcal{X}_a \times \{0\},$$

which proves the second claim. □

It is not always easy to find a mapping π that satisfies the assumptions of Lemma 5.32. The following example may be useful in the presence of a composite structure often found in applications.

Example 5.33 *Let*

$$f(x, u, \omega) = g(x, A(\omega)x + u, \omega),$$

where g *is a convex normal integrand on* $\mathbb{R}^n \times \mathbb{R}^m \times \Omega$ *and* A *is a random matrix with* $A\mathcal{X} \subseteq \mathcal{U}$. *Let* $\bar{x} \in \mathcal{X}_a$ *be such that* $(\bar{x}, A\bar{x} + \bar{u}) \in \operatorname{rcore}\operatorname{dom} Eg$. *Then* $\bar{x}$ *is strictly feasible. If* $\operatorname{aff}\operatorname{dom} Eg$ *is strongly closed, then* $\operatorname{aff}\operatorname{dom} Ef$ *is strongly closed. If, in addition, there exists a strongly continuous linear idempotent mapping* π' *on* $\mathcal{X} \times \mathcal{U}$ *such that* $\pi' \operatorname{aff}\operatorname{dom} Eg \subseteq \operatorname{aff}\operatorname{dom} Eg$ *and*

$$\operatorname{aff}\operatorname{dom} Eg - (I, \mathcal{A})\mathcal{X}_a = \operatorname{aff}\operatorname{dom} Eg - \operatorname{rge} \pi', \tag{5.12}$$

then $\operatorname{aff}\operatorname{dom} Ef - \mathcal{X}_a \times \{0\}$ *is strongly closed. Equality* (5.12) *holds if*

$$(I, \mathcal{A})\mathcal{X}_a \subseteq \operatorname{rge} \pi' \subseteq (I, \mathcal{A})\mathcal{X}_a + (\bar{x}, A\bar{x} + \bar{u}) - \operatorname{aff}\operatorname{dom} Eg. \tag{5.13}$$

This, in turn, holds if $(I, \mathcal{A})\mathcal{X}_a \subseteq \operatorname{rge} \pi' \subseteq \mathcal{X}_a \times \mathcal{U}$ *and*

$$(0, u) \in (I, \mathcal{A})\mathcal{X}_a + (\bar{x}, A\bar{x} + \bar{u}) - \operatorname{aff}\operatorname{dom} Eg$$

for all $u \in \mathcal{U}$ *such that* $(0, u) \in \operatorname{rge} \pi'$.

Proof We have

$$\operatorname{dom} Ef = \{(x, u) \in \mathcal{X} \times \mathcal{U} \mid (x, Ax + u) \in \operatorname{dom} Eg\} = \bar{\mathcal{A}}^{-1} \operatorname{dom} Eg,$$

where $\bar{\mathcal{A}} : \mathcal{X} \times \mathcal{U} \to \mathcal{X} \times \mathcal{U}$ denotes the linear mapping defined pointwise by $\bar{\mathcal{A}}(x,u) := \bar{A}(x,u)$, where

$$\bar{A}(\omega) := \begin{bmatrix} I & 0 \\ A(\omega) & I \end{bmatrix}.$$

The inverse of $\bar{\mathcal{A}}$ is given by a pointwise application of the matrix

$$(\bar{A})^{-1}(\omega) := \begin{bmatrix} I & 0 \\ -A(\omega) & I \end{bmatrix}.$$

By Lemma A.5, $\operatorname{rcore}\operatorname{dom} Ef = \bar{\mathcal{A}}^{-1} \operatorname{rcore}\operatorname{dom} Eg$, which proves the first claim. By Lemma 5.5, $\mathcal{A} : X \to \mathcal{U}$ is continuous, so $\bar{\mathcal{A}}$ is as well. By Lemma A.9, $\operatorname{aff}\operatorname{dom} Ef = \bar{\mathcal{A}}^{-1} \operatorname{aff}\operatorname{dom} Eg$. This proves the second claim.

To complete the proof, we apply Lemma 5.32 with the mapping $\pi = \bar{\mathcal{A}}^{-1}\pi'\bar{\mathcal{A}}$ which is idempotent since π' is so and since $\bar{\mathcal{A}}$ is invertible. Since $\operatorname{aff}\operatorname{dom} Ef = \bar{\mathcal{A}}^{-1} \operatorname{aff}\operatorname{dom} Eg$, we have $\pi \operatorname{aff}\operatorname{dom} Ef \subseteq \operatorname{aff}\operatorname{dom} Ef$. We have

$$\begin{aligned}
\operatorname{aff}\operatorname{dom} Ef - \mathcal{X}_a \times \{0\} &= \bar{\mathcal{A}}^{-1} \operatorname{aff}\operatorname{dom} Eg - \bar{\mathcal{A}}^{-1}\bar{\mathcal{A}}(\mathcal{X}_a \times \{0\}) \\
&= \bar{\mathcal{A}}^{-1}(\operatorname{aff}\operatorname{dom} Eg - (I, A)\mathcal{X}_a) \\
&= \bar{\mathcal{A}}^{-1}(\operatorname{aff}\operatorname{dom} Eg - \operatorname{rge}\pi') \\
&= \operatorname{aff}\operatorname{dom} Ef - \operatorname{rge}\pi,
\end{aligned}$$

where the second last equality holds by assumption and the last follows from $\operatorname{rge}\pi = \bar{\mathcal{A}}^{-1} \operatorname{rge}\pi'$. The inclusions in (5.13) give

$$\operatorname{aff}\operatorname{dom} Eg - (I, A)\mathcal{X}_a \subseteq \operatorname{aff}\operatorname{dom} Eg - \operatorname{rge}\pi' \subseteq \operatorname{aff}\operatorname{dom} Eg - (I, A)\mathcal{X}_a,$$

which is (5.12). To prove the last claim, let $(x', u') \in \operatorname{rge}\pi'$. We have $(I, \mathcal{A})x' \in \operatorname{rge}\pi'$, by assumption, so $(x', u') - (I, \mathcal{A})x' \in \operatorname{rge}\pi'$. Since $(x', u') - (I, \mathcal{A})x' = (0, u' - \mathcal{A}x')$, the last assumption gives

$$\begin{aligned}
(x', u') &= (x', u') - (I, \mathcal{A})x' + (I, \mathcal{A})x' \\
&\subseteq (I, \mathcal{A})\mathcal{X}_a + (\bar{x}, \mathcal{A}\bar{x} + \bar{u}) - \operatorname{aff}\operatorname{dom} Eg
\end{aligned}$$

so (5.13) holds. □

The following shows that $\operatorname{aff}\operatorname{dom} Ef - \mathcal{X}_a \times \{0\}$ may fail to be closed even if $\operatorname{aff}\operatorname{dom} Ef$ is strongly closed and $(P_{\mathcal{X}})$ is strictly feasible.

Counterexample 5.34 *Let $\mathcal{F}_0 = \{\emptyset, \Omega\}$, $T = 1$, $\mathcal{X} = L^\infty$ and $f(x, u, \omega) := \delta_{L(\omega)}(x)$, where $L(\omega) = \{x \mid \eta(\omega)x_0 = x_T\}$ with $\eta \notin L^\infty$ and $\eta \geq 1$. Then* dom $Ef = \mathcal{X}(L) \times \mathcal{U}$, *so* aff dom $Ef - \mathcal{X}_a \times \{0\}$ *is closed if and only if* $\mathcal{X}(L) - \mathcal{X}_a$ *is closed. Defining*

$$z^\nu := 1_{\{|\eta| \le \nu\}}(\eta^{-1/2}, \eta^{1/2})$$
$$x^\nu := 1_{\{|\eta| \le \nu\}}(0, \eta^{1/2})$$

we have $z^\nu \in \mathcal{X}(L)$, $x^\nu \in \mathcal{X}_a$ *and* $z^\nu - x^\nu \to z := (\eta^{-1/2}, 0)$ *in* L^∞*. However,* $z \notin \mathcal{X}(L) - \mathcal{X}_a$*. Indeed,* $z \in \mathcal{X}(L) - \mathcal{X}_a$ *would mean that there exists* $x \in \mathcal{X}_a$ *such that* $z + x \in \mathcal{X}(L)$*, or in other words, that* $(z_0 + x_0)\eta = (z_T + x_T)$*. This is impossible since* $z_T + x_T \in L^\infty$ *while* $(z_0 + x_0)\eta \notin L^\infty$ *since* x_0 *is constant. Thus,* $\mathcal{X}(L) - \mathcal{X}_a$ *is not closed.*

5.3.2 Relatively Complete Recourse

This section gives sufficient conditions for Assumption 5.25. We will denote the closure of a set $D \subset \mathcal{X}^t$ with respect to the weak topology $\sigma(\mathcal{X}^t, \mathcal{V}^t)$ by $\mathrm{cl}_\sigma D$. The following is an analogue of Lemma 5.26.

Lemma 5.35 *Under Assumption 5.25.2,*

$$E_t z^t \in \mathrm{cl}_\sigma \mathcal{D}^t \quad \forall z^t \in \mathrm{cl}_\sigma \mathcal{D}^t, \tag{5.14}$$

which is equivalent to

$$\sigma_{\mathcal{D}^t}(E_t p^t) \le \sigma_{\mathcal{D}^t}(p^t) \quad \forall p \in \mathcal{V}. \tag{5.15}$$

If dom $Ef(\cdot, \bar{u}) \neq \emptyset$*, then*

$$\sigma_{\mathcal{D}^t}(p^t) = E\sigma_{\mathrm{dom}\, f(\cdot,\bar{u})}(p^t, 0) \quad \forall p \in \mathcal{V}.$$

Proof By Lemma 3.6, E_t is weakly continuous, so Assumption 5.25.2 implies (5.14). The equivalence of (5.14) and (5.15) follows from Lemma A.47 and Theorem A.54. We have

$$\sigma_{\mathcal{D}^t}(p^t) = \sigma_{\mathrm{dom}\, Ef(\cdot,\bar{u})}(p^t, 0) = E\sigma_{\mathrm{dom}\, f(\cdot,\bar{u})}(p^t, 0),$$

where the last equality holds by the last identity in Theorem 3.15. □

It turns out that condition (5.14) holds if and only if the domain mapping $S(\omega) :=$ dom $f(\cdot, \bar{u}(\omega), \omega)$ is *adapted* in the sense that the projections

$$S^t(\omega) := \{x^t \in \mathbb{R}^{n^t} \mid x \in S(\omega)\}$$

are $\mathcal{F}_t$-measurable. Recall that problem (P) has *relatively complete recourse* if S is adapted and S^t is closed-valued for all t; see Sect. 2.2.3.

Theorem 5.36 *If* $\operatorname{dom} Ef(\cdot, \bar{u}) \neq \emptyset$*, then* (5.14) *is equivalent to the mapping*

$$S(\omega) := \operatorname{dom} f(\cdot, \bar{u}(\omega), \omega)$$

being adapted. In particular, if S^t *are closed-valued, Assumption 5.25 implies relatively complete recourse.*

Proof Let $(\operatorname{cl} S^t)(\omega) := \operatorname{cl} S^t(\omega)$. Since $\delta_{\mathcal{X}^t(\operatorname{cl} S^t)} = E\delta_{\operatorname{cl} S^t}$, Theorem 3.8 gives

$$\sigma_{\mathcal{X}^t(\operatorname{cl} S^t)}(p^t) = (E\delta_{\operatorname{cl} S^t})^*(p^t) = E\delta^*_{\operatorname{cl} S^t}(p^t) = E\sigma_{\operatorname{cl} S^t}(p^t) = E\sigma_S(p^t, 0)$$

so, by Lemma 5.35, $\sigma_{\mathcal{X}^t(\operatorname{cl} S^t)} = \sigma_{\mathcal{D}^t}$. By Theorem A.54, $\mathcal{X}^t(\operatorname{cl} S^t) = \operatorname{cl}_\sigma \mathcal{D}^t$ so (5.14) can be written as

$$E_t z^t \in \mathcal{X}^t(\operatorname{cl} S^t) \quad \forall z^t \in \mathcal{X}^t(\operatorname{cl} S^t).$$

By Theorem 3.14, this means that $\operatorname{cl} S^t$ is $\mathcal{F}_t$-measurable. By Lemma 1.1, this is equivalent to the $\mathcal{F}_t$-measurability of S^t, so S is adapted. □

The converse of Theorem 5.36 fails in general; see Counterexample 5.40 below. The following gives sufficient conditions for the converse.

Lemma 5.37 *Assumption 5.25.2 holds if* (P) *has relatively complete recourse and*

$$\operatorname{dom} Ef(\cdot, \bar{u}) = L^0(S),$$

where $S(\omega) := \operatorname{dom} f(\cdot, \bar{u}(\omega), \omega)$.

Proof If $x^t \in \mathcal{D}^t := \{x^t \mid x \in \operatorname{dom} Ef(\cdot, \bar{u})\}$, we clearly have $x^t \in \mathcal{X}^t(S^t)$. On the other hand, given $x^t \in \mathcal{X}^t(S^t)$, the mapping

$$\omega \mapsto \{z \in \mathbb{R}^n \mid z^t = x^t(\omega),\ z \in S(\omega)\}$$

is closed-valued and measurable so Corollary 1.28 gives the existence of a $z \in L^0(S)$ with $z^t = x^t$. By assumption, $z \in \operatorname{dom} Ef(\cdot, \bar{u})$ so $x^t \in \mathcal{D}^t$. Thus, $\mathcal{D}^t = \mathcal{X}^t(S^t)$. Relatively complete recourse means that S^t is closed-valued and $\mathcal{F}_t$-measurable. By Theorem 3.14, this is equivalent to Assumption 5.25.2. □

Lemma 5.37 can be extended as follows.

Lemma 5.38 *Assumption 5.25.2 holds if*

$$\operatorname{dom} Ef(\cdot, \bar{u}) = \bigcup_{j \in J} L^0(S_j),$$

where $\{S_j\}_{j\in J}$ is a collection of adapted mappings $S_j : \Omega \rightrightarrows \mathbb{R}^n$ with S_j^t closed-valued for all t.

Proof Much like in the proof of Lemma 5.37, we get

$$\mathcal{D}^t = \bigcup_{j\in J} \mathcal{X}^t(S_j^t).$$

By Theorem 3.14 again, $E_t x^t \in \mathcal{X}^t(S^t)$ for every t and $j \in J$ and $x^t \in \mathcal{X}^t(S^t)$. Thus, Assumption 5.25.2 holds. □

When $\mathcal{X} = L^\infty$, it is natural to construct the mappings S_j in Lemma 5.38 by truncating the domain mapping $S(\omega) := \operatorname{dom} f(\cdot, \bar{u}(\omega), \omega)$. If S is closed-valued and essentially bounded in the sense that there exists $r > 0$ with $S \subset \mathbb{B}_r$ almost surely, then S^t is automatically closed-valued for all t and $L^0(S) = L^\infty(S)$.

Example 5.39 (Bounded Recourse Condition) *Let $\mathcal{X} = L^\infty$, $S(\omega) := \operatorname{dom} f(\cdot, \bar{u}(\omega), \omega)$, and assume that there exists a sequence $(B^\nu)_{\nu\in\mathbb{N}}$ of bounded sets in $\mathbb{R}^n$ such that $\mathbb{B}_\nu \subseteq B^\nu$ and each $S_\nu(\omega) := S(\omega) \cap B^\nu$ is adapted and closed-valued with $L^\infty(S_\nu) \subseteq \operatorname{dom} Ef(\cdot, \bar{u})$. Then Assumption 5.25.2 holds.*

Relative complete recourse does not imply Assumption 5.25.2 in general.

Counterexample 5.40 *Let $T = 1$, $\mathcal{F}_0 = \{\Omega, \emptyset\}$, $\mathcal{X} = L^\infty$, $n = 2$ and $f(\cdot, \bar{u}(\omega), \omega) = \delta_S(\omega)$, where*

$$S(\omega) := \{x \in \mathbb{R}^2 \mid 0 \le x_0\xi(\omega) \le x_1\}$$

with $\xi \in L^1(\mathcal{F}_1)\backslash L^\infty$ such that $\xi \ge 1$. Since $S^0(\omega) = \mathbb{R}_+$ and S^1 is $\mathcal{F}_1$-measurable, S is adapted.

Let $z_1 \in L^\infty_+$ and $z_0 := z_1/\xi$ so that $z \in L^\infty(S)$. If $z_1 \neq 0$, then $E_0[z_0] > 0$ and $E_0[z_0]\xi \notin L^\infty$, so there is no $\bar{z} \in L^\infty(S)$ with $E_0 z_0 = \bar{z}_0$. Thus, Assumption 5.25.2 fails.

The following shows that, in Example 5.39, S may be adapted while its truncations S_r are not.

Counterexample 5.41 *Let S be as in Counterexample 5.40 and assume that $\xi \notin L^0(\mathcal{F}_0)$. We have*

$$S_r^0(\omega) = \{x_0 \in \mathbb{R} \mid |x_0| \le r,\ 0 \le x_0\xi \le r\} = \{x_0 \in \mathbb{R} \mid 0 \le x_0\xi(\omega) \le r\}$$

so the inverse images of open intervals $(\alpha, \infty) \subset \mathbb{R}_+$ can be expressed as

$$(S_r^0)^{-1}((\alpha, \infty)) = \{\omega \in \Omega \mid S_r^0(\omega) \cap (\alpha, \infty) \neq \emptyset\} = \{\omega \in \Omega \mid \xi(\omega) < r/\alpha\}.$$

Since $\xi \notin L^0(\mathcal{F}_0)$, this fails to be $\mathcal{F}_0$-measurable for some $\alpha > 0$.

We end this section with a quick observation about Assumption 5.25.1.

Lemma 5.42 *If* $(P_{\mathcal{X}})$ *is strictly feasible, then* $\bar{u} \in \operatorname{rcore} \operatorname{dom}_u Ef$ *and*

$$\operatorname{aff} \operatorname{dom}_u Ef = \{u \in \mathcal{U} \mid \exists x \in \mathcal{X} : (x, u) \in \operatorname{aff} \operatorname{dom} Ef\}.$$

If, in addition, $f(x, u, \omega) = g(x, A(\omega)x + u, \omega)$ *like in Example 5.33, then*

$$\operatorname{aff} \operatorname{dom}_u Ef = \{u \in \mathcal{U} \mid (0, u) \in \operatorname{aff} \operatorname{dom} Eg - \operatorname{gph} \mathcal{A}\}.$$

Proof Since $\operatorname{dom}_u Ef$ is the projection of $\operatorname{dom} Ef$, the first two claim follows from Lemmas A.5 and A.9. Recalling the proof of Example 5.33, we have $\operatorname{aff} \operatorname{dom} Ef = \bar{\mathcal{A}}^{-1} \operatorname{aff} \operatorname{dom} Eg$, so

$$\begin{aligned}
\operatorname{aff} \operatorname{dom}_u Ef &= \{u \in \mathcal{U} \mid \exists x \in \mathcal{X} : (x, u) \in \bar{\mathcal{A}}^{-1} \operatorname{aff} \operatorname{dom} Eg\} \\
&= \{u \in \mathcal{U} \mid \exists x \in \mathcal{X} : (x, \mathcal{A}x + u) \in \operatorname{aff} \operatorname{dom} Eg\} \\
&= \{u \in \mathcal{U} \mid (0, u) \in \operatorname{aff} \operatorname{dom} Eg - \operatorname{gph} \mathcal{A}\},
\end{aligned}$$

which completes the proof. □

5.4 Applications

This section applies the general results of this chapter to the examples studied in the previous chapters. In each application, we give concrete conditions that imply the sufficient conditions in Theorem 5.29. Throughout, the spaces $\mathcal{X}$ and $\mathcal{U}$ are assumed to satisfy the conditions of Sect. 5.3.

5.4.1 Mathematical Programming

Consider again problem (1.8)

$$\begin{aligned}
&\text{minimize} && Ef_0(x) \quad \text{over } x \in \mathcal{N}, \\
&\text{subject to} && f_j(x) \le 0 \quad j = 1, \dots, l \text{ a.s.}, \\
& && f_j(x) = 0 \quad j = l + 1, \dots, m \text{ a.s.},
\end{aligned}$$

from Sect. 1.3.1. This was embedded in the general duality framework in Sect. 3.4.1. Below, we will combine Theorem 3.41 with Theorem 5.29 in order to give necessary and sufficient conditions for the solutions of (1.8). In order to state the required assumptions, we first write problem (1.8) in the form

$$\begin{aligned} &\text{minimize} \quad && Ef_0(x) \quad \text{over} \ \ x \in \mathcal{N}, \\ &\text{subject to} \quad && F(x) \in \mathbb{R}^l_- \quad a.s., \\ & && Ax = b \quad a.s., \end{aligned}$$

where F is a random $\mathbb{R}^l_-$-convex function and A is a random $n \times (m-l)$-matrix. We assume that the matrix A has the block-diagonal form

$$A(\omega) = \begin{bmatrix} A_{0,0}(\omega) & 0 \cdots & 0 \\ \vdots & & \vdots \\ & & 0 \\ A_{T,0}(\omega) & \cdots & A_{T,T}(\omega) \end{bmatrix},$$

where $A_{s,t}(\omega) \in \mathbb{R}^{m^e_t \times n_s}$ with $m^e_1 + \cdots + m^e_T = m - l$. Accordingly, $b = (b_t)^T_{t=0}$ where $b_t \in \mathbb{R}^{m^e_t}$. Denoting $A^t(\omega) := (A_{t,0}(\omega), \ldots, A_{t,t}(\omega))$, we can write

$$A(\omega)x = (A^0(\omega)x^0, \ldots, A^T(\omega)x^T).$$

Analogously, we assume that

$$F(x,\omega) = (F^0(x^0,\omega), \ldots, F^T(x^T,\omega)),$$

where F^t is a random $\mathbb{R}^{m^{ie}_t}_-$-convex function on $\mathbb{R}^{n^t}$ with $m^{ie}_1 + \cdots + m^{ie}_T = l$.

With the above notation, we can write the normal integrand f in Sect. 3.4.1 as

$$\begin{aligned} f(x,u,\omega) &= \begin{cases} f_0(x,\omega) & \text{if } x \in \operatorname{dom} H(\cdot,\omega),\ H(x,\omega) + u \in K, \\ +\infty & \text{otherwise,} \end{cases} \\ &= \begin{cases} f_0(x,\omega) & \text{if } F(x,\omega) + u^{ie} \in \mathbb{R}^l_-,\ A(\omega)x + u^e = b(\omega), \\ +\infty & \text{otherwise,} \end{cases} \end{aligned}$$

where $u^{ie} := (u_j)^l_{j=1}$ and $u^e := (u_j)^m_{j=l+1}$ so that $u = (u^{ie}, u^e)$. We assume that $\mathcal{X} = L^\infty$. If the entries of the random matrix A were essentially bounded, it would be natural to choose $\mathcal{U} = L^\infty$. We allow for more general random matrices and assume that $\mathcal{U} = L^\infty(\mathbb{R}^l) \times \mathcal{U}^e$, where $\mathcal{U}^e$ is a space of $\mathbb{R}^{m-l}$-valued random variables satisfying the assumptions in Sect. 5.3.

Assumption 5.43 *Ef_0 is finite on $\mathcal{X}$ and*

1. *there exist $\bar{x} \in \mathcal{X}_a$ and $\epsilon > 0$ such that*

$$f_j(\bar{x}+x)+\epsilon \le 0 \quad \forall x \in \mathbb{B}_\epsilon, \quad j=1,\dots,l \; a.s.$$
$$A(\bar{x}) = b \quad a.s.$$

2. *$A\mathcal{X} \subseteq \mathcal{U}^e$ and $A\mathcal{X}_a$ and $A\mathcal{X}$ are strongly closed in $\mathcal{U}^e$,*
3. *For every t, A^t, b_t and F^t are $\mathcal{F}_t$-measurable, and*

$$\{x_t \in \mathcal{X}_t \mid F^t(x^{t-1}, x_t) \in \mathbb{R}^{m_t^1}_-, \; A^t(x^{t-1}, x_t) = b_t \; a.s.\}$$

is nonempty for every $\mathcal{F}_{t-1}$-measurable $x^{t-1} \in \mathcal{X}^{t-1}$ such that $F^{t'}(x^{t'}) \in \mathbb{R}^{m^1_{t'}}_-$ and $A^{t'}x^{t'} = b_{t'}$ for all $t' < t$.

Combining Theorem 3.41 with Theorem 5.29, yields the following.

Theorem 5.44 *Under Assumption 5.43,* $\inf$*(1.8)* $=$ $\sup$*(3.14) and the optimum in* (3.14) *is attained. If the optimum value is finite, a feasible $x \in \mathcal{N}$ solves* (1.8) *if and only if there exists $(p, y) \in \mathcal{N}^\perp \times \mathcal{Y}$ feasible in* (3.14) *such that*

$$p \in \partial_x[f_0 + y \cdot H](x),$$
$$H(x) \in K, \quad y \in K^\circ, \quad y \cdot H(x) = 0$$

almost surely.

Proof Recall that, (1.8) fits the general duality framework with $\bar{u} = 0$ and, since Assumption 5.43.1 implies $\operatorname{dom} Ef \cap (\mathcal{X} \times \mathcal{U}) \neq \emptyset$, the dual problem can be written as (3.14). By Theorem 3.41, it thus suffices to show that there is no duality gap and that the dual admits solutions. To this end, it suffices to verify the assumptions of Theorem 5.29.

Since Ef_0 is finite on $\mathcal{X}$, it is strongly lsc by Corollary 5.16. Assumption 5.43.1 implies properness of Ef, so Ef is proper and strongly lsc. We have

$$\begin{aligned}\operatorname{dom} Ef = \{(x,u) \in \mathcal{X} \times \mathcal{U} \mid F(x) + u^{ie} \in \mathbb{R}^l_-, \; a.s.\}\\ \cap \{(x,u) \in \mathcal{X} \times \mathcal{U} \mid Ax + u^e = b \; a.s.\}.\end{aligned}$$

By Assumption 5.43.1, the first set on the right has $(\bar{x}, 0)$ in its strong interior. By Lemmas A.5 and A.9,

$$\operatorname{rcore} \operatorname{dom} Ef = \{(x,u) \in \mathcal{X} \times \mathcal{U} \mid Ax + u^e = b \; a.s.\},$$
$$\operatorname{aff} \operatorname{dom} Ef = \{(x,u) \in \mathcal{X} \times \mathcal{U} \mid Ax + u^e = b \; a.s.\}.$$

In particular, $\bar{x}$ is strictly feasible. We have

$$\begin{aligned}\operatorname{aff}\operatorname{dom} Ef - \mathcal{X}_a \times \{0\} &= \{(x,u) \in \mathcal{X} \times \mathcal{U} \mid Ax + u^e = b\} - \mathcal{X}_a \times \{0\} \\ &= \{(x - x', u) \in \mathcal{X} \times \mathcal{U} \mid x' \in \mathcal{X}_a,\ Ax + u^e = b\} \\ &= \{(z,u) \in \mathcal{X} \times \mathcal{U} \mid Az + u^e \in A\mathcal{X}_a + b\}.\end{aligned}$$

Since A is continuous and $A\mathcal{X}_a$ is closed, $\operatorname{aff}\operatorname{dom} Ef - \mathcal{X}_a \times \{0\}$ is closed as the inverse image of a closed set under a continuous mapping. By Lemma A.9,

$$\operatorname{aff}\operatorname{dom}_u Ef = \{u \in \mathcal{U} \mid u^e \in A\mathcal{X} + b\},$$

which is strongly closed since $A\mathcal{X}$ is so by assumption.

As to the last assumption in Theorem 5.29, let $z \in \mathcal{X}$ with $(z,0) \in \operatorname{dom} Ef$. By Corollary 1.69 and Lemma B.13, the $\mathcal{F}_t$-measurability of A^t, b_t and F^t in Assumption 5.43.3 implies

$$\begin{aligned}F^{t'}(E_t z^{t'}) &\in \mathbb{R}^{t'}_- \quad \forall t' \le t, \\ A^{t'}(E_t z^{t'}) &= b_{t'} \quad \forall t' \le t.\end{aligned}$$

Under Assumption 5.43.3, $E_t z^t$ can be extended to a $\bar{z}$ such that $\bar{z}^t = E_t z^t$ and $(\bar{z}, 0) \in \operatorname{dom} Ef$. □

Recall from Example 1.25 that the scenariowise Moore-Penrose inverse $A^\dagger$ of a random matrix A is measurable. The following gives sufficient conditions for Assumption 5.43.2.

Example 5.45 *Assume that* $\operatorname{rge} A^t = \operatorname{rge} A_{t,t}$ *almost surely,* $A_{t,t}\mathcal{X}_t \subseteq \mathcal{U}^e_t$ *and that* $A^\dagger_{t,t}\mathcal{U}^e_t \subseteq \mathcal{X}_t$. *Then* $A\mathcal{X}$ *is closed in* $\mathcal{U}^e$. *If, in addition,* A^t *are* $\mathcal{F}_t$*-measurable, then* $A\mathcal{X}_a$ *is closed in* $\mathcal{U}^e$.

Proof Defining $\tilde{A}_{s,t} := 0$ for $s < t$ and $\tilde{A}_{t,t} := A_{t,t}$, the first assumption means that $\operatorname{gph} A^t = \operatorname{gph} \tilde{A}^t$ almost surely. By Theorem 1.72, this means

$$(\mathcal{X}^t \times \mathcal{U}^e_t)(\operatorname{gph} A^t) = (\mathcal{X}^t \times \mathcal{U}^e_t)(\operatorname{gph} \tilde{A}^t),$$

or, in other words,

$$A^t \mathcal{X}^t = A_{t,t}\mathcal{X}_t \quad \forall t. \tag{5.16}$$

We show next that

$$A\mathcal{X} = A_{1,1}\mathcal{X}_1 \times \cdots \times A_{T,T}\mathcal{X}_T. \tag{5.17}$$

The Eq. (5.16) implies that the left side is contained in the right side. We prove the converse by induction. The claim is clear when $T = 0$, so assume it holds for any $(T-1)$-period model. Given u with $u_t = A_{t,t}x_t$ for all $t = 0, \dots, T$, there exists $\tilde{x}^{T-1} \in \mathcal{X}^{T-1}$ with $u_t = A^t\tilde{x}^t$ for all $t = 0, \dots T-1$. Since $A^T\mathcal{X}^T = A_{T,T}\mathcal{X}_T$, there exists $\tilde{x}_T$ such that $A_{T,T}\tilde{x}_T = u_T - \sum_{t=0}^{T-1} A_{T,t}\tilde{x}_t$, so $A\tilde{x} = u$. Thus, (5.17) holds by induction. Since $A_{t,t}^\dagger\mathcal{U}_t^e \subseteq \mathcal{X}_t$, (5.17) and Lemma 3.5 imply that $A\mathcal{X}$ is closed.

Assume now that A^t are $\mathcal{F}_t$-measurable. Since $E_t\mathcal{X} \subseteq \mathcal{X}$ and $A^t\mathcal{X}^t = A_{t,t}\mathcal{X}_t$, by (5.16), Lemma B.13 gives $A^t\mathcal{X}^t(\mathcal{F}_t) = A_{t,t}\mathcal{X}_t(\mathcal{F}_t)$. An argument similar to the proof of (5.17) now gives

$$A\mathcal{X}_a = A_{1,1}\mathcal{X}_1(\mathcal{F}_1) \times \cdots \times A_{T,T}\mathcal{X}_T(\mathcal{F}_T). \tag{5.18}$$

Since $A_{t,t}^\dagger\mathcal{U}_t^e \subseteq \mathcal{X}_t$ implies $A_{t,t}^\dagger\mathcal{U}_t^e(\mathcal{F}_t) \subseteq \mathcal{X}_t(\mathcal{F}_t)$, (5.18) and Lemma 3.5 imply that $A\mathcal{X}_a$ is closed. □

5.4.2 Optimal Stopping

Recall the relaxed optimal stopping problem (1.10)

$$\begin{aligned} &\text{minimize} \quad && E\left[\sum_{t=0}^{T} R_t x_t\right] \quad \text{over } x \in \mathcal{N}, \\ &\text{subject to} \quad && x \geq 0, \ \sum_{t=0}^{T} x_t \leq 1 \quad a.s. \end{aligned}$$

from Sect. 1.3.2. As observed in Sect. 3.4.2, this fits the general duality framework with $n_t = 1, \mathcal{X} = L^\infty, \mathcal{V} = L^1, m = 1, \mathcal{U} = L^\infty, \mathcal{Y} = L^1$,

$$f(x,u,\omega) = \begin{cases} -\sum_{t=0}^{T} x_t R_t(\omega) & \text{if } x \geq 0 \text{ and } \sum_{t=0}^{T} x_t + u \leq 0, \\ +\infty & \text{otherwise} \end{cases}$$

and $\bar{u} = -1$.

We saw already in Remark 3.49 that the dual problem (3.17) admits solutions and that there is no duality gap. The optimal dual solution was constructed from the Doob decomposition of the Snell envelope of the reward process R. The general results of this chapter give an alternative proof. It turns out that the optimal stopping problem satisfies the sufficient conditions of this chapter without any additional assumptions on the problem.

Theorem 5.46 *If* $\max_t R_t \in L^1$, *then* $\inf(1.9) = \sup(3.17)$ *and the optimum in* (3.17) *is attained. In particular, a stopping time* $\tau \in \mathcal{T}$ *is optimal if and only if there exists* $(p, y) \in \mathcal{X}_a^\perp \times \mathcal{Y}$ *such that* $p_t + R_t \leq y$ *for all* t *and* $p_\tau + R_\tau = y$ *almost surely. This is equivalent to the existence of a martingale* y *such that* $R_t \leq y_t$ *for all* t *and* $R_\tau = y_\tau$ *almost surely.*

Proof Recall that in (1.10), $\bar{u} = -1$. It was shown in Sect. 3.4.2 that $(p, y) \in \partial f(x, \bar{u})$ can be written as the scenariowise optimality conditions given here. It thus suffices to verify the assumptions of Theorem 5.29.

We have

$$\operatorname{dom} Ef = \{(x, u) \in L^\infty \times L^\infty \mid x \geq 0, \sum_{t=0}^{T} x_t + u \leq 0\}.$$

There is a feasible solution $\bar{x} \in \mathcal{X}_a$ such that $\bar{x}_t \geq \epsilon$ and $\sum_{t=0}^{T} x_t \leq 1 - \epsilon$ for some constant $\epsilon > 0$. Thus, $\bar{x}$ is strictly feasible, $\operatorname{aff} \operatorname{dom} Ef = \mathcal{X} \times \mathcal{U}$ and $\operatorname{aff} \operatorname{dom}_u Ef = \mathcal{U}$. The mapping

$$S(\omega) := \operatorname{dom} f(\cdot, \bar{u}(\omega), \omega) = \{x \in \mathbb{R}^n \mid x \geq 0, \sum_{t=0}^{T} x_t \leq 1\}$$

is deterministic and bounded, and, in particular, adapted so the last assumption in Theorem 5.29 holds by Example 5.39. The last claim follows from Example 3.46. □

5.4.3 Optimal Control

Consider again the optimal control problem (1.11)

$$\begin{aligned} &\text{minimize} \quad && E\left[\sum_{t=0}^{T} L_t(X_t, U_t)\right] \quad \text{over } (X, U) \in \mathcal{N}, \\ &\text{subject to} \quad && \Delta X_t = A_t X_{t-1} + B_t U_{t-1} + W_t \quad t = 1, \dots, T \text{ a.s.} \end{aligned}$$

from Sect. 1.3.3 and recall from Sect. 3.4.3 that (1.11) fits the general duality framework with $x = (X, U)$, $\bar{u} = W$ and

$$f(x, u, \omega) = \sum_{t=0}^{T} L_t(X_t, U_t, \omega) + \sum_{t=1}^{T} \delta_{\{0\}}(\Delta X_t - A_t(\omega) X_{t-1} - B_t(\omega) U_{t-1} - u_t).$$

As in Assumption 3.53 in Sect. 3.4.3, we assume that

$$\mathcal{X}_t = \mathcal{S} \times \mathcal{C}, \qquad \mathcal{U}_t = \mathcal{S},$$
$$\mathcal{V}_t = \mathcal{S}' \times \mathcal{C}', \qquad \mathcal{Y}_t = \mathcal{S}',$$

where $\mathcal{S}$ and $\mathcal{C}$ are solid decomposable spaces with Köthe duals $\mathcal{S}'$ and $\mathcal{C}'$, respectively, and that, for all t,

1. $E_t\mathcal{S} \subseteq \mathcal{S}$ and $E_t\mathcal{C} \subseteq \mathcal{C}$,
2. $A_t\mathcal{S} \subseteq \mathcal{S}$ and $B_t\mathcal{C} \subseteq \mathcal{S}$.

In order to apply the results of this chapter, we also assume that $\mathcal{S}$ and $\mathcal{C}$ are endowed with Fréchet topologies under which their topological duals can be expressed as

$$\mathcal{S}^* = \mathcal{S}' \oplus (\mathcal{S}')^s \quad \text{and} \quad \mathcal{C}^* = \mathcal{C}' \oplus (\mathcal{C}')^s,$$

where $(\mathcal{S}')^s$ and $(\mathcal{C}')^s$ are the singular elements of $\mathcal{S}^*$ and $\mathcal{C}^*$, respectively. It follows that the spaces $\mathcal{X}$ and $\mathcal{U}$ satisfy the assumptions made in Sect. 5.3.

Assumption 5.47 *The functionals EL_t are proper and strongly lsc on $\mathcal{S} \times \mathcal{C}$,*

1. *there exists a feasible $(\bar{X}, \bar{U}) \in \mathcal{X}_a$ such that $(\bar{X}_t, \bar{U}_t) \in \operatorname{rcore} \operatorname{dom} EL_t$,*
2. $\operatorname{aff} \operatorname{dom} EL_t$ *is strongly closed,*
3. *(a)* $E_t \operatorname{dom} EL_{t'} \subseteq \operatorname{dom} EL_{t'}$ *for all $t' \le t$,*
   (b) for every $X_t \in \mathcal{S}$, there exists $U_t \in \mathcal{C}$ with $(X_t, U_t) \in \operatorname{dom} EL_t$

for all t.

Combining Theorem 3.51 and Corollary 3.56 with Theorem 5.29, yields the following.

Theorem 5.48 *Under Assumption 5.47, problem* (1.11) *is feasible for any $W \in \mathcal{U}_a$,* $\inf$*(1.11)* $=$ $\sup$*(3.23) and the optimum in the dual problem* (3.23) *is attained. If the optimum value is finite, a feasible (X, U) is optimal in* (1.11) *if and only if there exists a dual feasible $(p, y) \in \mathcal{N}^\perp \times \mathcal{Y}$ such that*

$$p_t - (\Delta y_{t+1} + A^*_{t+1} y_{t+1}, B^*_{t+1} y_{t+1}) \in \partial L_t(X_t, U_t)$$

almost surely for all t. If, in addition, each L_t is $\mathcal{F}_t$-measurable, this is equivalent to the existence of a $y \in \mathcal{Y}_a$ feasible in the reduced dual in Remark 3.56 such that

$$-E_t(\Delta y_{t+1} + A^*_{t+1} y_{t+1}, B^*_{t+1} y_{t+1}) \in \partial L_t(X_t, U_t)$$

almost surely.

Proof Recall from Sect. 3.4.3 that (1.11) fits the general duality framework with $\bar{u} = W$ and that, since Assumption 5.47.1 implies $\operatorname{dom} Ef \cap (\mathcal{X} \times \mathcal{U}) \neq \emptyset$, the dual problem can be written as (3.23). By Theorem 3.51 and Remark 3.56, it thus suffices to show that there is no duality gap and that the dual admits solutions. To this end, it suffices to verify the assumptions of Theorem 5.29.

Problem (1.11) fits the format $f(x, u) = g(x, Ax + u)$ of Example 5.33 with $g(x, u) := \sum L_t(x_t) + \delta_{\{0\}}(u)$ and $A(x) := (-\Delta X_t + A_t X_{t-1} + B_t U_{t-1})_{t=1}^T$. We have

$$\operatorname{dom} Eg = \{(x, u) \in \mathcal{X} \times \mathcal{U} \mid x_t \in \operatorname{dom} EL_t \ \forall t,\ u = 0\}.$$

By Lemmas A.5 and A.9, Assumption 5.47.1 implies

$$\operatorname{rcore} \operatorname{dom} Eg = \{(x, u) \in \mathcal{X} \times \mathcal{U} \mid x_t \in \operatorname{rcore} \operatorname{dom} EL_t \ \forall t,\ u = 0\}$$

and

$$\operatorname{aff} \operatorname{dom} Eg = \{(x, u) \in \mathcal{X} \times \mathcal{U} \mid x_t \in \operatorname{aff} \operatorname{dom} EL_t \ \forall t,\ u = 0\}. \tag{5.19}$$

By Example 5.33, Assumption 5.47.1 implies that $\bar{x}$ is strictly feasible.

Let $\pi'(x, u) := ({}^a x, {}^a u)$. To prove that $\operatorname{aff} \operatorname{dom} Ef - \mathcal{X}_a \times \{0\}$ is closed, it suffices, by Example 5.33, to show that $\operatorname{aff} \operatorname{dom} Eg$ is closed, $\pi' \operatorname{aff} \operatorname{dom} Eg \subseteq \operatorname{aff} \operatorname{dom} Eg$, $(I, \mathcal{A})\mathcal{X}_a \subseteq \operatorname{rge} \pi' \subseteq \mathcal{X}_a \times \mathcal{U}$, and

$$(0, u) \in (I, \mathcal{A})\mathcal{X}_a + (\bar{x}, A\bar{x} + \bar{u}) - \operatorname{aff} \operatorname{dom} Eg \tag{5.20}$$

for all $u \in \mathcal{U}$ with $(0, u) \in \operatorname{rge} \pi'$. Clearly, $\pi' = \mathcal{X}_a \times \mathcal{U}_a$. The closedness of $\operatorname{aff} \operatorname{dom} Eg$ follows from (5.19) and Assumption 5.47.2 while the inclusion $\pi' \operatorname{aff} \operatorname{dom} Eg \subseteq \operatorname{aff} \operatorname{dom} Eg$ follows from Assumption 5.47.3.a and Lemma A.9.1. The condition $(I, \mathcal{A})\mathcal{X}_a \subseteq \operatorname{rge} \pi' \subseteq \mathcal{X}_a \times \mathcal{U}$ is clear. Let $u \in \mathcal{U}_a$. Inclusion (5.20) holds if there exists $x \in \mathcal{X}_a$ with $(x, Ax + u - \bar{u}) \in \operatorname{dom} Eg$, i.e. if for all t, $(X_t, U_t) \in \operatorname{dom} EL_t$ and $\Delta X_t = A_t X_{t-1} + B_t U_{t-1} + \tilde{u}_t$, where $\tilde{u} := u - \bar{u} \in \mathcal{U}_a$. Assume that $(X^{t-1}, U^{t-1}) \in \mathcal{X}_a^{t-1}$ with $(X_{t'}, U_{t'}) \in \operatorname{dom} EL_{t'}$ and

$$\Delta X_{t'} = A_{t'} X_{t'-1} + B_{t'} U_{t'-1} + \tilde{u}_{t'}$$

for $t' = 0, \dots, t-1$. Since $A_t \mathcal{S} \subseteq \mathcal{S}$ and $B_t \mathcal{C} \subseteq \mathcal{S}$, by assumption, the random vector X_t defined by

$$\Delta X_t := A_t X_{t-1} + B_t U_{t-1} + \tilde{u}_t$$

belongs to $\mathcal{S}$. Since $\tilde{u} \in \mathcal{U}_a$, X_t is $\mathcal{F}_t$-measurable. By Assumption 5.47.3.b, there exists $\tilde{U}_t$ such that $(X_t, \tilde{U}_t) \in \operatorname{dom} EL_t$. Since X_t is $\mathcal{F}_t$-measurable, Assumption 5.47.3.a gives $(X_t, U_t) \in \operatorname{dom} L_t$, where $U_t := E_t \tilde{U}_t$. By induction on t, we thus find an $x \in \mathcal{X}_a$ with $(x, Ax + u - \bar{u}) \in \operatorname{dom} Eg$.

To prove that $\operatorname{aff} \operatorname{dom}_u Ef$ is closed, it suffices, by Lemma 5.42, to show that $\{u \in \mathcal{U} \mid (0, u) \in \operatorname{aff} \operatorname{dom} Eg - \operatorname{gph} \mathcal{A}\} = \mathcal{U}$. This means that, for every $u \in \mathcal{U}$, there exists $x \in \mathcal{X}$ with $(X_t, U_t) \in \operatorname{aff} \operatorname{dom} EL_t$ and $\Delta X_t = A_t X_{t-1} + B_t U_{t-1} + u_t$

for all t. The existence of such x can be established by an argument similar to that in the previous paragraph.

It remains to verify the last assumption in Theorem 5.29. Let $(z, \bar{u}) \in \operatorname{dom} Ef$ and $t' \leq t$. Assumption 5.47.3.a gives $E_t z_{t'} \in \operatorname{dom} EL_{t'}$ while, by Lemma 3.7,

$$E_t X_{t'} = E_t X_{t'-1} + A_{t'} E_t X_{t'-1} + B_{t'} E_t U_{t'-1} + \bar{u}_{t'}.$$

An induction argument similar to two paragraphs above gives $\bar{z} \in \mathcal{X}$ with $(\bar{z}, \bar{u}) \in \operatorname{dom} Ef$ and $\bar{z}^t = E_t z^t$, so the last assumption in Theorem 5.29 holds. □

Clearly, conditions 1–3 hold in Assumption 5.47 if the functions EL_t are finite on $\mathcal{S} \times \mathcal{C}$. Assumption 5.47.3.a holds automatically if each L_t is $\mathcal{F}_t$-measurable like in Remark 3.56. By Corollary 1.68, Assumption 5.47.3.a holds also if $\operatorname{dom} L_t$ are closed-valued and $\mathcal{F}_t$-measurable and if EL_t is finite at any $(X_t, U_t) \in \mathcal{S} \times \mathcal{C}$ such that $(X_t, U_t) \in \operatorname{dom} L_t$ almost surely.

Remark 5.49 *Assume that, for all t,* $\operatorname{dom} L_t$ *is closed-valued,* $\operatorname{dom} EL_t \neq \emptyset$ *and*

$$\operatorname{dom} EL_t = \{(X_t, U_t) \in \mathcal{S} \times \mathcal{C} \mid (X_t, U_t) \in \operatorname{dom} L_t \ a.s.\}.$$

Then $\operatorname{dom} L_t$ *is* $\mathcal{F}_t$*-measurable for all t if and only if Assumption 5.47.3.a holds. If, for all t and almost every* ω, $\operatorname{dom} L_t(X_t, \cdot, \omega) \neq \emptyset$ *for every* $X_t \in \mathbb{R}^N$*, then Assumption 5.47.3.b holds.*

Proof The first claim follows from Theorem 3.14. As to the second claim, let $X_t \in \mathcal{S}$. By assumption,

$$D_t(\omega) := \{U_t \in \mathbb{R}^M \mid L_t(X_t(\omega), U_t, \omega) < \infty\}$$

is closed nonempty-valued. By the measurable selection theorem, Corollary 1.28, there exists $U_t \in L^0(D_t)$. By the first assumption, $(X_t, U_t) \in \operatorname{dom} EL_t$, so Assumption 5.47.3.b holds. □

Example 5.50 (Bounded Strategies) *Let* $\mathcal{S} = L^\infty$ *and* $\mathcal{C} = L^\infty$ *and endow the spaces with the usual norm-topologies. If* EL_t *is finite on*

$$\mathcal{D}_t := \{(X_t, U_t) \in L^\infty \times L^\infty \mid \exists \epsilon > 0 : \mathbb{B}_\epsilon(X_t, U_t) \subset \operatorname{dom} L_t \quad a.s.\}$$

and if there is a feasible point $(\bar{X}, \bar{U})$ *such that* $(\bar{X}_t, \bar{U}_t) \in \mathcal{D}_t$ *for all t, then, by Remark 5.19, the conditions 1 and 2 in Assumption 5.47 hold.*

Example 5.51 (Linear-Quadratic Control) *Consider Example 2.104, let* $\mathcal{S}$ *and* $\mathcal{C}$ *be the Fréchet spaces of finite moments from Example 5.12 and assume that the entries of all the involved matrices have finite moments. Since* $E_t L^p \subseteq L^p$ *for every* $p \geq 1$*, we have* $E_t \mathcal{S} \subseteq \mathcal{S}$ *and* $E_t \mathcal{C} \subseteq \mathcal{C}$*. Since* $\mathcal{S}$ *and* $\mathcal{C}$ *are closed under componentwise products, we also have* $A_t \mathcal{S} \subseteq \mathcal{S}$ *and* $B_t \mathcal{C} \subseteq \mathcal{S}$ *and that the functionals* EL_t *are finite on* $\mathcal{S} \times \mathcal{C}$*. In particular, Assumption 5.47 holds.*

5.4.4 Problems of Lagrange

Consider again problem (1.13)

$$\text{minimize} \quad E\left[\sum_{t=0}^{T} K_t(x_t, \Delta x_t + u_t)\right] \quad \text{over } x \in \mathcal{N},$$

from Sect. 1.3.4 and recall from Sect. 3.4.4 that this fits the duality framework with $\bar{u} = 0$ and

$$f(x, u, \omega) = \sum_{t=0}^{T} K_t(x_t, \Delta x_t + u_t, \omega).$$

As in Sect. 3.4.4, we assume that

$$\mathcal{X}_t = \mathcal{S}, \quad \mathcal{V}_t = \mathcal{S}', \quad \mathcal{U} = \mathcal{X}, \quad \mathcal{Y} = \mathcal{V},$$

where $\mathcal{S}$ is solid decomposable space in separating duality with $\mathcal{S}'$. In addition, we assume that $E_t\mathcal{S} \subseteq \mathcal{S}$ for all t and that $\mathcal{S}$ is endowed with a Fréchet topology under which its topological dual can be expressed as

$$\mathcal{S}^* = \mathcal{S}' \oplus (\mathcal{S}')^s,$$

where $(\mathcal{S}')^s$ are singular elements of $\mathcal{S}^*$. The spaces $\mathcal{X}$ and $\mathcal{U}$ then satisfy the assumptions made in Sect. 5.3.

Assumption 5.52 *The functionals EK_t are proper and strongly lsc on $\mathcal{S} \times \mathcal{S}$,*

1. *there exists $\bar{x} \in \mathcal{X}_a$ such that, $(\bar{x}_t, \Delta\bar{x}_t) \in \operatorname{rcore}\operatorname{dom} EK_t$,*
2. $\operatorname{aff}\operatorname{dom} EK_t$ *is strongly closed,*
3. *(a) $E_t \operatorname{dom} EK_{t'} \subseteq \operatorname{dom} EK_{t'}$ for all $t' \leq t$,*
   (b) for every $x_{t-1} \in \mathcal{S}$, there exists $x_t \in \mathcal{S}$ such that $(x_t, \Delta x_t) \in \operatorname{aff}\operatorname{dom} EK_t$,
   (c) for every $\mathcal{F}_{t-1}$-measurable $x^{t-1} \in \mathcal{X}^{t-1}$ such that $(x_{t'}, \Delta x_{t'}) \in \operatorname{dom} EK_{t'}$ for all $t' \leq t-1$, there exists $x_t \in \mathcal{S}$ with $(x_t, \Delta x_t) \in \operatorname{dom} EK_t$

for all t.

Combining Theorem 3.62 and Corollary 3.64 with Theorem 5.29 yields the following.

Theorem 5.53 *Under Assumption 5.52,* $\inf$*(1.13)* $=$ $\sup$*(3.31) and the optimum in the dual problem* (3.31) *is attained. If the optimum value is finite, a feasible x is* *optimal in* (1.13) *if and only if there exists a dual feasible* $(p, y) \in \mathcal{N}^\perp \times \mathcal{Y}$ *such that*

$$p_t + \Delta y_{t+1} \in \partial_x H_t(x_t, y_t),$$
$$\Delta x_t \in \partial_y[-H_t](x_t, y_t)$$

almost surely. If, in addition, each K_t *is* $\mathcal{F}_t$ *measurable, this is equivalent to the existence of a* $y \in \mathcal{Y}_a$ *feasible in the reduced dual in Corollary 3.64 such that*

$$E_t \Delta y_{t+1} \in \partial_x H_t(x_t, y_t),$$
$$\Delta x_t \in \partial_y[-H_t](x_t, y_t),$$

almost surely.

Proof Recall from Sect. 3.4.4 that (1.13) fits the general duality framework with $\bar{u} = W$ and that, since Assumption 5.52.1 implies $\operatorname{dom} Ef \cap (\mathcal{X} \times \mathcal{U}) \neq \emptyset$, the dual problem can be written as (3.23). By Theorem 3.62 and Corollary 3.64, it thus suffices to show that there is no duality gap and that the dual admits solutions. To this end, it suffices to verify the assumptions of Theorem 5.29.

Problem (1.11) fits the format $f(x, u) = g(x, Ax + u)$ of Example 5.33 with $g(x, u) := \sum_{t=0}^T K_t(x_t, u_t)$ and $A(x) := (\Delta x_t)_{t=0}^T$. We have

$$\operatorname{dom} Eg = \{(x, u) \in \mathcal{X} \times \mathcal{U} \mid (x_t, u_t) \in \operatorname{dom} EK_t \ \forall t\}.$$

By Lemmas A.5 and A.9, Assumption 5.52.1 implies

$$\operatorname{rcore} \operatorname{dom} Eg = \{(x, u) \in \mathcal{X} \times \mathcal{U} \mid (x_t, u_t) \in \operatorname{rcore} \operatorname{dom} EK_t \ \forall t\}$$

and

$$\operatorname{aff} \operatorname{dom} Eg = \{(x, u) \in \mathcal{X} \times \mathcal{U} \mid (x_t, u_t) \in \operatorname{aff} \operatorname{dom} EK_t \ \forall t\}. \tag{5.21}$$

By Example 5.33, Assumption 5.52.1 implies that $\bar{x}$ is strictly feasible.

Let $\pi'(x, u) := ({}^a x, {}^a u)$. To prove that $\operatorname{aff} \operatorname{dom} Ef - \mathcal{X}_a \times \{0\}$ is closed, it suffices, by Example 5.33, to show that $\operatorname{aff} \operatorname{dom} Eg$ is closed, $\pi' \operatorname{aff} \operatorname{dom} Eg \subseteq \operatorname{aff} \operatorname{dom} Eg$, $(I, \mathcal{A})\mathcal{X}_a \subseteq \operatorname{rge} \pi' \subseteq \mathcal{X}_a \times \mathcal{U}$, and

$$(0, u) \in (I, \mathcal{A})\mathcal{X}_a + (\bar{x}, A\bar{x} + \bar{u}) - \operatorname{aff} \operatorname{dom} Eg \tag{5.22}$$

for all $u \in \mathcal{U}$ with $(0, u) \in \operatorname{rge} \pi'$. Clearly, $\pi' = \mathcal{X}_a \times \mathcal{U}_a$. The closedness of $\operatorname{aff} \operatorname{dom} Eg$ follows from (5.21) and Assumption 5.52.2 while the inclusion $\pi' \operatorname{aff} \operatorname{dom} Eg \subseteq \operatorname{aff} \operatorname{dom} Eg$ follows from Assumption 5.52.3.a and Lemma A.9.1. The condition $(I, \mathcal{A})\mathcal{X}_a \subseteq \operatorname{rge} \pi' \subseteq \mathcal{X}_a \times \mathcal{U}$ is clear. Let $u \in \mathcal{U}_a$. Since $\bar{u} = 0$, inclusion (5.22) holds if there exists $x \in \mathcal{X}_a$ with $(x, Ax + u) \in \operatorname{aff} \operatorname{dom} Eg$, i.e. if $(x_t, \Delta x_t + u_t) \in \operatorname{aff}\operatorname{dom} EK_t$ for all t. Assume that $x^{t-1} \in \mathcal{X}_a^{t-1}$ with $(x_{t'}, \Delta x_{t'} + u_{t'}) \in \operatorname{aff}\operatorname{dom} EK_{t'}$ for $t' \le t-1$. By Assumption 5.52.3.b, there exists $\tilde{x}_t$ such that $(\tilde{x}_t, \tilde{x}_t - x_{t-1} + u_t) \in \operatorname{aff}\operatorname{dom} EK_t$. Assumption 5.52.3.a implies that $E_t \operatorname{aff}\operatorname{dom} EK_t \subseteq \operatorname{aff}\operatorname{dom} EK_t$. Since u is adapted and x_{t-1} is $\mathcal{F}_{t-1}$-measurable, we get $(x_t, \Delta x_t + u_t) \in \operatorname{aff}\operatorname{dom} EK_t$, where $x_t := E_t \tilde{x}_t$. By induction on t, we thus find an $x \in \mathcal{X}_a$ with $(x, Ax + u) \in \operatorname{aff}\operatorname{dom} Eg$.

To prove that $\operatorname{aff}\operatorname{dom}_u Ef$ is closed, it suffices, by Lemma 5.42, to show that $\{u \in \mathcal{U} \mid (0, u) \in \operatorname{aff}\operatorname{dom} Eg - \operatorname{gph} \mathcal{A}\} = \mathcal{U}$. By (5.21), this means that, for every $u \in \mathcal{U}$, there exists $x \in \mathcal{X}$ with $(x_t, \Delta x_t + u_t) \in \operatorname{aff}\operatorname{dom} EK_t$ for all t. The existence of such x can be established by an argument similar to that in the previous paragraph.

It remains to verify the last assumption in Theorem 5.29. Let $(z, 0) \in \operatorname{dom} Ef$ and $t' \le t$. Assumption 5.52.3.a gives $E_t(z_{t'}, \Delta z_{t'}) \in \operatorname{dom} EK_{t'}$. Applying Assumption 5.52.3.c and 3.a, an induction argument similar to two paragraphs above gives $\bar{z} \in \mathcal{X}$ with $(\bar{z}, 0) \in \operatorname{dom} Ef$ and $\bar{z}^t = E_t z^t$, so the last assumption in Theorem 5.29 holds. □

If each EK_t is finite on $\mathcal{S} \times \mathcal{S}$, then Assumption 5.52 holds. If EK_t is proper on $\mathcal{S} \times \mathcal{S}$, K_t is $\mathcal{F}_t$-measurable and

$$\operatorname{dom} EK_t = \{(x_t, u_t) \in \mathcal{S} \times \mathcal{S} \mid (x_t, u_t) \in \operatorname{dom} K_t \ a.s.\},$$

then Jensen's inequality in Theorem 1.67 gives Assumption 5.52.3.a. Moreover, Assumption 5.52.3.b and 3.c hold if $\operatorname{dom}_1 EK_t$ or $\operatorname{dom}_2 EK_t$ is the whole space.

Example 5.54 (Bounded Strategies) *Assume that $\mathcal{S} = L^\infty$ is endowed with the usual norm-topology. If EK_t is finite on*

$$\mathcal{D}_t := \{(x_t, u_t) \in L^\infty \times L^\infty \mid \exists \epsilon > 0 : \mathbb{B}_\epsilon(x_t, u_t) \subseteq \operatorname{dom} K_t \quad a.s.\}$$

and if there is a feasible $\bar{x}$ such that $(\bar{x}_t, \Delta \bar{x}_t) \in \mathcal{D}_t$ for all t, then, by Remark 5.19, parts 1,2 and 3.b in Assumption 5.52 hold. More generally, if $K_t = K_t^0 + \delta_{H_t}$ where K_t^0 has the above properties and H_t is an affine-valued measurable mapping, and if there is a feasible $\bar{x}$ such that $(\bar{x}_t, \Delta \bar{x}_t) \in \mathcal{D}_t$ for all t then, by Example 5.20, Assumption 5.52.1 and 2 hold.

Example 5.55 (Bounded Recourse) *Let $\mathcal{S} = L^\infty$ and assume that, for every t, $\operatorname{dom} K_t$ is $\mathcal{F}_t$-measurable, $\operatorname{dom} EK_t = L^\infty(\operatorname{dom} K_t)$ and that, for every $r_t > 0$, there exists $r_{t+1} \ge r_t$ such that*

$$\begin{aligned}\{x_t \in \mathbb{R}^d \mid x_t \in \mathbb{B}_{r_t}, \exists x_{t-1} : K_t(x_t, x_t - x_{t-1}, \omega) < \infty\} \\ \subseteq \{x_t \in \mathbb{R}^d \mid \exists x_{t+1} \in \mathbb{B}_{r_{t+1}} : K_{t+1}(x_{t+1}, x_{t+1} - x_t, \omega) < \infty\}.\end{aligned} \tag{5.23}$$

Then the bounded recourse condition in Example 5.39 as well as parts a and c in Assumption 5.52.3 hold.

Proof Let $r_0^\nu := \nu$. By assumption, there exists r_t^ν such that (5.23) holds. Let $B^\nu := \mathbb{B}_{r_0^\nu} \times \cdots \times \mathbb{B}_{r_T^\nu}$, $S_\nu(\omega) := \operatorname{dom} f(\cdot, 0, \omega)) \cap B^\nu$ and $S_\nu^t(\omega) := \{x^t \in \mathbb{R}^{(t+1)d} \mid x \in S_\nu(\omega)\}$. Clearly, $S_\nu^t(\omega)$ is contained in

$$\{x^t \in \mathbb{R}^{(t+1)d} \mid (x_{t'}, \Delta x_{t'}) \in \operatorname{dom} K_{t'}(\cdot, \cdot, \omega),\ x_{t'} \in \mathbb{B}_{r_{t'}^\nu}\ \forall t' \le t\},$$

while (5.23) gives the converse. Thus, $\mathcal{F}_t$-measurability of dom K_t is implies that S_ν is adapted. This shows that the bounded recourse condition in Example 5.39 holds. Part a in Assumption 5.52.3 holds by Jensen's inequality and the assumption $\operatorname{dom} EK_t = L^\infty(\operatorname{dom} K_t)$ while part c follows from an application of the measurable selection theorem. □

5.4.5 Financial Mathematics

Consider again problem (1.15)

$$\text{minimize} \quad EV\left(c - \sum_{t=0}^{T-1} x_t \cdot \Delta s_{t+1}\right) \quad \text{over} \quad x \in \mathcal{N}_D,$$

from Sect. 1.3.5 and recall from Sect. (3.4.5) that it fits the general duality framework with $\bar{u} = c$ and

$$f(x, u, \omega) = V\left(u - \sum_{t=0}^{T} x_t \cdot \Delta s_t(\omega), \omega\right) + \sum_{t=0}^{T} \delta_{D_t(\omega)}(x_t).$$

We assume that $\mathcal{X} = L^\infty$ and that $\mathcal{U}$ is a space of real-valued random variables satisfying the assumptions in Sect. 5.3. By Example 5.11. $\mathcal{X}$ satisfies the assumptions too. The set

$$\operatorname{lin}\operatorname{core}\operatorname{dom} EV := (\operatorname{core}\operatorname{dom} EV)^\infty \cap [-(\operatorname{core}\operatorname{dom} EV)^\infty]$$

is the largest linear space contained in the recession cone of core dom EV. More concrete expressions for lin core dom EV will be given in Remark 5.58 below.

Assumption 5.56 *$EV : \mathcal{U} \to \overline{\mathbb{R}}$ is proper and strongly lsc and there exist $\bar{x} \in \mathcal{X}_a$ and $\epsilon > 0$ such that, for all t,*

1. *$c \in$ core dom EV and $\Delta s_t^j \in$ lin core dom EV for all $j \in J$,*
2. *$\mathbb{B}_\epsilon(\bar{x}_t) \subseteq D_t$ almost surely.*

Combining Theorem 3.68 and Example 3.71 with Theorem 5.29 yields the following.

Theorem 5.57 *Under Assumption 5.56,* $\inf(1.15) = \sup(3.33)$ *and the optimum in the dual problem* (3.33) *is attained. If the optimum value is finite, a feasible* $x \in \mathcal{N}$ *solves* (1.15) *if and only if there exists* $(p, y) \in \mathcal{N}^\perp \times \mathcal{Y}$ *feasible in* (3.33) *such that*

$$y \in \partial V(c - \sum_{t=0}^{T-1} x_t \cdot \Delta s_{t+1}),$$

$$p_t + y\Delta s_{t+1} \in N_{D_t}(x_t) \quad t = 0, \dots, T$$

almost surely. This holds if and only if there exists y feasible in the reduced dual in Example 3.71 such that

$$y \in \partial V(c - \sum_{t=0}^{T-1} x_t \cdot \Delta s_{t+1}),$$

$$E_t[y\Delta s_{t+1}] \in N_{D_t}(x_t) \quad t = 0, \dots, T$$

almost surely.

Proof Recall from Sect. 3.4.5 that (1.15) fits the general duality framework with $\bar{u} = c$ and that the dual problem can be written as (3.33). By Theorem 3.68 and Example 3.71, it thus suffices to show that there is no duality gap and that the dual admits solutions. To this end, it suffices to verify the assumptions of Theorem 5.29.

Problem (1.11) fits the format $f(x, u) = g(x, Ax + u)$ of Example 5.33 with $g(x, u) := V(u) + \sum_{t=0}^{T} \delta_{D_t}(x_t)$ and $Ax := -\sum_{t=0}^{T-1} x_t \cdot \Delta s_{t+1}$. We have

$$\operatorname{dom} Eg = \{(x, u) \in \mathcal{X} \times \mathcal{U} \mid x_t \in D_t \ \forall t \ a.s.,\ u \in \operatorname{dom} EV\}.$$

By Lemmas A.5 and A.9, Assumption 5.56.1 and 2 imply

$$\operatorname{core} \operatorname{dom} Eg = \{(x, u) \in \mathcal{X} \times \mathcal{U} \mid x_t \in \operatorname{core} L^\infty(D_t) \ \forall t,\ u \in \operatorname{core} \operatorname{dom} EV\}$$

and

$$\begin{aligned}\operatorname{aff} \operatorname{dom} Eg &= \{(x, u) \in \mathcal{X} \times \mathcal{U} \mid x_t \in \operatorname{aff} L^\infty(D_t) \ \forall t,\ u \in \operatorname{aff} \operatorname{dom} EV\} \\ &= \mathcal{X} \times \mathcal{U}.\end{aligned}$$

The monotonicity of V implies that $zu \in \operatorname{lin} \operatorname{core} \operatorname{dom} EV$ for every $u \in \operatorname{lin} \operatorname{core} \operatorname{dom} EV$ and $z \in L^\infty$. Indeed, if $u \in (\operatorname{core} \operatorname{dom} EV)^\infty$, then

$$EV(\bar{u} + \alpha \|z\|_{L^\infty} u) < +\infty \quad \forall \bar{u} \in \operatorname{core} \operatorname{dom} EV,\ \alpha > 0$$

and the monotonicity gives

$$EV(\bar{u} + \alpha z u) < +\infty \quad \forall \bar{u} \in \operatorname{core} \operatorname{dom} EV,\ \alpha > 0$$

which means, by Lemma A.4, that $zu \in (\operatorname{core}\operatorname{dom} EV)^\infty$. Similarly, $-u \in (\operatorname{core}\operatorname{dom} EV)^\infty$ implies $-zu \in (\operatorname{core}\operatorname{dom} EV)^\infty$. It follows that the second condition in Assumption 5.56.1 gives

$$\sum_{t=0}^{T-1} x_t \cdot \Delta s_{t+1} \in \operatorname{lin}\operatorname{core}\operatorname{dom} EV \quad \forall x \in \mathcal{X}. \tag{5.24}$$

By definition of the recession cone (see Sect. A.3), the first condition in Assumption 5.56.1 thus gives

$$c - \sum_{t=0}^{T-1} x_t \cdot \Delta s_{t+1} \in \operatorname{core}\operatorname{dom} EV \quad \forall x \in \mathcal{X}. \tag{5.25}$$

Thus, by Example 5.33, Assumption 5.56.2 implies that $\bar{x}$ is strictly feasible. Since $\operatorname{aff}\operatorname{dom} Eg = \mathcal{X} \times \mathcal{U}$, we have $\operatorname{dom} Ef = \mathcal{X} \times \mathcal{U}$ and thus, $\operatorname{aff}\operatorname{dom} Ef - \mathcal{X}_a \times \{0\} = \mathcal{X} \times \mathcal{U}$ and, by Lemma A.9, $\operatorname{aff}\operatorname{dom}_u Ef = \mathcal{U}$. Thus, assumption 3 of Theorem 5.29 holds.

It remains to verify the last assumption in Theorem 5.29. Let $(z, \bar{u}) \in \operatorname{dom} Ef$ and define $\bar{z} \in \mathcal{X}$ by $\bar{z}^t = E_t z^t$ and $\bar{z}_s = z_s$ for $s > t$. Since $z_t \in D_t$ almost surely, the $\mathcal{F}_t$-measurability of D_t implies, by Corollary 1.68, that $\bar{z}_t \in D_t$ almost surely. Thus, by (5.25), $(\bar{z}, \bar{u}) \in \operatorname{dom} Ef$. □

If the price process belongs to the Orlicz heart generated by the loss function V, it is natural to choose $\mathcal{U}$ as the corresponding Orlicz space.

Remark 5.58 *Let V nonrandom and real-valued. Then $\Phi(u) := V(|u|)$ defines a Young function in the sense of Sect. 5.1.1. Assume that EV is proper and lsc on the associated Orlicz space $\mathcal{U} := L^\Phi$; see Example 5.7. The set* $\operatorname{lin}\operatorname{dom} EV$ *coincides with the Orlicz heart*

$$M^\Phi = \{u \in L^1 \mid E\Phi(\lambda u) < \infty \ \forall \lambda > 0\};$$

see Example 5.8. In particular, $M^\Phi \subseteq$ $\operatorname{core}\operatorname{dom} EV$ *and the second condition in Assumption 5.56.1 means that $\Delta s_t \in M^\Phi$. Note also that if $c \in M^\Phi$, then we may chose $\mathcal{U} = M^\Phi$ so that $\mathcal{Y}^s = \{0\}$, by Example 5.8.*

Proof By definition of L^Φ, $0 \in \operatorname{core}\operatorname{dom} EV$. Thus,

$$(\operatorname{core}\operatorname{dom} EV)^\infty \subseteq \{u \in \mathcal{U} \mid \alpha u \in \operatorname{dom} EV \ \forall \alpha > 0\} \subseteq (\operatorname{cl}_s \operatorname{dom} EV)^\infty,$$

where the first inclusion follows from the definition of the recession cone and the second one from Theorem A.15. On the other hand, by Theorem A.15, $(\operatorname{cl}_s \operatorname{dom} EV)^\infty = (\operatorname{core}\operatorname{cl}_s \operatorname{dom} EV)^\infty$, where, by Theorem A.28, $\operatorname{core}\operatorname{cl}_s \operatorname{dom} EV = \operatorname{core}\operatorname{dom} EV$, since $\operatorname{int}\operatorname{dom} EV \neq \emptyset$ by Theorem A.33. Thus,

$$(\operatorname{core}\operatorname{dom} EV)^\infty = \{u \in \mathcal{U} \mid \alpha u \in \operatorname{dom} EV \ \forall \alpha > 0\} = (\operatorname{cl}_s \operatorname{dom} EV)^\infty$$

and consequently,

$$\operatorname{lin}\operatorname{dom} EV = \{u \in \mathcal{U} \mid \alpha u \in \operatorname{dom} EV \ \forall \alpha \in \mathbb{R}\}.$$

Since $V \leq \Phi$, we have $M^\Phi \subseteq \operatorname{lin}\operatorname{dom} EV$. On the other hand, since V is nondecreasing and $V(0) = 0$ we have, for any $u \in \operatorname{lin}\operatorname{dom} EV$, that $EV(\alpha u^+) < +\infty$ and $EV(\alpha u^-) < +\infty$ for all $\alpha > 0$ so $u \in M^\Phi$. □

If EV is finite throughout $\mathcal{U}$ and if there are no portfolio constraints, then by Corollary 5.17, Ef is $\tau(\mathcal{X} \times \mathcal{U}, \mathcal{V} \times \mathcal{Y})$-continuous as soon as $\Delta s_t \in \mathcal{U}$ for all t. By Theorem A.29 and Corollary A.62, this implies the weak subdifferentiability of φ as soon as it is proper. In that case, one can thus avoid going through the arguments in Sect. 5.3. In the presence of portfolio constraints, the Mackey continuity fails but we still find a dual solution in $\mathcal{V} \times \mathcal{Y}$ under Assumption 5.56. Note also that, the proof of Theorem 5.57 gives the existence of a solution in the strong dual (D_s) even without part 3 in Assumption 5.56.

The following applies Theorem 5.29 to the problem from Example 3.74.

Theorem 5.59 (Optimized Certainty Equivelent) *Let Assumption 5.56 hold and assume that* $L^\infty \subseteq \operatorname{core}\operatorname{dom} EV$. *Then* $\inf(3.36) = \sup(3.38)$ *and the optimum in the dual problem* (3.38) *is attained. If the optimum value is finite, a feasible* (α, x) *solves* (3.37) *if and only if there exists* $(p, y) \in \mathcal{N}^\perp \times \mathcal{Y}$ *feasible in* (3.38) *such that*

$$y \in \partial V(c - \sum_{t=0}^{T-1} x_t \cdot \Delta s_{t+1} - \alpha),$$

$$p_t + y\Delta s_{t+1} \in N_{D_t}(x_t) \quad t = 0, \ldots, T$$

almost surely. This holds if and only if there exists y feasible in the reduced dual in (3.39) *such that*

$$y \in \partial V(c - \sum_{t=0}^{T-1} x_t \cdot \Delta s_{t+1} - \alpha),$$

$$E_t[y\Delta s_{t+1}] \in N_{D_t}(x_t) \quad t = 0, \ldots, T$$

almost surely.

Proof As observed in Example 3.74 and the proof of Example 4.31, this fits the general framework with time t running from -1 to T, $x_{-1} = \alpha$, $\mathcal{F}_{-1} = \{\emptyset, \Omega\}$, $\bar{u} = c$ and

$$\hat{f}(\hat{x}, u, \omega) := \alpha + V(u - \alpha - \sum_{t=0}^{T-1} x_t \cdot \Delta s_{t+1}(\omega), \omega) + \sum_{t=0}^{T} \delta_{D_t(\omega)}(x_t),$$

where $\hat{x} = (\alpha, x)$. As in Example 3.74, we denote $\hat{\mathcal{N}} := L^0(\Omega, \mathcal{F}_{-1}, P; \mathbb{R}) \times \mathcal{N}$, $\hat{\mathcal{X}} := L^\infty \times \mathcal{X}$ and $\hat{\mathcal{X}}_a := \hat{\mathcal{X}} \cap \hat{\mathcal{N}}$. The topological dual of $\hat{\mathcal{X}}$ can be identified with $\mathcal{X}^* = (L^\infty)^* \times \mathcal{X}^*$. By Theorem 3.68 and Example 3.71, it thus suffices to show that there is no duality gap and that the dual admits solutions. To this end, it suffices to verify the assumptions of Theorem 5.29.

Recalling that

$$\hat{f}(\hat{x}, u, \omega) = \alpha + f(x, u - \alpha),$$

where f is the normal integrand corresponding to (1.15), we get

$$\operatorname{dom} E\hat{f} = \{(\hat{x}, u) \in \hat{\mathcal{X}} \times \mathcal{U} \mid (x, u - \alpha) \in \operatorname{dom} Ef\}.$$

As observed in the proof of Theorem 5.57, $\operatorname{aff} \operatorname{dom} Ef = \mathcal{X} \times \mathcal{U}$ and $\bar{x}$ is strictly feasible. By Lemmas A.5 and A.9,

$$\operatorname{rcore} \operatorname{dom} E\hat{f} = \{(\hat{x}, u) \in \hat{\mathcal{X}} \times \mathcal{U} \mid (x, u - \alpha) \in \operatorname{rcore} \operatorname{dom} Ef\}$$

and

$$\operatorname{aff} \operatorname{dom} E\hat{f} = \{(\hat{x}, u) \in \hat{\mathcal{X}} \times \mathcal{U} \mid (x, u - \alpha) \in \operatorname{aff} \operatorname{dom} Ef\}.$$

Thus, $(0, \bar{x}) \in \hat{\mathcal{X}}$ is strictly feasible in (3.37) and $\operatorname{aff} \operatorname{dom} E\hat{f} = \mathcal{X} \times \mathcal{U}$. By Lemma A.9, $\operatorname{dom}_u E\hat{f} = \mathcal{U}$. Thus, assumptions 1–3 of Theorem 5.29 hold.

It remains to verify the last assumption in Theorem 5.29. Since $L^\infty \subseteq \operatorname{core} \operatorname{dom} EV$, (5.24) implies

$$\sum_{t=0}^{T-1} x_t \cdot \Delta s_{t+1} + \alpha \in \operatorname{lin} \operatorname{core} \operatorname{dom} EV \quad \forall (\alpha, x) \in \hat{\mathcal{X}}.$$

By definition of the recession cone (see Sect. A.3), the first condition in Assumption 5.56.1 thus gives

$$c - \sum_{t=0}^{T-1} x_t \cdot \Delta s_{t+1} - \alpha \in \operatorname{core} \operatorname{dom} EV \quad \forall (\alpha, x) \in \hat{\mathcal{X}}. \tag{5.26}$$

Let $(\hat{z}, \bar{u}) \in \operatorname{dom} E\hat{f}$ and define $\bar{z} \in \mathcal{X}$ by $\bar{z}^t = E_t z^t$ and $\bar{z}_s = z_s$ for $s > t$. Since $z_t \in D_t$ almost surely, the $\mathcal{F}_t$-measurability of D_t implies, by Corollary 1.68, that $\bar{z}_t \in D_t$ almost surely. Thus, by (5.26), $(\bar{z}, \bar{u}) \in \operatorname{dom} E\hat{f}$. □

The superhedging problem from Example 1.88 fails the relatively complete recourse condition and its dual problem does not admit solutions, in general.

Counterexample 5.60 (Superhedging) *Consider Example 1.88 and assume that* $\mathcal{X} = L^\infty$, $\mathcal{U} = L^\infty$, $D_t = \mathbb{R}^J$, $T = 1$ *and* $\mathcal{F}_0 = \{\Omega, \emptyset\}$ *so that the problem becomes*

$$\begin{array}{ll} \text{minimize} & \alpha \quad \text{over } \alpha \in \mathbb{R},\ x_0 \in \mathbb{R} \\ \text{subject to} & \bar{u} \leq \alpha + x_0 \cdot \Delta s_1 \quad a.s. \end{array}$$

We have $\operatorname{dom}\varphi = \mathcal{X} \times \mathcal{U}$ *so the strong dual admits solutions, by Theorem 5.1. However, the last condition in Theorem 5.29 fails, in general.*

Indeed, assume that Δs_1 *is uniformly distributed on* $[-2, 1]$ *and that* $\bar{u} = (\Delta s_1)^+$. *It is easy to verify that the optimal solution is given by* $\bar{\alpha} := 2/3$ *and* $\bar{x}_0 := 1/3$. *By Example 3.74, an optimal dual solution thus has to satisfy*

$$E[y] = 1,$$
$$y(\bar{\alpha} + \bar{x}_0\Delta s_1 - \bar{u}) = 0.$$

Since $\bar{u} < \bar{\alpha} + \bar{x}_0\Delta s_1$ *almost surely, the latter implies* $y = 0$ *almost surely, which contradicts the former. Thus, there are no dual solutions in* $\mathcal{V} \times \mathcal{Y}$.

In the following counterexample, there is no duality gap and the dual problem has a solution but the primal problem does not have any. Interestingly, all the other assumptions of Example 4.31 are satisfied except that the dual feasible solution y satisfies the scaling condition only for $\lambda \in [1 - \epsilon, 1]$.

Counterexample 5.61 *Consider problem* (3.36) *with* $V(u) = \gamma \max\{u, 0\}$, *where* $\gamma > 1$, *so that the optimized certainty equivalent* $\mathcal{V}$ *becomes the Conditional value at Risk; see Remark 1.77. Assume that there are no portfolio constraints,* $|J| = 1$ *(only one risky asset),* $T = 1$, $\mathcal{F}_0 := \{\emptyset, \Omega\}$ *and* $c, \Delta s_T \in L^1$. *Since* $\operatorname{dom} EV = L^1$, *Assumption 5.56 is satisfied with* $\mathcal{U} = L^1$ *and* $\mathcal{Y} = L^\infty$. *Thus, as soon as the optimum value is finite, Theorem 5.59 implies that a solution* y *to the reduced dual in* (3.39) *exists and that* $(\alpha, x) \in \mathbb{R} \times \mathbb{R}$ *is primal optimal if and only if*

$$y \in \partial V(u - x\Delta s_T - \alpha)\ a.s. \tag{5.27}$$

Assume, in addition, that there is an $A \in \mathcal{F}$ *such that* $P(A) \in (0, 1/\gamma)$, $\Delta s_T = -1$ *on* $\Omega \setminus A$, $\Delta s_T \geq 0$ *on* A *and that*

$$E[\gamma(1 + \Delta s_T)1_A] = 1. \tag{5.28}$$

We will show below that the reduced dual is feasible and that its feasible solutions y *satisfy* $y = \gamma$ *on* A. *The optimum value is then finite and* (5.27) *implies* $u - x\Delta s_T - \alpha \geq 0$ *almost surely on* A. *Clearly, no such* $(x, \alpha) \in \mathbb{R} \times \mathbb{R}$ *exist if* s *is such that* $\Delta s_T 1_A \notin L^\infty$ *and* $c := -|\Delta s_T|^2$.

A $y \in \mathcal{Y}$ is feasible in the reduced dual (3.39) *if and only if* $y \in \operatorname{dom} V^* = [0, \gamma]$ *almost surely,* $E[y\Delta s] = 0$ *and* $E[y] = 1$. *The equations can be written as*

$$\begin{aligned} -E[y 1_{\Omega \setminus A}] + E[y \Delta s_T 1_A] &= 0, \\ E[y 1_{\Omega \setminus A}] + E[y 1_A] &= 1, \end{aligned}$$

which give

$$E[y(1 + \Delta s) 1_A] = 1.$$

Thus, (5.28) *and the constraint* $y \in [0, \gamma]$ *imply that a feasible solution* y *to the reduced dual has to equal* γ *on* A. *It now suffices to note that*

$$y(\omega) := \begin{cases} \gamma & \omega \in A, \\ (1 - \gamma P(A))/(1 - P(A)) & \omega \in \Omega \setminus A \end{cases}$$

is feasible in the reduced dual.

5.4.6 Subdifferentials and Conditional Expectations

Theorem 5.62 below uses the general theory of this chapter to give sufficient conditions for Theorem 3.79 on commutation of conditional expectation and subdifferentiation of convex normal integrands. We assume that, the spaces $\mathcal{X}$ and $\mathcal{V}$ satisfy the assumptions of Sect. 5.3, Eh is defined on $\mathcal{X}$ and Eh^* on $\mathcal{V}$.

Theorem 5.62 *Let* h *be a convex normal integrand and* $\mathcal{G} \subseteq \mathcal{F}$ *such that* Eh^* *is proper and* $E^{\mathcal{G}} x \in \operatorname{dom} Eh$ *for every* $x \in \operatorname{dom} Eh$. *If* $\operatorname{aff} \operatorname{dom} Eh$ *is strongly closed and* $(\operatorname{rcore} \operatorname{dom} Eh) \cap \mathcal{X}(\mathcal{G}) \neq \emptyset$, *then*

$$\partial(E^{\mathcal{G}} h)(x) = E^{\mathcal{G}}[\partial h(x)]$$

for every $x \in L^0(\mathcal{G})$ *such that* $h(x) \in L^1$ *and* $\mathcal{V}(\partial h(x)) \neq \emptyset$.

Proof Let $\bar{x} \in (\operatorname{rcore} \operatorname{dom} Eh) \cap \mathcal{X}(\mathcal{G})$. By Theorem 3.79, it suffices to show that $\mathcal{V}(\partial h(\bar{x})) \neq \emptyset$ and that, for every $v \in \mathcal{V}$, the function $\phi_v : \mathcal{X} \to \overline{\mathbb{R}}$ given by

$$\phi_v(z) := \inf_{x \in \mathcal{X}} \{E[h(x) - x \cdot v] \mid x - z \in \mathcal{X}(\mathcal{G})\}$$

is either subdifferentiable or equal to $-\infty$ at the origin. Since $\operatorname{aff} \operatorname{dom} Eh$ is strongly closed, Corollary A.64 implies strong subdifferentiability of Eh at $\bar{x}$. By Theorem 5.15, this implies weak subdifferentiability which, by Theorem 3.8, gives $\mathcal{V}(\partial h(\bar{x})) \neq \emptyset$.

It remains to show that, for every $v \in \mathcal{V}$, the function ϕ_v is either subdifferentiable or equal to $-\infty$ at the origin. As in the proof of Theorem 3.79, the function ϕ_v can be identified with the optimum value function of $(P_{\mathcal{X}})$ when $T = 0$, $\mathcal{F}_0 = \mathcal{G}$ and $f(x, u, \omega) := h(x, \omega) - x \cdot v(\omega)$. By Theorem 3.18, it suffices to show that there is no duality gap and the dual optimum is attained. To this end, it suffices to verify the assumptions of Theorem 5.29.

We have $\operatorname{dom} Ef = \operatorname{dom} Eh \times \mathcal{U}$, so $\operatorname{aff}\operatorname{dom} Ef = \operatorname{aff}\operatorname{dom} Eh \times \mathcal{U}$. Thus, $\bar{x}$ is strictly feasible, so the second assumption of Theorem 5.29 holds. Clearly $\operatorname{dom}_u Ef = \mathcal{U}$ is closed, so the second part of assumption 3 in Theorem 5.29 holds. Since we have assumed that $E^{\mathcal{G}} x \in \operatorname{dom} Eh$ for every $x \in \operatorname{dom} Eh$, the first part of assumption 3 follows from Lemma 5.32 with $\pi(x, u) := (E^{\mathcal{G}} x, u)$. The last assumption in Theorem 5.29 means that $E^{\mathcal{G}} x \in \operatorname{dom} Eh$ for every $x \in \operatorname{dom} Eh$.

□

The condition that $E^{\mathcal{G}} x \in \operatorname{dom} Eh$ for every $x \in \operatorname{dom} Eh \cap \mathcal{X}$ holds, in particular, if $x \in \operatorname{dom} Eh$ for every $x \in \mathcal{X}(\operatorname{dom} h)$ and if $\operatorname{dom} h$ is closed and $\mathcal{G}$-measurable. Indeed, by Corollary 1.68, $\mathcal{G}$-measurability of $\operatorname{dom} h$ implies that $E^{\mathcal{G}} x \in \operatorname{dom} h$ for every $x \in \mathcal{X}(\operatorname{dom} h)$.

5.5 Bibliographical Notes

Most results of this chapter are from [108] which extends the techniques of [131] (see also [103]) in two important says. First, we include the parameter $u \in \mathcal{U}$ which allows for dualization of different formats of stochastic optimization problems in a unified manner much like the conjugate duality framework of [127] unifies more specific problem formats in convex optimization. Second, we go beyond spaces of essentially bounded strategies and parameters. Allowing for more general Fréchet spaces $\mathcal{X}$ and $\mathcal{U}$, widens the scope of applicability of the theory.

Fréchet spaces of random variables satisfying the assumptions of Sect. 5.1.1 were studied in [106]. Theorem 5.6 recalls some sufficient conditions that imply the required structure for the topological dual. Besides the three examples given in Sect. 5.1.1, the theorem also covers e.g. Lorentz spaces, Orlicz-Lorentz spaces and generalized Musielak-Orlicz spaces; see [106, Section 6] for details. Such spaces have been recently studied in the context of risk measures; see e.g. [74] and the references there. Conjugates of integral functionals in spaces with the structure of Sect. 5.1.1 were first studied in [125] in the case of the Banach space L^∞ and later in [78] in the case of Orlicz spaces. Theorem 5.6 and the main result of [108] unify and extend the two settings.

The results of Sects. 5.2 and 5.3 are essentially from [108] which was largely inspired by Rockafellar and Wets [131]. The arguments of Sect. 5.3 generalize and simplify those of [131] which were based on an inductive argument involving the dynamic programming recursion. Section 5.3.2 extends the recourse-type conditions of [131, 138] to the general convex stochastic optimization format without requiring that the set of feasible solutions be uniformly bounded. In particular, the conditions in Example 5.55 extend "bounded recourse condition" from [138, Definition 1]. The assumptions in Example 5.54 extend the "interior feasibility condition" of [138, Definition 2]. The study of relatively complete recourse and induced constraints in convex stochastic programming goes back to [133, 157]. Remark 2.81 extends the definition of induced constraints from the single period framework of [133] to general convex stochastic optimization in finite discrete time.

Section 5.4.1 extends the main results of [136] by allowing for equality constraints and unbounded feasible sets. The strategy of [136] was to first relax the nonanticipativity constraint by using a shadow price of information p and then to construct a dual variable y via measurable selection arguments. The general theory of this chapter yields the existence of a dual optimal pair (p, y) directly. Theorem 5.46 is a discrete-time version of [105, Corollary 15] the proof of which was based on penalty representation of the constraints rather than going through the strong dual.

Section 4 of [136] gave sufficient conditions for dual existence in discrete-time stochastic control by applying the general existence results of [136] to the reduced formulation in Remark 1.86; see also Chapter 5 of [29]. Section 5.4.3 extends the results of [136] by allowing for unbounded control sets. This is essential e.g. in the classical models of linear-quadratic control and financial mathematics. The main results of this chapter could be applied also to the reduced formulation of stochastic optimal control much like in [136].

The results of Sect. 5.4.4 are closely related to those of [138, Section 4]. While the problem formulation in [138] involves an additional objective term, our conditions relax, as noted above, the bounded recourse condition of [138]. The sufficient conditions for the existence of dual solutions in Sect. 5.4.5 extend earlier results by the inclusion of portfolio constraints. Section 5 of [108] gives a further extension by including statically held derivative assets in the optimization problem as in Example 1.90. The idea of constructing Orlicz spaces from loss functions in Remark 5.58 is from [15]; see also [16]. Theorem 5.59 on the optimized certainty equivalent is new.

Theorem 5.62 extends the main result of [137] by relaxing the continuity and integrability assumptions made there. Earlier results on the commutation of conditional expectation and subdifferentiation can be found in [20] and [30, Theorem VIII.37].

Appendix A
Primer on Convex Analysis

This appendix collects, mostly well-known, results from convex analysis that have been used in this book. Most of the results below can be found, e.g. in the books of Moreau [87], Rockafellar [123, 127], Ekeland and Temam [50], Kelley and Namioka [75] and Zalinescu [160].

A set C in a vector space U is *convex* if

$$\alpha_1 x_1 + \alpha_2 x_2 \in C$$

for all $x_1, x_2 \in C$ and $\alpha_1, \alpha_2 > 0$ such that $\alpha_1 + \alpha_2 = 1$. A set C is a *cone* if $\lambda C \subseteq C$ for every $\lambda > 0$. An extended real-valued function $g : U \to \overline{\mathbb{R}} := \mathbb{R} \cup \{+\infty, -\infty\}$ is *convex* if its *epigraph*

$$\operatorname{epi} g := \{(x, \alpha) \in U \times \mathbb{R} \mid g(u) \leq \alpha\}$$

is a convex set in $U \times \mathbb{R}$. Equivalently, convexity of g means that

$$g(\alpha_1 u_1 + \alpha_2 u_2) \leq \alpha_1 g(u_1) + \alpha_2 g(u_2)$$

for all $\alpha_1, \alpha_2 > 0$ with $\alpha_1 + \alpha_2 = 1$ and all u_1 and u_2 in the *effective domain*

$$\operatorname{dom} g := \{u \in U \mid g(u) < \infty\}$$

of g. Convexity of g implies that its *lower level-sets*

$$\operatorname{lev}_\alpha g := \{u \in U \mid g(u) \leq \alpha\}$$

T. Pennanen, A.-P. Perkkiö, *Convex Stochastic Optimization*, Probability Theory and Stochastic Modelling 107, https://doi.org/10.1007/978-3-031-76432-5

are convex for every $\alpha \in \mathbb{R}$. The function g is *sublinear* if its epigraph is a convex cone. Equivalently, sublinearity of g means that

$$g(\alpha_1 u_1 + \alpha_2 u_2) \leq \alpha_1 g(u_1) + \alpha_2 g(u_2)$$

for all $u_i \in U$ and $\alpha_i > 0$. The *indicator function*

$$\delta_C(u) := \begin{cases} 0 & \text{if } u \in C \\ +\infty & \text{otherwise} \end{cases}$$

of a set C is convex (sublinear) if and only if C is convex (convex cone).

Given extended real numbers $\xi_1, \xi_2 \in \overline{\mathbb{R}}$, their sum $\xi_1 + \xi_2$ is well-defined except when one of the numbers is $+\infty$ and the other one $-\infty$. In this exceptional case, we define the sum as $\xi_1 + \xi_2 := +\infty$. It follows that a convex function g satisfies

$$g(\alpha_1 u_1 + \alpha_2 u_2) \leq \alpha_1 g(u_1) + \alpha_2 g(u_2)$$

for all $\alpha_1, \alpha_2 > 0$ with $\alpha_1 + \alpha_2 = 1$ and any u_1 and u_2 in U, not necessarily in the domain of g.

Given extended real numbers $\xi_1, \xi_2 \in \overline{\mathbb{R}}$, their product $\xi_1 \xi_2$ is well-defined except when one of them is zero and the other one is $+\infty$ or $-\infty$. In this exceptional case, we define the product as zero.

A.1 Convexity in Algebraic Operations

Convexity is preserved under many algebraic operations. It is easy to check, for example, that the sum of a finite collection of convex functions are convex. Also, the pointwise supremum of an arbitrary collection of convex functions is convex.

Theorem A.1 (Infimal Projection) *Given a convex function f on the product $X \times U$ of two vector spaces, the function*

$$\varphi(u) := \inf_{x \in X} f(x, u)$$

is a convex function on U.

Proof Convexity of f gives

$$\begin{aligned} \varphi(\alpha_1 u_1 + \alpha_2 u_2) &= \inf_{x_1, x_2 \in X} f(\alpha_1 x_1 + \alpha_2 x_2, \alpha_1 u_1 + \alpha_2 u_2) \\ &\leq \inf_{x_1, x_2 \in X} \{\alpha_1 f(x_1, u_1) + \alpha_2 f(x_2, u_2)\} \end{aligned}$$

$$= \alpha_1 \inf_{x_1 \in X} f(x_1, u_1) + \alpha_2 \inf_{x_2 \in X} f(x_2, u_2)$$

$$= \alpha_1 \varphi(u_1) + \alpha_2 \varphi(u_2)$$

for all $u_1, u_2 \in \operatorname{dom} \varphi$ and $\alpha_1, \alpha_2 > 0$ with $\alpha_1 + \alpha_2 = 1$. □

Let X and U be vector spaces and $K \subset U$ a convex cone. A function H from a subset $\operatorname{dom} H \subseteq X$ to U is said to be a *K-convex function from X to U* if the set

$$\operatorname{epi}_K H := \{(x, u) \in X \times U \mid x \in \operatorname{dom} H,\ H(x) - u \in K\}$$

is convex or, equivalently, if

$$H(\alpha_1 x_1 + \alpha_2 x_2) - \alpha_1 H(x_1) - \alpha_2 H(x_2) \in K$$

for all $x_i \in \operatorname{dom} H$ and $\alpha_i > 0$ with $\alpha_1 + \alpha_2 = 1$. An extended real-valued function is convex if and only if it is $\mathbb{R}_-$-convex.

Theorem A.2 (Composition) *If H is K-convex function from X to U and g is an extended real-valued convex function on U such that*

$$H(x) - u \in K \implies g(H(x)) \leq g(u) \quad \forall x \in \operatorname{dom} H$$

then the composition

$$(g \circ H)(x) := \begin{cases} g(H(x)) & \text{if } x \in \operatorname{dom} H, \\ +\infty & \text{if} x \notin \operatorname{dom} H \end{cases}$$

is an extended real-valued convex function on X.

Proof The function

$$f(x, u) := g(u) + \delta_{\operatorname{epi}_K H}(x, u)$$

is the sum of convex functions on $X \times U$ so it is convex. The growth condition gives

$$(g \circ H)(x) = \inf_{u \in U} f(x, u),$$

so the claim follows from Theorem A.1. □

A linear mapping $A : X \to U$ is $\{0\}$-convex function with $\operatorname{dom} A = X$ so Theorem A.2 implies that

$$(g \circ A)(x) := g(Ax)$$

is convex for any convex function g on U.

A.2 Positive Hulls and Separation Theorems

Let U be a vector space and $C \subseteq U$ convex. The *positive hull*

$$\operatorname{pos} C := \bigcup_{\lambda>0} \lambda C$$

of C is the intersection of all cones containing C. If $0 \in C$ and $0 < \lambda_1 < \lambda_2$, convexity implies $\lambda_1 C \subseteq \lambda_2 C$.

Lemma A.3 *Given a linear $A : X \to U$ and convex sets $C, C' \subseteq X$ and $D \subseteq U$, we have*

1. $\operatorname{pos}(AC) = A \operatorname{pos} C$,
2. $\operatorname{pos}(A^{-1}D) = A^{-1} \operatorname{pos} D$,
3. $\operatorname{pos}(C \times D) = \operatorname{pos} C \times \operatorname{pos} D$ *if* $0 \in C$ *and* $0 \in D$,
4. $\operatorname{pos}(C \cap C') = \operatorname{pos} C \cap \operatorname{pos} C'$ *if* $0 \in C$ *and* $0 \in C'$,
5. $\operatorname{pos}(C + C') = \operatorname{pos} C + \operatorname{pos} C'$ *if* $0 \in C$ *and* $0 \in C'$.

Proof The first two claims are clear. As to 3, we have $C \times D \subseteq \operatorname{pos} C \times \operatorname{pos} D$ so $\operatorname{pos}(C \times D) \subseteq \operatorname{pos} C \times \operatorname{pos} D$. If $(x, u) \in \operatorname{pos} C \times \operatorname{pos} D$, we have $x \in \lambda_1 C$ and $u \in \lambda_2 D$ for some $\lambda_1, \lambda_2 > 0$. Since C and D are convex sets containing the origins, we have $x \in \max\{\lambda_1, \lambda_2\}C$ and $u \in \max\{\lambda_1, \lambda_2\}D$, and thus, $(x, u) \in \operatorname{pos}(C \times D)$. Defining $A : X \to X \times X$ by $Ax = (x, x)$, we have $C \cap C' = A^{-1}(C \times C')$, so 4 follows from 2 and 3. Defining $A : X \times X \to X$ by $A(x, x') = x + x'$, we have $A(C \times C') = C + C'$, so 5 follows from 1 and 3. □

The *core* of a set $C \subseteq U$, denoted by $\operatorname{core} C$, is the set of points $u \in C$ for which $\operatorname{pos}(C-u) = U$. The *relative core* of C, denoted by $\operatorname{rcore} C$, is the core of C relative to the affine hull of C. More precisely,

$$\operatorname{rcore} C := \{u \in C \mid \operatorname{pos}(C - u) = \operatorname{aff}(C - u)\}.$$

A set C is *affine* if $\lambda u + (1 - \lambda)u' \in C$ for all $u, u' \in C$ and $\lambda \in \mathbb{R}$. The *affine hull*

$$\operatorname{aff} C := \left\{ \sum_{j \in J} \lambda^j C \;\middle|\; |J| < \infty, \sum_{j \in J} \lambda_j = 1 \right\}$$

of a set C is the intersection of all affine sets containing C.

Lemma A.4 *Given a nonempty convex set C,*

1. $\operatorname{aff}(C - u) = \operatorname{aff}(C - u')$ *for every* $u, u' \in C$,
2. $\operatorname{pos}(C - u) = \operatorname{pos}(C - u')$ *for every* $u, u' \in \operatorname{rcore} C$,
3. $\operatorname{rcore} C = \{u \in U \mid \operatorname{pos}(C - u)$ *is linear*$\}$,
4. $\alpha u + (1 - \alpha)\bar{u} \in \operatorname{rcore} C$ *for every* $u \in C$, $\bar{u} \in \operatorname{rcore} C$ *and* $\alpha \in (0, 1)$.

Proof The first claim follows from the fact that if a linear set contains $C-u$, then it also contains $C-u'$. The second claim follows from the first one and the definition of relative core. As to 3, let $u \in U$ and $u' \in C$. The linearity of $\text{pos}(C-u)$ gives the existence of $\alpha > 0$ and $u'' \in C$ such that $u' - u = -\alpha(u'' - u)$, or equivalently,

$$u = \frac{1}{\alpha+1}u' + \frac{\alpha}{\alpha+1}u''.$$

Thus, by convexity, $u \in C$ so 3 follows from the definition of relative core. As to 4, we have $C = \alpha C + (1-\alpha)C$, by convexity, so Lemma A.3.5 and the definition of relative core give

$$\begin{aligned}\text{pos}(C - \alpha u - (1-\alpha)\bar{u}) &= \text{pos}(\alpha(C-u) + (1-\alpha)(C-\bar{u}))\\ &= \text{pos}(C-u) + \text{pos}(C-\bar{u})\\ &= \text{pos}(C-u) + \text{aff}(C-\bar{u})\\ &= \text{aff}(C-\bar{u}),\end{aligned}$$

so 4 follows from 1 and the definition of relative core. □

Lemma A.5 *Given a linear* $A : X \to U$ *and convex sets* $C, C' \subseteq X$ *and* $D \subseteq U$, *the following hold provided the right side of the equation is nonempty:*

1. $\text{rcore}(AC) = A\,\text{rcore}\,C$;
2. $\text{rcore}(A^{-1}D) = A^{-1}\,\text{rcore}\,D$;
3. $\text{rcore}(C \times D) = \text{rcore}\,C \times \text{rcore}\,D$;
4. $\text{rcore}(C \cap C') = \text{rcore}\,C \cap \text{rcore}\,C'$;
5. $\text{rcore}(C + C') = \text{rcore}\,C + \text{rcore}\,C'$.

Proof Let $\bar{x} \in \text{rcore}\,C$. By Lemma A.3,

$$\text{pos}(AC - A\bar{x}) = A\,\text{pos}(C-\bar{x}) = A\,\text{aff}(C-\bar{x})$$

so $A\bar{x} \in \text{rcore}(AC)$, by Lemma A.4.3. To prove the converse, let $u \in \text{rcore}(AC)$. By definition, there is an $\epsilon > 0$ such that $u + \epsilon(u - A\bar{x}) \in AC$. Let $x \in C$ such that $u + \epsilon(u - A\bar{x}) = Ax$. This can be written as

$$u = A\left[\frac{1}{1+\epsilon}x + \frac{\epsilon}{1+\epsilon}\bar{x}\right]$$

so $u \in A(\text{rcore}\,C)$, by Lemma A.4.4. This proves 1. Claims 2 and 3 follow by an analogous argument. Defining $A : X \to X \times X$ by $Ax = (x, x)$, we have $C \cap C' = A^{-1}(C \times C')$, so 4 follows from 2 and 3. Defining $A : X \times X \to X$ by $A(x, x') = x + x'$, we have $A(C \times C') = C + C'$, so 5 follows from 1 and 3. □

The following was given in [123, Theorem 6.8] for Euclidean spaces.

Lemma A.6 *Let $C \subseteq X \times U$ be convex. If either* $\operatorname{rcore} C \neq \emptyset$ *or* $X = \mathbb{R}^n$*, then* $(x, u) \in \operatorname{rcore} C$ *if and only if* $u \in \operatorname{rcore} D$ *and* $x \in \operatorname{rcore} C_u$*, where*

$$D := \{u \in U \mid \exists x : (x, u) \in C\},$$
$$C_u := \{x \in X \mid (x, u) \in C\}.$$

Proof Assume first that $\operatorname{rcore} C \neq \emptyset$. Applying Lemma A.5.1 with $A(x, u) = u$, implies that $u \in \operatorname{rcore} D$ if and only if there exists x with $(x, u) \in \operatorname{rcore} C$. It thus suffices to show that, given $u \in \operatorname{rcore} D$, $(x, u) \in \operatorname{rcore} C$ if and only if $x \in \operatorname{rcore} C_u$. Denoting $M_u := \{(x, u) \in X \times U \mid x \in X\}$, this follows from

$$M_u \cap \operatorname{rcore} C = \operatorname{rcore}(M_u \cap C) = \{(x, u) \in X \times U \mid x \in \operatorname{rcore} C_u\},$$

where the first equality holds by Lemma A.5.4.

Assume now that $X = \mathbb{R}^n$, $u \in \operatorname{rcore} D$ and $x \in \operatorname{rcore} C_u$. Given $(x', u') \in \operatorname{aff} C$, define

$$G := C \cap (\mathbb{R}^n \times \{u + \lambda(u' - u) \mid \lambda \in \mathbb{R}\})$$

and $D_G := \{\bar{u} \in U \mid \exists x : (x, \bar{u}) \in G\}$. The set G is finite-dimensional, so $\operatorname{rcore} G \neq \emptyset$. Since

$$\begin{aligned} D_G &= \{\bar{u} \in U \mid \exists x : (x, \bar{u}) \in C \cap (\mathbb{R}^n \times \{u + \lambda(u' - u) \mid \lambda \in \mathbb{R}\}) \\ &= D \cap \{u + \lambda(u' - u) \mid \lambda \in \mathbb{R}\}, \end{aligned}$$

Lemma A.5 gives

$$\begin{aligned} \operatorname{rcore} D_G &= \operatorname{rcore}(D \cap \{u + \lambda(u' - u) \mid \lambda \in \mathbb{R}\}) \\ &= \operatorname{rcore} D \cap \{u + \lambda(u' - u) \mid \lambda \in \mathbb{R}\}, \end{aligned}$$

so $u \in \operatorname{rcore} D_G$. Since $G_u = C_u$, we have $x \in \operatorname{rcore} G_u$. By the first part of the proof, $(x, u) \in \operatorname{rcore} G$. Since $(x', u') \in \operatorname{aff} G$, there thus exists $\lambda > 0$ such that $(x + \lambda(x' - x), u + \lambda(u' - u)) \in G$. In particular, $(x + \lambda(x' - x), u + \lambda(u' - u)) \in C$. Since $(x', u') \in \operatorname{aff} C$ was arbitrary, $(x, u) \in \operatorname{rcore} C$. □

Lemma A.7 *Given a convex set C, we have* $\operatorname{rcore} \operatorname{pos} C = \operatorname{pos} \operatorname{rcore} C$.

Proof We have $\operatorname{pos} C = AG$, where $G := \{(u, \alpha) \in U \times \mathbb{R} \mid \alpha > 0,\ u \in \alpha C\}$ and $A(u, \alpha) := u$. Let $(u_i, \alpha_i) \in G$. We have

$$u_1 + u_2 \in \alpha_1 C + \alpha_2 C = (\alpha_1 + \alpha_2)\left(\frac{\alpha_1}{\alpha_1 + \alpha_2} C + \frac{\alpha_2}{\alpha_1 + \alpha_2} C\right) = (\alpha_1 + \alpha_2) C,$$

where the last equality holds by convexity. Thus, $(u_1+u_2, \alpha_1+\alpha_2) \in G$ so G convex since it is also a cone. By Lemma A.5.1, $\operatorname{rcore} \operatorname{pos} C = A \operatorname{rcore} G$. By Lemma A.6,

$$\operatorname{rcore} G = \{(u, \alpha) \mid \alpha > 0,\ u \in \operatorname{rcore}(\alpha C)\} = \{(u, \alpha) \mid \alpha > 0,\ u \in \alpha \operatorname{rcore} C\},$$

and thus, $A \operatorname{rcore} G = \operatorname{pos} \operatorname{rcore} C$. □

Lemma A.8 *If C is convex and $x \in C$, then*

$$\operatorname{aff}(C - x) = \operatorname{pos}(C - x) - \operatorname{pos}(C - x).$$

Proof Since $\operatorname{pos}(C - x) \subseteq \operatorname{aff}(C - x)$ and $\operatorname{aff}(C - x)$ is linear, the right side is contained in the left side. The converse follows from the facts that $\operatorname{aff}(C - x)$ is the smallest linear set containing $(C - x)$ and the set on the right is linear. □

Combining Lemma A.8 with Lemma A.3 gives the following.

Lemma A.9 *Given a linear $A : X \to U$ and convex sets $C, C' \subseteq X$ and $D \subseteq U$, we have*

1. $\operatorname{aff}(AC) = A \operatorname{aff} C$,
2. $\operatorname{aff}(A^{-1}D) = A^{-1} \operatorname{aff} D$ *if* $A^{-1} \operatorname{rcore} D \neq \emptyset$,
3. $\operatorname{aff}(C \times D) = \operatorname{aff} C \times \operatorname{aff} D$,
4. $\operatorname{aff}(C \cap C') = \operatorname{aff} C \cap \operatorname{aff} C'$ *if* $\operatorname{rcore} C \cap \operatorname{rcore} C' \neq \emptyset$,
5. $\operatorname{aff}(C + C') = \operatorname{aff} C + \operatorname{aff} C'$.

Proof Part 1 follows directly from the definition. To prove 2, let $\bar{x} \in A^{-1} \operatorname{rcore} D$. We have $A^{-1}D - \bar{x} = A^{-1}(D - A\bar{x})$. By Lemma A.5, $\bar{x} \in \operatorname{rcore} A^{-1}D$. Thus, by Lemmas A.4 and A.3,

$$\begin{aligned}
\operatorname{aff}(A^{-1}D) - \bar{x} &= \operatorname{aff}(A^{-1}D - \bar{x}) = \operatorname{pos}(A^{-1}D - \bar{x}) \\
&= \operatorname{pos}(A^{-1}(D - A\bar{x})) = A^{-1} \operatorname{pos}(D - A\bar{x}) \\
&= A^{-1}(\operatorname{aff}(D - A\bar{x})) = A^{-1}(\operatorname{aff} D - A\bar{x}) \\
&= A^{-1} \operatorname{aff} D - \bar{x}.
\end{aligned}$$

To prove 3, let $(x, u) \in C \times D$. By Lemmas A.8 and A.3,

$$\begin{aligned}
&\operatorname{aff}(C \times D) - (x, u) \\
&= \operatorname{pos}(C \times D - (x, u)) - \operatorname{pos}(C \times D - (x, u)) \\
&= \operatorname{pos}((C - x) \times (D - u)) - \operatorname{pos}((C - x) \times (D - u)) \\
&= \operatorname{pos}(C - x) \times \operatorname{pos}(D - u) - \operatorname{pos}(C - x) \times \operatorname{pos}(D - u) \\
&= (\operatorname{pos}(C - x) - \operatorname{pos}(C - x)) \times (\operatorname{pos}(D - u) - \operatorname{pos}(D - u)) \\
&= (\operatorname{aff} C - x) \times (\operatorname{aff} D - u) \\
&= \operatorname{aff} C \times \operatorname{aff} D - (x, u).
\end{aligned}$$

Defining $A : X \to X \times X$ by $Ax = (x, x)$, we have $C \cap C' = A^{-1}(C \times C')$, so 4 follows from 2 and 3. Defining $A : X \times X \to X$ by $A(x, x') = x + x'$, we have $A(C \times C') = C + C'$, so 5 follows from 1 and 3. □

Example A.10 *Given a convex function* $g : U \to \overline{\mathbb{R}}$,

$$\begin{aligned} \operatorname{aff} \operatorname{epi} g &= \operatorname{aff} \operatorname{dom} g \times \mathbb{R}, \\ \operatorname{rcore} \operatorname{epi} g &= \{(u, \alpha) \in U \times \mathbb{R} \mid u \in \operatorname{rcore} \operatorname{dom} g,\ g(u) < \alpha\}. \end{aligned}$$

Proof Since $\operatorname{epi} g = \operatorname{epi} g + \{0\} \times \mathbb{R}_+$, Lemma A.9 implies

$$\begin{aligned} \operatorname{aff} \operatorname{epi} g &= \operatorname{aff} \operatorname{epi} g + \operatorname{aff}(\{0\} \times \mathbb{R}_+) \\ &= \operatorname{aff} \operatorname{epi} g + (\{0\} \times \mathbb{R}) \\ &= \operatorname{aff}(\operatorname{epi} g + (\{0\} \times \mathbb{R})) \\ &= \operatorname{aff}(\operatorname{dom} g \times \mathbb{R}) \\ &= \operatorname{aff}(\operatorname{dom} g) \times \mathbb{R} \end{aligned}$$

which proves the first claim. The second claim follows from Lemma A.6. □

We now come to the classical separation theorems that give sufficient conditions for the existence of hyperplanes that separate convex sets. Recall that a *hyperplane* is an affine set of the form $\{u \in U \mid l(u) = \alpha\}$ where $l : U \to \mathbb{R}$ is a nonzero linear functional and $\alpha \in \mathbb{R}$.

Remark A.11 *Given a linear functional* l *on a linear subspace* $U' \subset U$, *there exists a linear functional* $\bar{l}$ *on* U *that coincides with* l *on* U'. *Given an affine set* $A \subset U$ *such that* $0 \notin A$, *there exists a hyperplane* $H \supseteq A$ *such that* $0 \notin H$.

Proof To prove the first claim, consider the set $\mathcal{P}$ of all linear extensions of l to subspaces of U larger than U' and endow $\mathcal{P}$ with the partial order defined by inclusion of the graphs of the linear functionals. The maximal element $\bar{l} \in \mathcal{P}$, provided by Zorn's lemma, necessarily has full domain. Indeed, if there existed an element u_0 outside of the domain of $\bar{l}$, then one could extend $\bar{l}$ further to the linear span of the domain of $\bar{l}$ and u_0, contradicting the maximality of $\bar{l}$.

As to the second claim, consider the set $\mathcal{P}$ of affine sets containing A but not the origin. Endow $\mathcal{P}$ with the partial order given by set inclusion. Zorn's lemma gives a maximal element H which is necessarily a hyperplane not containing the origin. Indeed, if it was not, then the affine span of H and the origin would not be all of U so one could find a point u_0 outside of the span. The affine span of H and u_0 would then be an affine set disjoint of the origin, contradicting the maximality of H. It now suffices to note that a maximal affine set is necessarily a hyperplane; see page 7 of [75]. □

Theorem 3.2 of [75] yields the following.

Theorem A.12 *Given a convex set C with* $\operatorname{rcore} C \neq \emptyset$*, we have* $0 \notin \operatorname{rcore} C$ *if and only if there exists a linear functional l such that*

$$\sup_{u \in C} l(u) \leq 0 \quad \textit{and} \quad \inf_{u \in C} l(u) < 0.$$

Proof Assume first that $0 \in \operatorname{aff} C$. Applying [75, Theorem 3.2] in $\operatorname{aff} C$ to the positive hull of C gives a nonzero linear functional $\tilde{l}$ on $\operatorname{aff} C$ such that

$$\sup_{u \in C} \tilde{l}(u) \leq 0 \quad \text{and} \quad \inf_{u \in C} \tilde{l}(u) < 0.$$

By Zorn's lemma, there is an linear extension l of $\tilde{l}$ to all of U; see Remark A.11. This proves the claim when $0 \in \operatorname{aff} C$. If $0 \notin \operatorname{aff} C$, the claim follows from the second claim of Remark A.11. □

Applying Theorem A.12 to the difference of two sets gives the following.

Corollary A.13 *Let C_1 and C_2 be convex sets such that* $0 \notin \operatorname{rcore}(C_1 - C_2) \neq \emptyset$*. Then there exists a linear functional l such that*

$$\sup_{u \in C_1} l(u) \leq \inf_{u \in C_2} l(u) \quad \textit{and} \quad \inf_{u \in C_1} l(u) < \sup_{u \in C_2} l(u).$$

A function is said to be *sublinear* if it is convex and positively homogeneous.

Corollary A.14 (Hahn–Banach) *Let p be a finite sublinear function on U and k a linear functional on a linear subspace $L \subset U$ such that $k \leq p$ on L. Then there exists a linear functional $\bar{k}$ on U such that $\bar{k} = k$ on L and $\bar{k} \leq p$ on U.*

Proof We apply Corollary A.13 to the sets

$$C_1 := \{(u, \alpha) \in U \times \mathbb{R} \mid u \in L,\ \alpha = k(u)\}$$

and $C_2 := \{(u, \alpha) \in U \times \mathbb{R} \mid p(u) < \alpha\}$. By Example A.10, $\operatorname{core} C_2 = C_2$, so $\operatorname{core}(C_1 - C_2) \neq \emptyset$. Thus, by Corollary A.13, there exists a nonzero linear functional l on $U \times \mathbb{R}$ such that

$$\sup_{u \in C_1} l(u) \leq \inf_{u \in C_2} l(u).$$

The functional l can be expressed as $l(u, \alpha) = \tilde{k}(u) + \beta\alpha$ for a linear functional $\tilde{k}$ on U and $\beta \in \mathbb{R}$. Since both sets contain the origin, both sides of the inequality equal zero. Thus,

$$\tilde{k}(u) + \beta k(u) \leq 0 \quad \forall u \in L$$

and

$$\tilde{k}(u) + \beta\alpha \geq 0 \quad \forall (u, \alpha) \in U \times \mathbb{R} : \ p(u) < \alpha.$$

By linearity of L, the first inequality must hold as an equality. Since l is nonzero, the last condition implies that $\beta \neq 0$. The functional $\bar{k}(u) := -\tilde{k}(u)/\beta$ thus has the claimed properties. □

A.3 Recession Analysis

This section collects some basic results on recession analysis from [121] and Sections 8 and 9 of [123]. While [123] was concerned with finite-dimensional spaces, many of its arguments go through in a general vector space U. The *recession cone* of a nonempty convex set $C \subseteq U$ is defined by

$$C^\infty := \{u \in U \mid \bar{u} + \lambda u \in C \ \forall \bar{u} \in C, \ \lambda > 0\}.$$

If $C = \emptyset$, we define $C^\infty := \emptyset$. Clearly, C^∞ is a cone and $C + C^\infty = C$. If C is a cone, then $C^\infty = C$. Note also that C^∞ is invariant under translations of C. For a closed convex set in $\mathbb{R}^n$, the recession cone coincides with the 'asymptotic cone' from [55] and the 'horizon cone' from [139].

We say that a set C is *algebraically closed* if

$$\{\lambda \in \mathbb{R} \mid \bar{u} + \lambda u \in C\}$$

is closed in $\mathbb{R}$ for every $\bar{u}$ and u in U. A closed subset in a topological vector space is algebraically closed, but the converse fails, in general.

Theorem A.15 *Given a nonempty algebraically closed convex set C,*

1. $C^\infty = \{u \mid \bar{u} + \lambda u \in C \ \forall \lambda > 0\}$ *for any* $\bar{u} \in C$,
2. $C^\infty = \bigcap_{\nu \in \mathbb{N}} \frac{1}{\nu}(C - \bar{u})$ *for any* $\bar{u} \in C$,
3. *if* $\operatorname{rcore} C \neq \emptyset$, *then* $(\operatorname{rcore} C)^\infty = C^\infty$.

Proof Let $\bar{u} \in C$ and $u \neq 0$ be such that $\bar{u} + \lambda u \in C$ for all $\lambda > 0$. To prove 1, it suffices to show that $u \in C^\infty$. Let $u' \in C$ and $\lambda' > 0$. For any $\lambda \geq \lambda'$,

$$u' + \lambda' u + \frac{\lambda'}{\lambda}(\bar{u} - u') = (1 - \frac{\lambda'}{\lambda})u' + \frac{\lambda'}{\lambda}(\bar{u} + \lambda u) \in C$$

by convexity. Letting $\lambda \nearrow \infty$ gives $u' + \lambda' u \in C$ since C is algebraically closed. Since $u' \in C$ was arbitrary, $u \in C^\infty$. Part 2 follows from 1 and the fact that, by convexity, the set $\lambda(C - \bar{u})$ grows with $\lambda > 0$. As to 3, let $u \in C^\infty$, $\bar{u} \in \operatorname{rcore} C$ and $\lambda > 0$. By 1, $\bar{u} + \lambda u \in C$ for every $\lambda > 0$. By Lemma A.4.4, $\bar{u} + \lambda u \in \operatorname{rcore} C$. Thus, $C^\infty \subseteq (\operatorname{rcore} C)^\infty$. The reverse inclusion follows from 1. □

The *recession function* g^∞ of a convex function g is defined by

$$\operatorname{epi} g^\infty := (\operatorname{epi} g)^\infty.$$

Clearly, g^∞ is sublinear. If g is sublinear, then $g^\infty = g$. Also, for any $\bar{u} \in U$ and $\alpha \in \mathbb{R}$, the function $g'(u) := g(u + \bar{u}) + \alpha$ has the same recession function as g.

A function is *proper* if it is finite at some point and never takes the value $-\infty$. A proper function is *algebraically closed* if its epigraph is algebraically closed.

Theorem A.16 *If g is a proper algebraically closed convex function, then*

$$g^\infty(u) = \sup_{\lambda>0} \frac{g(\bar{u} + \lambda u) - g(\bar{u})}{\lambda} = \lim_{\lambda \nearrow \infty} \frac{g(\bar{u} + \lambda u) - g(\bar{u})}{\lambda}$$

for every $\bar{u} \in \operatorname{dom} g$.

Proof The first expression follows by applying Theorem A.15.1 to the epigraph of g at the point $(\bar{u}, g(\bar{u}))$. The second follows from the fact that the convexity of g implies that the difference quotients are nondecreasing in λ. □

Theorem A.17 (Recession Calculus)

1. *If $(C_i)_{i\in I}$ is a family of algebraically closed convex sets with nonempty intersection, then*

$$\left(\bigcap_{i\in I} C_i\right)^\infty = \bigcap_{i\in I} C_i^\infty.$$

2. *If g is a proper algebraically closed convex function, then*

$$(\operatorname{lev}_{\le\beta} g)^\infty = \operatorname{lev}_{\le 0} g^\infty$$

for every $\beta \in \mathbb{R}$ such that $\operatorname{lev}_{\le\beta} g \neq \emptyset$.

3. *If g is a proper algebraically closed convex function, $A : X \to U$ is linear, $u \in U$ and $h(x) := g(Ax + u)$ is proper, then*

$$h^\infty(x) = g^\infty(Ax).$$

4. *If $(g_i)_{i\in I}$ is a finite family of proper algebraically closed convex functions, then*

$$\left(\sum_{i\in I} g_i\right)^\infty = \sum_{i\in I} g^\infty$$

as soon as $\sum_{i\in I} g_i$ is proper.

5. *If $(g_i)_{i\in I}$ is a family of proper algebraically closed convex functions such that their pointwise supremum is proper, then*

$$(\sup_{i\in I} g_i)^\infty = \sup_{i\in I} g_i^\infty.$$

Proof Let $\bar{u} \in \bigcap_{i \in I} C_i$. By Theorem A.15.1,

$$\begin{aligned}\left(\bigcap_{i\in I} C_i\right)^{\infty} &= \bigcap_{\lambda>0} \lambda\left(\bigcap_{i\in I} C_i - \bar{u}\right)\\ &= \bigcap_{\lambda>0} \lambda \bigcap_{i\in I}(C_i - \bar{u})\\ &= \bigcap_{i\in I}\bigcap_{\lambda>0} \lambda(C_i - \bar{u})\\ &= \bigcap_{i\in I} C_i^{\infty}.\end{aligned}$$

This proves 1. To prove 2, let $\bar{u} \in \mathrm{lev}_{\le\beta}\, g$. By Theorem A.16, $u \in \mathrm{lev}_{\le 0}\, g^{\infty}$ if and only if

$$g(\lambda u + \bar{u}) - g(\bar{u}) \le 0 \quad \forall \lambda > 0$$

while, by Theorem A.15, $u \in (\mathrm{lev}_{\le\beta}\, g)^{\infty}$ if and only if

$$g(\lambda u + \bar{u}) \le \beta \quad \forall \lambda > 0.$$

Hence $\mathrm{lev}_{\le 0}\, g^{\infty} \subseteq (\mathrm{lev}_{\le\beta}\, g)^{\infty}$. When $u \in (\mathrm{lev}_{\le\beta}\, g)^{\infty}$, the last inequality gives, by subtracting $g(\bar{u})$, dividing by λ and then taking the limit, that $u \in \mathrm{lev}_{\le 0}\, g^{\infty}$. This proves 2. Parts 3 and 4 follow directly from Theorem A.16 while 5 follows by applying 1 to the epigraphs of the functions. □

Remark A.18 *Let g be a proper algebraically closed convex function, $u \in L := \{u \mid g^{\infty}(u) \le 0\}$ and $\bar{u} \in \mathrm{dom}\, g$. Then $\lambda \to g(\lambda u + \bar{u})$ is a nonincreasing function on $\mathbb{R}_+$. If L is a linear space, then $g(u + \bar{u}) = g(\bar{u})$.*

Proof By Theorem A.16, $g(\lambda u + \bar{u}) \le g(\bar{u})$ for every $\lambda \ge 0$, which, by convexity, proves the first claim. When L is linear, the function $\lambda \mapsto g(-\lambda u + \bar{u})$ is nonincreasing as well. By convexity, this function is thus constant on the whole $\mathbb{R}$, which proves the second claim. □

Remark A.19 *Let g be a convex differentiable function on $\mathbb{R}^m$. Then*

$$g^{\infty}(u) = \lim_{\lambda \nearrow \infty} \nabla g(\lambda u) \cdot u,$$

where ∇g is the gradient of g.

Proof By convexity,

$$g(u^2) \ge g(u^1) + \nabla g(u^1) \cdot (u^2 - u^1) \quad \forall u^1, u^2 \in \mathbb{R}^m.$$

It follows that, for any $u \in \mathbb{R}^m$ and $\lambda, \lambda' > 0$,

$$\frac{g(\lambda' u + \lambda u) - g(\lambda' u)}{\lambda} \geq \nabla g(\lambda' u) \cdot u \geq \frac{g(u + (\lambda' - 1)u) - g(u)}{\lambda' - 1}.$$

Letting $\lambda \nearrow \infty$ and then $\lambda' \nearrow \infty$, the claim follows from Theorem A.16. □

Remark A.20 *If C_1 and C_2 are convex sets such that $C_1 \subseteq C_2$ and C_2 is algebraically closed, then $C_1^\infty \subseteq C_2^\infty$. If g_1 and g_2 are convex functions such that $g_1 \geq g_2$ and g_2 is algebraically closed, then $g_1^\infty \geq g_2^\infty$. In particular, if g is a convex function bounded from below by a constant, then $g^\infty \geq 0$.*

Proof The first claim follows from Theorem A.15.1. The second follows by applying the first one to epigraphs of functions. □

A.4 Lower Semicontinuity

Let U be a topological vector space. A function $g : U \to \overline{\mathbb{R}}$ is *lower semicontinuous* (lsc) if its epigraph

$$\operatorname{epi} g := \{(u, \alpha) \in U \times \mathbb{R} \mid g(u) \leq \alpha\}$$

is closed in the product topology of $U \times \mathbb{R}$. A function g is lower semicontinuous if and only if the lower level-sets $\operatorname{lev}_\alpha g$ are closed for every $\alpha \in \mathbb{R}$. In a metric space, lower semicontinuity means that

$$\liminf_{\nu \nearrow \infty} g(u^\nu) \geq g(u)$$

for every $u \in U$ and every sequence $(u^\nu)_{\nu \in \mathbb{N}}$ converging to u.

Example A.21 *Let C be a closed convex set in a topological vector space such that $0 \in C$. The* gauge

$$\gamma_C(u) := \inf\{\alpha > 0 \mid u/\alpha \in C\}$$

of C is a proper lsc convex function and the infimum in the definition is attained whenever $\gamma_C(u) > 0$.

Proof As observed at the beginning of Sect. A.2, the set αC is increasing with respect to $\alpha > 0$. Given $\beta \geq 0$, we thus have

$$\operatorname{lev}_\beta \gamma_C = \bigcap_{\alpha > \beta} \{u \in U \mid \gamma_C(u) < \alpha\} = \bigcap_{\alpha > \beta} \alpha C$$

which is closed. When $\gamma_C(u) > 0$, the sequence $(u/\alpha^\nu)_{\nu\in\mathbb{N}}$ converges for any minimizing sequence $(\alpha^\nu)_{\nu\in\mathbb{N}}$. □

We say that a function g is *inf-compact* if $\operatorname{lev}_\alpha g$ is compact for every $\alpha \in \mathbb{R}$.

Lemma A.22 *If g is inf-compact, then the set*

$$\operatorname{argmin} g := \{u \in U \mid g(u) = \inf g\}$$

of its minimizers is nonempty and compact.

Proof The collection of sets

$$\{\operatorname{lev}_\alpha g \mid \alpha > \inf g\}$$

has the finite intersection property so, by compactness, their intersection is nonempty. □

The following extends [139, Theorem 1.17] which was concerned with finite-dimensional spaces.

Theorem A.23 *Let f be a lsc function on $X \times U$ such that, for every $\bar{u} \in U$ and $\alpha \in \mathbb{R}$, there exists a neighborhood $U_{\bar{u}}$ of $\bar{u}$ and a compact set $K_{\bar{u}} \subset X$ such that*

$$\{x \in X \mid \exists u \in U_{\bar{u}} : f(x,u) \leq \alpha\} \subseteq K_{\bar{u}}.$$

Then, $\varphi(u) = \inf_x f(x,u)$ is lsc and the infimum is attained for every $u \in U$. If, in addition, $f(x,\cdot)$ is continuous for every $x \in X$, then φ is continuous.

Proof The attainment of the infimum follows from Lemma A.22 since the lower semicontinuity of f implies the lower semicontinuity of $x \mapsto f(x,u)$ for each fixed u. Let $\alpha \in \mathbb{R}$. If $\bar{u} \notin \operatorname{lev}_\alpha \varphi$, then $X \times \{\bar{u}\} \cap \operatorname{lev}_\alpha f = \emptyset$. Since $\operatorname{lev}_\alpha f$ is closed, there exist, for every $x \in X$, neighborhoods $X_x \ni x$ and a $U_x \ni \bar{u}$ such that $(X_x \times U_x) \cap \operatorname{lev}_\alpha f = \emptyset$. Since $K_{\bar{u}}$ is compact, there is a finite set of points $x_1, \ldots, x_n$ such that $K_{\bar{u}} \subset \bigcup_{i=1}^n X_{x_i}$. Then $V = \bigcap_{i=1}^n (U_{x_i} \cap U_{\bar{u}})$ is a neighborhood of $\bar{u}$ and $V \cap \operatorname{lev}_\alpha \varphi = \emptyset$. Indeed, if $u \in V \cap \operatorname{lev}_\alpha \varphi$, there is an $x \in X$ such that $(x,u) \in \operatorname{lev}_\alpha f$. But this is impossible since $u \in V \subset U_{\bar{u}}$ and $(x,u) \in \operatorname{lev}_\alpha f$ imply $x \in K_{\bar{u}}$ while

$$K_{\bar{u}} \times V \subset \bigcup_{i=1}^n (X_{x_i} \times U_{x_i}),$$

which is disjoint from $\operatorname{lev}_\alpha f$.

Assume now that f is continuous in u and let $\bar{u} \in U$. By the first part, there is an $\bar{x} \in X$ such that $\varphi(\bar{u}) = f(\bar{x},\bar{u})$ while $\varphi(u) \leq f(\bar{x},u)$ for all $u \in U$. The upper semicontinuity of f in u thus implies that φ is upper semicontinuous at $\bar{u}$. Since

$\bar{u}$ was arbitrary, φ is upper semicontinuous throughout U. Combining this with the lower semicontinuity obtained in the first part, gives the continuity. □

A finite-dimensional version of the following can be found in [139, Example 9.11].

Lemma A.24 *Let g be a proper lsc function on a metric space (X, d) such that $g(x) \geq -\rho d(x, 0) - m$ for some $\rho > 0$ and $m \geq 0$. The functions*

$$g^\nu(x) := \inf_{x' \in X} \{g(x') + \nu\rho d(x, x')\} \quad \nu \in \mathbb{N}$$

are $(\nu\rho)$-Lipschitz with $g^\nu(x) \geq -\rho d(x, 0) - m$ and as ν increases, they increase pointwise to g. If (X, d) is a normed space and g is convex then g^ν are convex.

Proof For any $x_1, x_2 \in X$,

$$\begin{aligned} g^\nu(x_1) &\leq \inf_{x'}\{g(x') + \nu\rho d(x_2, x') + \nu\rho d(x_1, x_2)\} \\ &= g^\nu(x_2) + \nu\rho d(x_1, x_2). \end{aligned}$$

By symmetry, g^ν is $\nu\rho$-Lipschitz continuous. Clearly, $g \geq g^\nu$. For every ν and $\epsilon > 0$, there exists an x^ν such that

$$\begin{aligned} g^\nu(x) &\geq g(x^\nu) + \nu\rho d(x, x^\nu) - \epsilon \\ &\geq -\rho d(x^\nu, 0) - m + \nu\rho d(x, x^\nu) - \epsilon \\ &\geq -\rho d(x, 0) - m + (\nu - 1)\rho d(x, x^\nu) - \epsilon. \end{aligned}$$

Thus, either $g^\nu(x) \to \infty$ or $x^\nu \to x$ as $\nu \to \infty$. Since g is lower semicontinuous, $\liminf g^\nu(x) \geq g(x)$ in either case. The last claim follows from Theorems A.2 and A.1. □

We end this section with some additional results on recession cones in topological vector spaces. The following is from [123, Theorem 8.2].

Theorem A.25 *If C is a nonempty sequentially closed convex set, then*

$$C^\infty = \{u \in U \mid \exists \lambda^\nu \searrow 0,\ u^\nu \in C : \lambda^\nu u^\nu \to u\}.$$

Proof Let $\bar{u} \in C$. If $u \in C^\infty$, Theorem A.15 implies that, for any $\lambda^\nu \searrow 0$, there exist $u^\nu \in C$ such that $u = \lambda^\nu(u^\nu - \bar{u})$ so $\lambda^\nu u^\nu \to u$. Assume now that $\lambda^\nu \searrow 0$, $u^\nu \in C$ and $\lambda^\nu u^\nu \to u$. Let $\lambda > 0$. For any $\lambda^\nu \leq \lambda$, convexity of C implies

$$\lambda^\nu(u^\nu - \bar{u}) \in \lambda^\nu(C - \bar{u}) \subseteq \lambda(C - \bar{u}).$$

Since C is closed, we get $u \in \lambda(C - \bar{u})$. Since $\lambda > 0$ was arbitrary, we have $u \in C^\infty$. □

Corollary A.26 *If g is a proper lsc convex function, then*

$$g^\infty(u) = \liminf_{\substack{\lambda^\nu \nearrow \infty \\ w^\nu \to u}} \frac{g(\lambda^\nu w^\nu)}{\lambda^\nu} = \liminf_{\substack{\lambda^\nu \nearrow \infty \\ \tilde{w}^\nu \to u}} \frac{g(\bar{u} + \lambda^\nu \tilde{w}^\nu) - g(\bar{u})}{\lambda^\nu}$$

for any $\bar{u} \in \operatorname{dom} g$.

Proof By Theorem A.25,

$$\operatorname{epi} g^\infty = \{(u, \alpha) \mid \exists \mu^\nu \searrow 0,\ (u^\nu, \alpha^\nu) \in \operatorname{epi} g : \ \mu^\nu(u^\nu, \alpha^\nu) \to (u, \alpha)\}.$$

Defining $\lambda^\nu = 1/\mu^\nu$, $w^\nu = \mu^\nu u^\nu$ and $\tilde{\alpha}^\nu = \mu^\nu \alpha^\nu$, this can be written as

$$\operatorname{epi} g^\infty = \{(u, \alpha) \mid \exists \lambda^\nu \nearrow \infty,\ (\lambda^\nu w^\nu, \lambda^\nu \tilde{\alpha}^\nu) \in \operatorname{epi} g : \ (w^\nu, \tilde{\alpha}^\nu) \to (u, \alpha)\}.$$

In other words, $\alpha \geq g^\infty(u)$ means that there exist sequences $\lambda^\nu \nearrow \infty$, $(w^\nu, \tilde{\alpha}^\nu) \to (u, \alpha)$ such that

$$\frac{g(\lambda^\nu w^\nu)}{\lambda^\nu} \leq \tilde{\alpha}^\nu.$$

This gives the first expression. The second expression is obtained with the change of variables $\tilde{w}^\nu = w^\nu - \bar{u}/\lambda$. □

The following is [123, Theorem 8.4].

Theorem A.27 *A nonempty closed convex set in $\mathbb{R}^m$ is bounded if and only if its recession cone only contains the origin.*

Proof It is clear that a set is unbounded if its recession cone is nonempty. On the other hand, if a convex closed set C is not bounded, it contains a sequence $(u^\nu)_{\nu \in \mathbb{N}}$ with $\lambda^\nu := |u^\nu|^{-1} \searrow 0$. Since $|\lambda^\nu u^\nu| = 1$, there is a further subsequence (still denoted by $(u^\nu)_{\nu \in \mathbb{N}}$) for which $\lambda^\nu u^\nu \to u$ for some $u \in \mathbb{R}^n$ with $|u| = 1$. By Theorem A.25, $u \in C^\infty$, so $C^\infty \neq \{0\}$. □

A.5 Continuity of Convex Functions

Let U be a topological vector space. The *interior* and the *closure* of a set $C \subseteq U$ will be denoted by $\operatorname{int} C$ and $\operatorname{cl} C$, respectively. The *relative interior* $\operatorname{rint} C$ of a set C is its interior with respect to its affine hull $\operatorname{aff} C$. By Rockafellar [123, Theorem 6.2], convex subsets of $\mathbb{R}^n$ always have nonempty relative interior.

Theorem A.28 *Let $C \subseteq U$ be a convex set with $\operatorname{aff} C$ is closed, $x \in \operatorname{rint} C$ and $\bar{x} \in \operatorname{cl} C$. Then*

$$\alpha x + (1-\alpha)\bar{x} \in \operatorname{rint} C \quad \forall \alpha \in (0,1].$$

If $\operatorname{rint} C \neq \emptyset$, *then* $\operatorname{cl} \operatorname{rint} C = \operatorname{cl} C$ *and*

$$\operatorname{rint} \operatorname{cl} C = \operatorname{rint} C = \operatorname{rcore} C.$$

Proof When $\operatorname{aff} C$ is closed, we have $\operatorname{cl} C \subseteq \operatorname{aff} C$, so $\operatorname{aff} \operatorname{cl} C = \operatorname{aff} C$. Thus, applying [1, Lemma 5.28] relative to $\operatorname{aff} C$ proves all the claims except the last equality. In the last one, the inclusion $\operatorname{rcore} C \supseteq \operatorname{rint} C$ is clear. Given $x' \in \operatorname{rcore} C$, we have $x' - \alpha(x - x') \in C$ for $\alpha > 0$ small enough. Since

$$\frac{\alpha}{1+\alpha} x + (1 - \frac{\alpha}{1+\alpha})(x' - \alpha(x - x')) = x',$$

$x' \in \operatorname{rint} C$ by the first claim. □

We say that an extended real-valued function is *continuous at a point* $u \in U$ if it is real-valued in a neighbourhood of u and continuous at u. The following is fundamental.

Theorem A.29 *A convex function that is bounded from above on a nonempty open set is either continuous or identically $-\infty$ throughout the core of its domain.*

Proof Let g be a convex function. Assume first that g is proper and that there is $M \in \mathbb{R}$ such that $g \leq M$ on an open set O. By [27, page II.18], g is continuous on O. Let $u \in \operatorname{core} \operatorname{dom} g$ and $\bar{u} \in O$. There exists $\alpha \in (0,1)$ and $u' \in \operatorname{dom} g$ such that $u = \alpha u' + (1-\alpha)\bar{u}$. By convexity, for every $w \in O$

$$g(\alpha u' + (1-\alpha)w) \leq \alpha g(u') + (1-\alpha) g(w) \leq \alpha g(u') + (1-\alpha) M.$$

The set $\alpha u' + (1-\alpha)O$ is a open set containing u, so g is continuous at u by the first part. This finishes the proof when g is proper. A simple line segment argument shows that, if a convex function equals $-\infty$ at some point, then it equals $-\infty$ throughout the core of its domain. □

We say that a function g is *relatively continuous* at $u \in \operatorname{dom} g$ if it is continuous at u relative to $\operatorname{aff} \operatorname{dom} g$. The following is a straightforward consequence of Theorem A.29.

Corollary A.30 *A convex function which is bounded from above on a relatively open set of* $\operatorname{aff} \operatorname{dom} g$ *is either relatively continuous or identically $-\infty$ throughout* $\operatorname{rcore} \operatorname{dom} g$.

If the seminorm p in the Hahn-Banach theorem in Corollary A.14 is continuous, Theorem A.29 implies that the linear extension $\bar{k}$ is continuous as well. This implies the following.

Corollary A.31 *If $L \subset U$ is a linear space and k a continuous linear functional on L, then there exists a continuous linear functional $\bar{k}$ on U such that $\bar{k} = k$ on L.*

Proof The continuity on L means that there exists a neighborhood $O \subset U$ of the origin and a constant $M \in \mathbb{R}$ such that $k(u) \leq M$ for $u \in O \cap L$. By linearity of k, we can take the convex hull of O and thus assume that O is convex. Given any $u \in L$ and $\alpha > 0$ such that $u \in \alpha O$, linearity of k gives $k(u) \leq M\alpha$. Minimizing over such α gives

$$k(u) \leq p(u) \quad \forall u \in L,$$

where $p := M\gamma_O$, see Example A.21. By Corollary A.14, there is a linear extension $\bar{k}$ of k such that $\bar{k} \leq p$ on U. Since $p \leq M$ on O, the continuity follows from Theorem A.29, □

The following gives a topological version of the separation theorem in Corollary A.13.

Corollary A.32 *Let C_1 and C_2 be convex sets with $0 \notin \operatorname{rint}(C_1 - C_2) \neq \emptyset$. There exists a continuous linear functional l such that*

$$\sup_{u \in C_1} l(u) \leq \inf_{u \in C_2} l(u) \quad \text{and} \quad \inf_{u \in C_1} l(u) < \sup_{u \in C_2} l(u).$$

Proof Translating, we may assume that $L := \operatorname{aff}(C_1 - C_2)$ is a linear space. Applying Theorem A.12 with respect to L gives the existence of a linear functional $l : L \to \mathbb{R}$ such that

$$\sup_{u \in (C_1 - C_2)} l(u) \leq 0.$$

This is the inequality in the statement. This also implies that l is bounded from above on $\operatorname{rint}(C_1 - C_2)$ so, by Theorem A.29, l is continuous on L. The claim now follows from Corollary A.31. □

Simpler conditions for continuity are available if the underlying space is *barreled* in the sense that $\operatorname{core} C = \operatorname{int} C$ for every closed convex set C. Every complete metrizable topological vector space U is barreled. Indeed, if $B \subseteq U$ is closed and convex with $0 \in \operatorname{core} B$, we have $U = \bigcup_{\lambda \in \mathbb{N}} \lambda B$, so the Baire category theorem (see, e.g. [75, Theorem 9.4]) says that $\operatorname{int} \lambda B = \emptyset$ for some $\lambda \in \mathbb{N}$. It follows that Fréchet spaces and, in particular Banach spaces, are barreled. Recall that a *Fréchet space* is a complete metrizable locally convex topological vector space.

Theorem A.33 *In a barreled space, a lsc convex function is either continuous or identically* $-\infty$ *throughout the core of its domain.*

Proof Let g be a lsc convex function on a barreled space U and let $u \in \operatorname{core} \operatorname{dom} g$ and $\alpha > g(u)$. Let $u' \in U$ and $\phi(\lambda) := g(u + \lambda u')$. We have $0 \in \operatorname{int} \operatorname{dom} \phi$ so, by convexity, ϕ is either continuous or identically $-\infty$ throughout the core of its domain, by Theorem A.29. It follows that $u \in \operatorname{core} \operatorname{lev}_\alpha g$. Since g is lsc and convex, $\operatorname{lev}_\alpha g$ is closed and convex, so $\operatorname{int} \operatorname{lev}_\alpha g \neq \emptyset$. The continuity of g now follows from Theorem A.29. □

Example A.34 (Banach–Steinhaus) *If* $\mathcal{A}$ *is a family of continuous linear operators from a Banach space* X *to a normed space such that*

$$\sup_{A \in \mathcal{A}} \|A(x)\| < \infty$$

for all $x \in X$*, then* $\sup_{A \in \mathcal{A}} \|A\| < \infty$.

Proof By Theorem A.33, the function

$$g(x) := \sup_{A \in \mathcal{A}} \|A(x)\|$$

is continuous. In particular, there exist $M > 0$ and $r > 0$ such that $g \leq M$ on $\mathbb{B}_r$. In other words,

$$M \geq \sup_{x \in \mathbb{B}_r} g(x) = \sup_{A \in \mathcal{A}} \sup_{x \in \mathbb{B}_r} \|A(x)\| = \sup_{A \in \mathcal{A}} r\|A\|,$$

which completes the proof. □

We say that an affine set is *barreled* if its translation to the origin is barreled relative to itself. If C is closed and convex and $\operatorname{aff} C$ is barreled, then applying the argument given just before Theorem A.33 on $\operatorname{aff} C$ gives

$$\operatorname{rint} C = \operatorname{rcore} C.$$

Remark A.35 *Closed subspaces of barreled spaces need not be barreled but closed linear subspaces of Fréchet spaces are Fréchet, and thus, barreled. In particular, if* C *is closed and convex subset of a Fréchet space and* $\operatorname{aff} C$ *is closed, then*

$$\operatorname{rint} C = \operatorname{rcore} C.$$

If g is a lsc convex function with $\operatorname{aff} \operatorname{dom} g$ barreled, then by Theorem A.33, g is either relatively continuous or identically $-\infty$ throughout $\operatorname{rcore} \operatorname{dom} g$. By Remark A.35 this implies the following.

Corollary A.36 *Let g be a lsc proper convex function on a Fréchet space such that* aff dom g *closed. Then g is either identically $-\infty$ or relatively continuous throughout* rcore dom g.

The following shows that the closedness condition in Corollary A.36 is necessary.

Remark A.37 *If g is a proper lsc convex function on a Fréchet space such that* rcore dom $g \neq \emptyset$ *and g is relatively continuous throughout* rcore dom g, *then* aff dom g *is closed.*

Proof Translating if necessary, we may assume that $0 \in \operatorname{rcore} g$. Given any $\beta > g(0)$, relative continuity gives a neighborhood N of the origin such that $N \cap \operatorname{aff}\operatorname{dom} g \subset \operatorname{lev}_\beta g$. Let $(u^\nu)_{\nu\in\mathbb{N}}$ be a sequence in aff dom g converging to a $\bar{u}$. The sequence is bounded so there is an $\alpha > 0$ such that $(u^\nu)_{\nu\in\mathbb{N}} \subset \alpha N$. It follows that $g(u^\nu/\alpha) \le \beta$ for all ν. The lower semicontinuity of g gives $g(\bar{u}/\alpha) \le \beta$ so $\bar{u} \in \operatorname{aff}\operatorname{dom} g$. □

The following far reaching extension of the open mapping theorem is from [160, Theorem 1.3.7].

Theorem A.38 (Ursescu) *Let X be a Fréchet space and let D be the projection of a closed convex set $C \subseteq X \times U$ on U. Assume that* aff D *is barreled. If $(x_0, u_0) \in C$ with $u_0 \in \operatorname{rcore} D$, then $u_0 \in \operatorname{int}_{\operatorname{aff} D}\{u \mid \exists x \in N : (x, u) \in C\}$ for every neighborhood N of x_0. In particular,* $\operatorname{rcore} D = \operatorname{rint} D$.

Proof This is essentially [160, Theorem 1.3.7] where, instead of Fréchet, X was assumed to be merely a complete and semi-metrizable locally convex topological vector space. We have denoted the graph of the mapping $\mathcal{R}$ in [160, Theorem 1.3.7] by C and used the fact, if aff D is barreled, then the set ${}^{ib}(\operatorname{Im}\mathcal{R})$ in [160, Theorem 1.3.7] is equal to rcore D. Thus, if this set is nonempty, the claim is that of [160, Theorem 1.3.7]. If $\operatorname{rcore} D = \emptyset$, the claim holds trivially since we always have $\operatorname{rint} D \subseteq \operatorname{rcore} D$. □

Theorem A.38 yields the following result which extends Corollary A.36 by relaxing the lower semicontinuity assumption to the structural assumption that the function be the infimal projection of a lsc convex function on a product space.

Theorem A.39 *Let X be a Fréchet space, let F be a lsc convex function on $X \times U$ and let*

$$\varphi(u) = \inf_{x\in X} F(x, u).$$

If aff dom φ *is barreled, then φ is either relatively continuous or identically $-\infty$ throughout* rcore dom φ. *The set* aff dom φ *is barreled, in particular, if it is closed and U is Fréchet.*

Proof We apply Theorem A.38 with $C = \operatorname{epi} F$ and

$$D := \{(u, \alpha) \in U \times \mathbb{R} \mid \exists x \in X : \; F(x, u) \le \alpha\},$$

the projection of C to $U \times \mathbb{R}$. By Example A.10,

$$\operatorname{aff}\operatorname{epi}\varphi = \operatorname{aff}\operatorname{dom}\varphi \times \mathbb{R},$$
$$\operatorname{rcore}\operatorname{epi}\varphi = \{(u,\alpha) \in U \times \mathbb{R} \mid u \in \operatorname{rcore}\operatorname{dom}\varphi,\ \alpha > \varphi(u)\},$$

so it follows that

$$\operatorname{rcore}\operatorname{epi}\varphi \subseteq D \subseteq \operatorname{epi}\varphi.$$

Thus, $\operatorname{aff} D = \operatorname{aff}\operatorname{dom}\varphi \times \mathbb{R}$ and $\operatorname{rcore}\operatorname{epi}\varphi = \operatorname{rcore} D$. By Scott Osborne [93, Proposition 4.3], products of barreled spaces are barreled, so $\operatorname{aff} D$ is barreled. Theorem A.38 now implies that, for any $u \in \operatorname{rcore}\operatorname{dom}\varphi$ and $\alpha > \varphi(u)$, there is a relatively open neighborhood of (u,α) in $\operatorname{aff}\operatorname{epi}\varphi$ that is contained in $\operatorname{epi}\varphi$. The first claim now follows from Corollary A.30. The second claim follows from Remark A.35. □

Theorem A.39 is from [161]; see also [160, Theorem 2.7.1]. It extends various earlier continuity results for the infimal projection; see, e.g. [127, Theorem 18].

Example A.40 (Closed Graph Theorem) *A linear mapping from a barreled space to a Fréchet space is continuous if it has a closed graph.*

Proof Let p be a continuous seminorm on X. If $A : U \to X$ is a linear mapping with a closed graph, the function

$$F(x,u) := p(x) + \delta_{\operatorname{gph} A}(u,x)$$

is lsc and convex and the function

$$\varphi(u) := \inf_{x \in X} F(x,u) = p(Au)$$

has $\operatorname{dom}\varphi = U$. By Theorem A.39, φ is continuous. Since p was an arbitrary continuous seminorm and since the topology of X is generated by the lower level-sets of such functions, the mapping A must be continuous. □

The following criterion was used in the proof of Lemma 5.32.

Lemma A.41 *Let $C \subseteq X$ and $A : X \to U$ a continuous linear mapping. If AC is closed, then* $\ker A + C$ *is closed. The converse holds if X and U are Frechét and* $\operatorname{rge} A$ *is closed.*

If C is closed and π is a continuous linear idempotent mapping on X such that $\pi C \subseteq C$, then the sets $\ker \pi + C$ *and* $\operatorname{rge}\pi + C$ *are closed.*

Proof We have

$$\begin{aligned}\ker A + C &= \{z + x' \mid Az = 0,\ x' \in C\}\\ &= \{x \mid A(x - x') = 0,\ x' \in C\}\\ &= \{x \mid Ax \in AC\}\\ &= A^{-1}(AC),\end{aligned}$$

which proves the first claim. The converse is from [160, Corollary 1.3.15] which is an application of Theorem A.38. Applying the first claim to $A = I - \pi$ gives the closedness of $\operatorname{rge}\pi + C$, since $AC = C - \pi C = C$ and $\ker A = \operatorname{rge}\pi$. Applying the last claim to $\pi' := 1 - \pi$, gives the closedness of $\ker\pi + C$. □

A.6 Conjugates and Subgradients

Let U and Y be vector spaces in *separating duality* under a real-valued bilinear form

$$(u, y) \mapsto \langle u, y\rangle.$$

This means that, for every nonzero $u \in U$, there is a $y \in Y$ such that $\langle u, y\rangle \neq 0$ and similarly for Y. The *conjugate* of a function $g : U \to \overline{\mathbb{R}}$ is the extended real-valued convex function on $\mathcal{Y}$ defined by

$$g^*(y) = \sup_{u\in U}\{\langle u, y\rangle - g(u)\}.$$

By definition, a function and its conjugate satisfy the *Fenchel's inequality*

$$g(u) + g^*(y) \geq \langle u, y\rangle$$

for all $u \in U$ and $y \in Y$. When $g(u)$ is finite, the inequality holds as an equality if and only if

$$g(u') \geq g(u) + \langle u' - u, y\rangle \quad \forall u' \in U.$$

We then say that y is a *subgradient* of g at u. The set of subgradients of g at u is known as the *subdifferential* of g at u and denoted by $\partial g(u)$. The subdifferential is defined as the empty set unless $g(u)$ is finite. If g is subdifferentiable at u, we have $g(u) + g^*(y) = \langle u, y\rangle$ for every $y \in \partial g(u)$ so $g(u) = g^{**}(u)$.

Given a convex set $D \subseteq U$, its *support function* $\sigma_D : Y \to \overline{\mathbb{R}}$ is the conjugate of its indicator function δ_D, i.e.

$$\sigma_D(y) := \sup_{u \in D} \langle u, y \rangle.$$

The support function of the epigraph of a function g can be written as

$$\sigma_{\text{epi}\, g}(y, \beta) = \sup_{(u,\alpha) \in U \times \mathbb{R}} \{\langle u, y \rangle + \alpha\beta \mid g(u) \le \alpha\}$$

so the conjugate of g can be expressed as

$$g^*(y) = \sigma_{\text{epi}\, g}(y, -1).$$

Recall that the *gauge* $\gamma_D : U \to \overline{\mathbb{R}}$ of a set $D \subseteq U$ is given by

$$\gamma_D(u) := \inf\{\alpha > 0 \mid u/\alpha \in D\}.$$

The *polar* of D is defined by

$$D^\circ := \{y \in Y \mid \langle u, y \rangle \le 1\ \forall u \in D\}.$$

Gauges and polars of convex sets in Y are defined analogously. Note that if D is a cone, then

$$D^\circ = \{y \in Y \mid \langle u, y \rangle \le 0\ \forall u \in D\}.$$

Example A.42 *Given a set $D \subseteq U$ with $0 \in D$, we have*

$$\gamma_{D^\circ} = \sigma_D \quad \textit{and} \quad \gamma_D^* = \delta_{D^\circ}.$$

Proof By definition,

$$\begin{aligned}
\gamma_{D^\circ}(y) &= \inf_{\alpha > 0} \{\alpha \mid y/\alpha \in D^\circ\} \\
&= \inf_{\alpha > 0} \{\alpha \mid \langle u, y/\alpha \rangle \le 1\ \forall u \in D\} \\
&= \inf_{\alpha > 0} \{\alpha \mid \langle u, y \rangle \le \alpha\ \forall u \in D\} \\
&= \sigma_D(y)
\end{aligned}$$

and

$$\begin{aligned}\gamma_D^*(y) &= \sup_{u\in U,\alpha>0} \{\langle u, y\rangle - \alpha \mid u/\alpha \in D\}\\ &= \sup_{u\in U,\alpha>0} \{\alpha(\langle u/\alpha, y\rangle - 1) \mid u/\alpha \in D\}\\ &= \sup_{\alpha>0} \alpha(\sigma_D(y) - 1)\\ &= \delta_{D^\circ}(y)\end{aligned}$$

for all $y \in Y$. □

The *normal cone* $N_D(u)$ of D at a point $u \in D$ is the subdifferential of δ_D at u, i.e.

$$\begin{aligned}N_D(u) :=&\{y \in Y \mid \delta_D(u) + \sigma_D(y) \le \langle u, y\rangle\}\\ =&\{y \in Y \mid \langle u' - u, y\rangle \le 0\ \forall u' \in D\}.\end{aligned}$$

Example A.43 *If D is a cone, then $y \in N_D(u)$ if and only if*

$$u \in D,\ \ y \in D^\circ,\ \langle u, y\rangle = 0.$$

Let (X, V) be another pair of vector spaces in separating duality and consider a linear mapping $A : X \to U$. A linear mapping $A^* : Y \to V$ is the *adjoint* of A if

$$\langle Ax, y\rangle = \langle x, A^* y\rangle \quad \forall x \in X,\ y \in Y;$$

see Lemma A.53 below. The following facts follow directly from definitions.

Remark A.44 *Given extended real-valued functions g_1 and g_2 on U,*

$$\partial g_1(u) + \partial g_2(u) \subseteq \partial(g_1 + g_2)(u) \quad \forall u \in U.$$

Given an extended real-valued function g on U and a linear mapping $A : X \to U$ with adjoint $A^* : Y \to V$,

$$A^*\partial g(Ax) \subseteq \partial(g \circ A)(x) \quad \forall x \in X.$$

Sufficient conditions for equalities to hold in Remark A.44 will be given in Corollary A.80 and Corollary A.81 below.

Lemma A.45 *If g^* is proper, then*

$$(g^*)^\infty = \sigma_{\operatorname{dom} g}.$$

Proof By definition,

$$g^*(y) = \sup_{u \in \operatorname{dom} g} \{\langle u, y\rangle - g(u)\}$$

so the claim follows from Theorem A.17.5 since the recession function of $\langle u, \cdot\rangle - g(u)$ is $\langle u, \cdot\rangle$ for every $u \in \operatorname{dom} g$. □

Remark A.46 *The subdifferential of a convex function g on U defines a set-valued mapping from U to Y associating to each point $u \in U$ the set $\partial g(u)$ of subgradients. This mapping, denoted by $\partial g : U \rightrightarrows Y$, is* monotone *in the sense that*

$$\langle u_1 - u_2, y_1 - y_2\rangle \geq 0 \quad \forall (u_1, y_1), (u_2, y_2) \in \operatorname{gph} \partial g.$$

This follows simply by adding together the subdifferential inequalities at u_1 and u_2.

Lemma A.47 *Let $\pi : U \to U$ be a linear mapping with adjoint $\pi^* : Y \to Y$ and let $g : U \to \overline{\mathbb{R}}$ such that $g \circ \pi \leq g$. Then $g^* \circ \pi^* \leq g^*$.*

Proof We have

$$\begin{aligned} g^*(\pi^* y) &= \sup_u \{\langle u, \pi^* y\rangle - g(u)\} \\ &\leq \sup_u \{\langle \pi u, y\rangle - g(\pi u)\} \\ &\leq \sup_u \{\langle u, y\rangle - g(u)\} \\ &= g^*(y), \end{aligned}$$

which proves the claim. □

A.7 Compatible Topologies

As in the previous section, we assume that U and Y are vector spaces in separating duality. The *weak topology* $\sigma(U, Y)$ on U is the weakest topology under which the linear functionals $u \mapsto \langle u, y\rangle$, $y \in Y$ are all continuous. The weak topology $\sigma(Y, U)$ on Y is defined similarly. Weak topologies are automatically locally convex vector topologies. The *Mackey topology* $\tau(U, Y)$ on U is the weakest topology under which the *support functions*

$$\sigma_D(u) := \sup_{y \in D} \langle u, y\rangle$$

of $\sigma(Y, U)$-compact sets $D \subseteq Y$ are continuous. By the Mackey-Arens theorem, the Mackey topology is the strongest locally convex vector topology under which all continuous linear functionals can be expressed as $u \mapsto \langle u, y \rangle$ for some $y \in Y$; see, e.g. [147, Section 4.3]. Similarly for the Mackey topology on Y.

Given a topological vector space U, its topological dual U^* is the space of all continuous linear functionals on U.

Remark A.48 *If U is a barreled locally convex topological vector space under a given topology τ and $Y = U^*$, then τ coincides with the Mackey topology generated by the pairing $\langle u, y \rangle := u^*(u)$. Indeed, by the Mackey-Arens theorem, the Mackey topology is at least as strong as τ. On the other hand, the Mackey topology is at most as strong as the topology generated by all finite seminorms which are τ-lsc. By Theorem A.33, such functions are τ-continuous so the generated topology cannot be stronger than τ.*

A set $C \subseteq U$ is *bounded* if, for every neighborhood O of the origin, there exists $\lambda > 0$ such that $C \subseteq \lambda O$.

Remark A.49 *A set in U is $\sigma(U, Y)$-bounded if and only if its support function is finite throughout Y. Combined with Remark A.48, this implies that a closed convex set in the topological dual U^* of a barreled locally convex topological vector space is $\sigma(U^*, U)$-bounded if and only if it is $\sigma(U^*, U)$-compact.*

Proof Let $C \subseteq U$ be bounded. Given $y \in Y$, the set $\{u \mid |\langle u, y \rangle| \leq 1\}$ is a neighborhood of the origin, so boundedness of C implies $\sigma_C(y) < \infty$. As to the converse, any $\sigma(U, Y)$-neighborhood of the origin contains a set of the form $D = \{u \in U \mid \langle u, y_i \rangle \leq 1\ i = 1, \ldots, \nu\}$. Clearly, $\langle u, y_i \rangle \leq \sigma_C(y_i)$ for every $u \in C$ so, if σ_C is finite everywhere, we get $C \subset \max_i \sigma_C(y_i) D$. □

Local convexity of a topology yields the following result on strong separation of convex sets.

Theorem A.50 *Let C_1 and C_2 be disjoint convex sets in U such that the origin does not belong to the Mackey-closure of $C_1 - C_2$. There exists a nonzero $y \in Y$ such that*

$$\sup_{u \in C_1} \langle u, y \rangle < \inf_{u \in C_2} \langle u, y \rangle.$$

Proof Since origin does not belong to the Mackey-closure of $C_1 - C_2$, local convexity gives the existence of a convex Mackey-neighborhood $\bar{C}_2$ of the origin that is disjoint from $\bar{C}_1 := C_1 - C_2$. Corollary A.32 gives the existence of a nonzero $y \in Y$ such that

$$\inf_{u \in \bar{C}_2} \langle u, y \rangle \geq \sup_{u \in \bar{C}_1} \langle u, y \rangle.$$

It now suffices to note that

$$\sup_{u\in\bar{C}_1}\langle u, y\rangle = \sup_{u\in C_1}\langle u, y\rangle - \inf_{u\in C_2}\langle u, y\rangle$$

and, since $0 \in \operatorname{int}\bar{C}_2$, we get $\inf_{u\in\bar{C}_2}\langle u, y\rangle < 0$. □

The condition that the origin does not belong to the closure of $C_1 - C_2$ holds, in particular, if C_1 is closed and C_2 is a singleton.

Corollary A.51 *The Mackey closed convex hull of any set $C \subseteq U$ is the intersection of all closed half-spaces containing C. In particular, a convex set is weakly closed if and only if it is Mackey closed.*

Proof We may assume that $\operatorname{cl}\operatorname{co} C \neq U$ since the intersection of an empty collection of sets is U. Applying Theorem A.50 with $C_1 = \operatorname{cl}\operatorname{co} C$ and C_2 a singleton not contained in C_1 shows that $\operatorname{cl}\operatorname{co} C$ is the intersection of all closed half-spaces containing $\operatorname{cl}\operatorname{co} C$. It now suffices to note that a closed half space contains C if and only it contains $\operatorname{cl}\operatorname{co} C$. □

The *lower semicontinuous (lsc) hull* of a function is the greatest lsc function it dominates. The lsc hull of a function g is denoted by $\operatorname{lsc} g$. We have

$$\operatorname{epi}(\operatorname{lsc} g) = \operatorname{cl}\operatorname{epi} g.$$

Indeed, the function $\bar{g}(u) = \inf\{\alpha \mid (u,\alpha) \in \operatorname{cl}\operatorname{epi} g\}$ has $\operatorname{epi}\bar{g} = \operatorname{cl}\operatorname{epi} g$, and, given another lsc function $\tilde{g} \leq g$, we have $\operatorname{epi} g \subseteq \operatorname{epi}\tilde{g}$ and thus $\operatorname{epi}\bar{g} = \operatorname{cl}\operatorname{epi} g \subseteq \operatorname{cl}\operatorname{epi}\tilde{g}$, so $\bar{g} \geq \tilde{g}$. Thus, Corollary A.51 gives the following.

Corollary A.52 *The weak and the Mackey lsc-hulls of a convex function coincide. In particular, a convex function is weakly lower semicontinuous if and only if it is Mackey lower semicontinuous.*

The following establishes the existence of the adjoint of a linear mapping $A : X \to U$ defined in Sect. A.6.

Lemma A.53 *Let (X, V) and (U, Y) be two pairs of vector spaces in separating duality and let $A : X \to U$ be linear. The following are equivalent:*

1. *A is continuous with respect to $\sigma(X, V)$ and $\sigma(U, Y)$;*
2. *the adjoint A^* exists,*

and imply that A^* is continuous with respect to $\sigma(U, Y)$ and $\sigma(V, X)$.

Proof Assume 1 and let $y \in Y$. The mapping $x \mapsto \langle Ax, y\rangle$ is $\sigma(X, V)$-continuous and linear, so there exists $v \in V$ such that $\langle Ax, y\rangle = \langle x, v\rangle$ and we may set $A^*y := v$. Thus, 1 implies 2. Assume that 2 holds. Given x and $\epsilon > 0$, we have $\{y \mid |\langle Ax, y\rangle| < \epsilon\} = \{y \mid |\langle x, A^*y\rangle| < \epsilon\}$, so the preimages under A of $\sigma(V, X)$-open sets are $\sigma(Y, U)$-open. Thus, 2 implies 1. Symmetrically, 2 implies the continuity of A^*. □

A.8 Biconjugate Theorem

Again, we assume that U and Y are vector spaces in separating duality. As observed in the previous section, a convex function on U is lsc with respect to the weak topology $\sigma(U, Y)$ if and only if it is lsc with respect to the Mackey topology $\tau(U, Y)$. From now on, we will simply say that a convex function is lower semicontinuous (lsc) if it is lsc with respect to a topology compatible with the pairing of U with Y.

The *closure* of a convex function $g : U \to \overline{\mathbb{R}}$ is the function $\operatorname{cl} g : U \to \overline{\mathbb{R}}$ defined by

$$\operatorname{cl} g = \begin{cases} \operatorname{lsc} g & \text{if } \operatorname{lsc} g(u) > -\infty \text{ for all } u \in U, \\ -\infty & \text{otherwise.} \end{cases}$$

A function $g : U \to \overline{\mathbb{R}}$ is *closed* at $u \in U$ if $g(u) = (\operatorname{cl} g)(u)$. A function is *closed* if it is closed at every point.

The following fundamental result goes back to [28, 54, 87].

Theorem A.54 (Biconjugate Theorem) *Given a function* $g : U \to \overline{\mathbb{R}}$,

$$g^{**} = \operatorname{cl} \operatorname{co} g.$$

Proof We have $(u, \alpha) \in \operatorname{epi} g^{**}$ if and only if

$$\alpha \geq \langle u, y \rangle - \beta \quad \forall (y, \beta) \in \operatorname{epi} g^*.$$

Here $(y, \beta) \in \operatorname{epi} g^*$ if and only if $g(u) \geq \langle u, y \rangle - \beta$ for every u. Thus, $\operatorname{epi} g^{**}$ is the intersection of the epigraphs of all continuous affine functionals dominated by g.

On the other hand, by Theorem A.51, $\operatorname{epi} \operatorname{lsc} \operatorname{co} g$ is the intersection of all closed half-spaces

$$H_{y,\beta,\gamma} := \{(u, \alpha) \in U \times \mathbb{R} \mid \langle u, y \rangle + \alpha\beta \leq \gamma\}$$

containing $\operatorname{epi} g$. We have $(\operatorname{lsc} \operatorname{co} g)(u) > -\infty$ for every $u \in U$ if and only if one of the half-spaces has $\beta \neq 0$, or in other words, there is an affine function h_0 dominated by g.

It thus suffices to show that if there is a half-space $H_{y,\beta,\gamma}$ containing $\operatorname{epi} g$ but not a point $(\bar{u}, \bar{\alpha})$, then there is an affine function h such that $g \geq h$ but $h(\bar{u}) > \bar{\alpha}$. If $\operatorname{epi} g \subseteq H_{y,\beta,\gamma}$, then necessarily $\beta \leq 0$. If $\beta < 0$, then the function $h(u) = \langle u, y/(-\beta) \rangle + \gamma/\beta$ will do. If $\beta = 0$, then $\operatorname{dom} g$ is contained in $\{u \mid \langle u, y \rangle \leq \gamma\}$ while $\bar{u}$ is not. It follows that g dominates the affine function $h(u) = h_0(u) + \lambda(\langle u, y \rangle - \gamma)$ for any $\lambda \geq 0$. Since $\langle \bar{u}, y \rangle > \gamma$, we have $h(\bar{u}) > \bar{\alpha}$ for λ large enough. □

Corollary A.55 *Given a convex function g closed at u, we have $y \in \partial g(u)$ if and only if $u \in \partial g^*(y)$.*

Proof By definition, $y \in \partial g(u)$ if and only if

$$g(u) + g^*(y) = \langle u, y \rangle.$$

If g is closed at u, Theorem A.54 says that this is equivalent to

$$g^{**}(u) + g^*(y) = \langle u, y \rangle,$$

which means that $u \in \partial g^*(u)$. □

The above implies that if g is a closed convex function, then the set-valued mappings $\partial g : U \rightrightarrows Y$ and $\partial g^* : Y \rightrightarrows U$ are inverses of each other.

Corollary A.56 *A closed proper sublinear function is the support function of a closed convex set.*

Proof The conjugate of a sublinear function only takes values 0 or $+\infty$, so the conjugate is an indicator function of a closed convex set. Thus, the claim follows from the biconjugate theorem. □

Corollary A.57 *Given a function $g : U \to \overline{\mathbb{R}}$, its* polar

$$g^\circ(y) := \sup_u \{\langle u, y \rangle \mid g(u) \leq 1\}$$

is a closed and sublinear function on Y. If g is sublinear, then $g^{\circ\circ} = \operatorname{cl} g$.

Proof The first claim is trivial. As to the second, note first that $g^* = \delta_{\operatorname{dom} g^*}$, by sublinearity of g, so $\operatorname{cl} g = \sigma_{\operatorname{dom} g^*}$, by Theorem A.54. Since $g^{\circ\circ} = \sigma_{\operatorname{lev}_1 g^\circ}$, it suffices to show that $\operatorname{dom} g^* = \operatorname{lev}_1 g^\circ$. Here $y \in \operatorname{dom} g^*$ means that $\langle u, y \rangle \leq g(u)$ for all $u \in U$ while $y \in \operatorname{lev}_1 g^\circ$ means that $\langle u, y \rangle \leq 1$ for all $u \in U$ with $g(u) \leq 1$. Thus, $\operatorname{dom} g^* \subseteq \operatorname{lev}_1 g^\circ$. If $y \notin \operatorname{dom} g^*$, there exists $\bar{u} \in U$ such that $\langle y, \bar{u} \rangle > g(\bar{u})$. Since g is sublinear, there exists $\alpha > 0$ such that $\langle y, \alpha\bar{u} \rangle > g(\alpha\bar{u}) = 1$, so $y \notin \operatorname{lev}_1 g^\circ$. □

Applying the biconjugate theorem to the support function of a convex set gives the following. Recall that polar of a set $D \subseteq U$ is defined by

$$D^\circ := \{y \in Y \mid \langle u, y \rangle \leq 1 \; \forall u \in D\}.$$

Corollary A.58 (Bipolar Theorem) *If D is a set with $0 \in \operatorname{cl} \operatorname{co} D$, then $D^{\circ\circ} = \operatorname{cl} \operatorname{co} D$.*

Proof Since $\sigma_D = \delta_D^*$, by definition, the biconjugate theorem gives $\sigma_D^* = \delta_{\mathrm{cl\,co}\,D}$ and $\sigma_D = \sigma_{\mathrm{cl\,co}\,D}$. On the other hand, Example A.42 gives

$$\sigma_D^* = \gamma_{D^\circ}^* = \delta_{D^{\circ\circ}}$$

so $\delta_{D^{\circ\circ}} = \delta_{\mathrm{cl}\,D}$. □

Given a convex function g and a scalar $\alpha \geq 0$, we define

$$(\alpha g)(u) := \begin{cases} \alpha g(u) & \text{if } \alpha > 0, \\ \delta_{\mathrm{cl\,dom}\,g}(u) & \text{if } \alpha = 0 \end{cases}$$

and

$$(g\alpha)(u) := \begin{cases} \alpha g(u/\alpha) & \text{if } \alpha > 0, \\ g^\infty(u) & \text{if } \alpha = 0. \end{cases}$$

For $\alpha > 0$, the above definitions coincide with the definitions of αg and $g\alpha$ in [123]. Note that, in this case, $\mathrm{epi}(g\alpha) = \alpha\, \mathrm{epi}\, g$. The cases where $\alpha = 0$ can be thought of as limits of αg and $g\alpha$, respectively, when $\alpha \searrow 0$; see [123, Section 8].

Corollary A.59 *Given a closed proper convex function g,*

$$\sigma_{\mathrm{epi}\,g^*}(u, -\alpha) = \begin{cases} (g\alpha)(u) & \textit{if } \alpha \geq 0, \\ +\infty & \textit{otherwise} \end{cases}$$

and, in particular, $g(u) = \sigma_{\mathrm{epi}\,g^*}(u, -1)$. For $\alpha \geq 0$, the functions $g\alpha$ and αg^* are conjugates of each other.

Proof Since g is closed and proper, Theorem A.54 implies that g^* is proper as well. By definition,

$$\sigma_{\mathrm{epi}\,g^*}(u, -\alpha) = \sup_{y,\beta}\{\langle u, y\rangle - \alpha\beta \mid g^*(y) \leq \beta\}.$$

If $\alpha < 0$, the right side is $+\infty$. When $\alpha = 0$, Lemma A.45 gives

$$\begin{aligned} \sigma_{\mathrm{epi}\,g^*}(u, -\alpha) &= \sup_{y,\beta}\{\langle u, y\rangle \mid g^*(y) \leq \beta\} \\ &= \sup_{y}\{\langle u, y\rangle \mid y \in \mathrm{dom}\, g^*\} = g^\infty(u). \end{aligned}$$

When $\alpha > 0$, Theorem A.54 gives

$$\sigma_{\mathrm{epi}\,g^*}(u, -\alpha) = \sup_{y}\{\langle u, y\rangle - \alpha g^*(y)\}$$

$$= \sup_y \alpha\{\langle u/\alpha, y\rangle - g^*(y)\}$$

$$= \alpha g(u/\alpha).$$

The above equalities also show that $(\alpha g^*)^* = g\alpha$. Since αg^* is closed, we have $\alpha g^* = (g\alpha)^*$, by the biconjugate theorem. □

We say that a function g is *weakly inf-compact* if $\operatorname{lev}_\alpha g$ is $\sigma(U, Y)$-compact for every $\alpha \in \mathbb{R}$. The following fundamental result goes back to [87].

Theorem A.60 *If a function $g : U \to \overline{\mathbb{R}}$ is Mackey continuous at u, then the function*

$$g_u^*(y) := g^*(y) - \langle u, y\rangle$$

is weakly inf-compact. Conversely, a closed proper convex function g is Mackey continuous at u if the function g_u^* is weakly inf-compact.

Proof By translation, we may assume that $u = 0$. The continuity implies the existence of a scalar α and a Mackey neighborhood $C \in \tau(U, Y)$ of the origin such that $g \le \delta_C + \alpha$. By definition of the Mackey topology, there exists a $\sigma(Y, U)$-compact convex set $D \subset Y$ such that $0 \in D$ and $D^\circ \subseteq C$ and thus,

$$g^*(y) \ge \sup_u\{\langle u, y\rangle - \delta_C(u) - \alpha\} \ge \sup_u\{\langle u, y\rangle - \delta_{D^\circ}(u) - \alpha\} = \sigma_{D^\circ}(y) - \alpha.$$

In particular, $\inf g^* \ge -\alpha$, so if $\beta > \inf g^*$, we have $\alpha + \beta > 0$ and then, by Corollary A.58,

$$\begin{aligned}
\operatorname{lev}_\beta g^* &\subseteq \{y \in Y \mid \sigma_{D^\circ}(y) \le \alpha + \beta\} \\
&= (\alpha + \beta)\{y \in Y \mid \sigma_{D^\circ}(y) \le 1\} \\
&= (\alpha + \beta)D
\end{aligned}$$

so g^* is inf-compact.

To prove the converse, we assume temporarily that $g^*(0) = \inf g^* = 0$. Let $D := \operatorname{lev}_\beta g^*$. If $y \notin D$, we have

$$\begin{aligned}
\gamma_D(y) :&= \inf\{\eta > 0 \mid y \in \eta D\} \\
&= \inf\{\eta > 1 \mid y \in \eta D\} \\
&= \inf\{\eta > 1 \mid g^*(y/\eta) \le \beta\} \\
&\le \inf\{\eta > 1 \mid g^*(y)/\eta \le \beta\} \\
&= g^*(y)/\beta,
\end{aligned}$$

where the inequality holds by convexity. If $y \in D$, we have $\gamma_D(y) \leq 1$ so $g^*(y) \geq 0 \geq \beta\gamma_D(y) - \beta$. Thus, $g^* \geq \beta\gamma_D - \beta$. By Theorem A.54,

$$g(u) \leq \beta\gamma_D^*(u/\beta) + \beta = \beta\delta_{D^\circ}(u/\beta) + \beta,$$

where the equality holds by Example A.42. Weak compactness of D implies that D° is Mackey neighborhood of the origin. By Theorem A.29, this implies that g is Mackey continuous the origin. This proves the claim in the case $g^*(0) = \inf g^* = 0$. The general case follows by noting that $\operatorname{argmin} g^* \neq \emptyset$, by Lemma A.22 (recall that we may assume that $u = 0$), and considering the function $\tilde{g}(u) := g(u) - \langle u, \bar{y}\rangle + g^*(\bar{y})$, where $\bar{y} \in \operatorname{argmin} g^*$. Indeed, $\tilde{g}^*(0) = \inf \tilde{g}^* = 0$. while g is continuous at the origin if and only if $\tilde{g}$ is so. □

Corollary A.61 (Banach-Alaoglu) *A closed convex set is a Mackey-neighborood of the origin if and only if its polar is weakly compact.*

Proof Let C be a closed convex and set $g := \delta_C$. The set C is a neighborhood of the origin if and only if g is continuous at the origin. By Theorem A.60, this is equivalent to σ_C being inf-compact. Since σ_C is positively homogeneous, it is inf-compact if and only if $\operatorname{lev}_1 \sigma_C$ is compact. □

Corollary A.62 *If g is a convex function Mackey continuous at u, then $\partial g(u)$ is nonempty and $\sigma(Y, U)$-compact.*

Proof By definition,

$$\partial g(u) = \operatorname*{argmin}_{y \in Y}\{g^*(y) - \langle u, y\rangle\},$$

which is nonempty and compact by Theorem A.60 and Lemma A.22. □

Corollary A.63 *If g is a convex function relatively Mackey continuous at u, then $\partial g(u) \neq \emptyset$.*

Proof Translating if necessary, we may assume that $u = 0$. We equip $\operatorname{aff}\operatorname{dom} g$ with the relative topology and pair $\operatorname{aff}\operatorname{dom} g$ with its topological dual. By Corollary A.62, there exists a continuous linear functional $\tilde{y}$ on $\operatorname{aff}\operatorname{dom} g$ such that

$$g(u') \geq g(u) + \langle u' - u, \tilde{y}\rangle \quad \forall u' \in \operatorname{aff}\operatorname{dom} g.$$

By Corollary A.14, there exists a $y \in Y$ that coincides with $\tilde{y}$ on $\operatorname{aff}\operatorname{dom} g$. Since $\operatorname{dom} g \subseteq \operatorname{aff}\operatorname{dom} g$, we have $y \in \partial g(u)$. □

Combining Corollaries A.63 and A.36 gives the following.

Corollary A.64 *If g is a lsc convex function such that* $\operatorname{aff}\operatorname{dom} g$ *is barreled, then g is subdifferentiable or identically $-\infty$ throughout* $\operatorname{rcore}\operatorname{dom} g$.

Combining Corollary A.63 and Theorem A.39 gives the following.

Theorem A.65 *Let X be a Fréchet space, let F be a lsc convex function on $X \times U$ and let*

$$\varphi(u) = \inf_{x \in X} F(x, u).$$

If $\operatorname{aff}\operatorname{dom}\varphi$ *is barreled, then* φ *is either subdifferentiable or identically* $-\infty$ *throughout* $\operatorname{rcore}\operatorname{dom}\varphi$*. The set* $\operatorname{aff}\operatorname{dom}\varphi$ *is barreled, in particular, if it is closed and U is Fréchet.*

Lemma A.66 *Let g be a convex function such that g^* is proper and* $\operatorname{aff}\operatorname{dom} g$ *is closed. Then* $0 \in \operatorname{rcore}\operatorname{dom} g$ *if and only if* $\operatorname{lev}_0(g^*)^\infty$ *is linear.*

Proof By Lemma A.4, $0 \in \operatorname{rcore}\operatorname{dom} g$ means that $\operatorname{pos}\operatorname{dom} g$ is linear. By Corollary A.58, a closed convex cone is linear if and only if its polar is so. We have

$$\begin{aligned}(\operatorname{pos}\operatorname{dom} g)^\circ &= \{y \in Y \mid \sigma_{\operatorname{pos}\operatorname{dom} g}(y) \leq 0\}\\ &= \{y \in Y \mid \sigma_{\operatorname{dom} g}(y) \leq 0\}\\ &= \{y \in Y \mid (g^*)^\infty(y) \leq 0\},\end{aligned}$$

where the last equality holds by Lemma A.45. □

We end this section with the Kreps-Yan theorem which was used in the proof of the "fundamental theorem of asset pricing" in Theorem 4.29. The theorem provides a strict form of separation of two convex cones. A topological space is *Lindelöf* if every collection of open sets that covers the whole space has a countable subcollection that still covers the space. The following is essentially [141, Theorem 1.1] but we slightly relax the assumption that the space Y be Banach.

Theorem A.67 (Kreps-Yan) *Assume that U is Lindelöf in the $\sigma(U, Y)$-topology and Y is Banach in a topology at least as strong as $\sigma(Y, U)$. Let K and C be $\sigma(U, Y)$-closed convex cones in U such that $C \cap K = \{0\}$, $-K \subseteq C$ and there exists y_0 such that $\langle u, y_0\rangle > 0$ for every $u \in K \setminus \{0\}$. Then there exists $y \in Y$ such that*

$$\begin{aligned}\langle u, y\rangle &\leq 0 \quad \forall u \in C,\\ \langle u, y\rangle &> 0 \quad \forall u \in K \setminus \{0\}.\end{aligned}$$

Proof Let $u \in K \setminus \{0\}$. By Theorem A.50, there exists $y_u \in Y$ with $\langle u', y_u\rangle < \langle u, y_u\rangle$ for every $u' \in C$. Since C is a cone, this means that $\langle u', y_u\rangle \leq 0$ for every $u' \in C$ and $\langle u, y_u\rangle > 0$. In other words, $y_u \in C^\circ$ and $\langle u, y_u\rangle > 0$. Since $-K \subset C$, we also have $y_u \in -K^\circ$. Let

$$A_u := \{u' \in U \mid \langle u', y_u\rangle > 0\} \quad \text{for } u \in K \setminus \{0\},$$

and $A_0 := \{u' \in U \mid |\langle u', y_0\rangle| < 1\}$. Then the family $(A_u)_{u\in K}$ is an open cover of K. Since closed subsets of a Lindelöf space are Lindelöf, there exists a sequence $(u^\nu)_{\nu\in\mathbb{N}}$ in K such that $(A_{u^\nu})_{\nu\in\mathbb{N}}$ is an open cover of K.

Since Y is Banach in a stronger topology, the series $\sum_{\nu=1}^{\infty} \frac{2^{-\nu}}{\|y_{u^\nu}\|} y_{u^\nu}$ converges to some $y \in Y$. Since $y_u \in C^\circ \cap (-K^\circ)$ for all $u \in K \setminus \{0\}$, we have $y \in C^\circ \cap (-K^\circ)$. Given $u \in K \setminus \{0\}$, we have $\langle u, y_0\rangle > 0$, by assumption, so there exists $\lambda > 0$ such that $\lambda u \notin A_0$. Thus, $\lambda u \in A_{u^\nu}$ for some $\nu \geq 1$, so $\langle \lambda u, y\rangle > 0$ and thus, $\langle u, y\rangle > 0$. □

Remark A.68 *A Banach space is said to be* weakly compactly generated *if it contains a weakly compact set whose linear hull is dense. A space is said to be* σ-compact *if it is a countable union of compact sets.*

1. *A weakly compactly generated Banach space is Lindelöf in the weak topology.*
2. *The Cartesian product of a σ-compact space and a Lindelöf space is Lindelöf.*

Proof The first claim is the main result of [151]. As to the second, we may assume that the first space is compact, since, clearly, countable unions of Lindelöf spaces are Lindelöf. Thus, the claim follows from the exercise in [88, p. 194]. □

Example A.69 *The space $L^\infty \times L^1$ is $\sigma(L^\infty \times L^1, L^1 \times L^\infty)$-Lindelöf.*

Proof By Banach-Alaoglu, L^∞ is σ-compact with respect to $\sigma(L^\infty, L^1)$. By Remark A.68, it suffices to show that L^1 is weakly compactly generated. The unit ball of L^∞ is weakly compact in L^1 and its linear span is dense in L^1. Indeed, the support function of the unit ball is $y \mapsto E[|y|]$, which is $\tau(L^\infty, L^1)$-continuous, by Corollary 5.17, so compactness follows Theorem A.60. The linear span is L^∞, which is clearly dense in L^1. □

A.9 Duality in Optimization

This section reviews the conjugate duality framework of Rockafellar (see [123, Part VI] and [127]) which provides the foundation for the duality theory developed in Chaps. 3–5 of this book. Conjugate duality is the most general duality framework in convex optimization. Well-known duality results, e.g. on Fenchel's model or linear programming are easy consequences of the general theory of conjugate duality. Our formulation below deviates slightly from [127] by adding an auxiliary parameter v which is in the dual space of the primal optimization variables. This extra parameter does not add to the generality of the theory, but it allows for convenient derivation of various calculus rules of conjugation and subdifferentiation of convex functions; see [127, Section 9] and Sect. A.10 below.

Consider a general convex optimization problem

$$\text{minimize} \quad f(x) \quad \text{over } x \in X, \tag{A.1}$$

where f is an extended real-valued convex function on a vector space X. We assume throughout that X is in separating duality with a vector space V. Given another pair (U, Y) of vector spaces in separating duality, we say, following [142] and [39], that a convex function $F : X \times U \to \overline{\mathbb{R}}$ is a *Rockafellian* for problem (A.1) if

$$f(x) = F(x, \bar{u}) - \langle x, \bar{v} \rangle$$

for some $\bar{u} \in U$ and $\bar{v} \in V$. Problem (A.1) can then be written as

$$\text{minimize} \quad F(x, \bar{u}) - \langle x, \bar{v} \rangle \quad \text{over } x \in X.$$

Within the duality framework described below, this will be called the *primal problem*. The *dual problem* associated with F is

$$\text{maximize} \quad \langle \bar{u}, y \rangle - F^*(\bar{v}, y) \quad \text{over } y \in Y.$$

The primal *optimum value function* is defined by

$$\varphi_v(u) := \inf_{x \in X} \{F(x, u) - \langle x, v \rangle\}$$

and the dual optimum value function by

$$\gamma_u(v) := \inf_{y \in Y} \{F^*(v, y) - \langle u, y \rangle\}.$$

By Theorem A.1, these functions are convex. By Fenchel's inequality,

$$F(x, u) + F^*(v, y) \geq \langle x, v \rangle + \langle u, y \rangle \quad \forall (x, u) \in X \times U, \ (v, y) \in V \times Y,$$

so

$$\varphi_v(u) \geq -\gamma_u(v) \quad \forall u \in U, v \in V.$$

If $\varphi_{\bar{v}}(\bar{u}) > -\gamma_{\bar{u}}(\bar{v})$, a *duality gap* is said to exist. Note that $\varphi_v^*(y) = F^*(v, y)$ and $\gamma_u^*(x) = F^{**}(x, u)$, where $F^{**} = \operatorname{cl} F$, by Theorem A.54. In particular, the dual problem can be written as

$$\text{maximize} \quad \langle \bar{u}, y \rangle - \varphi_{\bar{v}}^*(y) \quad \text{over } y \in Y.$$

The properties of conjugates and subgradients from the previous section thus imply that the absence of a duality gap and existence of dual solutions come down to lower semicontinuity and subdifferentiability of $\varphi_{\bar{v}}$ at $\bar{u}$; see Theorems A.71 and A.72 below.

The *Lagrangian* associated with F is the convex-concave function on $X \times Y$ given by

$$L(x, y) := \inf_{u \in U} \{F(x, u) - \langle u, y \rangle\}.$$

Clearly, the conjugate of F can be expressed as

$$F^*(v, y) = \sup_{x \in X} \{\langle x, v \rangle - L(x, y)\}.$$

The *Lagrangian minimax problem* to find a saddle value and/or a saddle point of the convex-concave function

$$L_{\bar{v},\bar{u}}(x, y) := L(x, y) - \langle x, \bar{v} \rangle + \langle \bar{u}, y \rangle.$$

We always have

$$\inf_x \sup_y L_{\bar{v},\bar{u}}(x, y) \geq \sup_y \inf_x L_{\bar{v},\bar{u}}(x, y).$$

When the equality holds, the common value is called the *minimax* or the *saddle value* of $L_{\bar{v},\bar{u}}$. A pair $(\bar{x}, \bar{y})$ is called a *saddle point* of $L_{\bar{v},\bar{u}}$ if

$$L_{\bar{v},\bar{u}}(x, \bar{y}) \geq L_{\bar{v},\bar{u}}(\bar{x}, \bar{y}) \geq L_{\bar{v},\bar{u}}(\bar{x}, y) \quad \forall x \in X, y \in Y.$$

When a saddle point $(\bar{x}, \bar{y})$ exists, the saddle value exists and equals $L_{\bar{v},\bar{u}}(\bar{x}, \bar{y})$.

We have

$$\langle \bar{u}, y \rangle - F^*(\bar{v}, y) = \inf_x L_{\bar{v},\bar{u}}(x, y) \tag{A.2}$$

so the dual problem can be viewed as the maximization-half of the Lagrangian minimax problem. When F is closed in u in the sense that $F(x, \cdot) = \operatorname{cl} F(x, \cdot)$ for all x, the biconjugate theorem gives

$$F(x, \bar{u}) - \langle x, \bar{v} \rangle = \sup_y L_{\bar{v},\bar{u}}(x, y),$$

so the primal problem is the minimization-half of the minimax problem. In general,

$$(\operatorname{cl}_u F)(x, \bar{u}) - \langle x, \bar{v} \rangle = \sup_y L_{\bar{v},\bar{u}}(x, y), \tag{A.3}$$

where $(\operatorname{cl}_u F)(x, \cdot) := \operatorname{cl} F(x, \cdot)$. Clearly,

$$\operatorname{cl} F \leq \operatorname{cl}_u F \leq F.$$

Recalling that $\varphi_v^*(y) = F^*(v, y)$ and $\gamma_u^*(x) = (\mathrm{cl}\, F)(x, u)$, Theorem A.54 yields the following.

Lemma A.70 *We have*

$$\begin{aligned}
(\mathrm{cl}\,\varphi_v)(u) &= \sup_y\{\langle u, y\rangle - \varphi_v^*(y)\} \\
&= \sup_y\{\langle u, y\rangle - F^*(v, y)\} \\
&= -\gamma_u(v) \\
&\le -(\mathrm{cl}\,\gamma_u)(v) \\
&= \inf_x\{\gamma_u^*(x) - \langle x, v\rangle\} \\
&= \inf_x\{(\mathrm{cl}\, F)(x, u) - \langle x, v\rangle\} \\
&\le \inf_x\{(\mathrm{cl}_u\, F)(x, \bar{u}) - \langle x, v\rangle\} \\
&\le \inf_x\{F(x, u) - \langle x, v\rangle\} \\
&= \varphi_v(u).
\end{aligned}$$

The following is an immediate consequence of Lemma A.70 and Eqs. (A.2) and (A.3).

Theorem A.71 *The implications* $1 \Leftrightarrow 2 \Rightarrow 3 \Rightarrow 4$ *hold among the following conditions:*

1. *there is no duality gap;*
2. $\varphi_{\bar{v}}$ *is closed at* $\bar{u}$*;*
3. $L_{\bar{v},\bar{u}}$ *has a saddle value;*
4. $\gamma_{\bar{u}}$ *is closed at* $\bar{v}$.

If F *is closed in* u*, then* $1 \Leftrightarrow 2 \Leftrightarrow 3$. *If* F *is closed, then* $1 \Leftrightarrow 2 \Leftrightarrow 3 \Leftrightarrow 4$.

Theorem A.72 *Assume that* $\varphi_{\bar{v}}(\bar{u}) < \infty$. *The implications* $1 \Leftrightarrow 2 \Rightarrow 3 \Rightarrow 4$ *hold among the following conditions:*

1. *there is no duality gap and* y *solves the dual;*
2. *either* $y \in \partial\varphi_{\bar{v}}(\bar{u})$ *or* $\varphi_{\bar{v}}(\bar{u}) = -\infty$*;*
3. $\inf_x \sup_y L_{\bar{v},\bar{u}}(x, y) = \inf_x L_{\bar{v},\bar{u}}(x, y)$*;*
4. $\gamma_{\bar{u}}$ *is closed at* $\bar{v}$ *and* y *solves the dual.*

If F *is closed in* u*, then* $1 \Leftrightarrow 2 \Leftrightarrow 3$. *If* F *is closed, then* $1 \Leftrightarrow 2 \Leftrightarrow 3 \Leftrightarrow 4$.

Proof Condition 1 means that either $\varphi_{\bar{v}}(\bar{u}) + \varphi_{\bar{v}}^*(y) = \langle \bar{u}, y\rangle$ or $\varphi_{\bar{v}}(\bar{u}) = -\infty$. Indeed, in the latter case, every $y \in Y$ solves the dual. This proves the equivalence of 1 and 2. The remaining claims follow from Lemma A.70 and Eqs. (A.2) and (A.3). □

Theorem A.73 *The implications* $1 \Leftrightarrow 2 \Leftrightarrow 3 \Rightarrow 4 \Leftrightarrow 5 \Rightarrow 6 \Leftrightarrow 7$ *hold among the following conditions:*

1. *There is no duality gap,* x *solves the primal,* y *solves the dual and both problems are feasible;*
2. $y \in \partial\varphi_{\bar{v}}(\bar{u})$ *and* x *solves the primal;*
3. $(\bar{v}, y) \in \partial F(x, \bar{u})$*;*
4. (x, y) *is a saddle point of* $L_{\bar{u},\bar{v}}$*;*
5. $\bar{v} \in \partial_x L(x, y)$ *and* $\bar{u} \in \partial_y[-L](x, y)$*;*
6. $x \in \partial\gamma_{\bar{u}}(\bar{v})$ *and* y *solves the dual;*
7. $(x, \bar{u}) \in \partial F^*(\bar{v}, y)$.

If F *is closed in* u*, then* $1 \Leftrightarrow 2 \Leftrightarrow 3 \Leftrightarrow 4 \Leftrightarrow 5$*. If* F *is closed, then* $1 \Leftrightarrow 2 \Leftrightarrow 3 \Leftrightarrow 4 \Leftrightarrow 5 \Leftrightarrow 6 \Leftrightarrow 7$.

Proof By Lemma A.70, 1 means that

$$\langle \bar{u}, y\rangle - \varphi_{\bar{v}}^*(y) = \langle \bar{u}, y\rangle - F^*(\bar{v}, y) = F(x, \bar{u}) - \langle x, \bar{v}\rangle = \varphi_{\bar{v}}(\bar{u}),$$

which is equivalent to both 2 and 3. Condition 5 is simply a reformulation of 4. By Lemma A.70, 7 means that

$$\langle \bar{u}, y\rangle - F^*(\bar{v}, y) = -\gamma_{\bar{u}}(\bar{v}) = \gamma_{\bar{u}}^*(x) - \langle x, \bar{v}\rangle = (\operatorname{cl} F)(x, \bar{u}) - \langle x, \bar{v}\rangle.$$

Since $\gamma_{\bar{u}}^*(x) = (\operatorname{cl} F)(x, \bar{u})$, this is equivalent to 6. By Lemma A.70, 1 implies

$$\langle \bar{u}, y\rangle - F^*(\bar{v}, y) = (\operatorname{cl}_u F)(x, \bar{u}) - \langle x, \bar{v}\rangle \tag{A.4}$$

which is 4. If $(\operatorname{cl}_u F) = F$, this is, by Lemma A.70, equivalent to 1. The equality (A.4) clearly implies

$$\langle \bar{u}, y\rangle - F^*(\bar{v}, y) = (\operatorname{cl} F)(x, \bar{u}) - \langle x, \bar{v}\rangle$$

which is 7. If $\operatorname{cl} F = F$, this is, by Lemma A.70, equivalent to 1. □

In [123, 127], condition 5 of Theorem A.73 was referred to as "Kuhn-Tucker conditions", although condition 5 above is far more general than the original Kuhn-Tucker conditions which only apply to optimization problems with explicit equality and inequality constraints; see Example A.76 below. It has now become customary to refer to the optimality conditions in Example A.76 as *Karush-Kuhn-Tucker (KKT) conditions* to also credit the work of Karush who introduced the conditions a few years before Kuhn and Tucker. The generalization in condition 5 of Theorem A.73 is due to Rockafellar [123, 127]. Accordingly, we will refer to condition 5 in Theorem A.73 as the KKTR-conditions.

Combining Theorem A.72 with general properties of convex functions on locally convex spaces, we obtain the following.

Lemma A.74 *Assume that F is closed and proper and that, for every $\bar{v} \in V$, $\varphi_{\bar{v}}$ is either relatively Mackey continuous at $\bar{u} \in U$ or $\varphi_{\bar{v}}(\bar{u}) = -\infty$. Then, for every $\bar{v} \in V$, there exists y such that all the conditions in Theorem A.72 hold and*

1. *$\gamma_{\bar{u}}$ is closed and proper and the infimum in its definition is attained,*
2. *the recession function of $\gamma_{\bar{u}}$ is given by*

$$\gamma_{\bar{u}}^{\infty}(v) = \inf_{y \in Y}\{(F^*)^{\infty}(v, y) - \langle \bar{u}, y \rangle\},$$

where the infimum is attained.

In particular, $\gamma_{\bar{u}} = F(\cdot, \bar{u})^*$.

Proof Since F is closed, all the conditions in Theorem A.72 are equivalent. By Corollary A.63, $\partial\varphi_{\bar{v}}(\bar{u}) \neq \emptyset$ or $\varphi_{\bar{v}}(\bar{u}) = -\infty$, so there exists y such that condition 2 in Theorem A.72 holds. This proves the first claim. In particular, $\gamma_{\bar{u}}$ is closed and the infimum in its definition is attained. Since $F(\cdot, \bar{u})$ is proper and, by definition, $F(\cdot, \bar{u})^* = \gamma_{\bar{u}}$, Theorem A.54 now implies that $\gamma_{\bar{u}}$ is proper.

Since F is closed and proper, Theorem A.54 implies that there exists $(\bar{v}, \bar{y}) \in \operatorname{dom} F^*$. By Fenchel's inequality,

$$F(x, u) + F^*(\bar{v}, \bar{y}) \geq \tilde{F}(x, u),$$

where $\tilde{F}(x, u) := \delta_{\operatorname{dom} F}(x, u) + \langle x, \bar{v} \rangle + \langle u, \bar{y} \rangle$. Thus,

$$\varphi_v(u) + F^*(\bar{v}, \bar{y}) \geq \tilde{\varphi}_v(u),$$

where

$$\tilde{\varphi}_v(u) := \inf_x\{\tilde{F}(x, u) - \langle x, v \rangle\}.$$

We have $\operatorname{dom} \tilde{\varphi}_v = \operatorname{dom} \varphi_v$ and $\varphi_v(\bar{u}) < \infty$, by assumption. When $\tilde{\varphi}_v(\bar{u})$ is finite, the above inequality implies that $\varphi_v(\bar{u})$ is finite as well, and then, by Corollary A.30, the relative continuity of φ_v at $\bar{u}$ implies that of $\tilde{\varphi}_v$. Thus, by Corollary A.63, either $\partial\tilde{\varphi}_v(\bar{u}) \neq \emptyset$ or $\varphi_v(\bar{u}) = -\infty$. By Theorem A.72,

$$\inf_{y \in Y}\{\tilde{F}^*(v, y) - \langle \bar{u}, y \rangle\} = \sup_{x \in X}\{\langle x, v \rangle - \tilde{F}(x, \bar{u})\} \quad \forall v \in V, \tag{A.5}$$

where the infimum is attained. By Theorem A.54, properness of F implies that of F^*, so Lemma A.45 gives

$$\begin{aligned}
\tilde{F}^*(v, y) &= \sup_{x,u}\{\langle x, v \rangle + \langle u, y \rangle - \tilde{F}(x, u)\} \\
&= \sup_{(x,u) \in \operatorname{dom} F}\{\langle x, v - \bar{v} \rangle + \langle u, y - \bar{y} \rangle\}
\end{aligned}$$

$$
\begin{aligned}
&= \sigma_{\mathrm{dom}\, F}(v-\bar v, y-\bar y)\\
&= (F^*)^\infty(v-\bar v, y-\bar y).
\end{aligned}
$$

Since $\gamma_{\bar u}$ is the conjugate of $F(\cdot, \bar u)$, Lemma A.45 also gives

$$
\begin{aligned}
\sup_{x\in X}\{\langle x, v\rangle - \tilde F(x, \bar u)\} &= \sup_{x\in \mathrm{dom}\, F(\cdot, \bar u)}\{\langle x, v-\bar v\rangle - \langle \bar u, \bar y\rangle\}\\
&= \sigma_{\mathrm{dom}\, F(\cdot, \bar u)}(v-\bar v) - \langle \bar u, \bar y\rangle\\
&= \gamma_{\bar u}^\infty(v-\bar v) - \langle \bar u, \bar y\rangle.
\end{aligned}
$$

Substituting into (A.5), we get

$$
\begin{aligned}
\gamma_{\bar u}^\infty(v-\bar v) - \langle \bar u, \bar y\rangle &= \inf_{y\in Y}\{(F^*)^\infty(v-\bar v, y-\bar y) - \langle \bar u, y\rangle\}\\
&= -\langle \bar u, \bar y\rangle + \inf_{y\in Y}\{(F^*)^\infty(v-\bar v, y) - \langle \bar u, y\rangle\}.
\end{aligned}
$$

Since $v \in V$ was arbitrary, this gives the expression for $\gamma_{\bar u}^\infty$. □

Note that $\mathrm{dom}\, \varphi_v$ does not depend on v. We denote this set by $\mathrm{dom}\, \varphi$.

Theorem A.75 *Let X and U be Fréchet and let V and Y, respectively, be the corresponding topological duals. Assume that F is closed and proper, $\bar u \in \mathrm{rcore}\,\mathrm{dom}\, \varphi$ and that $\mathrm{aff}\,\mathrm{dom}\, \varphi$ is closed. Then, for every $\bar v \in V$, $\varphi_{\bar v}$ is either relatively Mackey continuous at $\bar u$ or $\varphi_{\bar v}(\bar u) = -\infty$, there exists y such that all the conditions in Theorem A.72 hold and*

1. *$\gamma_{\bar u}$ is closed and proper and the infimum in its definition is attained,*
2. *the recession function of $\gamma_{\bar u}$ is given by*

$$
\gamma_{\bar u}^\infty(v) = \inf_{y\in Y}\{(F^*)^\infty(v, y) - \langle \bar u, y\rangle\},
$$

where the infimum is attained.

In particular, $\gamma_{\bar u} = F(\cdot, \bar u)^*$.

Proof By Remark A.48, Mackey topologies $\tau(X, V)$ and $\tau(U, Y)$ coincide with the Fréchet topologies of X and U, respectively. We apply Theorem A.39 to the function $(x, u) \mapsto F(x, u) - \langle x, v\rangle$. The assumptions here imply that, for every $\bar v \in V$, $\varphi_{\bar v}$ is either relatively Mackey continuous at $\bar u$ or $\varphi_{\bar v}(\bar u) = -\infty$. Thus, the claims follow from Lemma A.74. □

By Lemma A.5, the condition $\bar u \in \mathrm{rcore}\,\mathrm{dom}\, \varphi$ in Theorem A.75 holds if the primal problem is *strictly feasible* in the sense that there is an $\bar x \in X$ with $(\bar x, \bar u) \in \mathrm{rcore}\,\mathrm{dom}\, F$. If U is finite dimensional, then the closedness assumption in Theorem A.75 is redundant. The closedness assumption holds also if $\bar u \in \mathrm{int}\,\mathrm{dom}\, \varphi$

which also implies $\bar{u} \in \operatorname{core} \operatorname{dom} \varphi$. The interiority condition generalizes classical Slater-type conditions which ask that $\bar{u} \in \operatorname{int} \operatorname{dom} F(\bar{x}, \cdot)$ for some $\bar{x} \in X$. In general, aff dom φ is the projection of aff dom F to U. The closedness of projections of linear spaces is a classical question in functional analysis and various sufficient conditions have been given. Section 5.3.1 gives examples in the context of stochastic optimization.

Example A.76 (Lagrangian Duality) *Let X be a Fréchet space, $U = \mathbb{R}^m$, f_j, $j = 0, \dots, l$ lsc convex and f_j, $j = l+1, \dots, m$ continuous affine functions on X. Consider the problem*

$$
\begin{array}{lll}
\text{minimize} & f_0(x) & \text{over } x \in X, \\
\text{subject to} & f_j(x) \le 0 & j = 1, \dots, l, \\
& f_j(x) = 0 & j = l+1, \dots, m.
\end{array}
$$

This fits the general duality framework with V the topological dual of X, $Y = \mathbb{R}^m$, the Rockafellian

$$
F(x, u) = \begin{cases} f_0(x) & \text{if } f_j(x) + u_j \le 0 \text{ for } j = 1, \dots, l, \\ & \quad f_j(x) + u_j = 0 \text{ for } j = l+1, \dots, m, \\ +\infty & \text{otherwise,} \end{cases}
$$

$\bar{v} = 0$ and $\bar{u} = 0$. The Lagrangian becomes

$$
L(x, y) = \begin{cases} +\infty & \text{if } x \notin \bigcap_{j=0}^{l} \operatorname{dom} f_j, \\ f_0(x) + \sum_{j=1}^{m} y_j f_j(x) & \text{if } x \in \bigcap_{j=0}^{l} \operatorname{dom} f_j \text{ and } y \in \mathbb{R}^l_+ \times \mathbb{R}^{m-l}, \\ -\infty & \text{otherwise.} \end{cases}
$$

Assume that there exists a feasible $\bar{x} \in X$ such that $f_j(\bar{x}) < 0$ for $j = 1, \dots, l$ and either $m = l$ (no equality constraints) or $\bar{x} \in \bigcap_{j=0}^{l} \operatorname{rcore} \operatorname{dom} f_j$. Then the primal optimum value equals $\sup_y \inf_x L(x, y)$ where the supremum is attained. Moreover, an $x \in X$ solves the primal problem if and only if it is feasible and there exists $y \in \mathbb{R}^m$ such that

$$
\partial_x [f_0 + \sum_{j=1}^{m} y_j f_j](x) \ni 0,
$$

$$
y_j \ge 0, \quad y_j f_j(x) = 0 \quad j = 1, \dots, l.
$$

Proof To prove the first claim, it suffices, by Theorem A.75, to show that $0 \in \operatorname{rcore} \operatorname{dom} \varphi$, since aff dom φ is automatically closed in $\mathbb{R}^m$. If there are no equality constraints, $\varphi(u)$ is bounded from above by the constant $f_0(\bar{x})$ on the set $\{u \in \mathbb{R}^m \mid$

$f_j(\bar{x}) + u_j \le 0,\ j = 1, \ldots, m\}$ which is a neighborhood of the origin in $\mathbb{R}^m$, so $0 \in \operatorname{rcore} \operatorname{dom} \varphi$.

In general, $\operatorname{dom} \varphi$ is the projection to $\mathbb{R}^m$ of $\operatorname{dom} F = (\operatorname{dom} f_0 \times U) \bigcap C$, where

$$C := \{(x, u) \in X \times \mathbb{R}^m \mid (x, -u_j) \in \operatorname{epi} f_j,\ j = 1, \ldots, l,$$
$$f_j(x) + u_j = 0,\ j = l + 1, \ldots, m\}.$$

By Example A.10,

$$\operatorname{rcore} \operatorname{epi} f_j = \{(x, u) \in X \times \mathbb{R} \mid x \in \operatorname{rcore} \operatorname{dom} f_j,\ f_j(x) < u\},$$

so, by Lemma A.5, the assumptions on $\bar{x}$ give

$$\operatorname{rcore} C = \{(x, u) \in X \times \mathbb{R}^m \mid x \in \operatorname{rcore} \operatorname{dom} f_j,\ f_j(x) + u_j < 0,\ j = 1, \ldots, l,$$
$$f_j(x) + u_j = 0,\ j = l + 1, \ldots, m\}.$$

By Lemma A.5 again, $\operatorname{rcore} \operatorname{dom} F = (\operatorname{rcore} \operatorname{dom} f_0 \times U) \bigcap \operatorname{rcore} C$, so $(\bar{x}, 0) \in \operatorname{rcore} \operatorname{dom} F$, by assumption. By Lemma A.5, $0 \in \operatorname{rcore} \operatorname{dom} \varphi$. By Theorem A.73, the first claim implies the second one. □

Theorem A.77 *Assume that $f : \mathbb{R}^n \times \mathbb{R}^m \to \overline{\mathbb{R}}$ is a closed proper convex function such that*

$$N := \{x \in \mathbb{R}^n \mid f^\infty(x, 0) \le 0\}$$

is a linear space. Then

$$p(u) := \inf_{x \in \mathbb{R}^n} f(x, u)$$

is a closed proper convex function,

$$p^\infty(u) = \inf_{x \in \mathbb{R}^n} f^\infty(x, u),$$

and, for every $u \in \mathbb{R}^m$, both infimums are attained by an $x \in N^\perp$.

Proof We apply Theorem A.75 with $F(x, u) := f^*(u, x)$ and $\bar{u} = 0$. In a finite-dimensional space, linear sets are closed, while, by Lemma A.66, $0 \in \operatorname{rcore} \operatorname{dom} \varphi$ if and only if $\operatorname{lev}_0(\varphi^*)^\infty$ is linear. By Lemma A.45,

$$(\varphi^*)^\infty(y) = \sigma_{\operatorname{dom} \varphi}(y) = \sigma_{\operatorname{dom} F}(0, y) = (F^*)^\infty(0, y) = f^\infty(y, 0),$$

so $\operatorname{lev}_0(\varphi^*)^\infty$ is linear, by assumption. By Theorem A.75, the function p is closed and proper and both infimums are attained. By Remark A.18, the function $f(u, \cdot)$

is constant with respect to N, so the minimizers can be projected to $N^{\perp}$ without changing the function value. □

We close this section by some general remarks on saddle functions. The Lagrangian L is not necessarily closed in x but $-L$ is always closed in y. By the biconjugate theorem,

$$(\mathrm{cl}_x L)(x, y) = \sup_v \{\langle x, v\rangle - F^*(v, y)\},$$

where $(\mathrm{cl}_x L)$ is defined by $(\mathrm{cl}_x L)(\cdot, y) := \mathrm{cl}\, L(\cdot, y)$ for every $y \in Y$. When F is closed in u,

$$F(x, u) = \sup_y \{L(x, y) + \langle u, y\rangle\}$$

and, when F is closed,

$$F(x, u) = \sup_y \{(\mathrm{cl}_x L)(x, y) + \langle u, y\rangle\}.$$

The following can be found in [126].

Theorem A.78 *If F is closed, then all saddle functions between* $\mathrm{cl}_x L$ *and L have the same saddle value and saddle points.*

Proof Given a function $(x, y) \mapsto \tilde{L}(x, y)$, we define

$$(\mathrm{cl}_y \tilde{L})(x, \cdot) := -\mathrm{cl}(-\tilde{L})(x, \cdot) \quad \forall x \in X.$$

Since the infimums of a function and of its closure coincide, Theorem A.54 gives

$$\begin{aligned}\sup_y \{\langle u, y\rangle + (\mathrm{cl}_y \mathrm{cl}_x L)(x, y)\} &= \sup_y \{\langle u, y\rangle + (\mathrm{cl}_x L)(x, y)\} \\ &= \sup_{v,y} \{\langle x, v\rangle + \langle u, y\rangle - F^*(v, y)\} \\ &= F(x, u).\end{aligned}$$

As already noted, $F(x, u) = \sup_y \{L(x, y) + \langle u, y\rangle\}$. Thus, by Theorem A.54 again, $\mathrm{cl}_y \mathrm{cl}_x L = L$. It follows that

$$\inf_x L(x, y) = \inf_x (\mathrm{cl}_x L)(x, y) \quad \forall y \in Y$$

and

$$\sup_y L(x, y) = \sup_y (\mathrm{cl}_y \, \mathrm{cl}_x \, L)(x, y) = \sup_y (\mathrm{cl}_x \, L)(x, y) \quad \forall x \in X.$$

The claims now follow from the fact that saddle values and saddle points are characterized by the expressions on the left. □

A.10 Calculating Conjugates and Subgradients

As shown in [127, Section 9], the general duality theory yields powerful rules for calculating conjugates and subgradients of functions that have been constructed by algebraic operations from other functions. This section extends some of the calculus rules given in [127] by employing the slightly more general sufficient conditions given in the previous section. The following simple consequence of Theorems A.72 and A.73 will be useful in this.

Theorem A.79 *Assume that, for all v, either $\partial \varphi_v(\bar{u}) \neq \emptyset$ or $\varphi_v(\bar{u}) = -\infty$. Then*

$$F(\cdot, \bar{u})^*(v) = \inf_y \{F^*(v, y) - \langle \bar{u}, y\rangle\},$$

where the infimum is attained, and

$$\partial_x F(x, \bar{u}) = \{v \in V \mid \exists y : \ (v, y) \in \partial F(x, \bar{u})\} \quad \forall x.$$

If F is closed in u, then

$$\partial_x F(x, \bar{u}) = \left\{v \in V \;\middle|\; \exists y : \ v \in \partial_x L(x, y), \ \bar{u} \in \partial_y[-L](x, y)\right\} \quad \forall x.$$

If F is closed, then

$$\partial_x F(x, \bar{u}) = \left\{v \in V \;\middle|\; \exists y : \ (x, \bar{u}) \in \partial F^*(v, y)\right\} \quad \forall x.$$

Proof By definition,

$$\varphi_v(u) = \inf_x \{F(x, u) - \langle x, v\rangle\} = F(\cdot, u)^*(v),$$

so the expression for $F(\cdot, \bar{u})^*$ and the attainment of the infimum follow from the equivalences of 1 and 2 in Theorem A.72. We always have

$$\partial_x F(x, u) \supseteq \{v \in V \mid \exists y : \ (v, y) \in \partial F(x, u)\}$$

so if the left side is empty, so too is the expression on the right. We have $v \in \partial_x F(x,u)$ if and only if x achieves the infimum in the definition of $\varphi_v(u)$. By Theorem A.73, this is equivalent to the existence of a y such that $(v,y) \in \partial F(x,u)$. The remaining two expressions follow from the equivalent formulations of the condition $(v,y) \in \partial F(x,u)$ in Theorem A.73. □

The following was proved for Banach spaces in [3] and extended to Fréchet spaces in [160, Theorem 2.8.7].

Corollary A.80 *Let f_1, f_2 be closed proper convex functions on a Fréchet space X such that* $\operatorname{aff}(\operatorname{dom} f_1 - \operatorname{dom} f_2)$ *is closed and*

$$0 \in \operatorname{rcore}(\operatorname{dom} f_1 - \operatorname{dom} f_2).$$

Then

$$(f_1+f_2)^*(v) = \inf_{y\in X^*}\{f_1^*(v-y)+f_2^*(y)\} \quad \forall v \in X^*,$$

where the infimum is attained, and

$$\partial(f_1+f_2)(x) = \partial f_1(x) + \partial f_2(x) \quad \forall x \in X.$$

Proof We apply Theorem A.75 with $U = X$, $Y = V = X^*$, $F(x,u) = f_1(x) + f_2(x+u)$ and $\bar{u} = 0$. We have

$$\operatorname{dom}\varphi = \{u \in U \mid \exists x \in \operatorname{dom} f_1 : u + x \in \operatorname{dom} f_2\} = \operatorname{dom} f_2 - \operatorname{dom} f_1,$$

so $\operatorname{aff}\operatorname{dom}\varphi$ is closed and $0 \in \operatorname{rcore}\operatorname{dom}\varphi$, by assumption. Thus, Theorem A.75 says that the assumption of Theorem A.79 holds. We have $F(\cdot,0) = f_1 + f_2$ and

$$F^*(v,y) = f_1^*(v-y) + f_2^*(y),$$

so the expression for $(f_1+f_2)^*$ follows from the first claim in Theorem A.79. Noting that

$$L(x,y) = f_1(x) + \langle x,y\rangle - f_2^*(y),$$

the third claim in Theorem A.79 gives

$$\begin{aligned}
\partial(f_1+f_2)(x) &= \left\{v \in V \mid \exists y : v \in \partial_x L(x,y),\ \bar{u} \in \partial_y[-L](x,y)\right\}\\
&= \{\partial f_1(x) + y \mid x \in \partial f_2^*(y)\}\\
&= \{\partial f_1(x) + y \mid y \in \partial f_2(x)\}\\
&= \partial f_2(x) + \partial f_2(x),
\end{aligned}$$

where the third equality comes from Corollary A.55. □

The following can be found, e.g. in [160, Theorem 2.8.1].

Corollary A.81 *Let X and U be Fréchet spaces, $g : U \to \overline{\mathbb{R}}$ closed proper and convex and $A : X \to U$ continuous and linear such that* $\operatorname{aff}(\operatorname{dom} g - \operatorname{rge} A)$ *is closed and*

$$0 \in \operatorname{rcore}(\operatorname{dom} g - \operatorname{rge} A).$$

Then

$$(g \circ A)^*(v) = \inf_{y \in Y}\{g^*(y) \mid A^*y = v\} \quad \forall v \in X^*,$$

where the infimum is attained, and

$$\partial(g \circ A)(x) = A^*\partial g(Ax) \quad \forall x \in X.$$

Proof We apply Theorem A.75 with $V = X^*$, $Y = U^*$, $F(x, u) = g(Ax + u)$ and $\bar{u} = 0$. We have

$$\operatorname{dom} \varphi = \{u \in U \mid \exists x : u + Ax \in \operatorname{dom} g\} = \operatorname{dom} g - \operatorname{rge} A$$

and

$$F^*(v, y) = g^*(y) + \delta_{\{0\}}(v - A^*y),$$
$$L(x, y) = \langle Ax, y\rangle - g^*(y),$$

so the claims follow from Theorem A.79 much like in the proof of Corollary A.80.

□

The conditions on the domains in the above results are sufficient but not necessary for the validity of the calculus rules. In Euclidean spaces, for example, the rules are valid for proper polyhedral functions without any extra conditions. The same holds for the more general class of piecewise linear-quedratic functions; see [139, Theorem 11.42].

The following is a simple application of Corollary A.80.

Corollary A.82 *Given a closed proper convex function on $g : \mathbb{R}^n \to \overline{\mathbb{R}}$, the associated* proximal mapping *defined by*

$$P_g(\bar{u}) := \operatorname*{argmin}_{u \in \mathbb{R}^n}\{g(u) + \frac{1}{2}|u - \bar{u}|^2\}$$

is a single-valued contraction. In particular, the minimum norm projection to a nonempty closed convex set in $\mathbb{R}^n$ is a single-valued contraction.

Proof By Theorem A.77, the set $P_g(\bar{u})$ is nonempty for all $\bar{u}$. Properness and strict convexity give uniqueness. By Corollary A.80, the elements of $P_g(\bar{u})$ are the solutions u of

$$\partial g(u) + u - \bar{u} \ni 0.$$

Equivalently,

$$P_g(\bar{u}) = (\partial g + I)^{-1}(\bar{u}).$$

The monotonicity of ∂g (see Remark A.46) implies

$$\langle u_1 - u_2, \bar{u}_1 - \bar{u}_2 \rangle \geq |u_1 - u_2|^2 \quad \forall (u_1, \bar{u}_1), (u_2, \bar{u}_2) \in \operatorname{gph}(\partial g + I),$$

which implies

$$|\bar{u}_1 - \bar{u}_2| \geq |u_2 - u_2| \quad \forall (u_1, \bar{u}_1), (u_2, \bar{u}_2) \in \operatorname{gph}(\partial g + I).$$

Thus, the proximal mapping is a contraction. □

The following is concerned with proximal mappings associated with a parametric family of closed convex functions. The proximal mappings are defined as in Corollary A.82.

Corollary A.83 (Tikhonov Regularization) *Given a closed proper convex function on $g : \mathbb{R}^n \to \overline{\mathbb{R}}$, we have* argmin $g \neq \emptyset$ *if and only if $P_{g/\epsilon}(\bar{u})$ is bounded when $\epsilon \searrow 0$ and, in this case, $P_{g/\epsilon}(\bar{u})$ converges to the minimum norm projection of $\bar{u}$ to* argmin g.

Proof As in Corollary A.82, $P_{g/\epsilon}(\bar{u})$ is the unique solution u_ϵ to

$$\partial g(u_\epsilon) + \epsilon(u_\epsilon - \bar{u}) \ni 0.$$

If $(u_\epsilon)_{\epsilon \searrow 0}$ is bounded, then the closedness of gph g implies that every cluster point u_0 of $(u_\epsilon)_{\epsilon \searrow 0}$ satisfies $\partial g(u_0) \ni 0$ which means $u_0 \in \operatorname{argmin} g$. On the other hand, if $\hat{u} \in \operatorname{argmin} g$, then $\partial g(\hat{u}) \ni 0$ and the monotonicity of ∂g gives

$$\begin{aligned}
& \langle \hat{u} - u_\epsilon, \epsilon(u_\epsilon - \bar{u}) \rangle \geq 0 \\
\iff \quad & \langle \hat{u} - \bar{u} - (u_\epsilon - \bar{u}), u_\epsilon - \bar{u} \rangle \geq 0 \\
\iff \quad & |u_\epsilon - \bar{u}|^2 \leq \langle u_\epsilon - \bar{u}, \hat{u} - \bar{u} \rangle \\
\implies \quad & |u_\epsilon - \bar{u}| \leq |\hat{u} - \bar{u}|.
\end{aligned}$$

Since $\hat{u} \in \operatorname{argmin} g$ was arbitrary, we get

$$|u_\epsilon - \bar{u}| \leq \inf\{|u - \bar{u}| \mid u \in \operatorname{argmin} g\}. \tag{A.6}$$

It follows that $(u_\epsilon)_{\epsilon \searrow 0}$ is bounded and the closedness of $\operatorname{gph} g$ implies again that all cluster points of $(u_\epsilon)_{\epsilon \searrow 0}$ are in $\operatorname{argmin} g$. The bound (A.6) implies that all the cluster points are minimum norm projections of $\bar{u}$ on $\operatorname{argmin} g$. The uniqueness of the projection in Corollary A.82 now implies that the whole sequence converges to the projection. □

Example A.84 (Moore-Penrose Inverse) *Given any $A \in \mathbb{R}^{m\times n}$, the limit*

$$A^\dagger := \lim_{\epsilon \searrow 0} (A^*A + \epsilon I)^{-1} A^*$$

exists in $\mathbb{R}^{m\times n}$, the set

$$\operatorname*{argmin}_{x\in\mathbb{R}^n} |Ax - b|^2$$

is nonempty and, for any $b \in \mathbb{R}^m$, its minimum norm element is given by $A^\dagger b$.

Proof We apply Corollary A.83 to the function $g_b(x) = \frac{1}{2}|Ax - b|^2$ and $\bar{u} = 0$. The proximal mapping becomes

$$P_{g_b/\epsilon}(0) = \operatorname*{argmin}_{x\in\mathbb{R}^n}\{\frac{1}{2}|Ax - b|^2 + \frac{\epsilon}{2}|x|^2\} = (A^*A + \epsilon I)^{-1} A^* b$$

as is easily checked. Since, by Theorem A.77, $\operatorname{argmin} g_b \neq \emptyset$, Corollary A.83 says that the limit

$$\lim_{\epsilon \searrow 0} (A^*A + \epsilon I)^{-1} A^* b$$

exists and equals the minimum norm element of $\operatorname{argmin} g_b$. Since $b \in \mathbb{R}^m$ was arbitrary, it now suffices to note that the above limit is linear in b so it can be represented by an element of $\mathbb{R}^{m\times n}$. □

A.11 Scaling Properties

Given $p \in \mathbb{R}$, we say that a function g on a linear space U is *p-homogeneous* if there is a constant $C \in \mathbb{R}$ such that

$$g(\alpha u) = \begin{cases} \alpha^p g(u) - C\frac{\alpha^p - 1}{p} & p \neq 0, \\ g(u) - C \ln \alpha & p = 0. \end{cases}$$

for all $u \in U$ and $\alpha > 0$. This generalizes the classical notion of *positive homogeneity of degree p* which means that

$$g(\alpha u) = \alpha^p g(u) \quad \forall u \in U,\ \alpha > 0;$$

see, e.g. [123, Section 15] for the case $p \geq 1$. Many classical utility functions are p-homogeneous with $p < 1$; see Remark A.85 below. If g is p-homogeneous, then $u \mapsto ag(u) + b$ is p-homogeneous for any $a > 0$ and $b \in \mathbb{R}$.

The following characterizes convex nondecreasing p-homogeneous functions on the real line.

Remark A.85 *Given a lsc proper convex nondecreasing function $\phi : \mathbb{R} \to \overline{\mathbb{R}}$, the following are equivalent:*

1. *ϕ is p-homogeneous;*
2. $\operatorname{dom}\phi$ *is a cone, ϕ is twice differentiable on* $\operatorname{dom}\phi \setminus \{0\}$ *with*

$$\phi''(u) = \frac{p-1}{u}\phi'(u) \quad u \in \operatorname{dom}\phi \setminus \{0\};$$

3. *there exist constants $a > a' > 0$ and b such that either $\phi = b$, $\phi = \delta_{\mathbb{R}_-} + b$ or*

$$\phi(u) = \begin{cases} a(u^+)^p + b \quad \text{with } \operatorname{dom}\phi = \mathbb{R} & \text{if } p > 1, \\ au^+ - a'u^- + b \quad \text{with } \operatorname{dom}\phi = \mathbb{R} & \text{if } p = 1, \\ -a(-u)^p + b \quad \text{with } \operatorname{dom}\phi = \mathbb{R}_- & \text{if } p \in (0,1), \\ -a\ln(-u) + b \quad \text{with } \operatorname{dom}\phi = \mathbb{R}_- \setminus \{0\} & \text{if } p = 0, \\ a(-u)^p + b \quad \text{with } \operatorname{dom}\phi = \mathbb{R}_- \setminus \{0\} & \text{if } p < 0. \end{cases}$$

Proof Given any $u_0 \in \operatorname{dom}\phi$, 1 implies

$$\phi(\alpha) = \begin{cases} (\alpha/u_0)^p\phi(u_0) - C\frac{(\alpha/u_0)^p - 1}{p} & p \neq 0, \\ \phi(u_0) - C\ln(\alpha/u_0) & p = 0. \end{cases}$$

Calculating the derivatives gives 2. The functions in 3 are the only convex increasing solutions of the differential equation in 2. Clearly, 3 implies 1. □

Nondecreasing convex functions $\phi : \mathbb{R} \to \overline{\mathbb{R}}$ correspond to utility functions $U : \mathbb{R} \to \overline{\mathbb{R}}$ through the identity $U(u) = -\phi(-u)$. In this context, the number

$$|u|U''(u)/U'(u)$$

is known as the *relative risk aversion*. By Remark A.85, a utility function has *constant relative risk aversion* (CRRA) of $1 - p$ if and only if it is p-homogeneous.

Recall from the beginning of this chapter that the composition of a function H from a vector space X to another U with a function $g : U \to \overline{\mathbb{R}}$ is the function $g \circ H : X \to \overline{\mathbb{R}}$ defined by

$$(g \circ H)(x) := \begin{cases} g(H(x)) & \text{if } x \in \operatorname{dom} H, \\ +\infty & \text{otherwise.} \end{cases}$$

We say that H is *K-positively homogeneous* if its *K-epigraph* is a convex cone.

Theorem A.86 *The following functions are p-homogeneous:*

1. *$g \circ H$, where g is p-homogeneous and H is K-positively homogeneous,*
2. *$g \circ A$, where g is p-homogeneous and A is linear,*
3. *Sum of p-homogeneous functions on U,*
4. *$\varphi(u) := \inf_x f(x, u)$, where $f : X \times U \to \overline{\mathbb{R}}$ is p-homogeneous. In this case,*

$$\operatorname*{argmin}_x f(x, \alpha u) = \alpha \operatorname*{argmin}_x f(x, u)$$

for all $\alpha > 0$.

Proof In each case, the p-homogeneity follows directly from the definition. Given $\alpha > 0$, we have for any $\bar{x} \in X$ that

$$\begin{aligned} & \alpha\bar{x} \in \operatorname*{argmin}_x f(x, \alpha u) \\ \iff & f(\alpha\bar{x}, \alpha u) = \inf_x f(\alpha x, \alpha u) \\ \iff & \alpha^p f(\bar{x}, u) = \alpha^p \inf_x f(x, u) \\ \iff & \bar{x} \in \operatorname*{argmin}_x f(x, u), \end{aligned}$$

which proves the last claim. □

Given a differentiable convex function $\phi : \mathbb{R} \to \overline{\mathbb{R}}$, the number

$$\frac{u\phi'(u)}{\phi(u)}$$

is known as the *elasticity* of ϕ at u. If ϕ is twice differentiable with constant elasticity p, then ϕ is p-homogeneous, by Remark A.85. On the other hand, if $p < 1$ and ϕ is p-homogeneous, then there is a constant C such that the elasticity is given by

$$\frac{u\phi'(u)}{\phi(u)} = p - \frac{C}{\phi(u)}$$

for every u with $\phi(u) \neq 0$. This follows from the identity in the proof of Remark A.85 by differentiating at $\alpha = u_0$.

The numbers

$$AE_+(\phi) := \liminf_{u\to+\infty} \frac{u\phi'(u)}{\phi(u)} \quad \text{and} \quad AE_-(\phi) := \limsup_{u\to-\infty} \frac{u\phi'(u)}{\phi(u)}$$

are called the *asymptotic elasticities* of ϕ at $+\infty$ and $-\infty$, respectively. The following criteria were used in Remark 2.121 and Sect. 4.3.5.

Theorem A.87 (Asymptotic Elasticity) *Let $\phi : \mathbb{R} \to \overline{\mathbb{R}}$ be convex, nondecreasing and differentiable with* $\inf \phi < 0$.

1. *If $AE_-(\phi) < p \in (0, 1)$, then there exists a $\bar{y} \in \operatorname{dom}\phi^*$ such that*

$$\phi^*(\lambda y) \le \lambda^{\frac{p}{p-1}} \phi^*(y) \quad \forall \lambda \in (0, 1),\ y \in [0, \bar{y}].$$

2. *If $AE_+(\phi) > p > 1$, then there exists a $\bar{y} \in \operatorname{dom}\phi^*$ such that*

$$\phi^*(\lambda y) \le \lambda^{\frac{p}{p-1}} \phi^*(y) \quad \forall \lambda \ge 1,\ y \ge \bar{y}.$$

Proof The condition $AE_-(\phi) < p$ means that there exists $\bar{u} < 0$ such that

$$\phi'(u) \le p\phi(u)/u \quad \forall u \le \bar{u}.$$

Given $y \le \bar{y} := \phi'(\bar{u})$, we have $(\phi^*)'(y) \le (\phi^*)'(\bar{y}) = \bar{u}$, so

$$\begin{aligned}
\phi^*(y) &= \sup_u \{uy - \phi(u)\} \\
&= (\phi^*)'(y)y - \phi((\phi^*)'(y)) \\
&\ge (\phi^*)'(y)y - (\phi^*)'(y)\phi'((\phi^*)'(y))/p \\
&= (\phi^*)'(y)y - (\phi^*)'(y)y/p \\
&= \frac{p-1}{p}(\phi^*)'(y)y
\end{aligned}$$

and thus,

$$(\phi^*)'(y) \ge \frac{p}{p-1}\frac{1}{y}\phi^*(y) \quad \forall y \in (0, \bar{y}].$$

Since $\bar{y} \in \operatorname{dom} \phi^*$, the first claim holds trivially if $\bar{y} = 0$. Otherwise, let $y \in (0, \bar{y}]$ and $\lambda \in (0, 1)$. By Grönwall's inequality,

$$\phi^*(y) \geq \phi^*(\lambda y) \exp\left(\int_{\lambda y}^{y} \frac{p}{p-1}\frac{1}{s}ds\right) = \phi^*(\lambda y)\lambda^{\frac{p}{1-p}}$$

which gives the first claim.

The condition $AE_+(\phi) > p$ means that there exists $\bar{u} > 0$ such that

$$\phi'(u) \geq p\phi(u)/u \quad \forall u \geq \bar{u}.$$

Given $y \geq \bar{y} := \phi'(\bar{u})$, we have $(\phi^*)'(y) \geq (\phi^*)'(\bar{y}) = \bar{u}$, so

$$\begin{aligned}
\phi^*(y) &= \sup_u\{uy - \phi(u)\} \\
&= (\phi^*)'(y)y - \phi((\phi^*)'(y)) \\
&\geq (\phi^*)'(y)y - (\phi^*)'(y)\phi'((\phi^*)'(y))/p \\
&= (\phi^*)'(y)y - (\phi^*)'(y)y/p \\
&= \frac{p-1}{p}(\phi^*)'(y)y
\end{aligned}$$

and thus,

$$(\phi^*)'(y) \leq \frac{p}{p-1}\frac{1}{y}\phi^*(y) \quad \forall y > \bar{y}.$$

Given $y \geq \bar{y}$ and $\lambda \geq 1$, Grönwall's inequality gives

$$\phi^*(\lambda y) \leq \phi^*(y) \exp\left(\int_{y}^{\lambda y} \frac{p}{p-1}\frac{1}{s}ds\right) = \phi^*(y)\lambda^{\frac{p}{p-1}},$$

which is the second claim. □

The second condition in Theorem A.87 implies the Δ_2-condition: there exists $K > 0$ and $\bar{y}$ such that $\phi^*(2y) \leq K\phi^*(y)$ for all $y \geq \bar{y}$. By Rao and Ren [119, Corollary 2.3.4], the two conditions are in fact equivalent.

Example A.88 *The asymptotic elasticities and conjugates of classical utility/loss functions are easily calculated.*

1. Given $\beta > 0$, the exponential loss function

$$\phi(u) := (\exp(\beta u) - 1)/\beta$$

has $AE_-(\phi) = 0, \quad AE_+(\phi) = +\infty$ *and*

$$\phi^*(y) = \beta(y \log y - y + 1) + \delta_{\mathbb{R}_+}(y).$$

2. *Given* $p < 1$, *the loss function*

$$\phi(u) := \begin{cases} -\frac{1}{p}(-u)^p & \operatorname{dom}\phi = \mathbb{R}_- & \text{if } p \in (0,1), \\ -\ln(-u) & \operatorname{dom}\phi = \mathbb{R}_- \setminus \{0\} & \text{if } p = 0, \\ \frac{1}{p}(-u)^p & \operatorname{dom}\phi = \mathbb{R}_- \setminus \{0\} & \text{if } p < 0 \end{cases}$$

has $AE_-(\phi) = p$ *and*

$$\phi^*(y) = \begin{cases} \frac{1}{q}y^q & \operatorname{dom}\phi^* = \mathbb{R}_+ & \text{if } p \in (0,1), \\ -\ln y - 1 & \operatorname{dom}\phi^* = \mathbb{R}_+ \setminus \{0\} & \text{if } p = 0, \\ -\frac{1}{q}y^q & \operatorname{dom}\phi^* = \mathbb{R}_+ \setminus \{0\} & \text{if } p < 0, \end{cases}$$

where $1/p + 1/q = 1$.

Appendix B
Primer on Probability

B.1 Spaces of Random Variables

Let L^0 be the metric vector space of equivalence classes of $\mathbb{R}^n$-valued random variables defined in Sect. 1.1.5. We will denote the metric by

$$d(x, x') := E\rho(|x' - x|),$$

where $\rho : \mathbb{R} \to [0, 1]$ is a nondecreasing continuous function vanishing only at the origin and such that $\rho(\alpha_1 + \alpha_2) \leq \rho(\alpha_1) + \rho(\alpha_2)$ for all $\alpha_1, \alpha_2 > 0$.

Theorem B.1 *A sequence in L^0 converges in probability if and only if every subsequence has an almost surely convergent subsequence with a common limit. The space L^0 is a complete topological vector space where a sequence converges if and only if it converges in probability.*

Proof Let $(x^\nu)_{\nu\in\mathbb{N}}$ be a sequence in L^0 and $x \in L^0$. If $x^\nu \to x$ in probability, there is a subsequence $(x^{\nu_k})_{k\in\mathbb{N}}$ such that $P(\rho(|x^{\nu_k} - x|) \geq 2^{-k}) \leq 2^{-k}$. Let $A_k := \{\omega \mid \rho(|x^{\nu_k}(\omega) - x(\omega)|) \geq 2^{-k}\}$. By monotone convergence,

$$\begin{aligned}
E[\sum_{k=1}^{\infty} \rho(|x^{\nu_k} - x|)] &= \sum_{k=1}^{\infty} E[\rho(|x^{\nu_k} - x|)] \\
&= \sum_{k=1}^{\infty} E[1_{A_k}\rho(|x^{\nu_k} - x|) + 1_{\Omega\setminus A_k}\rho(|x^{\nu_k} - x|)] \\
&\leq \sum_{k=1}^{\infty} E[1_{A_k} + 1_{\Omega\setminus A_k}2^{-k}] \\
&\leq \sum_{k=1}^{\infty}(2^{-k} + 2^{-k}) < \infty.
\end{aligned}$$

T. Pennanen, A.-P. Perkkiö, *Convex Stochastic Optimization*, Probability Theory and Stochastic Modelling 107, https://doi.org/10.1007/978-3-031-76432-5

Thus, $\sum_{k=1}^{\infty} \rho(|x^{\nu_k} - x|) < \infty$ almost surely so $\rho(|x^{\nu_k} - x|) \to 0$ almost surely. For the converse, assume that x^ν does not converge to x in probability. Then there is an $\epsilon > 0$ and a subsequence such that $P(|x^{\nu_k} - x| \geq \epsilon) > \epsilon$. By dominated convergence, this cannot hold for almost surely converging subsequences.

Let $x^\nu \to x$ in probability. By the first claim, every subsequence has an almost surely converging subsequence $x^{\nu_k} \to x$. By dominated convergence, $d(x^{\nu_k}, x) \to 0$. This implies that the whole sequence convergences in the L^0 metric. If x^ν does not converge to x in probability, there is an $\epsilon > 0$, $\delta > 0$ and a subsequence such that $P(|x^{\nu'} - x| \geq \epsilon) \geq \delta$. Then $d(x^{\nu'}, x) \geq \delta\rho(\epsilon)$, so subsequences of $(x^{\nu'})$ cannot converge to x in L^0.

If $(x^\nu)_{\nu\in\mathbb{N}}$ is Cauchy in L^0, there is a subsequence $(x^{\nu_k})_{k\in\mathbb{N}}$ such that $d(x^{\nu_{k+1}}, x^{\nu_k}) \leq 2^{-k}$. By monotone convergence,

$$E[\sum_{k=1}^{\infty} \rho(|x^{\nu_{k+1}} - x^{\nu_k}|)] = \sum_{k=1}^{\infty} E[\rho(|x^{\nu_{k+1}} - x^{\nu_k}|)] \leq \sum_{k=1}^{\infty} 2^{-k} < \infty,$$

so $\sum_{k=1}^{\infty} \rho(|x^{\nu_{k+1}} - x^{\nu_k}|) < \infty$ almost surely. Thus, $(x^{\nu_k})_{k\in\mathbb{N}}$ is almost surely Cauchy in $\mathbb{R}^n$ so it converges almost surely to an $x \in L^0$. By dominated convergence, $x^{\nu_k} \to x$ in L^0. The triangle inequality now implies that the whole sequence converges to x. □

We now come to the proof of Theorem 5.6. We refer to Sect. 5.1.1 for the definitions.

Proof of Theorem 5.6 Since $\mathcal{P}$ is countable, $\mathcal{U}$ is metrizable. Let $(u^\nu)_{\nu\in\mathbb{N}}$ be a Cauchy sequence in $\mathcal{U}$. This means that, for every $\epsilon > 0$ and $p \in \mathcal{P}$, there is a $\bar{\nu}$ such that

$$p(u^\nu - u^\mu) \leq \epsilon \quad \forall \nu, \mu \geq \bar{\nu}.$$

By (A1), the sequence is Cauchy also in L^1. By completeness of L^1, the sequence L^1-converges to an $u \in L^1$. By (A3),

$$p(u^\nu - u) \leq \epsilon \quad \forall \nu \geq \bar{\nu}$$

so $u \in \mathcal{U}$, by triangle inequality, and $(u^\nu)_{\nu\in\mathbb{N}}$ converges in $\mathcal{U}$ to u. This proves that $\mathcal{U}$ is complete and thus Fréchet. Fréchet spaces are barreled, so $\mathcal{Y}$ is the Köthe dual of $\mathcal{U}$, by [106, Lemma 8].

Assumption (A1) implies that $L^\infty \subseteq \mathcal{U} \subseteq L^1$. Assumption (A2) then implies that $\mathcal{U}$ is solid and decomposable. By part 2 of [106, Lemma 8], (A1) and (A2) imply that $\mathcal{Y}$ is solid and $L^\infty \subseteq \mathcal{Y} \subseteq L^1$. Thus, $\mathcal{Y}$ is decomposable as well. Since both spaces contain L^∞, the duality is separating. Theorem 6 of [106] gives the expression for the dual space as well as the fact that condition (A4) implies $\mathcal{Y}^s = \{0\}$. Part 3 follows from that of [106, Lemma 8].

It remains to show that $\mathcal{Y}^s = \{0\}$ implies (A4). If condition (A4) fails, there is a $p \in \mathcal{P}$, $u \in \mathcal{U}$, a decreasing sequence $(A^\nu)_{\nu\in\mathbb{N}}$ with $P(A^\nu) \searrow 0$ and $\epsilon > 0$ with $p(1_{A^\nu}u) > \epsilon$ for all $\nu \in \mathbb{N}$. By Corollary A.57 $p = \sigma_B$, where

$$B := \{y \in \mathcal{U}^* \mid p^\circ(y) \le 1\}$$

is $\sigma(\mathcal{U}^*, \mathcal{U})$-compact by Corollary A.61. Thus, there is a sequence $(y^\nu)_{\nu\in\mathbb{N}} \subset B$ with $\langle 1_{A^\nu}u, y^\nu\rangle = p(1_{A^\nu}u)$. Denote the adjoint of the linear mapping $u \mapsto u1_{A^\nu}$ by $1^*_{A^\nu}$. Assumption (A2) and the decomposability of $\mathcal{U}$ imply

$$\begin{aligned} p^\circ(1^*_{A^\nu}y) &= \sup_u\{\langle u, 1^*_{A^\nu}y\rangle \mid p(u) \le 1\} \\ &= \sup_u\{\langle u1_{A^\nu}, y\rangle \mid p(u) \le 1\} \\ &\le \sup_u\{\langle u1_{A^\nu}, y\rangle \mid p(u1_{A^\nu}) \le 1\} \\ &\le \sup_u\{\langle u, y\rangle \mid p(u) \le 1\} \end{aligned}$$

so $1^*_{A^\nu}y^\nu \in B$. Since B is $\sigma(\mathcal{U}^*, \mathcal{U})$-compact, $(1^*_{A^\nu}y^\nu)_{\nu\in\mathbb{N}}$ has a subnet converging to y. This means that there is a directed set N, an order-preserving map $j : N \to \mathbb{N}$ and a $y \in B$ such that for every $\nu \in \mathbb{N}$, there exists $\mu \in N$ with $j(\mu) \ge \nu$, and $\lim_{\mu\in N} 1^*_{A^{j(\mu)}}y^{j(\mu)} = y$. By construction,

$$\langle u, y\rangle = \lim_{\mu\in N}\langle u, 1^*_{A^{j(\mu)}}y^{j(\mu)}\rangle = \lim_{\mu\in N}\langle 1_{A^{j(\mu)}}u, y^{j(\mu)}\rangle \ge \epsilon.$$

For every ν, the identity mapping from $N_\nu := \{\mu \in N \mid j(\mu) \ge \nu\}$ to N defines a further subnet, so $\langle u1_{A^\nu}, y\rangle = \langle u, y\rangle$. This cannot happen if $y \in \mathcal{Y}$. Thus, $\mathcal{Y}^s \ne \emptyset$. □

Besides the examples given in Sect. 5.1.1, assumptions (A1)–(A3) there are satisfied, e.g. by Lorentz and Marcinkiewicz spaces; see [106, Section 6].

The following classical result on weakly compact sets in L^1 was used in the proof of Lemma 5.3.

Theorem B.2 (Dunford-Pettis) *The $\sigma(L^1, L^\infty)$-closure of a set $C \subset L^1$ is $\sigma(L^1, L^\infty)$-compact if and only if $\sup_{u\in C} E[|u|] < \infty$ and, for every $\epsilon > 0$, there exists $\delta > 0$ such that $\sup_{u\in C} E[|u|1_A] \le \epsilon$ for all $A \in \mathcal{F}$ with $P(A) \le \delta$.*

Proof See, e.g. [42, Theorem II.19 and Theorem II.25]. □

B.2 Extended Real-Valued Random Variables

Recall that the sum $\xi_2 + \xi_2$ of extended real numbers $\xi_1, \xi_2 \in \overline{\mathbb{R}}$ is defined as $+\infty$ if one of them is $+\infty$. The expectation of an extended real-valued random variable ξ is defined as

$$E[\xi] := E[\xi^+] - E[\xi^-].$$

Lemma B.3 *Given extended real-valued random variables ξ_1 and ξ_2, we have*

$$E[\xi_1 + \xi_2] = E[\xi_1] + E[\xi_2]$$

under any of the following conditions:

1. $\xi_1^+, \xi_2^+ \in L^1$ *or* $\xi_1^-, \xi_2^- \in L^1$;
2. $\xi_1 \in L^1$ *or* $\xi_2 \in L^1$;
3. ξ_1 *or* ξ_2 *is* $\{0, +\infty\}$*-valued.*

Proof Claim 3 is clear. If both ξ_1 and ξ_2 are positive, the equality follows from the definition of the Lebesgue integral and the monotone convergence theorem. Assume now that $\xi_1, \xi_2 \in L^1$ and let $\xi = \xi_1 + \xi_2$. We have $\xi^+ - \xi^- = \xi_1^+ - \xi_1^- + \xi_2^+ - \xi_2^-$ so $\xi^+ + \xi_1^- + \xi_2^- = \xi^- + \xi_1^+ + \xi_2^+$ and

$$E\xi^+ + E\xi_1^- + E\xi_2^- = E\xi^- + E\xi_1^+ + E\xi_2^+.$$

Rearranging gives the equality. We will show that in the remaining cases, the equality holds with both sides being equal to either $+\infty$ or $-\infty$. By the sublinearity of the functions $\xi \mapsto \xi^+$ and $\xi \mapsto \xi^-$,

$$(\xi_1 + \xi_2)^+ \leq \xi_1^+ + \xi_2^+, \quad (\xi_1 + \xi_2)^- \leq \xi_1^- + \xi_2^-,$$
$$\xi_i^+ \leq (\xi_1 + \xi_2)^+ + \xi_j^-, \quad \xi_i^- \leq (\xi_1 + \xi_2)^- + \xi_j^+.$$

If $\xi_1^+, \xi_2^+ \in L^1$, the first inequality implies $(\xi_1 + \xi_2)^+ \in L^1$. If $\xi_i^- \notin L^1$, the last inequality gives $(\xi_1 + \xi_2)^- \notin L^1$ so the equality holds. The sufficiency of $\xi_1^-, \xi_2^- \in L^1$ is proved similarly. Assume now $\xi_1 \in L^1$. If $\xi_2^+ \in L^1$, the equation holds by 1 while if $\xi_2^+ \notin L^1$, the third inequality gives $(\xi_1 + \xi_2)^+ \notin L^1$, so the equation holds again. □

Corollary B.4 *The expectation is a sublinear function on L^0.*

Proof Positive homogeneity is clear so it remains to show that

$$E[\xi_1 + \xi_2] \leq E[\xi_1] + E[\xi_2] \quad \forall \xi_1, \xi_2 \in L^0.$$

By Lemma B.3, this holds as an equality if $\xi_1^+, \xi_2^+ \in L^1$ while in the complementary case the right side equals $+\infty$. □

Lemma B.5 *Given extended real-valued random variables* ξ_1 *and* ξ_2*, we have*

$$\xi_1 \leq \xi_2 \quad a.s.$$

if and only if

$$E[\xi_1 1_B] \leq E[\xi_2 1_B] \quad \forall B \in \mathcal{F}.$$

Proof It is clear that the first condition implies the second. To prove the converse, assume that the set $B = \{\omega \in \Omega \mid \xi_1(\omega) > \xi_2(\omega)\}$ has positive probability. For ν large enough, the set

$$B^\nu := \{\omega \in B \mid \xi_1(\omega) \geq -\nu,\ \xi_2(\omega) \leq \nu\}$$

has positive probability and $E[\xi_1 1_{B^\nu}] > E[\xi_2 1_{B^\nu}]$. □

B.3 Some Measurability Results

The following can be found, e.g. in [30, Theorem III.23].

Theorem B.6 *If* $(\Omega, \mathcal{F}, P)$ *is complete, then*

$$\{\omega \in \Omega \mid \exists x \in \mathbb{R}^n : (x, \omega) \in A\} \in \mathcal{F}$$

for every $A \in \mathcal{B}(\mathbb{R}^n) \otimes \mathcal{F}$.

Recall that a measurable space $(\Xi, \mathcal{A})$ is a *Borel space* if there exists a Borel subset A of the unit interval and a bijective mapping $\phi : \Xi \to A$ such that both ϕ and ϕ^{-1} are measurable. Every Borel subset of a completely metrizable space is a Borel space; see, e.g. [46, Theorem 2.1.22]. The following can be found, e.g. in [73, Lemma 1.13].

Lemma B.7 (Doob-Dynkin) *Let* ξ *and* $\hat{\xi}$ *be measurable functions from* $(\Omega, \mathcal{F})$ *to a measurable space* $(\Xi, \mathcal{A})$ *and to a Borel space* $(\hat{\Xi}, \hat{\mathcal{A}})$*, respectively. Then* $\hat{\xi}$ *is* $\sigma(\xi)$*-measurable if and only if there exists a measurable function* $\phi : (\Xi, \mathcal{A}) \to (\hat{\Xi}, \hat{\mathcal{A}})$ *such that* $\hat{\xi} = \phi \circ \xi$.

Remark B.8 *Given a probability measure* P *on* $(\Omega, \mathcal{F})$*, the function* ϕ *in Lemma B.7 is* P_ξ*-almost everywhere unique. This follows from the fact that* ϕ *is unique on the range of* ξ*. In particular, the linear mapping* $\Pi : L^0(\Xi, \mathcal{A}, P_\xi; \hat{\Xi}) \to L^0(\Omega, \sigma(\xi), P; \hat{\Xi})$ *given by*

$$\Pi\phi = \phi \circ \xi$$

is a bijection.

A collection $\mathcal{C}$ of sets in Ω is a

1. *monotone class* if $\bigcup_{\nu\in\mathbb{N}} A^\nu \in \mathcal{C}$ for every increasing sequence $(A^\nu)_{\nu\in\mathbb{N}}$ of sets $A^\nu \in \mathcal{C}$ and $\bigcap_{\nu\in\mathbb{N}} A^\nu \in \mathcal{C}$ for every decreasing sequence $(A^\nu)_{\nu\in\mathbb{N}}$ of sets $A^\nu \in \mathcal{C}$,
2. *π-class* if $A \cap B \in \mathcal{C}$ whenever $A, B \in \mathcal{C}$,
3. *λ-class* if i) $\Omega \in \mathcal{C}$, ii) $B \setminus A \in \mathcal{C}$ whenever $A, B \in \mathcal{C}$ with $A \subset B$, iii) $\bigcup_{\nu\in\mathbb{N}} A^\nu \in \mathcal{C}$ for every increasing sequence $(A^\nu)_{\nu\in\mathbb{N}}$ of sets $A^\nu \in \mathcal{C}$.

The following can be found, e.g. in [62, Theorem 1.2].

Theorem B.9 (Monotone Class Theorem) *Let $\mathcal{C}$ and $\mathcal{D}$ be collections of sets in Ω with $\mathcal{C} \subseteq \mathcal{D}$. Then $\sigma(\mathcal{C}) \subseteq \mathcal{D}$ under either of the following conditions:*

1. *$\mathcal{D}$ is a monotone class and $\mathcal{C}$ is an algebra;*
2. *$\mathcal{D}$ is a λ-class and $\mathcal{C}$ is a π-class.*

A function $\mu : \Omega \times \mathcal{A} \to \overline{\mathbb{R}}$ is said to be a *probability kernel* from $(\Omega, \mathcal{F})$ to a measurable space $(\Xi, \mathcal{A})$ if for each $\omega \in \Omega$, $\mu(\omega, \cdot)$ is a probability measure on $(\Xi, \mathcal{A})$ and, for each $A \in \mathcal{A}$, $\mu(\cdot, A)$ is a $\mathcal{F}$-measurable function on Ω. Given an extended real-valued $\mathcal{B}(\mathbb{R}^n) \otimes \mathcal{A}$-measurable function H, we will use the notation

$$(E^\mu H)(x, \omega) := \int_\Xi H(x, s)\mu(\omega, ds),$$

where $\int_\Xi H(x, s)\mu(\omega, ds) := \int_\Xi H^+(x, s)\mu(\omega, ds) - \int_\Xi H^-(x, s)\mu(\omega, ds)$. The following fact is well-known in the case of real-valued functions H; see, e.g. [73, Theorem 6.4].

Lemma B.10 *Given an extended real-valued $\mathcal{B}(\mathbb{R}^n) \otimes \mathcal{A}$-measurable function H and a probability kernel from $(\Omega, \mathcal{F})$ to $(\Xi, \mathcal{A})$, the function $E^\mu H$ is $\mathcal{B}(\mathbb{R}^n) \otimes \mathcal{F}$-measurable.*

Proof Let $\mathcal{D}$ be the collection sets $C \in \mathcal{B}(\mathbb{R}^n) \otimes \mathcal{A}$ such that $E^\mu 1_C$ is $\mathcal{B}(\mathbb{R}^N) \otimes \mathcal{F}$-measurable. The dominated convergence theorem implies that $\mathcal{D}$ is a monotone class. If $H = 1_{A\times B}$ for $A \in \mathcal{B}(\mathbb{R}^n)$ and $B \in \mathcal{A}$, we have $(E^\mu H)(x, \omega) = 1_A(x)\mu(\omega, B)$, so $A \times B \in \mathcal{D}$. By linearity of the operator E^μ on bounded measurable functions, $\mathcal{D}$ contains the collection $\mathcal{C}$ of all finite unions of sets $A \times B$ with $A \in \mathcal{B}(\mathbb{R}^n)$ and $B \in \mathcal{A}$. It is easy to check that $\mathcal{C}$ is an algebra so, by Theorem B.9, $\mathcal{D}$ contains $\sigma(\mathcal{C}) = \mathcal{B}(\mathbb{R}^n) \otimes \mathcal{A}$. By monotone convergence theorem, measurability holds for any nonnegative $\mathcal{B}(\mathbb{R}^n) \otimes \mathcal{A}$-measurable function H. The general case follows from by expressing H as the difference of its positive and negative parts. □

B.4 Conditional Expectation and Independence

Most of the results in this section can be found, e.g. in [149]. Recall that the product $\xi_1\xi_2$ of extended real numbers $\xi_1, \xi_2 \in \overline{\mathbb{R}}$ is defined as zero if one of the numbers is zero. An extended real-valued random variable ξ is *quasi-integrable* if either its positive part ξ^+ or the negative part ξ^- is integrable.

Theorem B.11 *Given a quasi-integrable extended real-valued random variable ξ and a σ-algebra $\mathcal{G} \subseteq \mathcal{F}$, there exists a $\mathcal{G}$-measurable quasi-integrable extended real-valued random variable $E^{\mathcal{G}}\xi$, unique almost surely, such that*

$$E\left[\alpha(E^{\mathcal{G}}\xi)\right] = E\left[\alpha\xi\right] \quad \forall \alpha \in L^{\infty}_{+}(\Omega, \mathcal{G}, P).$$

Proof When $\xi \in L^{\infty}$, the mapping $l(\eta) = E[\eta\xi]$ is linear and continuous on $L^1(\mathcal{F})$, by Example 5.11. The restriction of l to $L^1(\mathcal{G})$ is linear and continuous as well, so, by Example 5.11, it can be represented by an element of $L^{\infty}(\mathcal{G})$. This gives the existence of $E^{\mathcal{G}}\xi$ when $\xi \in L^{\infty}$. Given a nonnegative ξ, let $\xi^{\nu} := \min\{\xi, \nu\} \in L^{\infty}$. For any $\alpha \in L^{\infty}_{+}(\mathcal{G})$, monotone convergence theorem gives

$$E[\alpha\xi] = \lim E[\alpha\xi^{\nu}] = \lim E[\alpha E^{\mathcal{G}}\xi^{\nu}] = E[\alpha E^{\mathcal{G}}[\lim \xi^{\nu}]],$$

so we can take $E^{\mathcal{G}}\xi := \lim E^{\mathcal{G}}\xi^{\nu}$. In the general case, we can take $E^{\mathcal{G}}\xi := E^{\mathcal{G}}\xi^+ - E^{\mathcal{G}}\xi^-$. Indeed, when ξ is quasi-integrable, either $E^{\mathcal{G}}\xi^+$ or $E^{\mathcal{G}}\xi^-$ is integrable so, by Lemma B.3,

$$\begin{aligned} E\left[\alpha[(E^{\mathcal{G}}\xi)^+ - (E^{\mathcal{G}}\xi)^-]\right] &= E[\alpha(E^{\mathcal{G}}\xi)^+] - E^{\mathcal{G}}[\alpha(E^{\mathcal{G}}\xi)^-] \\ &= E[\alpha\xi^+] - E^{\mathcal{G}}[\alpha\xi^-] \\ &= E\left[\alpha\xi\right] \end{aligned}$$

for any $\alpha \in L^{\infty}_{+}(\Omega, \mathcal{G}, P)$. □

The random variable $E^{\mathcal{G}}\xi$ in Theorem B.11 is known as the *$\mathcal{G}$-conditional expectation* of ξ.

The claims in the following theorem are easy consequences of the ordinary Fatou's lemma and monotone and dominated convergence theorems.

Theorem B.12 *Let $(\xi^{\nu})_{\nu\in\mathbb{N}}$ be a sequence of quasi-integrable extended real-valued random variables converging almost surely to an extended real-valued random variable ξ.*

1. *If $\xi^{\nu} \geq 0$ for all ν, then* $\liminf_{\nu} E^{\mathcal{G}}[\xi^{\nu}] \geq E^{\mathcal{G}}[\xi]$ *almost surely.*
2. *If $0 \leq \xi^{\nu} \leq \xi^{\nu+1}$ for all ν, then* $\lim_{\nu} E^{\mathcal{G}}[\xi^{\nu}] = E^{\mathcal{G}}[\xi]$.
3. *If there exists $\eta \in L^1$ with $|\xi^{\nu}| \leq \eta$ for all ν, then* $\lim_{\nu} E^{\mathcal{G}}[\xi^{\nu}] = E^{\mathcal{G}}[\xi]$.

Lemma B.13 *Let ξ_1 and ξ_2 be extended real-valued random variables.*

1. *If ξ_1 and ξ_2 are quasi-integrable and satisfy any of the conditions in Lemma B.3, then $\xi_1 + \xi_2$ is quasi-integrable and*

$$E^{\mathcal{G}}[\xi_1 + \xi_2] = E^{\mathcal{G}}[\xi_1] + E^{\mathcal{G}}[\xi_2].$$

2. *If ξ_2 and $(\xi_1\xi_2)$ are quasi-integrable, and ξ_1 is $\mathcal{G}$-measurable, then*

$$E^{\mathcal{G}}[\xi_1\xi_2] = \xi_1 E^{\mathcal{G}}[\xi_2].$$

Proof Let $\alpha \in L^\infty_+(\mathcal{G})$. By Lemma B.3,

$$E[\alpha(\xi_1+\xi_2)] = E[\alpha\xi_1]+E[\alpha\xi_2] = E[\alpha E^{\mathcal{G}}\xi_1]+E[\alpha E^{\mathcal{G}}\xi_2] = E[\alpha(E^{\mathcal{G}}\xi_1+E^{\mathcal{G}}\xi_2)],$$

which proves 1. To prove 2, note first that the claim is clear if ξ_1 is bounded and either nonpositive or nonnegative. Let ξ_1^ν be the projection of ξ_1 to $[-\nu, \nu]$. Since $\xi_1\xi_2$ is quasi-integrable, 1 gives

$$\begin{aligned}E^{\mathcal{G}}[\xi_1^\nu\xi_2] &= E^{\mathcal{G}}[1_{\{\xi_1^\nu\geq 0\}}\xi_1^\nu\xi_2] + E^{\mathcal{G}}[1_{\{\xi_1^\nu<0\}}\xi_1^\nu\xi_2]\\ &= 1_{\{\xi_1^\nu\geq 0\}}\xi_1^\nu E^{\mathcal{G}}[\xi_2] + 1_{\{\xi_1^\nu<0\}}\xi_1^\nu E^{\mathcal{G}}[\xi_2]\\ &= \xi_1^\nu E^{\mathcal{G}}[\xi_2].\end{aligned}$$

The last term converges almost surely to $\xi_1 E^{\mathcal{G}}[\xi_2]$. On the other hand, by 1,

$$E^{\mathcal{G}}[\xi_1^\nu\xi_2] = E^{\mathcal{G}}[(\xi_1^\nu\xi_2)^+ - (\xi_1^\nu\xi_2)^-] = E^{\mathcal{G}}[(\xi_1^\nu\xi_2)^+] + E^{\mathcal{G}}[-(\xi_1^\nu\xi_2)^-].$$

Since both terms on the right are monotone in ν and one of them is bounded almost surely,

$$\lim E^{\mathcal{G}}[\xi_1^\nu\xi_2] = \lim E^{\mathcal{G}}[(\xi_1^\nu\xi_2)^+] + \lim E^{\mathcal{G}}[-(\xi_1^\nu\xi_2)^-]$$

almost surely. By Theorem B.12,

$$\lim E^{\mathcal{G}}[\xi_1^\nu\xi_2] = E^{\mathcal{G}}[(\xi_1\xi_2)^+] + E^{\mathcal{G}}[-(\xi_1\xi_2)^-],$$

where, by 1, the right side equals $E^{\mathcal{G}}[\xi_1\xi_2]$ and the left side equals $\xi_1 E^{\mathcal{G}}[\xi_2]$ as observed above. □

Lemma B.14 *Given a quasi-integrable extended real-valued random variable ξ and σ-algebras $\mathcal{G}' \subseteq \mathcal{G} \subseteq \mathcal{F}$, we have*

$$E^{\mathcal{G}'}\xi = E^{\mathcal{G}'}[E^{\mathcal{G}}\xi]$$

and

$$E\xi = E[E^{\mathcal{G}}\xi].$$

Proof The first claim follows directly from the definition. As to the second, apply Lemmas B.13 and B.3 to $\xi = \xi^+ - \xi^-$. □

Given another probability measure $Q \ll P$, a Q-quasi-integrable random variable has a Q-almost surely unique $\mathcal{G}$-conditional expectation with respect to Q. The following is a slight generalization of [57, Proposition A.16].

Theorem B.15 *Given a probability measure $Q \ll P$, a σ-algebra $\mathcal{G} \subseteq \mathcal{F}$ and a Q-quasi-integrable extended real-valued random variable ξ, we have*

$$E^{\mathcal{G}}\left[\frac{dQ}{dP}\xi\right] = E^{\mathcal{G}}\left[\frac{dQ}{dP}\right]E_Q^{\mathcal{G}}[\xi] \quad P\text{-a.s.},$$

where $E_Q^{\mathcal{G}}[\xi]$ denotes any random variable that is P-almost surely equal to the $\mathcal{G}$-conditional expectation of ξ with respect to Q. The random variable $E^{\mathcal{G}}[dQ/dP]$ is Q-almost surely strictly positive. In particular,

$$E_Q^{\mathcal{G}}[\xi] = \frac{1}{E^{\mathcal{G}}\left[\frac{dQ}{dP}\right]}E^{\mathcal{G}}\left[\frac{dQ}{dP}\xi\right] \quad Q\text{-a.s.}$$

Proof Let $y := \frac{dQ}{dP}$ and $\tilde{\xi} := E_Q^{\mathcal{G}}[\xi]$. For any $A \in \mathcal{G}$, Lemma B.13 implies

$$E[E^{\mathcal{G}}[y\xi]1_A] = E[y\xi 1_A] = E[y\tilde{\xi}1_A] = E[E^{\mathcal{G}}[y\tilde{\xi}]1_A] = E[E^{\mathcal{G}}[y]\tilde{\xi}1_A],$$

which proves the first claim. By the definition of conditional expectation,

$$E^Q[1_{\{E^{\mathcal{G}}\left[\frac{dQ}{dP}\right]=0\}}] = E^P[\frac{dQ}{dP}1_{\{E^{\mathcal{G}}\left[\frac{dQ}{dP}\right]=0\}}] = E^P[E^{\mathcal{G}}[\frac{dQ}{dP}]1_{\{E^{\mathcal{G}}\left[\frac{dQ}{dP}\right]=0\}}] = 0,$$

which proves the second claim. □

Given a σ-algebra $\mathcal{H} \subset \mathcal{F}$, σ-algebras $\mathcal{G}'$ and $\mathcal{G}$ are *$\mathcal{H}$-conditionally independent* if

$$E^{\mathcal{H}}[1_{A'}1_A] = E^{\mathcal{H}}[1_{A'}]E^{\mathcal{H}}[1_A]$$

for every $A' \in \mathcal{G}'$ and $A \in \mathcal{G}$. The smallest σ-algebra containing $\mathcal{G}$ and $\mathcal{H}$ is denoted by $\mathcal{G} \vee \mathcal{H}$. The following is essentially [73, Proposition 6.6 and Corollary 6.7]

Lemma B.16 *Given σ-algebras $\mathcal{G}'$, $\mathcal{G}$ and $\mathcal{H}$, the following are equivalent:*

1. *$\mathcal{G}'$ and $\mathcal{G}$ are $\mathcal{H}$-conditionally independent;*
2. *$E^{\mathcal{H}}[w'w] = E^{\mathcal{H}}[w']E^{\mathcal{H}}[w]$ for every $w' \in L^1(\mathcal{G}')$ and $w \in L^\infty(\mathcal{G})$;*
3. *$E^{\mathcal{G}\vee\mathcal{H}}[w'] = E^{\mathcal{H}}[w']$ for every $w' \in L^1(\mathcal{G}')$*

and imply that, if $\mathcal{G}$ and $\mathcal{H}$ are independent, then so too are $\mathcal{G}$ and $\mathcal{G}'$.

Proof Part 2 follows from 1 and the monotone convergence, by expressing the functions w' and w as limits of simple functions. When 2 holds, we have, for any $w' \in L^1(\mathcal{G}')$, $A \in \mathcal{G}$ and $B \in \mathcal{H}$,

$$E[E^{\mathcal{H}}[w']1_{A\cap B}] = E[E^{\mathcal{H}}[w'1_A]1_B] = E[w'1_A1_B] = E[E^{\mathcal{G}\vee\mathcal{H}}[w']1_{A\cap B}],$$

and, by the monotone class theorem Theorem B.9, this extends from sets of the form $A \cap B$ to any set in $\mathcal{G} \vee \mathcal{H}$. Thus, 2 implies 3. Assuming 3, we have, for $A' \in \mathcal{G}'$ and $A \in \mathcal{G}$,

$$\begin{aligned} E^{\mathcal{H}}[1_A1_{A'}] &= E^{\mathcal{H}}[E^{\mathcal{G}\vee\mathcal{H}}[1_A1_{A'}]] = E^{\mathcal{H}}[1_A E^{\mathcal{G}\vee\mathcal{H}}1_{A'}] \\ &= E^{\mathcal{H}}[1_A E^{\mathcal{H}}1_{A'}] = E^{\mathcal{H}}[1_A]E^{\mathcal{H}}[1_{A'}], \end{aligned}$$

so 1 holds.

Assume now in addition that $\mathcal{G}$ and $\mathcal{H}$ are independent. Given any $A' \in \mathcal{G}'$ and $A \in \mathcal{G}$, we get

$$\begin{aligned} E[1_{A'}1_A] &= E[E^{\mathcal{H}}[1_{A'}1_A]] = E[E^{\mathcal{H}}[1_{A'}]E^{\mathcal{H}}[1_A]] \\ &= E[E^{\mathcal{H}}[1_{A'}]E[1_A]] = E[1_{A'}]E[1_A], \end{aligned}$$

which implies the independence if $\mathcal{G}$ and $\mathcal{G}'$. □

For a version of the following result involving countably many σ-algebras; see [73, Proposition 6.8].

Lemma B.17 *Given σ-algebras $\mathcal{G}_1, \mathcal{G}_2, \mathcal{G}'$ and $\mathcal{G}_0$, the following are equivalent:*

1. *$\mathcal{G}_1 \vee \mathcal{G}_2$ is $\mathcal{G}_0$-conditionally independent of $\mathcal{G}'$;*
2. *$\mathcal{G}_1$ is $\mathcal{G}_0$-conditionally independent of $\mathcal{G}'$ and $\mathcal{G}_2$ is $\mathcal{G}_0 \vee \mathcal{G}_1$-conditionally independent of $\mathcal{G}'$.*

Proof By Lemma B.16.3, 1 means that

$$E^{\mathcal{G}^1\vee\mathcal{G}^2\vee\mathcal{G}_0}[w'] = E^{\mathcal{G}_0}[w'] \quad \forall w \in L^1(\mathcal{G}')$$

while 2 means that

$$E^{\mathcal{G}_1\vee\mathcal{G}_0}[w'] = E^{\mathcal{G}_0}[w'], \qquad E^{\mathcal{G}_1\vee\mathcal{G}_2\vee\mathcal{G}_0}[w'] = E^{\mathcal{G}_1\vee\mathcal{G}_0}[w'] \quad \forall w \in L^1(\mathcal{G}').$$

Thus, 2 implies 1. Assuming 1,

$$\begin{aligned} E^{\mathcal{G}_1\vee\mathcal{G}_0}[w'] &= E^{\mathcal{G}_1\vee\mathcal{G}_0}[E^{\mathcal{G}_1\vee\mathcal{G}_2\vee\mathcal{G}_0}[w']] \\ &= E^{\mathcal{G}_1\vee\mathcal{G}_0}[E^{\mathcal{G}_0}[w']] = E^{\mathcal{G}_0}[w'] = E^{\mathcal{G}^1\vee\mathcal{G}^2\vee\mathcal{G}_0}[w'] \end{aligned}$$

for all $w \in L^1(\mathcal{G}')$. □

The following gives a sufficient condition for the existence of the regular conditional distributions that were studied in Sects. 2.1.5 and 2.2.4.

Theorem B.18 *A random variable ξ with values in a Borel space $(\Xi, \mathcal{A})$ admits a regular $\mathcal{G}$-conditional distribution, i.e. a probability kernel μ such that*

$$E^{\mathcal{G}}[X(\xi)](\omega) = \int_{\Xi} X(s)\mu(\omega, ds)$$

for every $\mathcal{A}$-measurable extended real-valued function X such that $X(\xi)$ is quasi-integrable.

Proof By Theorem 5 in Section 7.II of [149], there exists a probability kernel μ such that the equation holds for functions X of the form $X = 1_A$, where $A \in \mathcal{A}$. By the monotone convergence theorem, this extends to nonnegative measurable functions X. If $X(\xi)$ is quasi-integrable, we have

$$\begin{aligned} E^{\mathcal{G}}[X(\xi)](\omega) &= E^{\mathcal{G}}[X^+(\xi)](\omega) - E^{\mathcal{G}}[X^-(\xi)](\omega) \\ &= \int_{\Xi} X^+(s)\mu(\omega, ds) - \int_{\Xi} X^-(s)\mu(\omega, ds) \\ &= \int_{\Xi} X(s)\mu(\omega, ds) \end{aligned}$$

which completes the proof. □

The equality in Theorem B.18 extends to random functions X.

Theorem B.19 *Let ξ be a random variable with values in a measurable space $(\Xi, \mathcal{A})$ and let μ be its regular $\mathcal{G}$-conditional distribution. Then*

$$E^{\mathcal{G}}[X(\xi)](\omega) = \int_{\Xi} X(s, \omega)\mu(\omega, ds)$$

for every $\mathcal{A} \otimes \mathcal{G}$-measurable extended real-valued function X such that $\omega \mapsto X(\xi)(\omega) := X(\xi(\omega), \omega)$ is quasi-integrable.

Proof Given $A \in \mathcal{A}$ and $B \in \mathcal{G}$,

$$E^{\mathcal{G}}[1_{A\times B}(\xi)](\omega) = 1_B(\omega)E^{\mathcal{G}}[1_A(\xi)](\omega) = \int_{\Xi} 1_{A\times B}(s,\omega)\mu(\omega, ds).$$

By Theorem B.9, this extends to functions of the form 1_C, where $C \in \mathcal{A} \otimes \mathcal{G}$. The extension to nonnegative X follows from the monotone convergence theorem. The general case follows by applying this to the positive and negative parts of $X(\xi)$. □

The following was used in the proof of Corollary 2.60.

Lemma B.20 *Let ξ_i be a $(\Xi_i, \mathcal{A}_i)$-valued random variable and let ν be a regular ξ_1-conditional distribution of ξ_2. Then the regular ξ_1-conditional distribution of (ξ_1, ξ_2) is given by*

$$\hat{\nu}(s_1, A) = \int_{\Xi_2} 1_A(s_1, s_2)\nu(s_1, ds_2)$$

for all $A \in \mathcal{A}_1 \otimes \mathcal{A}_2$.

Proof Since sections of product measurable sets are measurable, the function $\hat{\nu}(s_1, \cdot)$ is, for every $s_1 \in \Xi_1$, a measure on $\mathcal{A}_1 \otimes \mathcal{A}_2$. To show that $\hat{\nu}(\cdot, A)$ is measurable for all $A \in \mathcal{A}_1 \otimes \mathcal{A}_2$, we apply the monotone class theorem Theorem B.9 with $\mathcal{D}$ consisting of the sets $A \in \mathcal{A}_1 \otimes \mathcal{A}_2$ such that $\hat{\nu}(\cdot, A)$ is measurable. Given $A_i \in \mathcal{A}_i$, we have

$$\hat{\nu}(s_1, A_1 \times A_2) = 1_{A_1}(s_1) \int_{\Xi_2} 1_{A_2}(s_2)\nu(s_1, ds_2).$$

This is measurable in s_1. The measurability extends to finite unions of products of measurable sets. This collection $\mathcal{C}$ of sets thus belongs to $\mathcal{D}$. Since $\mathcal{C}$ is an algebra, the monotone class theorem gives $\sigma(\mathcal{C}) \subseteq \mathcal{D}$. Thus, $\hat{\nu}(\cdot, A)$ is measurable for all $A \in \mathcal{A}_1 \otimes \mathcal{A}_2$.

For any $A_i \in \mathcal{A}_i$,

$$\begin{aligned}
\hat{\nu}(\xi_1, A_1 \times A_2) &= \int_{\Xi_2} 1_{A_1\times A_2}(\xi_1, s_2)\nu(\xi_1, ds_2) \\
&= 1_{A_1}(\xi_1) \int_{\Xi_2} 1_{A_2}(s_2)\nu(\xi_1, ds_2) \\
&= 1_{A_1}(\xi_1)E^{\sigma(\xi_1)}[1_{A_2}(\xi_2)] \\
&= E^{\sigma(\xi_1)}[1_{A_1}(\xi_1)1_{A_2}(\xi_2)] \\
&= E^{\sigma(\xi_1)}[1_{A_1\times A_2}(\xi_1, \xi_2)].
\end{aligned}$$

By the monotone class argument again (see Theorem B.9), this extends to all $A \in \mathcal{A}_1 \otimes \mathcal{A}_2$. By monotone convergence this extend from indicators 1_A to nonnegative $\mathcal{A}_1 \otimes \mathcal{A}_2$-measurable random variables and then, by taking differences of positive and negative parts, to all quasi-integrable X. □

B.5 Essential Infimum

Given a collection C of random variables, its *essential infimum* $\operatorname{essinf} C$ is a random variable such that $\operatorname{essinf} C \leq \xi$ almost surely for every $\xi \in C$ and if $\underline{\xi}$ is another random variable with this property, then $\underline{\xi} \leq \operatorname{essinf} C$ almost surely. Equivalently, $\operatorname{essinf} C$ is an extended real-valued random variable such that

$$\underline{\xi} \leq \xi \quad \forall \xi \in C \iff \underline{\xi} \leq \operatorname{essinf} C$$

for every extended real-valued random variable $\underline{\xi}$. This notion was used in Remark 2.3 and Lemma 2.4. Lemma B.21 below says that every collection of random variables has a unique essential infimum. The proof is essentially that of [89, Proposition II.4.1] (see also [57, Theorem A.17]), but we allow for extended real-valued random variables.

Lemma B.21 *Any collection C of extended real-valued random variables admits an essential infimum and, moreover, there exists a sequence $(\xi^\nu)_{\nu\in\mathbb{N}}$ in C such that* $\inf_{\nu\in\mathbb{N}} \xi^\nu = \operatorname{essinf} C$ *almost surely.*

Proof Transforming by a strictly increasing $\phi : \overline{\mathbb{R}} \to [0, 1]$, if necessary, we may assume that the random variables in C take values in $[0, 1]$. Let $\mathcal{J}$ be the collection of all sequences of elements of C. For each $(\xi^\nu)_{\nu\in\mathbb{N}} \in \mathcal{J}$, let

$$g((\xi^\nu)_{\nu\in\mathbb{N}}) := E \inf_{\nu\in\mathbb{N}} \xi^\nu.$$

Let $(\xi^\nu_k)_{\nu\in\mathbb{N}} \in \mathcal{J}$, $k \in \mathbb{N}$, be a minimizing sequence for g so that

$$\inf_{k\in\mathbb{N}} g((\xi^\nu_k)_{\nu\in\mathbb{N}}) = \inf_{(\xi^\nu)_{\nu\in\mathbb{N}}\in\mathcal{J}} g((\xi^\nu)_{\nu\in\mathbb{N}}).$$

Let $(\bar{\xi}^\nu)_{\nu\in\mathbb{N}} \in \mathcal{J}$ be a sequence containing all ξ^ν_k. Clearly, $g((\bar{\xi}^\nu)_{\nu\in\mathbb{N}}) \leq g((\bar{\xi}^\nu_k)_{\nu\in\mathbb{N}})$ for all k so $(\bar{\xi}^\nu)_{\nu\in\mathbb{N}}$ minimizes g over $\mathcal{J}$.

We claim that $\underline{\xi} := \inf_{\nu\in\mathbb{N}} \bar{\xi}^\nu = \operatorname{essinf} C$. If ξ is a random variable such that $\xi \leq \xi'$ for all $\xi' \in C$, then clearly $\xi \leq \underline{\xi}$. If there existed a $\xi \in C$ with $P(\xi < \underline{\xi}) > 0$, then adding ξ to the sequence $(\bar{\xi}^\nu)_{\nu\in\mathbb{N}}$ would strictly reduce the value of g contradicting the minimality of $(\bar{\xi}^\nu)_{\nu\in\mathbb{N}}$. □

Given a collection C of extended real-valued random variables and a σ-algebra $\mathcal{G} \subseteq \mathcal{F}$, the *$\mathcal{G}$-conditional essential supremum* $\operatorname{esssup}^{\mathcal{G}} C$ of C is a $\mathcal{G}$-measurable random variable such that $\operatorname{esssup}^{\mathcal{G}} C \geq \xi$ almost surely for all $\xi \in C$ and, if $\bar{\xi}$ is another $\mathcal{G}$-measurable random variable with the same property, then $\bar{\xi} \geq \operatorname{esssup}^{\mathcal{G}} C$ almost surely. This notion was used in the characterization of the conditional essential supremum of a normal integrand in Theorem 2.44. The following establishes the existence of a $\mathcal{G}$-conditional essential supremum of a family of random variables. Its proof is a simple consequence of Lemma B.21. A different argument can be found in [5, Proposition 2.6].

Corollary B.22 *Given a σ-algebra $\mathcal{G} \subseteq \mathcal{F}$, any collection of extended real-valued random variables admis a $\mathcal{G}$-conditional essential supremum.*

Proof Let C be a collection of extended real-valued random variables. Applying Lemma B.21 to the collection

$$C' := \{\xi' \in L(\mathcal{G}) \mid \xi' \geq \xi \ a.s. \ \forall \xi \in C\}$$

gives a sequence $(\xi^\nu)_{\nu \in \mathbb{N}}$ in C' with $\bar{\xi} := \inf_{\nu \in \mathbb{N}} \xi^\nu = \operatorname{essinf} C'$. Since $\xi^\nu \in C'$ almost surely, we have $\bar{\xi} \in C'$ almost surely. Thus, $\bar{\xi} = \operatorname{esssup}^{\mathcal{G}} C$. □

Corollary B.23 *Let ξ and ξ' be extended real-valued random variables and let $A \in \mathcal{G}$. Then,*

$$\operatorname{esssup}^{\mathcal{G}} \xi = \operatorname{esssup}^{\mathcal{G}} (\xi 1_A) + \operatorname{esssup}^{\mathcal{G}} (\xi 1_{\Omega \setminus A})$$

and

$$\operatorname{esssup}^{\mathcal{G}} (\xi + \xi') \leq \operatorname{esssup}^{\mathcal{G}} \xi + \operatorname{esssup}^{\mathcal{G}} \xi',$$

where the inequality holds as an equality if ξ' is $\mathcal{G}$-measurable and $\xi' > -\infty$ almost surely.

Proof To prove the first claim, it suffices to show that $\operatorname{esssup}^{\mathcal{G}} (\xi 1_A) = \operatorname{esssup}^{\mathcal{G}} (\xi) 1_A$. Clearly, $\operatorname{esssup}^{\mathcal{G}} (\xi 1_A) \leq \operatorname{esssup}^{\mathcal{G}} (\xi) 1_A$. If the inequality were strict with positive probability, $\operatorname{esssup}^{\mathcal{G}} (\xi 1_A) 1_A + \operatorname{esssup}^{\mathcal{G}} (\xi) 1_{\Omega \setminus A}$ would majorize ξ and be smaller than $\operatorname{esssup}^{\mathcal{G}} \xi$, which is a contradiction.

The inequality in the second claim is clear, so assume that ξ' is $\mathcal{G}$-measurable and $\xi' > -\infty$ almost surely. On $\{\xi' = \infty\}$, we have $\xi + \xi' = \infty$, so $\operatorname{esssup}^{\mathcal{G}} ((\xi + \xi') 1_{\{\xi' = \infty\}}) = \infty$. On the complement $\{\xi' < \infty\}$, we have,

$$\begin{aligned} \operatorname{esssup}^{\mathcal{G}} (\xi) + \xi' &= \operatorname{esssup}^{\mathcal{G}} (\xi + \xi' - \xi') + \xi' \\ &\leq \operatorname{esssup}^{\mathcal{G}} (\xi + \xi'), \end{aligned}$$

which finishes the proof. □

B.6 Komlós' Theorem

This section gives two measure theoretic compactness-like results that were used in the proofs in Chap. 4. More on the topic can be found, e.g. in [40], [68, Section 5.2] and [154].

Lemma B.24 *Given a sequence* $(x^\nu)_{\nu\in\mathbb{N}}$ *of* $\overline{\mathbb{R}}_+$*-valued random variables, there is a sequence of convex combinations* $\bar{x}^\nu \in \operatorname{co}\{x^\mu \mid \mu \geq \nu\}$ *that converges almost surely to an* $\overline{\mathbb{R}}_+$*-valued random variable.*

Proof See Lemma 9.8.1 and Remark 9.8.2 in [40]. □

Theorem B.25 (Komlós) *Let* $(x^\nu)_{\nu\in\mathbb{N}}$ *be a sequence in* $L^0(\Omega, \mathcal{F}, P; \mathbb{R}^n)$ *and assume one of the following conditions:*

1. $(x^\nu)_{\nu\in\mathbb{N}}$ *is almost surely bounded in the sense that*

$$\sup_{\nu\in\mathbb{N}} |x^\nu(\omega)| < \infty \quad a.s.;$$

2. $(x^\nu)_{\nu\in\mathbb{N}}$ *is bounded in* L^1.

Then there exists a sequence $(\bar{x}^\nu)_{\nu\in\mathbb{N}}$ *of convex combinations* $\bar{x}^\nu \in \operatorname{co}\{x^\mu \mid \mu \geq \nu\}$ *that converges almost surely to an* $\mathbb{R}^n$*-valued random variable.*

Proof The first claim can be found, e.g. in [57, Lemma 1.70] while the second one is [40, Theorem 15.1.4]. □

References

1. C.D. Aliprantis, K.C. Border, *Infinite-Dimensional Analysis*, 2nd edn. (Springer, Berlin, 1999). A hitchhiker's guide
2. V.I. Arkin, I.V. Evstigneev, *Stochastic Models of Control and Economic Dynamics* (Academic Press, Cambridge, 1987)
3. H. Attouch, H. Brezis, Duality for the sum of convex functions in general Banach spaces, in *Aspects of Mathematics and its Applications*. North-Holland Mathematical Library, vol. 34, pp. 125–133 (North-Holland, Amsterdam, 1986)
4. K. Back, S.R. Pliska, The shadow price of information in continuous time decision problems. Stochastics **22**(2), 151–186 (1987)
5. E.N. Barron, P. Cardaliaguet, R. Jensen, Conditional essential suprema with applications. Appl. Math. Optim. **48**(3), 229–253 (2003)
6. N. Bäuerle, U. Rieder, *Markov Decision Processes with Applications to Finance*. Universitext (Springer, Heidelberg, 2011)
7. E.M.L. Beale, On minimizing a convex function subject to linear inequalities. J. R. Stat. Soc. Ser. B. **17**, 173–184, 194–203 (1955) (Symposium on linear programming.)
8. M. Beiglböck, P. Henry-Labordère, F. Penkner, Model-independent bounds for option prices—a mass transport approach. Financ. Stoch. **17**(3), 477–501 (2013)
9. R. Bellman, *Dynamic programming*. Princeton Landmarks in Mathematics (Princeton University Press, Princeton, 2010). Reprint of the 1957 edition, With a new introduction by Stuart Dreyfus
10. A. Ben-Tal, M. Teboulle, Expected utility, penalty functions, and duality in stochastic nonlinear programming. Manag. Sci. **32**(11), 1445–1466 (1986)
11. A. Ben-Tal, M. Teboulle, An old-new concept of convex risk measures: the optimized certainty equivalent. Math. Financ. **17**(3), 449–476 (2007)
12. D.P. Bertsekas, *Dynamic Programming and Stochastic Control* (Academic Press [Harcourt Brace Jovanovich Publishers], New York, 1976). Mathematics in Science and Engineering, 125
13. D. Bertsekas, *Reinforcement Learning and Optimal Control*. Athena Scientific Optimization and Computation Series (Athena Scientific, Nashua, 2019)
14. D.P. Bertsekas, S.E. Shreve, *Stochastic Optimal Control*. Mathematics in Science and Engineering, vol. 139 (Academic Press Inc. [Harcourt Brace Jovanovich Publishers], New York, 1978). The discrete time case
15. S. Biagini, M. Frittelli, A unified framework for utility maximization problems: an Orlicz space approach. Ann. Appl. Probab. **18**(3), 929–966 (2008)

T. Pennanen, A.-P. Perkkiö, *Convex Stochastic Optimization*, Probability Theory and Stochastic Modelling 107, https://doi.org/10.1007/978-3-031-76432-5

16. S. Biagini, A. Černý, Convex duality and Orlicz spaces in expected utility maximization. Math. Financ. **30**(1), 85–127 (2020)
17. S. Biagini, T. Pennanen, A.P. Perkkiö, Duality and optimality conditions in stochastic optimization and mathematical finance. J. Convex Anal. **25**, 403–420 (2018)
18. J.R. Birge, F. Louveaux, *Introduction to Stochastic Programming*. Springer Series in Operations Research and Financial Engineering, 2nd edn. (Springer, New York, 2011)
19. J.M. Bismut, Intégrales convexes et probabilités. C. R. Acad. Sci. Paris Sér. A-B **274**, A915–A917 (1972)
20. J.M. Bismut, Conjugate convex functions in optimal stochastic control. J. Math. Anal. Appl. **44**, 384–404 (1973)
21. J.M. Bismut, *Convex Analysis and Probability*. Theses, Université Paris VI (1973)
22. J.M. Bismut, Intégrales convexes et probabilités. J. Math. Anal. Appl. **42**, 639–673 (1973)
23. J.M. Bismut, B. Skalli, Temps d'arrêt optimal, théorie générale des processus et processus de Markov. Z. Wahrscheinlichkeitstheorie und Verw. Gebiete **39**(4), 301–313 (1977)
24. F. Black, M. Scholes, The pricing of options and corporate liabilities. J. Political Econ. **81**(3), 637–654 (1973)
25. J.F. Bonnans, *Convex and Stochastic Optimization*. Universitext (Springer, Cham, 2019)
26. B. Bouchard, N. Touzi, A. Zeghal, Dual formulation of the utility maximization problem: the case of nonsmooth utility. Ann. Appl. Probab. **14**(2), 678–717 (2004)
27. N. Bourbaki, *Topological Vector Spaces. Chapters 1–5*. Elements of Mathematics (Berlin) (Springer, Berlin, 1987). Translated from the French by H. G. Eggleston and S. Madan
28. A. Brøndsted, Conjugate convex functions in topological vector spaces. Mat.-Fys. Medd. Danske Vid. Selsk. **34**(2), 27pp. (1964)
29. P. Carpentier, J.P. Chancelier, G. Cohen, M. De Lara, *Stochastic Multi-Stage Optimization*. Probability Theory and Stochastic Modelling, vol. 75 (Springer, Cham, 2015). At the crossroads between discrete time stochastic control and stochastic programming
30. C. Castaing, M. Valadier, *Convex Analysis and Measurable Multifunctions* (Springer, Berlin, 1977). Lecture Notes in Mathematics, Vol. 580
31. P. Cheridito, M. Kupper, N. Vogelpoth, Conditional analysis on $\mathbb{R}^d$, in *Set Optimization and Applications—The State of the Art*. Springer Proceedings in Mathematics & Statistics, vol. 151, pp. 179–211 (Springer, Heidelberg, 2015)
32. C. Choirat, C. Hess, R. Seri, A functional version of the birkhoff ergodic theorem for a normal integrand: a variational approach. Ann. Probab. **31**(1), 63–92 (2003)
33. P.L. Combettes, Perspective functions: properties, constructions, and examples. Set-Valued Var. Anal. **26**(2), 247–264 (2018)
34. C. Czichowsky, J. Muhle-Karbe, W. Schachermayer, Transaction costs, shadow prices, and duality in discrete time. SIAM J. Financ. Math. **5**(1), 258–277 (2014)
35. R.C. Dalang, A. Morton, W. Willinger, Equivalent martingale measures and no-arbitrage in stochastic securities market models. Stoch. Stoch. Rep. **29**(2), 185–201 (1990)
36. G.B. Dantzig, Linear programming under uncertainty. Manag. Sci. **1**, 197–206 (1955)
37. M.H.A. Davis, G. Burstein, A deterministic approach to stochastic optimal control with application to anticipative control. Stoch. Stoch. Rep. **40**(3–4), 203–256 (1992)
38. M.H.A. Davis, I. Karatzas, A deterministic approach to optimal stopping, in *Probability, Statistics and Optimisation*, Wiley Series in Probability and Statistics—Applied Probability and Statistics Section (Wiley, Chichester, 1994), pp. 455–466
39. M. De Lara, Duality between Lagrangians and Rockafellians. J. Convex Anal. **30**(3), 887–896 (2023)
40. F. Delbaen, W. Schachermayer, *The Mathematics of Arbitrage*. Springer Finance (Springer, Berlin, 2006)
41. F. Delbaen, P. Grandits, T. Rheinländer, D. Samperi, M. Schweizer, C. Stricker, Exponential hedging and entropic penalties. Math. Financ. **12**(2), 99–123 (2002)
42. C. Dellacherie, P.A. Meyer, *Probabilities and Potential*. North-Holland Mathematics Studies (North-Holland, Amsterdam, 1978)

43. M.A.H. Dempster, (ed.), *Stochastic Programming*. Institute of Mathematics and its Applications Conference Series (Academic Press, [Harcourt Brace Jovanovich, Publishers], London, 1980)
44. M.A.H. Dempster, I.V. Evstigneev, M.I. Taksar, Asset pricing and hedging in financial markets with transaction costs: an approach based on the Von Neumann–Gale model. Ann. Financ. **2**(4), 327–355 (2006)
45. D. Dentcheva, A. Ruszczyński, *Risk-Averse Optimization and Control: Theory and Methods*. Springer Series in Operations Research and Financial Engineering (Springer, Cham, 2024)
46. R. Durrett, *Probability—Theory and Examples*. Cambridge Series in Statistical and Probabilistic Mathematics, vol. 49, 5th edn. (Cambridge University Press, Cambridge, 2019)
47. E.B. Dynkin, Probabilistic concave dynamic programming. Mat. Sb. (N.S.) **87**(129), 490–503 (1972)
48. E.B. Dynkin, I.V. Evstigneev, Regular conditional expectation of correspondences. Teor. Verojatnost. i Primenen. **21**(2), 334–347 (1976)
49. M.J. Eisner, P. Olsen, Duality for stochastic programming interpreted as L.P. in L_p-space. SIAM J. Appl. Math. **28**, 779–792 (1975)
50. I. Ekeland, R. Temam, *Convex Analysis and Variational Problems* (North-Holland, Amsterdam, 1976). Translated from the French, Studies in Mathematics and its Applications, Vol. 1
51. M. El Mansour, E. Lépinette, Conditional interior and conditional closure of random sets. J. Optim. Theory Appl. **187**(2), 356–369 (2020)
52. M. El Mansour, E. Lépinette, Robust discrete-time super-hedging strategies under AIP condition and under price uncertainty. Maths Action **11**(1), 193–212 (2022)
53. I.V. Evstigneev, Measurable selection and dynamic programming. Math. Oper. Res. **1**(3), 267–272 (1976)
54. W. Fenchel, On conjugate convex functions. Can. J. Math. **1**, 73–77 (1949)
55. W. Fenchel, *Convex Cones, Sets, and Functions*. Department of Mathematics, Logistics Research Project (Princeton University, Princeton, 1953)
56. W.H. Fleming, R.W. Rishel, *Deterministic and Stochastic Optimal Control*. Applications of Mathematics, vol. 1 (Springer, Berlin, 1975)
57. H. Föllmer, A. Schied, *Stochastic Finance*. De Gruyter Graduate. De Gruyter, Berlin (2016). An Introduction in Discrete Time, Fourth Revised and Extended Edition of [MR1925197]
58. M. Frittelli, The minimal entropy martingale measure and the valuation problem in incomplete markets. Math. Financ. **10**(1), 39–52 (2000)
59. S.J. Garstka, On duality in stochastic programming with recourse. Zeitschrift für Wahrscheinlichkeitstheorie und Verwandte Gebiete **29**(1), 21–24 (1974)
60. J.M. Harrison, D.M. Kreps, Martingales and arbitrage in multiperiod securities markets. J. Econ. Theory **20**(3), 381–408 (1979)
61. J.M. Harrison, S.R. Pliska, Martingales and stochastic integrals in the theory of continuous trading. Stoch. Process. Appl. **11**(3), 215–260 (1981)
62. S.W. He, J.G. Wang, J.A. Yan, *Semimartingale Theory and Stochastic Calculus* (Kexue Chubanshe (Science Press), Beijing, 1992)
63. F. Hiai, H. Umegaki, Integrals, conditional expectations, and martingales of multivalued functions. J. Multivariate Anal. **7**(1), 149–182 (1977)
64. A.D. Ioffe, On lower semicontinuity of integral functionals. I. SIAM J. Control Optim. **15**(4), 521–538 (1977)
65. A.D. Ioffe, Survey of measurable selection theorems: Russian literature supplement. SIAM J. Control Optim. **16**(5), 728–732 (1978)
66. J. Jacod, A.N. Shiryaev, Local martingales and the fundamental asset pricing theorems in the discrete-time case. Financ. Stoch. **2**(3), 259–273 (1998)
67. Y.M. Kabanov, Hedging and liquidation under transaction costs in currency markets. Financ. Stoch. **3**(2), 237–248 (1999)
68. Y.M. Kabanov, M. Safarian, *Markets with Transaction Costs*. Springer Finance (Springer, Berlin, 2009). Mathematical theory

69. Y.M. Kabanov, C. Stricker, The Harrison-Pliska arbitrage pricing theorem under transaction costs. J. Math. Econ. **35**(2), 185–196 (2001). Arbitrage and control problems in finance
70. Y.M. Kabanov, C. Stricker, A teachers' note on no-arbitrage criteria, in *Séminaire de Probabilités, XXXV*. Lecture Notes in Mathematics, vol. 1755 (Springer, Berlin, 2001), pp. 149–152
71. P. Kall, J. Mayer, *Stochastic Linear Programming*. International Series in Operations Research & Management Science, vol. 156, 2nd edn. (Springer, New York, 2011). Models, theory, and computation
72. P. Kall, S.W. Wallace, *Stochastic Programming*. Wiley-Interscience Series in Systems and Optimization (Wiley, Chichester, 1994)
73. O. Kallenberg, *Foundations of Modern Probability*. Probability and its Applications (New York), 2nd edn. (Springer, New York, 2002)
74. T. Kalmes, A. Pichler, On Banach spaces of vector-valued random variables and their duals motivated by risk measures. Banach J. Math. Anal. **12**(4), 773–807 (2018)
75. J.L. Kelley, I. Namioka, *Linear Topological Spaces* (Springer, New York, 1976). With the collaboration of W. F. Donoghue, Jr., Kenneth R. Lucas, B. J. Pettis, E. T. Poulsen, G. B. Price, W. Robertson, W. R. Scott, and K. T. Smith, Second corrected printing, Graduate Texts in Mathematics, No. 36
76. W.K. Klein Haneveld, *Duality in Stochastic Linear and Dynamic Programming*. Lecture Notes in Economics and Mathematical Systems, vol. 274 (Springer, Berlin, 1986)
77. W.K. Klein Haneveld, M.H. van der Vlerk, W. Romeijnders, *Stochastic Programming*. Graduate Texts in Operations Research (Springer, Cham, 2020). Modeling decision problems under uncertainty
78. A. Kozek, Convex integral functionals on Orlicz spaces. Comment. Math. Prace Mat. **21**(1), 109–135 (1980)
79. A. Kozek, Z. Suchanecki, Multifunctions of faces for conditional expectations of selectors and Jensen's inequality. J. Multivariate Anal. **10**(4), 579–598 (1980)
80. D. Kramkov, W. Schachermayer, The condition on the asymptotic elasticity of utility functions and optimal investment in incomplete markets. Ann. Appl. Probab. **9**(3), 904–950 (1999)
81. D. Kuhn, *Generalized Bounds for Convex Multistage Stochastic Programs*. Lecture Notes in Economics and Mathematical Systems (Springer, Berlin, 2005)
82. E. Lépinette, I. Molchanov, Conditional cores and conditional convex hulls of random sets. J. Math. Anal. Appl. **478**(2), 368–392 (2019)
83. A. Madansky, Dual variables in two-stage linear programming under uncertainty. J. Math. Anal. Appl. **6**, 98–108 (1963)
84. H.M. Markowitz, *Portfolio Selection: Efficient Diversification of Investments*. Cowles Foundation for Research in Economics at Yale University, Monograph, vol. 16 (Wiley, New York, 1959)
85. R.C. Merton, Lifetime portfolio selection under uncertainty: the continuous-time case. Rev. Econ. Stat. **3**, 373–413 (1969)
86. I. Molchanov, *Theory of Random Sets*. Probability and its Applications (New York) (Springer, London, 2005)
87. J.J. Moreau, *Fonctionelles Convexes*. Séminaire sur les Equations aux Dérivées Partielles (Collège de France, Paris, 1967)
88. J. Munkres, *Topology*, 2nd edn. (Prentice Hall, Hoboken, 2000)
89. J. Neveu, *Bases mathématiques du calcul des probabilités* (Masson et Cie, Éditeurs, Paris, 1964)
90. P. Olsen, Multistage stochastic programming with recourse as mathematical programming in an L_p space. SIAM J. Control Optim. **14**(3), 528–537 (1976)
91. P. Olsen, Multistage stochastic programming with recourse: the equivalent deterministic problem. SIAM J. Control Optim. **14**(3), 495–517 (1976)
92. P. Olsen, When is a multistage stochastic programming problem well-defined? SIAM J. Control Optim. **14**(3), 518–527 (1976)

93. M.S. Osborne, *Locally Convex Spaces*. Graduate Texts in Mathematics, vol. 269 (Springer, Cham, 2014)
94. T. Pennanen, Arbitrage and deflators in illiquid markets. Financ. Stoch. **15**(1), 57–83 (2011)
95. T. Pennanen, Convex duality in stochastic optimization and mathematical finance. Math. Oper. Res. **36**(2), 340–362 (2011)
96. T. Pennanen, Dual representation of superhedging costs in illiquid markets. Math. Financ. Econ. **5**, 233–248 (2011)
97. T. Pennanen, Superhedging in illiquid markets. Math. Financ. **21**(3), 519–540 (2011)
98. T. Pennanen, Optimal investment and contingent claim valuation in illiquid markets. Financ. Stoch. **18**(4), 733–754 (2014)
99. T. Pennanen, Erratum: "convex duality in stochastic optimization and mathematical finance". Math. Oper. Res. **41**(2), 732–733 (2016)
100. T. Pennanen, I. Penner, Hedging of claims with physical delivery under convex transaction costs. SIAM J. Financ. Math. **1**, 158–178 (2010)
101. T. Pennanen, A.P. Perkkiö, Stochastic programs without duality gaps. Math. Program. **136**(1), 91–110 (2012)
102. T. Pennanen, A.P. Perkkiö, Convex duality in optimal investment and contingent claim valuation in illiquid markets. Financ. Stoch. **22**(4), 733–771 (2018)
103. T. Pennanen, A.P. Perkkiö, Shadow price of information in discrete time stochastic optimization. Math. Program. **168**(1–2, Ser. B), 347–367 (2018)
104. T. Pennanen, A.P. Perkkiö, Duality in convex stochastic optimization. submitted (2022)
105. T. Pennanen, A.P. Perkkiö, Optimal stopping without snell envelopes, in *Proceedings of the American Mathematical Society* (2022)
106. T. Pennanen, A.P. Perkkiö, Topological duals of locally convex function spaces. Positivity **26**(1), 2, 38 (2022)
107. T. Pennanen, A.P. Perkkiö, Dynamic programming in convex stochastic optimization. J. Convex Anal. **30**(4), 1241–1283 (2023)
108. T. Pennanen, A.P. Perkkiö, Dual solutions in convex stochastic optimization. Math. Oper. Res. (to appear), https://doi.org/10.1287/moor.2022.0270
109. T. Pennanen, A.P. Perkkiö, M. Rásonyi, Existence of solutions in non-convex dynamic programming and optimal investment. Math. Financ. Econ. **11**(2), 173–188 (2017)
110. A.P. Perkkiö, Stochastic programs without duality gaps for objectives without a lower bound. Manuscript (2016)
111. G. Peskir, A. Shiryaev, *Optimal Stopping and Free-Boundary Problems*. Lectures in Mathematics ETH Zürich (Birkhäuser Verlag, Basel, 2006)
112. J. Pfanzagl, Convexity and conditional expectations. Ann. Probab. **2**, 490–494 (1974)
113. G.C. Pflug, A. Pichler, *Multistage Stochastic Optimization*. Springer Series in Operations Research and Financial Engineering (Springer, Cham, 2014)
114. S.R. Pliska, Duality theory for some stochastic control models, in *Stochastic Differential Systems (Bad Honnef, 1982)*. Lecture Notes in Control and Information Sciences, vol. 43 (Springer, Berlin, 1982), pp. 329–337
115. L.S. Pontryagin, V.G. Boltyanskii, R.V. Gamkrelidze, E.F. Mishchenko, *The Mathematical Theory of Optimal Processes* (Wiley, New York, 1962). Translated from the Russian by K. N. Trirogoff; edited by L. W. Neustadt
116. W.B. Powell, *Introduction to Markov Decision Processes*, chapter 3 (Wiley, Hoboken, 2007)
117. A. Prékopa, *Stochastic Programming*. Mathematics and its Applications, vol. 324 (Kluwer Academic, Dordrecht, 1995)
118. M. L. Puterman, Markov decision processes, in *Stochastic Models*. Handbooks in Operations Research and Management Science, vol. 2 (North-Holland, Amsterdam, 1990), pp. 331–434
119. M.M. Rao, Z.D. Ren, *Theory of Orlicz Spaces*. Monographs and Textbooks in Pure and Applied Mathematics, vol. 146 (Marcel Dekker, New York, 1991)
120. M. Rásonyi, L. Stettner, On utility maximization in discrete-time financial market models. Ann. Appl. Probab. **15**(2), 1367–1395 (2005)

121. R.T. Rockafellar, Level sets and continuity of conjugate convex functions. Trans. Am. Math. Soc. **123**, 46–63 (1966)
122. R.T. Rockafellar, Integrals which are convex functionals. Pac. J. Math. **24**, 525–539 (1968)
123. R.T. Rockafellar, *Convex Analysis*. Princeton Mathematical Series, vol. 28 (Princeton University Press, Princeton, 1970)
124. R.T. Rockafellar, Convex integral functionals and duality, in *Contributions to Nonlinear Functional Analysis (Proceedings of Symposium, Mathematics Research Center, University of Wisconsin, Madison)* (Academic Press, New York, 1971), pp. 215–236
125. R.T. Rockafellar, Integrals which are convex functionals. II. Pac. J. Math. **39**, 439–469 (1971)
126. R.T. Rockafellar, Saddle-points and convex analysis, in *Differential Games and Related Topics (Proceedings of International Summer School, Varenna, 1970)* (North-Holland, Amsterdam, 1971), pp. 109–127
127. R.T. Rockafellar, *Conjugate Duality and Optimization* (Society for Industrial and Applied Mathematics, Philadelphia, 1974)
128. R.T. Rockafellar, Integral functionals, normal integrands and measurable selections, in *Nonlinear Operators and the Calculus of Variations (Summer School, University of Libre Bruxelles, Brussels, 1975)*. Lecture Notes in Mathematics, vol. 543 (Springer, Berlin, 1976), pp. 157–207
129. R.T. Rockafellar, R.J.B. Wets, Continuous versus measurable recourse in N-stage stochastic programming. J. Math. Anal. Appl. **48**, 836–859 (1974)
130. R.T. Rockafellar, R.J.B. Wets, Stochastic convex programming: Kuhn-Tucker conditions. J. Math. Econ. **2**(3), 349–370 (1975)
131. R.T. Rockafellar, R.J.B. Wets, Nonanticipativity and L^1-martingales in stochastic optimization problems. Math. Program. Stud. **6**, 170–187 (1976). Stochastic systems: modeling, identification and optimization, II (Proc. Sympos., Univ Kentucky, Lexington, Ky., 1975)
132. R.T. Rockafellar, R.J.B. Wets, Stochastic convex programming: basic duality. Pac. J. Math. **62**(1), 173–195 (1976)
133. R.T. Rockafellar, R.J.B. Wets, Stochastic convex programming: relatively complete recourse and induced feasibility. SIAM J. Control Optim. **14**(3), 574–589 (1976)
134. R.T. Rockafellar, R.J.B. Wets, Stochastic convex programming: singular multipliers and extended duality singular multipliers and duality. Pac. J. Math. **62**(2), 507–522 (1976)
135. R.T. Rockafellar, R.J.B. Wets, Measures as Lagrange multipliers in multistage stochastic programming. J. Math. Anal. Appl. **60**(2), 301–313 (1977)
136. R.T. Rockafellar, R.J.B. Wets, The optimal recourse problem in discrete time: L^1-multipliers for inequality constraints. SIAM J. Control Optim. **16**(1), 16–36 (1978)
137. R.T. Rockafellar, R.J.B. Wets, On the interchange of subdifferentiation and conditional expectations for convex functionals. Stochastics **7**(3), 173–182 (1982)
138. R.T. Rockafellar, R.J.B. Wets, Deterministic and stochastic optimization problems of Bolza type in discrete time. Stochastics **10**(3–4), 273–312 (1983)
139. R.T. Rockafellar, R.J.B. Wets, *Variational Analysis*. Grundlehren der Mathematischen Wissenschaften [Fundamental Principles of Mathematical Sciences], vol. 317 (Springer, Berlin, 1998)
140. L.C.G. Rogers, Monte Carlo valuation of American options. Math. Financ. **12**(3), 271–286 (2002)
141. D.B. Rokhlin, The Kreps-Yan theorem for L^∞. Int. J. Math. Math. Sci. **17**, 2749–2756 (2005)
142. J.O. Royset, R.J.B. Wets, *An Optimization Primer*. Springer Series in Operations Research and Financial Engineering (Springer, Cham, 2021)
143. P.A. Samuelson, Lifetime portfolio selection by dynamic stochastic programming. Rev. Econ. Stat. **51**(3), 239–246 (1969)
144. W. Schachermayer, A Hilbert space proof of the fundamental theorem of asset pricing in finite discrete time. Insurance Math. Econ. **11**(4), 249–257 (1992)
145. W. Schachermayer, Optimal investment in incomplete markets when wealth may become negative. Ann. Appl. Probab. **11**(3), 694–734 (2001)

146. W. Schachermayer, The fundamental theorem of asset pricing under proportional transaction costs in finite discrete time. Math. Financ. **14**(1), 19–48 (2004)
147. H.H. Schaefer, M.P. Wolff, *Topological Vector Spaces*. Graduate Texts in Mathematics, vol. 3, 2nd edn. (Springer, New York, 1999)
148. A. Shapiro, D. Dentcheva, A. Ruszczyński, *Lectures on Stochastic Programming—Modeling and Theory*. MOS-SIAM Series on Optimization, vol. 28, 3rd edn. (Society for Industrial and Applied Mathematics, Philadelphia/Mathematical Optimization Society, Philadelphia, 2021). ©2021
149. A.N. Shiryaev, *Probability*. Graduate Texts in Mathematics, vol. 95, 2nd edn. (Springer, New York, 1996). Translated from the first (1980) Russian edition by R. P. Boas
150. C. Striebel, *Optimal Control of Discrete Time Stochastic Systems* (Springer, Berlin-New York, 1975). Lecture Notes in Economics and Math. Systems, Vol. 110
151. M. Talagrand, Sur une conjecture de H. H. Corson. Bull. Sci. Math. **99**(4), 211–212 (1975)
152. L. Thibault, Espérances conditionnelles d'intégrandes semi-continus. Ann. Inst. H. Poincaré Sect. B (N.S.) **17**(4), 337–350 (1981)
153. A. Truffert, Conditional expectation of integrands and random sets. Ann. Oper. Res. **30**(1–4), 117–156 (1991). Stochastic programming, Part I (Ann Arbor, MI, 1989)
154. G. Žitković, Convex compactness and its applications. Math. Financ. Econ. **3**(1), 1–12 (2010)
155. R.J.B. Wets, Programming under uncertainty: the equivalent convex program. SIAM J. Appl. Math. **14**, 89–105 (1966)
156. R.J.B. Wets, Problèmes duaux en programmation stochastique. C. R. Acad. Sci. Paris Sér. A-B **270**, A47–A50 (1970)
157. R. Wets, Induced constraints for stochastic optimization problems, in *Techniques of Optimization (Fourth IFIP Colloquium Optimization Techniques, Los Angeles,California, 1971)* (1972), pp. 433–443
158. R.J.B. Wets, Stochastic programs with fixed recourse: the equivalent deterministic program. SIAM Rev. **16**, 309–339 (1974)
159. R.J.B. Wets, On the relation between stochastic and deterministic optimization, in *Control Theory, Numerical Methods and Computer Systems Modelling*, ed. by A. Bensoussan, J.L. Lions. Lecture Notes in Economics and Mathematical Systems, vol. 107 (Springer, Berlin, 1975), pp. 350–361
160. C. Zălinescu, *Convex Analysis in General Vector Spaces* (World Scientific, River Edge, 2002)
161. C. Zălinescu, A comparison of constraint qualifications in infinite-dimensional convex programming revisited. J. Austral. Math. Soc. Ser. B **40**(3), 353–378 (1999)

Index

T. Pennanen, A.-P. Perkkiö, *Convex Stochastic Optimization*, Probability Theory and Stochastic Modelling 107, https://doi.org/10.1007/978-3-031-76432-5

GPSR Compliance

The European Union's (EU) General Product Safety Regulation (GPSR) is a set of rules that requires consumer products to be safe and our obligations to ensure this.

If you have any concerns about our products, you can contact us on ProductSafety@springernature.com

In case Publisher is established outside the EU, the EU authorized representative is:

Springer Nature Customer Service Center GmbH
Europaplatz 3
69115 Heidelberg, Germany

Batch number: 08378489

Printed by Printforce, the Netherlands